BEAUFORT
Prudhoe Bay
Mt Michelson
8855
Mt Chamberlin
9020
ROMANZOF MTNS
DAVIDSON MTNS
RANGE
PHILIP SMITH MTNS
RICHARDSON MTNS
Mackenzie River
Porcupine River
Fort Yukon
River
Peel River
Circle
Yukon River
Eagle
Fairbanks
Dawson
Klondike River
Stewart River
Tanana River
Delta Junction
Mt Deborah
12339
Mt Hayes
13832
Tok
Pelly River
Mt Mather
RANGE
TALKEETNA MTNS
Susitna River
White River
Mt Sanford
16237
Mt Drum
12010
Mt Wrangell
14163
Mt Blackburn
16390
Mt Bona
16421
Mt Bear
14831
Watson Lake
Whitehorse
Palmer
Mt Marcus Baker
13176
Anchorage
Valdez
Copper River
Port Valdez
College Fiord
CHUGACH MTNS
SAINT ELIAS MTNS
Mt Logan
19850
Whittier
Prince William Sound
Orca Bay
Cordova
Mt St Elias
18008
Robinson Mtns
Chilkoot Pass
White Pass
Skagway
Haines
Hinchinbrook Island
Wingham Island
Kayak Island
Cape Saint Elias
Icy Bay
Russell Fiord
Yakutat Bay
Yakutat
Seward
Montague Island
Resurrection Bay
Muir Inlet
Lynn Canal
Devils Paw
8584
Mt Fairweather
15300
Glacier Bay
Cape Fairweather
Juneau
Stikine River
Icy Strait
Hoonah
Stephens Passage
Chichagof Island
Admiralty Island
Seymour Canal
Peril Strait
Angoon
Kates Needle
10023
GULF OF ALASKA
ALEXANDER ARCHIPELAGO
Sitka
Mt Edgecumbe
3201
Baranof Island
Chatham Strait
Sitka Sound
Petersburg
Mitkof Island
Wrangell
Wrangell Island
Sumner Strait
Etolin Island
Behm Canal
Portland Canal
0 50 100 200 Miles
Cape Ommaney
Prince of Wales Island
Klawock
Craig
Ketchikan
Annette Island
Metlakatla
Bucareli Bay
Dall Island
Cape Muzon
Dixon Entrance
ALEUTIAN ISLANDS
Andreanof Islands
Korovin Volcano
5030
Atka Pass
Atka Island
Kiska Volcano
4004
Kanaga Volcano
4287
Adak Island
Amchitka Island
Island
Queen Charlotte Islands

Alaska

Also by Walter R. Borneman

A Climbing Guide to Colorado's Fourteeners
(with Lyndon J. Lampert)

Alaska

Saga of a Bold Land

Walter R. Borneman

HarperCollins*Publishers*

HarperCollins books may be purchased for educational, business, or sales promotional use. For information, please write: Special Markets Department, HarperCollins Publishers Inc., 10 East 53rd Street, New York, NY 10022.

FIRST EDITION

Designed by Nancy B. Field

Library of Congress Cataloging-in-Publication Data

Borneman, Walter R.
Alaska : saga of a bold land / Walter R. Borneman.— 1st ed.
p. cm.
Includes bibliographical references and index.
ISBN 0-06-050306-8
1. Alaska—History. I. Title.

F904.B75 2003
979.8—dc21 2002027271

03 04 05 06 07 ❖/RRD 10 9 8 7 6 5 4 3 2 1

For my son, Russ,

who has traveled this land with me
and who has his own frontiers yet to cross.

Contents

BOOK SEVEN
Postwar Rumblings: Statehood and Earthquake (1945–1964)

BOOK EIGHT
North Again: This Time the Gold Is Black (1964–1980)

BOOK NINE
Whose Land? Competing Claims (1980–2001)

PROLOGUE

Alaska—a Sense of Scale

In a bone-chilling drizzle we waded through the swollen waters of the Anaktuvuk River. Two days of steady rain and snowmelt had turned its clear waters a muddy and ominous brown. Somewhere up ahead lay Ernie Pass and the crest of the Brooks Range—the Arctic Divide. Even in June at the height of the brief Arctic summer, there was evidence everywhere of the brevity of the season and the harshness of the land's extremes.

Clouds obscured the highest peaks, but surely the summit of the pass must be close. For hours we made slow but steady progress across a trailless patchwork of toadstool-shaped tussocks and grassy bogs. That had to be the pass—right up ahead. The verdict was unanimous except for one lone dissenter. That was the pass, all right, he argued, but it was still four miles and a major drainage away—not within a hopeful thirty minutes.

A lively discussion consulting topo map and GPS soon confirmed the unpopular position. There were still miles to go. Under a quiet blanket of misty gray clouds, the rest had been lulled into a false hope by the sheer scale of the land itself. Here, one can walk for days and still see major landmarks seemingly unchanged.

Scale. That's what this land is all about.

INTRODUCTION

Crossing the Next Frontier

When I would mention to those only generally knowledgeable with Alaska and its literature that I was writing a history of that bold land, all too often the response would be, "Oh, you mean like James Michener." As a historian and not a novelist, I gradually became upset by this comparison. Closer examination of such responses, however, actually affirmed that I was indeed on the right track. What I was writing was in fact a "big picture" look at Alaska, and people seemed to understand what that meant.

Now to be sure, there have been many, many books written about Alaska, but few—among them Michener's novel—tell the entire Alaskan story from its first inhabitants to contemporary challenges. Given Alaska's wide-ranging geography, its ubiquitous cast of dynamic characters, and its diverse cultural complexities, such a task is a tall order. There are so many pieces that—just like a jigsaw puzzle—it is impossible to pick up the whole without the pieces falling apart. You can admire the assembled puzzle flat on a table, but it defies handling for close-up examination. Thus, you describe the puzzle in broad terms, all the while striving not to obscure the details of its individual pieces.

But if such a broad view is necessary, why is this entire sweep of events—this big picture of Alaska's history—important, or even insightful? Alaska's historic themes are surprisingly consistent and reoccurring: new land; new people; new riches—and ever-present competing views over their use. These themes have been played out by many different personalities, events, and responses over the years, but to better appreciate their place in the puzzle—to get one's hands around the whole story and pick up the puzzle—requires a step backward.

The view that results from such a vantage point is one of a succession of waves crashing across Alaska's landscape: Siberian emigrants crossing the Bering Land Bridge; Russian *promyshlenniki* (hunters) exploiting a fur empire; European flag bearers checking rival advances; Americans looking north in the aftermath of civil war; argonauts from across the world stampeding north to the clarion summons of "gold"; soldiers and sailors battling out a decisive chapter in a world war on the most bizarre of battlegrounds; another rush for a different kind of mineral wealth; and always at the core, the conflict over how the land will be used and by whom.

Throughout Alaska's history there has always been another wave cresting, another frontier to cross. Some have wished that the land would remain static, but it never has and never will. Others have been at the vanguard of change, leading the charge to extract tangible resources from its intangible beauty.

Some call Alaska the last frontier. But seen in the full spectrum of its recorded history, that appellation could be applied to any number of periods. It is not so much that Alaska is the last frontier, but rather that it is a land that has always had new frontiers thrust upon it. It is a land that always has been, and always will be, crossing the next frontier.

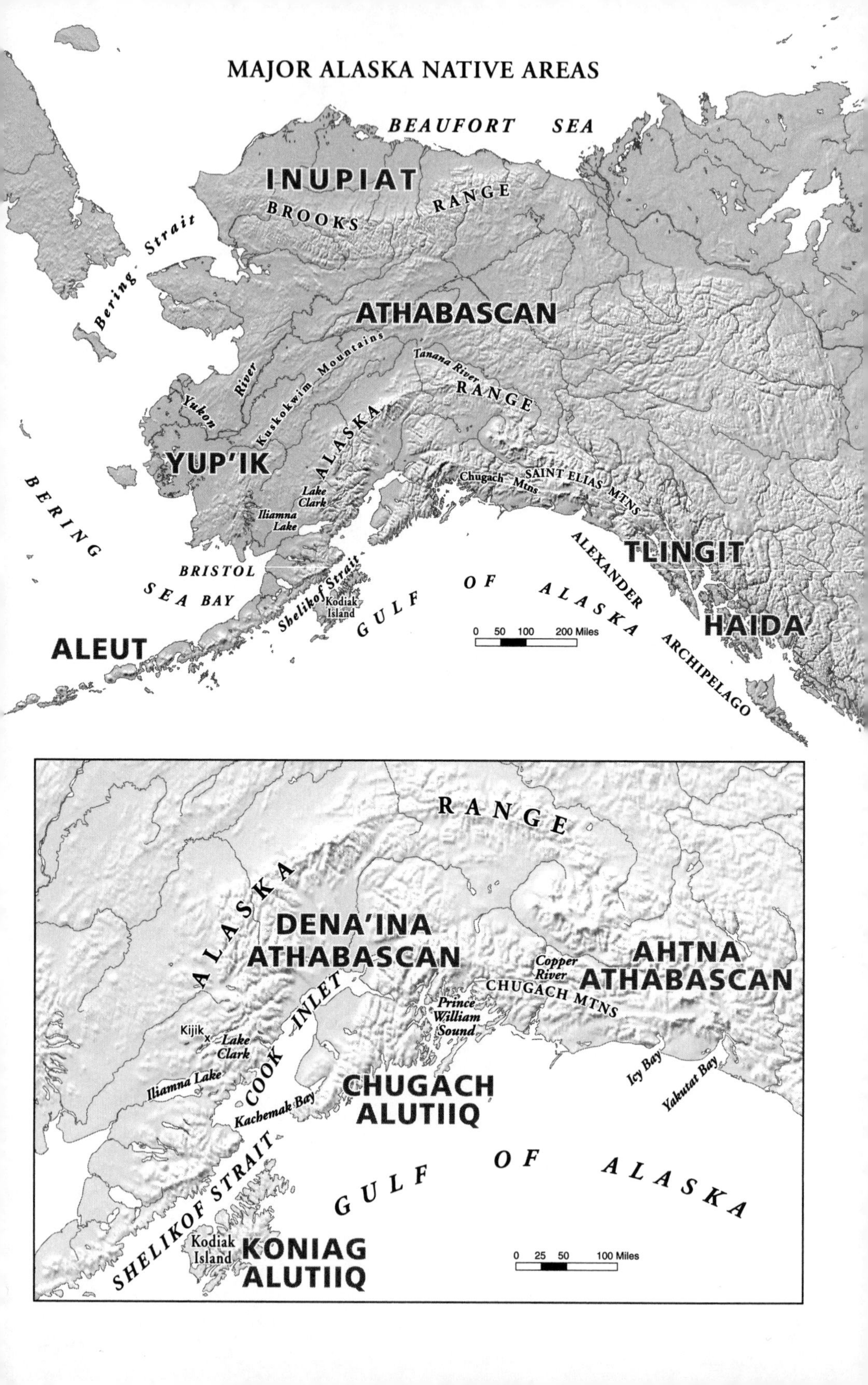

MAJOR ALASKA NATIVE AREAS
BEAUFORT SEA
INUPIAT
BROOKS RANGE
Bering Strait
ATHABASCAN
Kuskokwim Mountains
Tanana River
Yukon River
ALASKA RANGE
YUP'IK
BERING SEA
Lake Clark
Iliamna Lake
Chugach Mtns
SAINT ELIAS MTNS
BRISTOL BAY
Shelikof Strait
Kodiak Island
GULF OF ALASKA
ALEXANDER ARCHIPELAGO
TLINGIT
HAIDA
ALEUT
0 50 100 200 Miles
RANGE
ALASKA
DENA'INA ATHABASCAN
AHTNA ATHABASCAN
Copper River
CHUGACH MTNS
Prince William Sound
Kijik
Lake Clark
COOK INLET
Iliamna Lake
Kachemak Bay
CHUGACH ALUTIIQ
Icy Bay
Yakutat Bay
SHELIKOF STRAIT
GULF OF ALASKA
Kodiak Island
KONIAG ALUTIIQ
0 25 50 100 Miles

BOOK ONE

The Land before Time

Prehistory–1728

Raven came, releasing the Sun, Moon, and Stars.
His cunning creations, the elders say, changed the world.

—Tlingit oral history

Mountains, Glaciers, and Innumerable Rivers
First Steps, Continuing Traditions

Look at a map of Alaska. *What you notice first, and what remains with you long afterward, is the scale. Here is a land where superlatives abound and comparisons are few. Here is a land that dwarfs almost any wilderness you have known: Highest mountain in North America—Denali; third longest river system in North America—the Yukon; largest U.S. national park—Wrangell–St. Elias; southernmost tidewater glacier in North America—LeConte; northernmost town in the United States—Barrow; largest U.S. national forest—the Tongass; largest subpolar icefield in the world—Bagley Icefield.*

Alaska is 615,230 square miles of rugged mountains, grinding glaciers, seemingly endless tundra, broad rivers, and rushing streams—larger than all but seventeen of the world's countries.[1] *Everything that has happened or will happen here—from the first people migrating from Siberia, to the gold rushes and oil booms, to the competing issues of wilderness versus development—* everything *is inextricably tied to the land. One cannot understand the story without knowing something about the setting and the people who first set foot upon it.*

Mountains, Glaciers, and Innumerable Rivers

Rivers, creeks, and streams almost without number course through the extent of Alaska, but it is the mountain ranges that most define its landscape. Jutting southward from the main landmass, the Aleutian Range is the backbone of the narrow Alaska Peninsula. To the north, the Alaska Range sprawls more than 500 miles across the heart of the state, rising to 20,320 feet atop the icy crown of North America. The Chugach and Wrangell–St. Elias Mountains shadow the Alaska Range to the south and wrap around southcentral Alaska's turbulent rim of fire and ice. Southeast of the St. Elias Range, the Coast Mountains march down the southeast panhandle above the emerald waters of the Inside Passage. West of the Alaska Range, the Kuskokwim Mountains are mere foothills by comparison, but this range rises above the entangled streams and wetlands of the Yukon and Kuskokwim river systems. The Kuskokwim Mountains point north toward the Brooks Range, the northernmost mountain chain in the United States and the roof above Alaska's North Slope.

Aleutian Range, Alaska Range, Chugach Mountains, Wrangell–St. Elias Mountains, Coast Mountains, Kuskokwim Mountains, and Brooks Range; these are the seven great mountain systems that define Alaska.

The Aleutian Range dominates the sweeping arc of the Alaska Peninsula and the Aleutian Islands as they slice between the waters of the Bering Sea and the North Pacific Ocean. This wild, 1,600-mile tail of Alaska is a grand necklace of rugged peaks, lowland plains, and rocky beaches cast upon a restless and frequently rambunctious sea. The high-point of the range, 11,413-foot Mount Torbert, lies near its tangled juncture with the Alaska Range. Chakachamna Lake and the Chakachatna River slice through the range just south of Mount Torbert and Mount Spurr (11,070 feet) and tempt some to lump these peaks with the Alaska Range rather than the Aleutians. South of here, however, beyond Lake Clark Pass, there can be no doubt. The active volcanoes of Redoubt

(10,197 feet) and Iliamna (10,016 feet) rise above the waters of Cook Inlet to the east and sparkling Lake Clark to the west.

The Aleutian Range fades briefly near Iliamna Lake before regaining height in the peaks north and south of Mount Katmai. Had it not been for the 1912 eruption of nearby Novarupta that collapsed Katmai's summit cone, the mountain would be some 7,500 feet high. As it is, 6,715-foot Mount Katmai is one of eleven 6,000-foot-plus mountains in Katmai National Park and Preserve and one of fifteen active volcanoes lining Shelikof Strait between the peninsula and Kodiak Island.

Southwest of Katmai, Aniakchak caldera also bears stark witness to Alaska's rim of fire and ice. About 3,500 years ago, a cataclysmic eruption blew the top off Aniakchak Mountain. This caused its summit slopes to collapse, leaving a 2,000-foot-deep, six-mile-wide caldera. The Aniakchak River rises in Surprise Lake within the caldera and then cuts through its rim at the Gates, embarking on a rollicking thirty-two-mile journey to the sea.

Southwest of Aniakchak, the Alaska Peninsula ends opposite Unimak Island. This narrow waterway was called False Pass because passage on its northern end appeared blocked at low tide. Mount Shishaldin (9,372 feet) on Unimak Island towers above the strait and is one of those volcanoes with an almost perfect symmetry to its cone. Here, the terrain sweeps upward from sea level to above 9,000 feet in less than ten miles. Beyond watery Unimak Pass, the Aleutian Islands trail off across the North Pacific toward Asia's Kamchatka Peninsula. The islands get smaller as the chain bends westward, but mountains—many more than 4,000 feet tall—continue to dominate the landscape.

In the other direction from Lake Clark Pass, the rocky backbone of the Alaska Range curves northeastward for some 500 miles across the heart of Alaska, dividing the southcentral coast from the interior. No roads cross its crest in the western half, and it is a barrier even to the moisture-laden clouds that drop most of their load south of the range.

The U.S. Board of Geographic Names insists that the tallest mountain in the range—and the highpoint of North America—is called Mount McKinley. Athabascan Natives of the interior long called the mountain Denali, meaning "the high one." Charles Sheldon, who was instrumental in the creation of Mount McKinley National Park, also always referred to the mountain as Denali. Sheldon first arrived near the mountain from the north in 1906 and wrote: "Soon after starting again we caught glimpses of snowy peaks toward the south, and when we reached the top, Denali and the Alaska Range suddenly burst into view ahead, apparently very near. I can never forget my sensations at the sight. No description could convey any suggestion of it."[2] Sheldon was not the first—nor would he be the last—to be fooled by this country's scale. Denali was still a good thirty miles away, but Sheldon was certainly right about the mountain's name.

The highest summits of the Alaska Range cluster about Denali: Mounts Crosson (12,800 feet), Foraker (17,400 feet), Russell (11,670 feet), and Dall (8,756 feet) curving to the southwest; Hunter (14,573 feet) and Huntington (12,240 feet) forming a barrrier to the south; and Silverthrone (13,220 feet), Deception (11,826 feet), and Mather (12,123 feet) running eastward along the crest of the range. Today, all of these summits are within Denali National Park and Preserve.

The lowest crossing of the main Alaska Range is 2,300-foot Broad Pass, the route of both the Alaska Railroad and the George Parks Highway. Broad Pass is just that—almost flat and very broad—sending waters either north to the Nenana-Yukon drainage or south to the Susitna River. Both the railroad and the highway ease through the remainder of the range by following the canyon of the Nenana River.

East of where the Nenana River bisects the range, Mount Deborah (12,339 feet), Mount Hayes (13,832 feet), and Mount Moffit (13,020 feet) rise north of the Denali Highway. East of the Deborah massif, the Alaska Range is crossed by the Alaska Pipeline and the Richardson Highway over 3,000-foot Isabel Pass. The Glenn Highway cuts a third crossing of the range via Mentasta Pass (2,280 feet) at the headwaters of the Little Tok River.

Other than its gigantic mountains, the most striking features of the Alaska Range are the huge glaciers that flow from its spine. Warm moist air flowing north from the Gulf of Alaska drops most of its precipitation on the south side of the range, making the glaciers there considerably larger than those on the north side. The largest, the Kahiltna, is up to three miles wide and forty-five miles long. On the north side of the range, the glacial monarch is the Muldrow, flowing northeast from the upper slopes of Denali and fed by the Traleika and Brooks Glaciers. From Kahiltna Pass below the West Buttress of Denali, the Peters Glacier flows northeast beneath the heights of Wickersham Wall and Pioneer Ridge. During the summer of 1987, the Peters Glacier surged forward three miles, moving at an astonishing rate of up to seventy-five yards per day.

In 2001, the Tokositna Glacier on the south side of the range went on a similar rampage. These events, which occurred when most of the world's glaciers were receding, bear witness to the dynamic forces that continue to shape the Alaskan landscape.

South of the Alaska Range, the Chugach Mountains mirror the larger range's crescent arc and form a 250-mile-long divide along the Gulf of Alaska. Only the Copper River cuts a path completely through the range. The highpoint is 13,176-foot Mount Marcus Baker, but hundreds of snowy summits rise above Turnagain Arm, Prince William Sound, and the Copper River Delta. A southwestern extension of the range, the Kenai Mountains, dominates the Kenai Peninsula along with the Harding and Sargent Icefields.

At the eastern end of the range, between the Chugach Mountains and the Gulf of Alaska, the Bagley Icefield is the largest subpolar icefield in North America and extends for some eighty miles. The Martin River, Steller, Bering, Yahtse, and Guyot Glaciers flow south from it toward the Gulf of Alaska. Everywhere, there are braided rivers and streams laden with glacial silt carving away at the landscape.

East of the Chugach, the landscape intensifies even more. The Alaska Range may have Denali, but the Wrangell and St. Elias Mountains are the most extensive realm of towering mountains, raging rivers, and massive glaciers in North America. Twelve of the fifteen highest peaks in Alaska, and ten of the fifteen highest peaks in North America, rise in these ranges. The international boundary between Alaska and Canada's Yukon Territory bisects the St. Elias Mountains and makes 19,432-foot Mount Logan the highest point in Canada. Had the boundary been drawn one degree of longitude to the east—a mere thirty-four miles at this latitude—Mount Logan and the 16,000-foot-plus giants of Mounts Lucania and Steele and King Peak would also be in Alaska.

Much of this region is now Wrangell–St. Elias National Park and Preserve. With some 13 million acres, it is the largest national park in the United States—six times the size of Yellowstone. Together with neighboring Kluane National Park in the Yukon Territory, Tatshenshini-Alsek Wilderness Provincial Park in British Columbia, and Glacier Bay National Park and Preserve in Alaska's panhandle, Wrangell–St. Elias National Park and Preserve has been designated a World Heritage Area—evidence indeed that the interdependency and cohesiveness of a far-flung ecosystem cannot be severed by political boundaries.

Highest Mountains of North America

1.	Mount McKinley	Alaska Range	Alaska	20,320 feet
2.	Mount Logan	St. Elias Range	Canada	19,432 feet
3.	Pico de Orizaba	Mexico	Mexico	18,700 feet
4.	Mount St. Elias	St. Elias Range	Alaska/Canada	18,008 feet
5.	Popocatepetl	Mexico	Mexico	17,887 feet
6.	Mount Foraker	Alaska Range	Alaska	17,400 feet
7.	Iztaccihuatl	Mexico	Mexico	17,343 feet
8.	Mount Lucania	St. Elias Range	Canada	17,150 feet
9.	King Peak	St. Elias Range	Canada	16,972 feet
10.	Mount Steele	St. Elias Range	Canada	16,600 feet
11.	Mount Bona	St. Elias Range	Alaska	16,421 feet
12.	Mount Blackburn	Wrangell Mts.	Alaska	16,390 feet
13.	Mount Sanford	Wrangell Mts.	Alaska	16,237 feet
14.	Mount Wood	St. Elias Range	Canada	15,880 feet
15.	Mount Vancouver	St. Elias Range	Alaska/Canada	15,700 feet

Highest Mountains of Alaska

1.	Mount McKinley	Alaska Range	Alaska	20,320 feet
2.	Mount St. Elias	St. Elias Range	Alaska/Canada	18,008 feet
3.	Mount Foraker	Alaska Range	Alaska	17,400 feet
4.	Mount Bona	St. Elias Range	Alaska	16,421 feet
5.	Mount Blackburn	Wrangell Mts.	Alaska	16,390 feet
6.	Mount Sanford	Wrangell Mts.	Alaska	16,237 feet
7.	Mount Vancouver	St. Elias Range	Alaska/Canada	15,700 feet
8.	Mount Churchill	St. Elias Range	Alaska	15,638 feet
9.	Mount Fairweather	St. Elias Range	Alaska	15,330 feet
10.	Mount Hubbard	St. Elias Range	Alaska	15,015 feet
11.	Mount Bear	St. Elias Range	Alaska	14,831 feet
12.	Mount Hunter	Alaska Range	Alaska	14,573 feet
13.	Mount Alverstone	St. Elias Range	Alaska/Canada	14,565 feet
14.	University Peak	St. Elias Range	Alaska	14,470 feet
15.	Mount Wrangell	Wrangell Mts.	Alaska	14,163 feet
16.	Mount Augusta	St. Elias Range	Alaska/Canada	14,070 feet

The Wrangell Mountains extend eastward some 110 miles from the valley of the Copper River to the vicinity of Chitistone and Skolai Passes. The first Wrangell volcanoes formed about 26 million years ago when the northwest-moving Pacific plate began to push beneath the North American plate. Called the Wrangell volcanic field, this extensive volcanic terrain covers about 4,000 square miles and extends into the St. Elias Mountains.

The highpoint of the Wrangell Mountains is the 16,390-foot dome of Mount Blackburn, which is almost completely covered with icefields and glaciers. The best-known glacier is the Kennicott, which sweeps south from the mountain in a relatively straight line and now terminates near the towns of McCarthy and Kennecott.

Northwest of Blackburn, Mount Wrangell (14,163 feet) is an enormous shield volcano. Its ice-filled summit caldera is 3.5 miles long and 2.5 miles wide with three small craters containing active fumaroles situated along its rim. There have been at least three reports (1784, 1884–85, and 1900) of eruptions, including unconfirmed reports of flowing lava. One landmark on the mountain is Mount Zanetti (13,009 feet), a large cinder cone high on Mount Wrangell's northwest flank that may have erupted less than 25,000 years ago.

The highpoint in the western Wrangell Mountains is Mount Sanford (16,237 feet). Its soaring south face rises some 8,000 feet in about one mile, one of the steepest gradients in North America. The Sanford Glacier heads in the massive cirque at the base of the south face and carries with it tons of debris from the rockfall and avalanches that roar down the face.

Mount Drum (12,010 feet) is the westernmost of the Wrangell volcanoes. Although 4,000 feet lower than neighboring Mount Sanford, the mountain dominates the local landscape because of its vertical rise above the adjacent Copper River valley. Major eruptions on its southern flanks produced mudflows that poured into the Copper River valley and flowed downstream at least as far as the current site of Chitina. The Nadina Glacier is the largest of eleven glaciers on the mountain and flows southwest more than nine miles from the large amphitheater created by these mudflows.[3]

East of the Wrangells, the St. Elias Mountains arc southeastward from the vicinity of Chitistone and Skolai Passes to the waters of Glacier Bay. Denali is indeed taller, but no area in North America matches the St. Elias Mountains in any other category. It is the largest and highest concentration of snow-covered mountains in North America. The complexity and enormity of the range was best described by geologist Israel Cook Russell after an 1891 attempt to climb Mount St. Elias. Russell viewed the country to the north for the first time and wrote:

> I expected to see a comparatively low, wooded country stretching away to the north, with lakes and rivers and perhaps some signs of human habitation, but I was entirely mistaken. What did meet my eager gaze was a vast snow-covered region, limitless in its expanse, through which hundreds, and perhaps thousands, of barren angular mountain peaks projected. There was not a stream, not a lake, and not a trace of vegetation of any kind in sight. A more desolate or utterly lifeless land one never beheld. Vast, smooth snow surfaces, without crevasses or breaks, so far as I could judge, stretched away to unknown distances, broken only by jagged and angular mountain peaks.[4]

Mount St. Elias (18,008 feet), the highest peak in the range in Alaska, is located on the Alaskan-Yukon border above the waters of Icy Bay. The mountain rises dramatically—and perpendicularly—less than twenty miles from sea level at Icy Bay. Before a thorough survey of Denali was undertaken, St. Elias was briefly thought to be the tallest peak on the continent. Less than thirty miles northeast of the pointy pyramid of St. Elias, the broad plateau of Mount Logan (19,432 feet) is the highpoint of Canada. On a clear day, Logan's broad plateau and St. Elias's pointy pyramid are easily discernible to air travelers en route to and from Anchorage.

The Fairweather Range—many immediately call its name a misnomer—is a southern extension of the St. Elias Range and runs some seventy miles from Grand Pacific and Grand Plateau Glaciers south to Cross Sound and Icy Strait. The range separates Glacier Bay from the Gulf of Alaska, and along with Canada's Alsek and Takhinsha Mountains, it is the source of the glaciers that sculpted Glacier Bay. The many inlets of Glacier Bay have only recently been exposed from beneath massive glaciers. Mount Fairweather (15,330 feet), the highpoint of the range, sits squarely on yet another corner of the jagged United States–Canada boundary.

From the southern end of the St. Elias Range, the Coast Mountains stand like a towering picket fence above the waters of the North Pacific for some 900 miles, all the way south to the forty-ninth parallel and the

Cascade Range. Look at a relief map of North America, and here is a graphic example of the mountain-building forces of plate tectonics. In general, the Pacific plate has dug under the North American plate and pushed it upward, folding and buckling the earth's crust in a process called subduction.

For 350 of these miles, from the windswept slopes of Chilkoot and White Passes southward to the southern tip of Dall Island, the connecting thread at the foot of the range is the Inside Passage, the tangled, meandering system of waterways that runs the length of southeast Alaska. It is called the Inside Passage because a myriad of islands—some quite large, others mere piles of boulders—form a barrier between the unruly North Pacific Ocean "outside" and quieter waters "inside." Much of the land—almost 17 million acres—is Tongass National Forest, the largest in the United States, covering a combined area larger than the state of West Virginia.

No road crosses the Coast Mountains from White Pass south to Prince Rupert, British Columbia. By Alaska Range and St. Elias Range standards, Coast summits are not particularly lofty, but their dramatic rise within a comparatively short distance from sea level, combined with fickle southeast Alaska weather, crumbling sedimentary rock, and extensive glaciers, make them mountaineering challenges. The international boundary runs along the crest of the range, and the highpoint of the Alaskan portion is 10,023-foot Kates Needle above the Stikine Icefield.

As in the Alaska Range, warm moisture-laden air from the Pacific cools on the windward side of the range and deposits its cargo. The larger glaciers flow west from the crest. They are easily seen from the waters of the Inside Passage. The most famous are the Mendenhall Glacier north of Juneau, flowing from the Juneau Icefield, and the LeConte Glacier east of Petersburg, flowing from the Stikine Icefield.

The Mendenhall is one of thirty-eight glaciers flowing from the 1,500-square-mile (larger than Rhode Island) Juneau Icefield. In about twelve miles, the Mendenhall drops from an elevation of 4,500 feet at the icefield to 54 feet above sea level at its terminus at Mendenhall Lake. Ice formed at the glacier's head takes about eighty years to make the journey. Less than 250 years ago, the glacier's face stood 2.5 miles farther down the valley than its current position, and as recently as the 1930s, it still covered the rocks where the visitor center now stands.

The LeConte Glacier at the southern end of the 2,900-square mile Stikine Icefield is the southernmost tidewater glacier in North America.

Icebergs from the glacier frequently choke LeConte Bay and float with the tide into Frederick Sound. Tlingit Natives called LeConte Glacier "Hutli," meaning "thunder." The Tlingit knew well the thundering noise of ice calving from the glacier's face. A Tlingit legend tells that a mythical bird produced the thunder by flapping its giant wings. The mountain looking down on LeConte Bay is appropriately named "Thunder Mountain."[5]

If the mountains of southeast Alaska are draped in ice and snow much of the year, the cape thrown across much of the Kuskokwim Mountains is one of watery muskeg and wetlands. The Kuskokwim Mountains and their more rugged southwestern extensions, the Kilbuck, Ahklun, and Wood River Mountains, extend from Cape Newenham between Kuskokwim and Bristol Bays northeast some 550 miles to the Tanana River west of Fairbanks. This is the great divide between the Yukon and Kuskokwim river systems.

No roads cross the Kuskokwim Mountains, and the real route across the range is that of the Iditarod Trail. After cresting the Alaska Range at Rainy Pass, the Iditarod crosses the south fork of the Kuskokwim River at Rohn Roadhouse, jumps the Kuskokwim's main stem at Big River Roadhouse, and then runs up Fourth of July Creek before crossing the range and descending Bonanza Creek to the Iditarod River. The highpoints of the range are Von Frank Mountain (4,508 feet) west of Lake Minchumina, Mount Oratia (4,658 feet) in the Ahklun Mountains, and Mount Waskey (5,026 feet) in the Wood River Mountains northwest of Dillingham.

Dominating the entire northern third of Alaska, the Brooks Range is the northernmost mountain chain in the United States. Named for Alaskan explorer and geologist Alfred H. Brooks, the range spans the roof of Alaska for some 600 miles, from Cape Lisburne on the Chukchi Sea east into Canada. Along its sinuous crest, the Arctic Divide sends waters flowing either north across the tundra of the North Slope or south through boreal forests to the Yukon.

By its sheer immensity, the Brooks Range encompasses what anywhere else would be great ranges isolated unto themselves, including the DeLong, Baird, Waring, Schwatka, Endicott, Philip Smith, Franklin, Romanzof, and Davidson Mountains. Obtaining accurate measurements in this geographic maze was long problematic, but 9,020-foot Mount Chamberlain in the Franklin Mountains, less than fifty miles from the Arctic Ocean, is currently considered the highpoint of the range. Mount

Isto (8,975 feet) and Mount Michelson (8,855 feet) in the Romanzof Mountains are other highpoints before the Brooks chain heads into Canada and terminates in the British and Richardson Mountains west of the massive Mackenzie River delta.

The central part of the Brooks Range is largely Gates of the Arctic National Park and Preserve, dominated by the Schwatka and Endicott Mountains. Between these ranges rise the twin-turreted fortress of Igikpak Peak (8,510 feet), the park's highpoint, and the razor-thin arêtes and sweeping flatirons of the Arrigetch Peaks (7,190 feet). From this mountainous maze flow the innumerable tributaries of four great rivers: the Colville draining north to the Arctic, the Noatak and the Kobuk coursing west to the Chukchi Sea, and the Koyukuk, flowing south to the Yukon. Because of its relatively drier climate—at least when compared to the Chugach and Coast Ranges—the Brooks Range has no massive valley glaciers except in the higher reaches of the Franklin and Romanzof Mountains.

Aleutian Range, Alaska Range, Chugach Mountains, Wrangell–St. Elias Mountains, Coast Mountains, Kuskokwim Mountains, and Brooks Range. East, west, north, or south, these mountain ranges dominate Alaska, but it is the rivers that are the avenues into their domains, and it is the rivers that give character to the lands in between. Alaska's rivers are many, and their tributaries almost infinite, but four great systems intertwine the mountain ranges: the mighty Yukon drainage spanning the entire state; the Kuskokwim flowing southwest from the north side of the Alaska Range; the Susitna and Copper Rivers winding around and through the Chugach Mountains; and the Kobuk, Noatak, and Colville draining the Brooks Range.

The headwaters of the Yukon River rise in a collection of lakes just north of 2,915-foot White Pass on the Alaska–British Columbia border. Sea level is less than 40 miles away at the head of the Inside Passage, but the Yukon goes on a wild 1,500-mile sweep north and then west before finally reaching the Bering Sea. More than any other river, the Yukon and the tentacles of its many tributaries have been the highways into the interior of Alaska.

From its headwaters the Yukon flows north through Canada's Yukon Territory past Whitehorse and Dawson, being joined in the process by the Pelly, Stewart, White, and famous Klondike Rivers. Below Dawson, the Yukon swings west into Alaska and skirts the Tanana Hills before meeting the Porcupine and Chandalar Rivers in the swampy wetlands of the Yukon Flats. Ever westward now, the river cuts across the heart of interior Alaska and picks up the Tanana and Koyukuk Rivers. Then, blocked by the Nulato Hills from making a straight 20-mile dash to the Unalakleet River and Norton Sound, the river detours in a huge U-loop—south, west, and then north—of some 400 more miles before finally reaching Norton Sound and the Bering Sea.

The Yukon's three major tributaries are the Porcupine, Tanana, and Koyukuk. The Porcupine heads in Canada and flows west some 460 miles to join the Yukon at Fort Yukon. Its basin on both sides of the international boundary is home to the Porcupine caribou herd. The Tanana rises from the Chisana and Nabesna Rivers that flow from glaciers of the same names on the north side of the Wrangell Mountains. The two rivers force their way through the Mentasta and Nutzotin Mountains at the eastern end of the Alaska Range and then meet to form the Tanana. The Tanana flows generally northwest for 440 miles along the northern side of the Alaska Range before meeting the Yukon at Tanana, deep in the interior. The Koyukuk has three main branches, the North Fork, Middle Fork, and South Fork, the former two heading deep in the Brooks Range atop the Arctic Divide. From the confluence of the North and Middle Forks, the main stem of the Koyukuk flows generally southwest for 425 miles to join the Yukon at—where else?—Koyukuk. Anywhere else but Alaska, all three tributaries would be major rivers in their own right.

South of the Yukon, the Kuskokwim River shadows the big river for the lower third of its course. The Yukon flows north of the Kuskokwim Mountains and the Kuskokwim River south of the northern two-thirds of the range. At one point, the rivers are barely twenty-five miles apart before each meanders its own way through the marshy wetlands of their expansive deltas.

Both the Susitna and Copper Rivers flow from north of the Chugach Mountains into the Gulf of Alaska. The Susitna takes the easier route down the broad valley between the Alaska Range and the Talkeetna Mountains to Cook Inlet, while the Copper cuts straight through the range. The Susitna heads at the glaciers near Mount Deborah and begins to flow south. Thinking better of the idea, the river cuts west north of the Talkeetna Mountains and roars through Devils Canyon, perhaps the toughest whitewater in Alaska. Then, having been joined by the Chulitna River flowing south from Broad Pass, the "Big Su" swings wide and muddy through the lowlands north of Cook Inlet.

The Copper River rises from glaciers on the north side of Mount Wrangell and begins to flow north as if bound for the Yukon. But when confronted by the Alaska Range at Mentasta Pass, the river turns west and then loops south around the western Wrangell Mountains. Its major tributary, the Chitina, joins it just before the river carves a rocky path through the Chugach Mountains. Once the route of an amazing railroad that no road has dared to follow, the Copper River rolls through Wood and Baird Canyons before reaching the wide wetland delta at its mouth. Like so many wetlands in Alaska, the Copper River delta is a critical wildlife habitat, especially for migratory birds.

The Kobuk, Noatak, and Colville are the great rivers of northern Alaska and the Brooks Range. The Kobuk and Noatak each flow west for some 400 miles from sources within a few miles of each other between Mount Igikpak and the Arrigetch Peaks. Once out of the canyons at its headwaters, the Kobuk runs between the Waring and Baird Mountains and is wide and placid for most of its length. It flows through enormous oxbows and past the Kobuk sand dunes on its way to Kotzebue Sound on the Chukchi Sea. The Noatak runs north of the Baird Mountains, across the vast tundra of the Noatak Basin, and through the 65-mile-long Grand Canyon of the Noatak before swinging south to a labyrinthine wetland delta on Kotzebue Sound. Both river basins are critical habitat to the Western caribou herd.

North of the Brooks Range, the Colville River flows just the opposite way—east from the northern slopes of the DeLong Mountains across a full half of the North Slope. Although its many tributaries draining the northern reaches of the range descend in haste and with a sense of urgency, by the time their waters mingle in the Colville, it is a slow, meandering river curving across tundra flats to the Arctic Ocean.

All of the forces that built this diverse landscape are still very much at work. Cataclysmic, landscape-altering events continue to occur here, some with little or no warning. On July 9, 1958, an earthquake rocked Lituya Bay on the Fairweather Fault just west of Glacier Bay. During this one episode, the Pacific plate moved northwestward an estimated twenty-one feet. The quake triggered several large landslides that in turn created a giant tidal wave in the bay. As it surged toward the ocean, the wave tore mature trees from Lituya's shoreline up to an elevation of 1,740 feet, leaving a mountainside scar that is still visible. Of course, the Good Friday Earthquake of 1964 reshaped the coastline along hundreds of miles of southcentral Alaska.

Even more recent evidence of nature's capriciousness was the rambling surge of the Hubbard Glacier. The Hubbard heads in the Icefield Ranges of Canada's Kluane National Park and flows south for more than ninety miles to Disenchantment Bay. Glaciologists estimate that about A.D. 1000 the glacier extended all the way to the Gulf of Alaska, completely covering what is now Disenchantment *and* Yakutat Bays. Over time, Hubbard Glacier retreated more than thirty miles to uncover these bodies of water.

Then in the spring of 1986, the Hubbard advanced to cut off Russell Fjord at the head of Disenchantment Bay. The glacier pushed a thick plug of mud, gravel, and boulders across the mouth of the fjord, and runoff from rain and glacial melt soon turned the fjord into a freshwater lake. Meanwhile, the ice continued to build an increasingly taller dam at its mouth. By early October, the "lake" was more than eighty feet above sea level. On October 8, 1986, the ice dam burst and an estimated 3.5 million cubic feet of water per second, thirty-five times the flow of Niagara Falls, gushed out of the lake into Disenchantment Bay, creating standing waves ten to thirty feet high. Such is the power of Alaska's elements.

When the United States purchased Alaska from Russia in 1867, vehement cries against the transaction called the new acquisition "Seward's icebox." Alaska, it was said, was nothing more than a barren and frozen wasteland, locked in ice and cold as the proverbial witch's teat. In fact, Alaska's climate is as diverse as its varied geographic regions, and it is a land of extremes on both ends of the thermometer and rain gauge.

The warm Japanese current in the North Pacific keeps the southeast downright balmy in comparison to the interior. Cool summers with occasional days in the sixties or seventies blend into winters where temperatures rarely fall below twenty degrees along the Inside Passage. (All temperatures are given in Fahrenheit degrees.) It is a wet cold, however, and rain and clouds are the norm. Parts of southeast Alaska get upwards of 200 inches of rainfall per year. Kids in Ketchikan learned long ago how to answer the inevitable "How long has it been raining?" queries of cruise-ship tourists disembarking under rainy skies. Answer: "I don't know, I'm only twelve years old."

The Aleutians and the Alaska Peninsula are foggy and wet, too, but not nearly as warm as the southeast. The Japanese current flows too far south to be of much benefit here, and prevailing winds blow from cold

Siberia across the Bering Sea. Along the remainder of the southcentral coast, from Anchorage and Cook Inlet east to Prince William Sound, winters are hard, but with about five hours of daylight, and summers—well, it can be cool and misty or sunny and up to ninety-two degrees, the record for Anchorage. Just to the north of Anchorage, moderate temperatures and the daylight of long Arctic summers have made the Matanuska Valley Alaska's breadbasket.

Much of interior Alaska sits in the rain shadow of the Alaska Range. Here, rainfall averages less than a dozen inches per year, although what rain and snow does fall has little place to go given the spongelike muskeg and watery wetlands of the Yukon and Kuskokwim Valleys. Elevation plays a part, of course, and in June it can be below zero on the summit of Denali and ninety degrees on the banks of the Chena River in Fairbanks. Winter is another matter, however, with Fairbanks routinely reporting below-zero temperatures. At Barrow, the temperatures are just as low, but there the cold is masked in the total darkness that lasts from about November 18 until January 23.

The Brooks Range is even more arid than the interior, receiving less than six inches of rainfall a year. What snow does fall usually melts by May with the lengthening days of the Arctic summer. Of course, winter here and throughout the interior finds, as Robert Service wrote, "the white land locked tight as a drum."[6]

And, of course, within these generalities, there are the daily extremes. Changing sea breezes, high mountain environments, and fickle Arctic weather systems combine to make for rapid weather changes—sometimes within minutes. Perhaps Inga Sjolseth, then a single woman crossing the Chilkoot Pass in the spring of 1898, summed up Alaska's climate most succinctly when she confided to her diary: "The weather is very changeable here."[7] So it was, so it remains, but therein, too—along with its mountains, glaciers, and innumerable rivers—lies much of the mystique of this great land.

First Steps, Continuing Traditions

Perhaps no facet of Alaska is more difficult to get one's hands around than the diversities and complexities of the Alaska Natives. Limited archaeological evidence, conflicting oral history traditions, and the insensitivity of the first centuries of European contact are all parts of the puzzle. Even today, cultural and linguistic groupings are not universally accepted—particularly by Alaska Natives themselves.

What is without question is that Alaska was the gateway to the Americas. There may have been other avenues of settlement, but the majority of evidence supports a stream of migrations from Asia to Alaska that then spread throughout North and South America. So when did people first make this journey? Pick a date between 30,000 and 15,000 years ago, and there is apt to be an argument to support it. There is little agreement on dates, and some theories suggest that these migrations occurred in several phases, perhaps thousands of years apart.

These general dates coincide with the cold spell—if a "spell" can be said to last for millennia—of the Pleistocene Ice Age. The massive ice sheets that draped across the northern parts of North America and Asia locked up great quantities of the earth's water. With so much water trapped as ice, the sea levels dropped at least 250 feet and perhaps as much as 350 feet below contemporary levels. Much of the world's continental shelves were exposed, including a wide swath of land that in effect joined North America and Asia. Long called the Bering Land Bridge, this "bridge" was in fact over 1,000 miles wide at the lowest sea levels. It extended from the general area of the Pribilof Islands to more than 500 miles north of the current Bering Strait. Look at a map of the floors of the Pacific and Arctic Oceans, and the extent of these continental shelves becomes apparent. With lower sea levels, this "bridge" was a huge plain that is now called Beringia.

Changes in sea level occurred gradually over considerable time. Naturally, the "bridge" was also a gateway for land animals as well as a barrier between marine animals in the Pacific and Arctic Oceans. As the plain became exposed, hunters followed game onto it. Whether these first people were "just passing through" or actually settled in Alaska is debatable, but because they were of a hunter-gatherer culture, it is likely that any passage southward was gradual. Evidence of human occupation of Alaska dates from about 11,000 years ago—probably long after these migrations started—and comes in the form of small stone hunting tools, called microblades. Again, how "permanent" this occupation may have been is debatable. When the ice sheets began to melt and the sea level rose, this was also a gradual process. Even after waters formed the Bering Strait, probably about 12,000 years ago, there is evidence that there were cultural exchanges between the two continents, and there may even have been continuing smaller migrations from west to east.

When referring to Alaska's Native peoples, it is sometimes difficult to use terms that are both historically correct and universally accepted. Indeed, Native peoples sometimes argue among themselves about what they should be called. When possible, the name of a particular tribal affiliation or clan should be used, but sometimes generalizations and historical uncertainties do not permit this. The following general terminology is both respectful and historically accurate. *Alaska Natives* collectively refers to the special status of the Native peoples of Alaska, whether they are of Inupiat Eskimo, Yup'ik Eskimo, Aleut, Alutiiq, Athabascan, Tlingit, Haida, or Tsimshian heritage. (A *native Alaskan* is anyone born in Alaska.) *Eskimo* has been used historically to group Alaska Natives throughout the Arctic coastal region, but it is more correct to identify the specific languages and cultures of Inupiat or Yup'ik.

American Indian is a term that usually has been used to describe sovereign Native nations that have a treaty relationship with the United States. For a variety of reasons, which will become clear as this story of Alaska's history unfolds, Alaska Natives have historically had a different legal relationship with the United States than American Indian nations in the lower forty-eight United States. *Indian* is still an acceptable term—as long as it is not used derogatorily. Frequently, historical references give no greater detail than this generalization, and it is difficult to place a historical reference to "Indians" accurately into current terminology. *Native American* has become the politically correct term in the United States, but technically it does not distinguish between true Native peoples and those of any race who are "native" just because they were born in the United States. *Native peoples* is the more internationally accepted term. In Canada, the term *First Nations* has come to be used to distinguish the continuing sovereignty of Native peoples there.[8]

That said, it seems overly simplistic to say that there are four major cultural areas of Alaska Natives: Aleuts throughout the Aleutians; Inupiat and Yup'ik Eskimos along the Arctic coast; Athabascans in the interior; and the Tlingit and Haida of the southeast coast. But it is a place from which to begin.

The domain of the Aleuts is the Alaska Peninsula and the necklace of islands stretching westward toward Asia. *Aleut* is a term introduced by the Russians from the languages of Siberia and means something akin to "coastal people." The Aleut call themselves Unangan, "the original people." They are famous for their consummate skill in paddling a *baidarka* (a Siberian word for a form of kayak) and righting it when capsized, even in the most frigid of waters. Working their way westward from the peninsula, the Aleuts may have colonized some of the Aleutian Islands as long as 10,000 years ago. By the time the Russians arrived, the Aleuts inhabited all of the major islands of the chain, although their population was concentrated among the eastern islands and the peninsula where they had access to salmon and caribou.

Aleut life revolved around a principal village and seasonal fishing camps spread out among the smaller islands. The permanent settlements tended to be located on the southern side of the Alaska Peninsula to avoid the winter ice of the Bering Sea, but on the northern side of the larger islands to be somewhat sheltered from the prevailing southwest winds. A typical village might have five to ten dwellings, housing up to several hundred inhabitants. The Aleuts are a matrilineal society, where ancestry is traced through the females. While the man was generally the family leader, the houses in the village belonged to women, who shared them with a house group that frequently included the woman's brothers and their wives.

The Aleuts had a child-rearing procedure called *avunculate* that may have helped reduce adolescent friction between fathers and sons. When a boy approached puberty, he was sent to live in the home of his mother's oldest brother. This uncle became the boy's primary teacher and trainer. The role was a strict one undertaken to ensure the young man's competence in the skills he would need to survive as a hunter and paddler. Uncle and nephew were frequently at odds, but this arrangement left the boy's father to play a more nurturing and supportive role. One can almost hear an Aleut father sympathetically commiserating with his son: "Uncle made you practice rolling your *baidarka* that many times!"

And indeed, the Aleuts were preeminently adapted to the marine world. The most important animal in the culture was the sea lion—long before the appellation of "Steller" was attached to it. Like the buffalo to the Plains Indians, the sea lion provided many essentials: hide for boat covers, sinew for line and cord, blubber for oil, bones for tools, teeth for fishhooks, flippers for boot soles, and, of course, food. Whales and sea otters were also important, and halibut and cod supplemented their diet.

Aleut whaling was a highly ritualized exercise, undertaken alone with harpoons from *baidarkas*. Whales also provided many things, and frequently whale bone served in place of wood in their treeless domain. The houses, called *barabara*, another Siberian word, were oblong pit dwellings with grass and sod overlaying whale-bone frames and rafters.

While the men hunted, the women were busy sewing, weaving, and cooking. Aleut basketry made from the fine grasses found on the islands is among the best in the world, and long before GoreTex, the Aleuts were constructing waterproof boots and garments. Almost everything had a utilitarian function. The long hats and visors of the Aleut culture not only kept the frequent rain from one's face but also protected the hunter's eyes and reduced the glare of the seas. Sea otter furs were widely worn. The few extravagances were ceremonial cloaks adorned with the bright feathers of tufted puffins.

The social structure of the village was fairly low-key. The Russians coined the term *toyon* for the individual who was the leader among the house group leaders. While there is some evidence of warfare with the Yup'ik Eskimos and small raiding parties going between villages to right some alleged wrong, feasting and dancing between villages was more the norm. The Aleuts were a respectful and harmonious people until wooden ships from the west began to anchor off the shores of their islands.[9]

North of the Alaska Peninsula, the long Arctic coast is the domain of the Inupiat and the Yup'ik, historically identified collectively as Eskimos. The term *Eskimo* is a non-Native word that Europeans used to describe all of the Native peoples of northeastern Siberia, northern Alaska, and northern Canada. The word has been given various meanings, but the most common may come from the Micmac Indian word *Eskameege*, which was roughly translated by French Canadian trappers in the early 1800s as "to eat raw fish." Arctic Native peoples definitely do not use *Eskimo* to describe themselves, and it hardly does justice to the many different cultures in this area. It is far more accurate to speak generally of the Inupiat and Yup'ik cultures and then identify the main units within each.

The Inupiat occupy the Arctic coast from Norton Sound north and are divided into four main units: the Bering Strait people on the Seward Peninsula (around current Nome); the Kotzebue Sound people around Kotzebue Sound and the Kobuk and Noatak Rivers; the North Alaska Coast people or Tareumiut, meaning "people of the sea," along the Chukchi and Beaufort sea coasts (including Point Hope and Barrow); and the Interior North Alaska people or Nunamiut, meaning "people of the land," on the north slopes of the Brooks Range (current Anaktuvuk Pass). There are other smaller units recognized with the suffix *miut*, meaning "people of."

Far from living in solitary igloos, the Inupiat congregated in larger communities that might have hundreds of people. Present-day Point Hope, Wales, and Barrow started as Inupiat villages. The houses had underground tunnel entrances that served as cold traps so that cold air would not get in (cold air settles and does not rise), and were built into the ground as much as possible, using the earth as natural insulation. Frequently, the building material was sod over wood or whale-bone frames. The much-storied igloo was really used only as an emergency shelter when hunting parties were caught away from their villages.

The coastal units of the Inupiat depended heavily on bowhead and beluga whales, walruses, and seals. The inland Nunamiut relied on the Western caribou herd. Water travel was by *umiak*, a large, open skin boat that could be from fifteen to twenty feet long, and by kayak. On land, sleds were used. Dogs served as pack animals but apparently weren't put in harness pulling sleds until only a century or two before the Europeans arrived.

Unlike many Alaska Natives, the Inupiat did not have a matrilineal family organization. Rather, kinship was determined bilaterally with relatives on both sides being of equal importance. Within this family tree, kinship was very important, however, and there was a genuine distrust of anyone not somehow related to the extended family.

Disputes among men within local groups were settled by song duels—apparently something akin to a "battle of the bands." Each of the feuding parties took turns singing songs that sought to set the record straight and belittle their opponent in the process. These were sung before the group, and the group's laughter and cheering gave praise, bestowed ridicule, and helped determine which of the two crooners ultimately withdrew in shame. When that happened, the duel ended, and the matter was declared resolved. Conflicts outside the group were not settled so amicably. The Inupiat engaged in both warfare and trade with the Chukchi and Siberian Eskimos.

The major social event was the Messenger's Feast, where one group would invite another for several days of eating. This is similar to the potlatches of the Athabascans and Tlingit, but while gifts were given to guests, such giving did not progress so far as to bankrupt the host. Other feasts included highly ritualized preparations for bowhead whale hunts.

South of the Inupiat territory, the respective units of the Yup'ik stretched along the Arctic coast from St. Lawrence Island to Prince William Sound. Given the differences in terrain over this distance, the different Yup'ik units are very diverse. In the Bering Sea, the St. Lawrence Island Yup'ik lived on rugged St. Lawrence Island; the Yukon-Kuskokwim Delta Yup'ik occupied the wetland deltas of those two rivers; the few Nunivak Island Yup'ik inhabited Nunivak Island off the coast between the deltas; and the Bristol Bay Yup'ik surrounded Bristol Bay and the southwestern two-thirds of seventy-eight-mile-long Iliamna Lake. On the Pacific side of the Alaska Peninsula, the southernmost Yup'ik were the Koniag Alutiiq on Kodiak Island and its surrounding archipelago; the few Uneqkurmuit Yup'ik on the tip of the Kenai Peninsula (current Seldovia); and the Chugach Alutiiq along the coasts of Kenai Fjords and Prince William Sound.

Spirit poles are erected at Yup'ik graves to keep the spirits of the dead from disrupting the world of the living. This world includes *kashgees*, the ceremonial houses for men used by Yup'ik, Athabascans, and Alutiiq. Here the man was in charge, and adults taught traditional skills to boys. In the home, the woman ruled and taught their daughters the required homemaking skills. The Chugach in particular traded and inevitably fought with others, including the Athabascans.

The farther south a group was, the more it could rely on salmon runs. Fish-drying camps with racks and racks of salmon were a common sight. Conversely, Bering Sea groups were more dependent on seals and whales. The Yup'ik also held Messenger's Feasts, but an even more important occasion was the Bladder Feast. Yup'ik believe that the seal spirit is housed in its bladder. Seal bladders are saved from the previous hunt and then inflated and hung up during five days of dancing. After this, the bladders are returned to the sea. This ensures that the spirits of dead seals are returned to their homes so that they can be reborn and hunted again. The ceremony is a simple, yet profound, recognition of nature's cycle.

The broad expanse of interior Alaska—from the Brooks Range to the Alaska Range and along almost the entire length of the Yukon and its tributaries—and the valleys of the Copper and Susitna Rivers are the territory of the Athabascans. Within this wide geographic area, Athabascans make up nine ethnic-linguistic groupings, which are generally divided into riverine, upland, and Pacific categories based on their geographic range and their resulting hunting and fishing methods. The riverine are the Ingalik, Koyukon, Tanana, and Holikachuk along the lower and middle Yukon and the lower Koyukuk and Tanana Rivers. The upland are the Kutchin, Han, and Upper Tanana along the upper Yukon and upper Tanana Rivers. The Pacific are the Ahtna and Dena'ina through the valleys of the Copper and Susitna Rivers, with the Dena'ina territory extending around the Alaska Range and Cook Inlet all the way to the upper Kuskokwim and the northeast end of Iliamna Lake.

Salmon fishing, supplemented by moose and caribou hunting, was the mainstay of the riverine and Pacific groups. Salmon were caught in dip nets or a variety of stone traps or wooden weirs. The upland groups lacked the strong salmon runs and depended on caribou. The more plentiful and reliable the game—such as the salmon runs on the Copper River for the Ahtna—the more apt the group was to build more substantial dwellings and remain in one territory. The Dena'ina Athabascans built permanent winter villages throughout their realm, the largest of which comprised some 260 semisubterranean houses at Kijik west of Lake Clark. The upland groups were almost always moving in pursuit of game or to intercept the caribou migrations. Just as the sea lion was essential to the Aleuts, the upland Athabascans depended heavily on caribou for food, clothing, utensils, and necessary accoutrements.

Because they were hunters and gathers in a relatively spartan environment, the Athabascans spread out in small bands. There were no large villages, and there were only an estimated 10,000 Athabascans across this whole vast area by the mid-1800s. Such an existence made them highly adaptive, and they borrowed readily from the neighbors that they came in contact with through trade and occasional warfare. The Dena'ina on Cook Inlet, the only Athabascan group with direct access to the seacoast, may have borrowed the kayak design from their Aleut or Yup'ik neighbors or traded or purchased kayaks from Yup'ik along their shared boundaries on the Bering Sea coast, the Mulchatna River, and Iliamna Lake. The Tlingit may have influenced the Ahtna's use of large plank dwellings and clan symbols. Trade was important and was carried on outside of the Athabascan family. The copper of the Ahtna was particularly prized. Because of the rugged geography, the Ahtna clan held a lucrative trade monopoly between the coast and the interior, acting as middlemen for all trade along the Copper River. The Koyukon and Kutchin also traded with their Inupiat neighbors, and the Ahtna traded with the Tlingit.

There are some similarities between the Athabascan culture and that of the woodland American Indians. The Athabascans hunted with bow and arrows, wore fringed and beaded buckskin clothing, and built birch-bark canoes. In fact, they used birch bark for just about everything, including bowls, containers, and construction. In emergencies, they stretched moose hide over birch or willow frames and built coracles. These rounded shells were fine for floating downriver, but they were far too cumbersome to be paddled upstream with much success. Also, Athabascans were masters of the hunt with snares or deadfalls. Life was tenuous, particularly among the groups in the interior, and there was no energy to waste.

Perhaps in such an environment it is only natural that these peoples would harbor strong relationships with animals, believing in reincarnation as animals and vice versa. Raven was at the center of this culture and was the great trickster who constantly strove to upset the moral order with his cunning deceptions. There are many Athabascan legends about Raven, and they are frequently used to teach children right from wrong.

To varying degrees among the nine groups, potlatches were important ceremonial events. These may have been borrowed from the Tlingit, although how large the event was seems to have been dictated by the food supply. Potlatches might take the role of a few friends coming over for dinner, rather than the larger extravaganzas prevalent among the Tlingit. The matrilineal clans, divided into Raven and Seagull moieties, may also be the result of Tlingit influence—or, of course, vice versa. And then there is the question of language. The Athabascan tree of languages is closely related to that of the Navajo and Apache of the American Southwest. Who influenced whom? Or are they simply descendants of common ancestors, some of whom moved southward long ago while others stayed in the North?

In contrast to the Athabascans and other Alaska Natives, those living in the resource-rich southeast have been abundantly blessed by nature's bounty. This is the land of the Tlingit and the Haida. While much of interior Alaska was free of ice 20,000 years ago, the verdant coast of southeast Alaska was extensively covered by giant glaciers. As the glaciers slowly receded, the first steps were taken here 10,000 to 7,000 years ago by the Tlingit. Some Tlingit legends tell of migrations northward from the Skeena River in British Columbia. Others speak of journeys over the high Coast Mountains and down the river valleys to the sea. In time, the Tlingit inhabited all of southeast Alaska, settling in villages on the mainland from icy Yakutat Bay in the north to Portland Canal in the south and on the major islands of the Alexander Archipelago. The Haida are much more recent arrivals. In the seventeenth or eighteenth century, they moved northward from the Queen Charlotte Islands off British Columbia and forced the Tlingit out of southern Prince of Wales Island. Much later, in 1887, the Tsimshian moved from Canada into southeast Alaska and settled on a reservation on Annette Island near Ketchikan.

The Tlingit, Haida, and Tsimshian peoples are the northernmost groups representative of the Northwest Coast Native Culture. Although they differ in many ways from one another and have unrelated languages, the Tlingit, Haida, and Tsimshian share the basic social, economic, and cultural patterns characteristic of Northwest Coast Natives from Alaska to Washington. The Tlingit, however, also share a heritage with other Alaska Natives because their language shows a distinct relationship to Athabascan.

By the late 1700s, the Tlingit had a combined population of about 15,000, most heavily concentrated along the Stikine and Chilkat Rivers. Each clan—such as the Kiksadi Tlingit at Sitka—established permanent winter villages and then dispersed in small groups to seasonal hunting and fishing camps. Their winter dwellings were quite large and well built. Some of these cedar plank houses were as large as forty by sixty feet and were shared by four to six families. The houses had four large interior house posts supporting the roof system. The carvings on these house posts were the beginnings of the various types of totem poles that came to be placed in many villages.

Turning to the ocean for most of their food, the Tlingit fished for cod, halibut, and herring, plucking them from the water with spears and wooden hooks with barbs made from bones. Once on the surface, the larger halibut were clubbed to death. In summer, the Tlingit stretched traps across the rocky streams to harvest salmon swimming upstream to spawn. On the larger streams, they constructed elaborate devices, such as stone traps that worked in conjunction with the tidal action, as well as wooden weirs and nets. Dip nets, spears, and gaffhooks were also used. Villages on the islands also depended on seals and Sitka black-tailed deer, while villages on the mainland had more opportunities to hunt moose and mountain goats. Seals were usually clubbed on land, and hunting was done with bows, arrows, and spears.

The chief transportation vehicles were dugout canoes carved from the trunks of massive red cedar trees. The larger crafts might be up to sixty feet in length and could accommodate many people. They were used for long-distance travel, transporting goods, and occasional warfare. These were v-shaped vessels with deep drafts and large prows and sterns, frequently adorned with clan crests. A smaller model—the Tlingit version of the runabout—was used for local trips. It was ten to sixteen feet long with a u-shaped bottom and could carry only four to six people.

Southeast Alaska had a much warmer climate than that experienced by other Alaska Natives, but the prized article of clothing was the Chilkat robe. These robes were woven by the women from mountain goat wool and cedar bark strips based on totemic patterns designed by the men. A variety of dyes gave the robes a standard yellow-and-black coloration. Worn, or sometimes only displayed on ceremonial occasions, Chilkat robes were a major trading commodity.

Within both Tlingit and Haida societies, social organization and the inheritance of leadership and wealth are determined by matrilineal descent within the clans. These are named after animals or mythical beings, which are depicted in symbols or crests that are put on clothing, blankets, totem poles, and other clan property. Clans have always been very important within Tlingit and Haida societies, and historically it was the clan itself that held ownership to property, including houses, canoes, fishing grounds, ceremonial garments, crests, songs, dances, and stories.

Each clan had its exclusive fishing waters. Infringement on these waters by other clans was cause for war or, at the very least, some form of compensation. Each clan followed its own trade routes along the coast and into the interior, trading in particular with the Athabascans. The Tlingit traded dried fish, otter furs, and Chilkat robes for caribou skins, fox furs, jade, and copper—items they could not obtain on the coast.

Both the Tlingit and Haida led a rich cultural life, gathering for all types of occasions, celebrating births and weddings, mourning deaths, and dancing before long fishing and trading expeditions. The major ceremonial institution was the potlatch. The host clan of a potlatch invited guests to its house, often for days at a time. There was a strong competitive element to potlatches, and the hosts were expected to give gifts and dole out their many possessions one by one to their guests. Such hospitality left the host clan greatly elevated in social rank. Fortunately, they knew that their guests would be required to reciprocate the invitation and that they would soon be on the receiving end. Such generosity was recognized with special potlatch hats with rings that indicated the number of potlatches its wearer had sponsored.

The totem poles found among the Tlingit and Haida are an influence of the Northwest Coast Native Culture. Originally, the carvings were adornments to house posts, but the poles came to symbolize many different things. There are crest poles that give the ancestry of a particular family or clan; history poles that record an event in the history of a clan; legend poles that illustrate folklore or real-life experience; or memorial poles that commemorate a particular individual. There are also mortuary poles that contain the ashes of a person who was cremated. Contrary to popular belief based on early missionary reports, totem poles were never religious in nature or use.

The legends and history depicted by the poles come from the oral history of the Tlingit and Haida people. The best method for tracing them is to learn the clan affiliations of the residents of the villages at the time the poles were collected and then find corresponding legends on the totems. Interpretation of totem poles must be approached with caution. Many of the stories and events associated with them have been lost over time. Others are interpreted differently by different clans and communities.[10]

Perhaps the most important thing to remember about Alaska's Native peoples is that all believe that at the basis of all life is the land. Their existence, whether searching for a solitary bowhead whale off St. Lawrence Island or casting nets into a rush of thrashing salmon in the Stikine River, depends on nature's bounty. It has always been a tenuous balance. Alaska Natives took the first steps in this land, and their proud traditions continue.

Postscript to an Era (Prehistory–1728)

Athabascan legends *include a tale passed down for centuries that foretold of a new people who would have yellow hair and pale skin and who would drive them from their hunting grounds. The winds of change were blowing. A new frontier was on the horizon. The Athabascan legend was about to come true.*

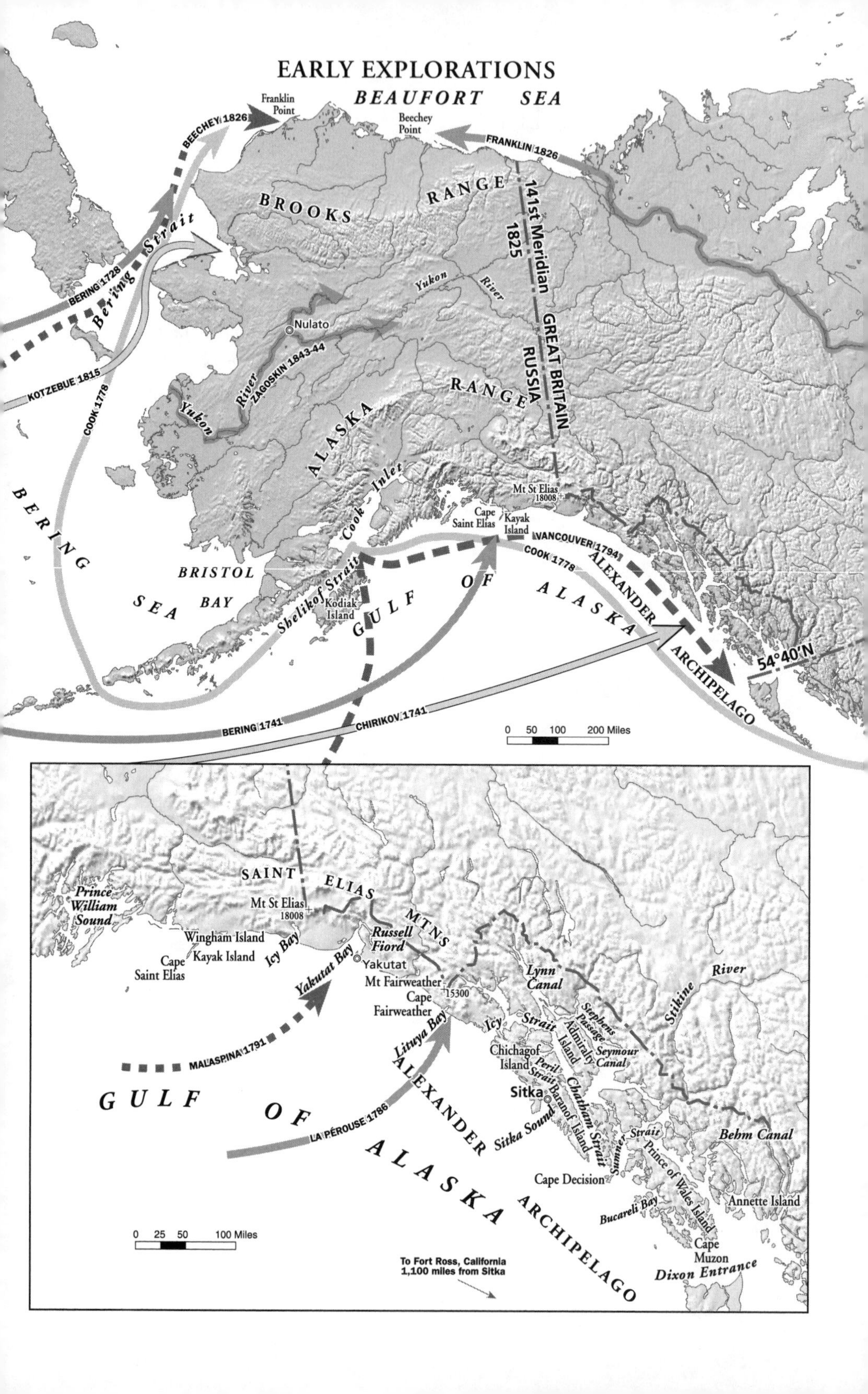
EARLY EXPLORATIONS
BEAUFORT SEA
Franklin Point
Beechey Point
BEECHEY 1826
FRANKLIN 1826
BROOKS RANGE
141st Meridian 1825
Bering Strait
BERING 1728
KOTZEBUE 1815
COOK 1778
Yukon River
Nulato
ZAGOSKIN 1843-44
GREAT BRITAIN
RUSSIA
ALASKA RANGE
Cook Inlet
Mt St Elias 18008
Cape Saint Elias
Kayak Island
VANCOUVER 1794
COOK 1778
ALEXANDER ARCHIPELAGO
54°40′N
BERING SEA
BRISTOL BAY
Shelikof Strait
Kodiak Island
GULF OF ALASKA
BERING 1741
CHIRIKOV 1741
0 50 100 200 Miles
SAINT ELIAS MTNS
Prince William Sound
Mt St Elias 18008
Wingham Island
Cape Saint Elias
Kayak Island
Icy Bay
Yakutat Bay
Yakutat
Russell Fiord
Mt Fairweather 15300
Cape Fairweather
Lituya Bay
Lynn Canal
Stephens Passage
Stikine River
Icy Strait
Admiralty Island
Seymour Canal
Chichagof Island
Peril Strait
Sitka
Baranof Island
Chatham Strait
Sitka Sound
Behm Canal
Sumner Strait
Prince of Wales Island
Cape Decision
Annette Island
Bucareli Bay
Cape Muzon
Dixon Entrance
MALASPINA 1791
LA PÉROUSE 1786
GULF OF ALASKA
ALEXANDER ARCHIPELAGO
0 25 50 100 Miles
To Fort Ross, California 1,100 miles from Sitka

BOOK TWO

Lifting the Veil: An Empire Up for Grabs 1728–1865

God is in His heaven, and the Czar is far away.

—Common lament in Russian America

The historical exploration of Alaska *is a tale interwoven with the same themes and players present in the broader saga of the colonization of North America. All of Europe's major powers explored here in an attempt to win greater empires—and the riches that came with them.*

In the early 1700s, Europe was embroiled in war. As was often the case, the provoking issue was royal succession, but the ramifications of the resulting treaty were geographic. From 1701 until 1713, the War of the Spanish Succession, called Queen Anne's War in the North American colonies, saw France and Spain face a Grand Alliance led by England to dispute the claim of Louis XIV's grandson to the throne of Spain. When the Treaty of Utrecht brought an end to hostilities, the world map was set for a century of further explorations and escalating colonial rivalries.

At Utrecht, Great Britain (as a unified England and Scotland was now called) solidified its dominance as a world sea power with the acquisition of Gibraltar from Spain. Great Britain also began its envelopment of French North America by acquiring Newfoundland, Nova Scotia, and territories to the west of Hudson Bay from France. In addition to Gibraltar, Spain lost territory in Italy to Austria. Spain and France, though weakened, nonetheless used the resulting peace and the tentative cordiality between each other to turn their attentions to furthering their individual North American empires.

Perhaps most significant to the future of Alaska, a major European power emerged that had not been a party to the Treaty of Utrecht. While western Europe fought among itself, Czar Peter the Great's Russia used the time to crush Charles XII of Sweden at Poltava in 1709 and secure Russia's long-sought "window on the west," a seaport on the Baltic. It slowly dawned on the other European powers that Russia would henceforth be a player on both European and global chessboards. So it was that Alaska—or what was then only a void upon the world map—came under the musings of these European powers. The Russians would be first to probe the mysterious veil and would make the most lasting inroads in this new frontier.

The Czar Looks East

Just as Czar Peter the Great led Russia to open its windows to the west, he also directed the exploration of Siberia and pushed Russia's influence to the east. In doing so, he called upon a Danish sailor named Vitus Bering. Born in Horsens on the eastern coast of Denmark's Jutland in 1680, Bering joined Peter's fledgling navy and served Russia ably against Sweden in the Great Northern War. In 1725, as one of the last official acts before his death, Czar Peter ordered Bering to organize an expedition to explore eastward from Siberia's Kamchatka Peninsula.

History's shorthand has long suggested that the goal of the expedition was to determine whether Asia and North America were connected. In truth, this question had apparently been settled some years before—although both Peter and Bering seem to have been oblivious of that fact. In 1648, Semen Ivanovich Dezhnev, a Siberian cossack, sailed eastward with ninety men in seven small boats from the Kolyma River on the Arctic coast of northeastern Siberia. They followed the coastline south through the strait at the easternmost tip of the Chukotsk Peninsula and then continued southwest, eventually reaching the vicinity of the Anadyr River. Although it took years for the report of this voyage to reach Moscow, it should have established that a watery passage separated Asia from North America.

For whatever reason, Peter's orders to Bering strongly implied that the continents were in fact connected. Bering was directed to reach the place where Asia "is joined with America" and then continue along the coast "to any city of European possession." If possible, he was to hail any European vessels he encountered and learn from their crews the names and mapping information applied to this new land. In other words, Peter wanted to know just how far his European rivals in the west were encroaching on the North American continent, and how long it would be before they were also knocking on his eastern door.[1]

Bering's first task was to move his command across the 4,000-mile expanse between St. Petersburg and the tiny Siberian outpost of Okhotsk. Every skilled worker that might be needed, and every item that the land or sea would not provide, had to be hauled along. At Okhotsk, the expedition built a small ship and sailed 600 miles across the Sea of Okhotsk to the western side of the Kamchatka Peninsula. Eastern Siberia and the Kamchatka region were still on the very fringes of the Russian Empire. Only within the last quarter century had Russian fur hunters, *promyshlenniki*, pushed eastward from Okhotsk and reached Kamchatka. Just as American mountain men and French-Canadian voyageurs pushed the frontiers of North America westward a century later, these *promyshlenniki* opened the avenues upon which later settlers would travel.

Unfortunately for Bering, the avenue he next chose was laborious at best. He and his men hauled their materials eastward across Kamchatka yet another 100-plus miles and finally set about building the sailing vessel *Gabriel* at Nizhne-Kamchatsk near the mouth of the Kamchatka River and the shores of the Pacific Ocean. Only later did Bering learn, when blown off course in a storm, that Kamchatka was indeed a peninsula and that he could have simply sailed around its southern tip, thus minimizing the final leg of a most arduous trek.

Finally, in July 1728, three and one-half years after Czar Peter's death, Bering and his crew nervously sailed the *Gabriel* northward along Kamchatka's eastern shores to carry out Peter's orders. Considering the years of preparation, it turned out to be a short trip. Prudently—some would say much too cautiously—Bering kept the shores of Asia within sight to port for much of the voyage. When land did appear off the starboard quarter, he named St. Lawrence Island, apparently the first non-Native name applied to what is now part of Alaska.

Then Bering sailed northward through the strait that today bears his name—never mind that Dezhnev had been there eighty years before. The comforting shoreline of Asia soon disappeared to the west, and no land was visible to the east. Bering sailed on to 67°18' north, about the latitude of Cape Krusenstern, and then, fearful of getting trapped in the ice pack, he turned south. By September 2, the ship and crew were safely back at Nizhne-Kamchatsk preparing to hunker down for the winter. An attempt to sail north again the following spring met with a storm that blew the *Gabriel* westward around the southern tip of the Kamchatka Peninsula and back into the Sea of Okhotsk, inking that passage on the map. A year later, Bering was back in St. Petersburg, having made a journey of gargantuan proportions, but failing even to sight the North American mainland.

That it took Bering almost a decade to mount another expedition says something about both the political intrigues of the Russian court and the complicated logistical requirements of the journey. Leaving once again for Okhotsk in 1737, Bering and his men completed the construction of the *St. Peter* and the *St. Paul* there in the summer of 1740. The vessels were identical, each eighty feet in length with a twenty-two-foot beam and a nine-and-a-half foot draft. They were rigged as brigs with two masts, and each was armed with fourteen two- and three-pound cannon. Bering commanded the *St. Peter* with seventy-five men on board, and Aleksei Chirikov, a veteran of Bering's 1728 voyage, commanded the *St. Paul* with a seventy-six-man crew. Needless to say, accommodations were cramped.

That fall, the two ships sailed from Okhotsk, rounded the southern tip of the Kamchatka Peninsula, and dropped anchor in Avacha Bay, a sheltered harbor on the peninsula's Pacific side. There, they settled in for a long winter. Bering named their collection of huts Petropavlovsk after his two vessels.

Theirs was a surprisingly international assemblage. Bering, the Dane, was still viewed as a foreigner by some of the Russians. Chirikov, a Russian, was one who held that view strongly, despite his years of service with Bering. Then, there was the hot-tempered German Georg Wilhem Steller. Listed as a "mineralogist," but schooled in both botany and medicine, he had his own ideas of what lands lay eastward. Add to the mix the Frenchman Louis de la Croyere, astronomer, well-dressed dandy, and half brother of geographer Joseph Delisle. De la Croyere had with him Delisle's map of the Pacific that had been sanctioned by the Russian Academy of Sciences in St. Petersburg. The Delisle map showed the lands of Yezo and Gama as extensions of Asia lying east of Japan in the North Pacific. It was to prove that the only thing worse than no map is a wrong map.

After wintering on Avacha Bay, the expedition made ready to sail east in the spring of 1741. Things began to go wrong from the start. May 1 found the two ships still tightly encrusted in heavy ice. Necessary supplies being shipped overland by dogsled from Bolsheretsk were only trickling in. When the ice finally cleared, the wind died for days on end. Anxious to get moving, Bering ordered the ships rowed toward the mouth of the bay. Finally, on June 4, a breeze freshened from the northwest, and they were on their way. But to where?

Bering left the operation of the *St. Peter* to his first mate, Lieutenant

Sven Waxell, a Swede, and retired to the cabin he shared with Steller. For starters, Bering was sixty, an old man by the life span of the time, and long worn out by the hardships of Siberian winters. Then there was this crazy notion of searching for Yezo and Gama Land. Bering doubted their existence, but his instructions demanded he search for them.

And so the tiny flotilla sailed southeast to where Gama Land was shown at a latitude of forty-six degrees north. This path skirted well south of the long tail of the Aleutians and led to nothing but empty ocean. Steller, for one, insisted that land lay just to the south, but a hurried council of officers agreed to alter course to the east. Delisle's map and the others Bering carried showed nothing but a void in that direction.

Clearly, the intent of each captain was to keep the other in sight, but the Alaskan weather was not conducive to such intentions. Intermittent rain gave way to gale-force winds, and then soupy thick fog hung over the rolling seas. On the morning of June 20, the *St. Paul* hove to in order to reduce her canvas, and the two ships soon became separated in the fog.

Standing orders called for each ship to put about and return to the last spot of contact. This the *St. Peter* did, combing the seas diligently for a period of several days. Then, unaccountably, Bering decided to sail south again in the hope of finding either Gama Land or the *St. Paul.* Neither was found, and by the time the ship's officers voted to continue the voyage to America, precious time had been wasted. For the better part of three weeks, the *St. Peter* sailed northeast—straight into the Gulf of Alaska. Finally, on July 16, 1741, Sofron Khitrov, who was keeping the expedition log for Bering, recorded the sighting of "high, snow-covered mountains" and "among them a high volcano." They were southwest of Yakutat Bay. Four days later, the *St. Peter* rounded a cape that Bering named for the patron saint of the day, St. Elias. Later explorers, including the English captain James Cook, would attach the name to the highest peak and the mountain range as well.

Cape St. Elias was actually the southern point of a small island jutting south from near the mainland. It still looks mysterious today—a long narrow slice of land, twenty-two miles long and one and a half miles wide. Dark, rain-filled clouds often obscure the island's rocky spine and shadow its spruce-hemlock forests. The Russians and all later visitors would come to call it Kayak Island because of its silhouette from afar. The *St. Peter* anchored off its western shores, and Bering prepared to send parties ashore to replenish freshwater supplies.

Steller was beside himself. This first step onto North American shores was what he had been waiting for—sometimes none too patiently—for years. Bering at first was not disposed to permit him ashore. Only after Steller threw a fit and told Bering that the captain was keeping him from doing his assigned duties did the old man relent. Had Steller been a naval officer, he probably would have been thrown in the brig; but he was a civilian scientist, and Bering chose to make light of his outburst and permitted him ashore. Thus, one watering party landed on Wingham Island just to the north, while Steller and several crewmen landed on Kayak Island, perhaps at a location now called Watering Place Creek. While the seamen fell to the task of filling the water casks, Steller determined to seize "every opportunity to accomplish as much as possible with the greatest possible dispatch." And indeed he did so, scurrying about the western shores of the island and gathering an impressive array of plant species.

If Bering had any doubts about whether he was indeed in North America, Steller's journal entry cast them aside. "Luck, through my hunter," Steller wrote, "placed in my hands only a single specimen, which I remember having seen painted in vivid colors and written about in the newest description of Carolina plants and birds published in French and English not long ago in London and whose author's name eludes me. This bird alone sufficiently convinced me that we were really in America."[2] Thus did Steller record his acquisition of what is still called a Steller's jay.

The scientist in Steller was taking over, and he asked Bering for extra time to continue his exploration of the island. For whatever reasons, Bering gruffly declined and told Steller to return to the ship quickly or be left behind. It was, Steller noted in his journal, as if ten years of preparation had led to but ten hours of exploration. Put in contemporary terms, it was something akin to arriving at the Grand Canyon or Yellowstone on a whirlwind vacation and then hopping out of the car for only five minutes before driving on another 300 miles.

But it was a start. Meanwhile, Aleksei Chirikov and the *St. Paul* had also made a landfall. After becoming separated from the *St. Peter*, Chirikov made a brief search and then held the *St. Paul*'s course east by northeast. Some would later say that he was all too glad to be separated from Bering. Whatever glory lay ahead would now belong solely to Russia. On July 15, one day before Bering's sighting of Mount St. Elias, Chirikov glimpsed the wooded mountains of what was probably Baranof Island, recording in his log: "This must be America, judging by the latitude and longitude."[3]

Cautiously, the *St. Paul* cruised northward along the fog-enshrouded coastline. Then on July 18, the high peaks of what would later be called the Fairweather Range came into view, and Chirikov saw a break in the shore, apparently the mouth of Lisianski Strait south of Cross Sound. Chirikov, too, was concerned about fresh water. He sent a detachment of eleven men ashore in a small boat at the mouth of the strait. Landing anywhere along the coast would have been problematic, but the entrances to Cross Sound and Lisianski Strait have some of the toughest and most erratic tides and currents in southeast Alaska.

The crew lining the rail of the *St. Paul* watched the little boat fight the surf into the narrow entrance and then disappear behind a row of foamy breakers. For five anxious days they waited, but the boat never returned. Chirikov dispatched his remaining boat and two volunteers to search for the first, but it suffered a similar fate, vanishing without a trace. When two Tlingit canoes approached the *St. Paul* on the morning of July 25, Chirikov concluded that their occupants had either killed or detained the shore parties. More likely, the two boats were caught in the jaws of a strong riptide and dashed to bits on the rocks.

Stunned by the loss of almost a fifth of his crew and now without small boats to go ashore and fetch fresh water anywhere else, Chirikov elected to head directly back to Avacha Bay. So much for glory. The *St. Paul* made good headway across the Gulf of Alaska, making landfall off the southern tip of the Kenai Peninsula on August 1. Then began the fight with the westerly winds. The little ship tacked this way and that and some days made no progress at all to the west. On September 9, the crew sighted Adak Island, two-thirds of the way out on the Aleutian chain. A freshwater stream cascaded down the hillside, but without boats there was no way to land. Aleuts appeared and approached in kayaks. The Russians were able to barter somewhat, exchanging knives for a meager supply of fresh water before a squall blew up and forced the *St. Paul* from the anchorage.

With the remaining water severely rationed and the autumn gales blowing fiercely, the dreaded scurvy made its appearance. Two of Chirikov's lieutenants, the navigator, and the dapper Frenchman de la Croyere were among those who died. At last the mountains of Kamchatka came into view, and the *St. Paul* sailed slowly into Avacha Bay on October 10, 1741. But where was her sister ship?

Off Kayak Island, Bering the seaman had sought to cool Steller the scientist's enthusiasm for further exploration with the time-honored "we're only halfway there" speech of sailors and mountaineers. Yes, they had quite a distance to cover in returning to Avacha Bay, but surely without the prior detours to search for Gama Land and the *St. Paul,* there was still time for another look. Bering said no, however, and in his haste to weigh anchor and leave the shores of Kayak Island, he also left twenty water casks unfilled. So Steller got only his ten hours ashore, and the *St. Peter* creaked and groaned her way into the same westerlies that were blasting the *St. Paul.* Ironically, in late August their tracks were quite similar and only a day or two apart. Neither ship saw the other, however.

Scurvy broke out on board the *St. Peter,* too. Then the error of the twenty empty water casks showed itself. Bering chose to detour northward toward land to find fresh water. On August 29, forty days after leaving Kayak Island, the *St. Peter* dropped anchor in the Shumagin Islands off the Alaska Peninsula. Steller was allowed to accompany the watering party without protest this time, but while he located several clear streams, the crew was busy filling the casks from brackish pools too close to the shore. Despite Steller's protests, the crew continued the procedure and, like the faulty map, salty water proved almost as bad as no water.

Westward they went, or at least tried to. The *St. Peter* bumped her way along the Aleutian chain while scurvy took a grim toll. Now well behind the *St. Paul,* the ship passed Kiska, and on October 30, with Waxell at the helm as he had been most of the trip, they sighted the Semichi Islands in the Near group at the extreme western end of the Aleutians. Bering was dying in his cabin. Avacha Bay was only a week's sail west, but wracked by scurvy Waxell mistook the Semichi Islands for the Kuriles and thought the ship to be to the south of Kamchatka. So he turned north—north into the icy teeth of winter.

Finally, during another attempt to get fresh water, the *St. Peter* was driven aground in the Komandorskie Islands east of Kamchatka, so close but still so far from her destination. Convinced the ship had in fact reached Kamchatka, Bering died of scurvy in December on what is now Bering Island. Steller and others eked out an existence through the winter, surviving on fish, seals, and the now-extinct Steller's sea cow. In the spring they salvaged enough timber from the *St. Peter* to build a small boat and finally made their way back to Petropavlovsk on Avacha Bay, arriving there on August 27, 1742.

Chirikov meanwhile was hailed as the discoverer of a great new land. (To his credit, he did sail the *St. Paul* back to Attu in the Aleutians in the spring of 1742 looking for Bering and the *St. Peter.*) Bering had been given up for dead, which in fact he now was. Steller never returned to St. Petersburg and died in Siberia in 1746.[4]

The immediate impacts of Bering's and Chirikov's voyages were few. Russian bureaucracy moved with glacial speed, and, as always, it was difficult to maintain interest—and the funding necessary to act on that interest—in lands thousands of miles to the east. There were always more pressing matters to be addressed at the doors of Europe. Upon the return of the survivors of these voyages, however, a seed was planted that was to grow into the tree of the Russian American Empire. That seed was furs. Bering's survivors, principally Chirikov's men, managed to return with a load of rich furs, particularly the pelts of sea otters. These caused a sensation, and slowly but steadily the *promyshlenniki* began to island-hop their way from Kamchatka to the Komandorskies and then eastward along the Aleutian chain.

During the half century after Bering, these efforts were largely private. The Russian government dispatched an occasional survey and kept a wary eye on reports of what the Spanish were up to along the North American coast, but by and large this was a fur rush of private entrepreneurs bent on individual fortune. What a contrast there was to be between this capitalistic approach, especially as championed under the laissez-faire policies of Catherine the Great from 1762 to 1796, and the strong involvement of both church and state in the later affairs of the Russian-American Company.

Spanish Entradas

So where were the Spanish? Large as it was, the vast Pacific Ocean had once been termed a Spanish lake, a mare clausum, closed to all but courtiers of Madrid until that Elizabethan sea dog Sir Francis Drake had dared to enter it in 1579. Spanish power reached its zenith with the defeat of its mighty armada in 1588, but a century and a half later Spain still held considerable sway over much of the Western Hemisphere. Tales of a Northwest Passage through North America intrigued the Spanish, just as they did other nations. The Spanish contributed their share of rumors to the quest, fired in part by the nonsensical—if not outright fraudulent—ramblings of seafarer Juan de Fuca. Slowly, however, it became apparent that Spain must strengthen its northern frontier in California to counter the pressures advancing upon it from Russian America and British North America. Spain's answer was to fortify its presence in California by establishing a series of missions from San Diego north, right up the California coast.

In 1774 as part of this overall plan, Antonio Bucareli, viceroy of New Spain, ordered Juan Josef Perez, a veteran captain who had been ferrying colonists and supplies to the California missions since the founding of San Diego five years earlier, to sail north from Mexico and show the Spanish flag. Perez was to sail the *Santiago* to a latitude of sixty degrees north and land here and there, claiming the best locations for Spanish settlements. This course would have brought him all the way to Steller's Kayak Island. But scurvy, as Steller well knew, was more of an enemy in those days than the unknown. Its outbreak and the North Pacific's typical weather conspired to turn Perez southward well short of his goal. His farthest northerly landfall was somewhere off the western shores of Prince of Wales Island, near what would later be named Bucareli Bay.

In the gray mists of the North Pacific, Perez missed the mouth of the mighty Columbia River and the wide strait that would soon be named for Juan de Fuca. But on the cruise southward, he put in at a mountain-encircled sound on the western shore of Vancouver Island. Perez named the place San Lorenzo. In the course of trading with Natives, two silver spoons were stolen from his ship's pilot, Estévan Martinez. The spoons would show up again and play a key role in a plot filled with international intrigue. By then, the placid little bay would be known as Nootka Sound.

Perez returned to Mexico but was ordered to take the *Santiago* north again the following year, this time under the command of his superior officer, Bruno Heceta. With the *Santiago* sailed a minuscule thirty-six-foot schooner, the *Sonora*, under the command of Juan Francisco de la Bodega y Quadra. Somewhere off the coast of present-day Washington, a landing party was dispatched from each ship. Heceta's party took formal possession of the Pacific Northwest for Spain. The party sent ashore by Bodega to fill water casks was massacred by Natives.

His command bloodied, Heceta held a council and ordered Bodega and the *Sonora* back to Mexico. How Bodega got away with it is unclear, but he ignored the order and sailed northward instead. The *Sonora* reached the vicinity of present-day Sitka, and Bodega landed long enough to claim the area for Spain. In the process, he also mapped and named Bucareli Bay on Prince of Wales Island.

Perhaps it was the naming of the bay for the viceroy, but Bodega y Quadra's career seems not to have suffered from his insubordination in heading north. In 1779, he sailed north again from San Blas, Mexico, in command of the *Favorita* on a second Alaskan voyage. Indeed, by 1780, he was a frigate captain, and later served out of Havana and Cadiz before being pulled once more into the politics of the converging frontiers of the Pacific Northwest.

Meanwhile, Perez's earnest lieutenant, Estévan Martinez, was busy supplying the California missions. Martinez happened to be in Monterey in 1786 when he lent a hand and piloted two French ships into the harbor. The Frenchman La Pérouse brought news from the north that the Russians were far from sleeping. Spain, it seemed, must continue to look to its northern borders. Consequently, the Spanish viceroy ordered Martinez north for another reconnaissance. In 1788, in command of the *Princesa*, Martinez reached Kodiak Island and Unalaska in the Aleutians. There he learned of Russian plans to establish a post at the quiet little inlet that he had first visited with Perez on Vancouver Island fourteen years before. Returning to Mexico, Martinez proposed a bold move to check the Russians: establish a Spanish outpost there before the Russians could do so themselves.

The viceroy concurred, and on May 5, 1789, Martinez sailed into Nootka Sound. But the neighborhood was already getting crowded. The Russians were not there, but an English ship was, along with a small schooner from that upstart nation of thirteen former colonies on the Atlantic coast. Martinez was not pleased.

Over the course of the following summer, the comings and goings of additional ships at Nootka Sound and the aggressive demonstrations of Spanish sovereignty by Martinez—including seizing a few ships when wining and dining their captains didn't work—resembled a twenty-first-century television soap opera rather than eighteenth-century gentlemen's diplomacy. Remember the two silver spoons? Perhaps the height of the drama came when Martinez quoted from the great Captain Cook's *A Voyage to the Pacific Ocean* about meeting a Native at Nootka Sound wearing two Spanish spoons on a cord around his neck. They were his spoons, Martinez asserted, lost a full four years before Cook's visit and proof positive set in English type of Spain's prior claim to the region.

When news of this confrontation in a once-quiet Northwest cove reached the governments of Europe, all rattled their swords and threatened war. At issue was far more than Nootka Sound as a base for lucrative fur-trading operations. The underlying international law principle was whether any nation—in this case Spain—could hold exclusive title to land by virtue of discovery alone. In a turn of events ominously predictive of the tangling alliances that later led to the First World War, Spain looked to aid from France, while England sought alliances with Prussia and the Netherlands. In the fledgling United States, President George Washington and his secretary of state, Thomas Jefferson, agonized over the prospect of England and Spain waging a global war centered upon the very landscape that they hoped to claim for themselves in due course.

In the end, tensions cooled when England and Spain signed the Nootka Sound Convention on October 28, 1790. The Nootka Sound Convention was another of those treaties to which Russia was not a party, but which nonetheless cast a shadow over its future moves. In effect, Spain gave up its exclusive claim by right of discovery to the Pacific Northwest, and Spain and Great Britain mutually agreed that each country was entitled to trade and build settlements on those parts of the coast not previously occupied. In other words, sovereignty still followed the flag, but it was made inviolate only by occupancy. Just about everything north of the Spanish mission at San Francisco was up for grabs. The consequence for Alaska was that as Spain, Great Britain, and the United

States each scrambled to stake out permanent settlements in the Northwest, Russia was forced to do likewise. In time, this would mean Russian outposts throughout southeast Alaska and one smack in the middle of northern California.

As for the staunch Martinez, the consensus of his superiors in light of the Nootka Sound Convention was that he had perhaps been a little too strident in asserting Spanish sovereignty. They appointed Juan Francisco de la Bodega y Quadra to take his place. By all accounts, Bodega y Quadra served in a far more diplomatic manner, and Martinez, still missing his two spoons, returned to Spain.[5]

So what lay north of Nootka Sound of interest to the Spanish? In 1791, Spain dispatched Alejandro Malaspina, an Italian in the service of the Spanish navy, to find out. Born in the Duchy of Parma in 1754, Malaspina entered the Spanish equivalent of Annapolis as a midshipman in November 1774. Completing two years of training, he was promoted to second lieutenant and saw action against Great Britain in the Atlantic and Mediterranean. By 1784, he was in command of the *Asuncion* on a cruise to the Philippines and back. By the time he returned to Spain after a round-the-world voyage in command of the *Astrea*, he was as competent a navigator as any under the Spanish flag—clearly Madrid's ready answer to James Cook and George Vancouver.

Spanish naval minister Antonio Valdes gave Malaspina new orders in 1789. He was to embark on a global voyage of discovery that would chart the coastline of the Americas and concentrate on astronomy and natural history. Malaspina repeated this theme in his own instructions to his second in command, José de Bustamante. "On this commission," Malaspina wrote, "the scientific aspect rather than the military is what will contribute to public usefulness."[6] Perhaps, but surely Valdes must have put his arm around his chosen captain and reminded him that his job had political and military purposes as well.

Malaspina oversaw the construction of two three-masted vessels for the expedition. Each was 109 feet in length with a loaded draft of 14 feet. Evidently, Malaspina was an eighteenth-century "techie." Not only did he consult the experts in Cadiz and Madrid for the latest in chronometers, sextants, and related navigational gear, but also he was well ahead of his time in understanding and implementing the food storage and preparation procedures necessary to maintaining healthy crews. Antiscorbutics to combat scurvy were a must, and he even experimented with distilling drinking water.

Malaspina captained the *Descubierta* (Discovery) and Bustamante commanded the *Atrevida* (Daring) when the expedition left Cadiz on July 30, 1789. They sailed south across the Atlantic, surveyed the Rio de la Plata and the eastern coast of Tierra del Fuego, and then rounded Cape Horn, swinging widely around it at sixty degrees south latitude. Then the two ships leapfrogged their way up the western coast of South America, making landfalls and charting harbors in places that included Valparaiso and Lima before arriving at Acapulco in the early spring of 1791. Here, fresh orders awaited them.

Mexico's newly appointed viceroy, the conde de Revillagigedo, had recently arrived in Mexico in the company of Francisco de la Bodega y Quadra with reports of the goings-on at Nootka Sound. Bodega was on his way there to relieve the overly zealous Martinez. Malaspina was directed to postpone his surveys of the Hawaiian Islands and instead sail straight north to sixty degrees north latitude. He was to conduct yet another search for the Northwest Passage and then chart the coastline before arriving in Nootka Sound from the north.

On May 1, 1791, the *Descubierta* and *Atrevida* left Acapulco, each with fourteen six-pounders, two four-pounders, and eight additional six-pound cannon in their holds. So much for science! The two ships stood well off the coast and raced north for the vicinity of Mount St. Elias. Making landfall, they dropped anchor in the icy waters of Yakutat Bay. Malaspina established an observation station on the south shore of the bay and proceeded to fix accurate bearings and also survey the height of Mount St. Elias and its neighboring peaks. "It is hard to believe," Malaspina wrote in his log, "that even in the months of June and July these mountains are covered with snow and . . . are destined to be forever inhabited only by bears."[7]

The entire area was much more heavily glaciated than it is today, but one arm of the bay extended northeastward and seemed to offer some encouragement as a possible passage. As on so many other voyages, hopes were raised that this waterway might lead eastward through the North American continent. Malaspina ordered a thorough investigation. The two vessels cautiously sailed up the slender arm but soon encountered floating ice, then larger bergs, and finally the massive face of a glacier blocking all hope of further advance. The ships turned about and Malaspina named this upper arm of the bay Puerto del Desengaño

(Disenchantment Bay). Later, the huge glacier dominating the north shores of Yakutat Bay would be named for Malaspina.

Disappointed in not finding the passage others spoke of with such certainty, Malaspina nonetheless lingered around Yakutat, recording and exploring. Despite his complement of cannon, he really did have with him a diverse and highly educated company of men who were committed to the scientific aspects of the voyage. There were astronomers, cartographers, botantists, naturalists, and painters, including prolific botanical collectors Tadeo Haenke and Luis Nee and reknowned artist Tomas de Suria. A great many specimens were collected, and the maps, drawings, and portraits produced on the expedition are a classic case of telling more by one picture than with a thousand words. Suria recorded glimpses of the Tlingit culture, as well as the landscape. One pencil-and-wash sketch of the *Descubierta* and *Atrevida* at anchor beneath Mount St. Elias is particularly grand.

In due course, the ships sailed west along the coast, cruising off the mouth of Icy Bay, then barely an indentation in the face of a gigantic glacial wall. They skirted Kayak Island and continued west as far as Hinchinbrook Island at the entrance to Prince William Sound. After trading with Chugach Natives, Malaspina turned south on July 27. He hugged the coast as far south as Cape Edgecombe west of present-day Sitka, charting this lesser-known territory, and then struck out for Nootka Sound, arriving there on August 12, 1791. The maps of the anchorage there made by Malaspina's cartographers are surprisingly accurate in retrospect, and here, too, the artists captured—with perhaps some poetic license—the assemblage of foreign ships and Native canoes. Then it was south to the presidio at San Francisco, a stop in Monterey, and back to Acapulco.

But Malaspina's voyage of discovery was far from over. After resting and taking on supplies, the expedition headed west across the Pacific to Guam, the Philippines, and eventually New Zealand, mapping and collecting at every opportunity. Then the two ships sailed westward to Australia and anchored in Sydney Harbor before starting the long return trip back to Spain by sailing east across the Pacific rather than westward across the Indian Ocean. One may surmise that Malaspina sought this easterly route because he had more mapping to do in the Pacific even if it meant rounding the Horn again. With his two ships and their crews in surprisingly good condition, Malaspina cruised into Cadiz on September 21, 1793, after four years at sea.

At first Malaspina was hailed by one and all, but he soon fell out of favor with the Spanish Crown. Much of his success had been premised on his enlightened ideas, but in the end, it may well have been these liberal views—particularly his criticisms of Spanish colonial administration around the globe—that got him into trouble. He spent time in prison before being banished to Italy. His records of the expedition were taken from him and much of its collections and drawings scattered. Surely, if his report had been published, it would have rivaled the accounts of the great Cook, whom he had sought to emulate. No account of Malaspina's four-year voyage was published until 1885—almost a century after his ships first sailed from Cadiz.[8]

Some writers have tended to downplay the importance of the Spanish to the exploration of Alaska, citing as evidence the lack of permanent settlements and only a handful of remaining Spanish place-names scattered across the map. Yet it was the Spanish reach north from Mexico into the North Pacific that concerned Russia from the time of Peter the Great. Yes, the English would sail these waters and interject themselves into the fray, and somewhere, far, far away over the rugged Coast Range, there were English colonies, but for the better part of a century, it was the Spanish who were on the most direct collision course with Russian America. In retrospect, the Spanish threat to Russian America proved more imagined than real, but Spanish and Russian interests stood face-to-face at Fort Ross in northern California as late as 1841. That open warfare never developed between Spain and Russia was more because each had overextended its respective empire and was worn out, rather than that Alaska was not worth winning.

Cook and Vancouver

Almost two centuries had passed since Sir Francis Drake's daring bolt across the Spanish Main in the Pacific, and Great Britain had been busy building its global empire of colonies. In July of 1776, while one band of those colonists was declaring its independence in Philadelphia, the British Admiralty instructed Captain James Cook to sail to the northwest coast of North America. Cook was the best of the Royal Navy's elite. He had already made two Pacific voyages, mapping the southern and central regions and disproving the theory of a large southern continent—Australia, after all, was on the small side. Now he was to sail the northern regions and search for the western outlet of the elusive Northwest Passage.

Just getting there was still an adventure. Cook left Plymouth, England, in command of the *Resolution* and was joined shortly by the *Discovery* under the command of Captain Charles Clerke. The two ships sailed the length of the Atlantic, rounded the Cape of Good Hope, and made landfalls at New Zealand, Tahiti, and the Hawaiian Islands, before arriving at Nootka Sound on March 30, 1778. There, Cook recorded meeting the Native wearing Martinez's spoons.

Over the course of the next five months, Cook cruised northward along the Alaskan coast, mapping key features but finding no continental passage. With Cook were two Englishmen who would later captain their own ships and leave their own names on Pacific charts, Midshipman George Vancouver and Master William Bligh. Together they cruised the Alexander Archipelago and made the entire circuit around the Gulf of Alaska, naming Cross Sound, Mount Fairweather, and Prince William Sound in the process.

Around the Kenai Peninsula they sailed before encountering a broad bay that Vancouver would later name Cook Inlet. It ended disappointingly in vast mud flats, but another, narrower estuary leading back eastward looked promising. Could this be the fabled passage? Cook sailed eastward and fought the tides for almost fifty miles before reaching the conclusion that this too was a dead end. Another "turnagain." Cook called it the River Turnagain. Vancouver would later label it Turnagain Arm. Then the two ships picked their way through the Aleutians and sailed as far north as Icy Cape on the Arctic Ocean, almost 400 miles north of the Bering Strait. It was a remarkable effort that tied Bering's geography with that of the known Northwest coast.

Turning around at Icy Cape, Cook sailed south to winter in the Hawaiian Islands and continue his mapping of that chain. When Cook met death at the hands of Hawaiian Natives in February 1779, the *Discovery*'s captain, Charles Clerke, took command of the expedition and led it back to Aleutian waters intent on completing the mapping. The *Discovery* and the *Resolution* again sailed north through the Bering Strait until confronted with impassable ice, and then bore west for Kamchatka. Clerke died there of tuberculosis, another victim who had paid the price of exploration.

From Kamchatka, the two ships started the long journey back to England. But first, they called at Canton for supplies. Lieutenant James King, now in command of the *Discovery*, took twenty medium-grade sea otter pelts ashore to barter for necessities. The Chinese merchants went crazy. They paid handsomely for King's pelts and then demanded to know where and how they could get more. King had all he could do to quell a potential mutiny among his crewmen, who were suddenly starstruck by the commercial power of what they had considered only incidental goods the day before. Some wanted to return immediately to the Aleutians for more pelts. King forestalled such actions, but once the *Resolution* and the *Discovery* docked in England, the word went out: There were great riches to be had in the fur trade. No wonder that Estévan Martinez found things a little crowded at Nootka Sound some years later.

Two of Cook's men from the *Discovery* were determined to be in on the rush. Sailing as a commercial venture and not as part of the Royal Navy, Captain Nathaniel Portlock and Captain George Dixon respectively sailed the *King George* and *Queen Charlotte* from England in 1785 bound for the North Pacific. They took the route around stormy Cape Horn, landed in the Hawaiian Islands, and arrived in Cook Inlet in July 1786. They traded with the Natives there—probably Dena'ina Athabascans—and in Prince William Sound—probably Chugach Alutiiq—and then made their way eastward down the coast, eventually calling at Nootka Sound before wintering in Hawaii.

When Portlock and Dixon returned to Prince William Sound in the spring of 1787, they met a fellow trader who had not gotten the word about wintering in the land of balmy breezes. Like Portlock and Dixon, John Meares had sailed to Prince William Sound in the summer of 1786, arriving there in August after the other two had sailed east. Meares unwisely decided to winter over on the sound and was in dire straits by the time Portlock and Dixon found him and his crew. After assisting Meares, Portlock and Dixon split up to cover more ground. Portlock worked Prince William Sound, while Dixon visited Yakutat Bay, Sitka Sound, and what would later be called Dixon Entrance between the Alexander Archipelago and the Queen Charlotte Islands. What became clear to both was that they were not the only ones with the idea of making money in the fur trade. The Dena'ina, Eyak, and Tlingit all were used to trading with men in tall ships. Portlock and Dixon reached Macao in November 1787, disposed of cargoes of furs, and then bought cargoes of tea to take back to England, arriving there separately in the fall of 1788.

As for John Meares, one winter on Prince William Sound was not enough to dissuade him from the potential profits of the fur trade. He organized several other trading ventures and managed to find himself at Nootka Sound when all hell broke loose there in 1789.

In fact, the Nootka Sound controversy delayed the sailing of Great Britain's second great Alaskan survey voyage. When the danger of war with Spain ebbed, the British Admiralty ordered one of its most experienced officers to fill in the blank spots on the charts of the Northwest coast. Having cut his teeth under Cook, George Vancouver was now a captain in command of a different *Discovery* and the *Chatham*. Leaving Falmouth on April 1, 1791, they sailed the long route south around the Cape of Good Hope, to Australia and New Zealand, and then north to Tahiti and the Hawaiian Islands. April 30, 1792, found the two vessels anchored in the Strait of Juan de Fuca and their crews ready to begin surveying. This they did along the coasts of northern California, Oregon, and Washington through the summer of 1792. They were not alone, of course, and there was plenty of uneasy discourse with the Spanish at Monterey and Nootka Sound.

Then in the summer of 1793, after a second winter in Hawaii, Vancouver pushed northward determined to reach the northern limits of Spanish exploration. Unlike some later travelers who would fancy themselves explorers with a carte blanche for names, Vancouver seems to have generally respected the Russian and Spanish names that were known to him as he mapped his way north. Sometimes, that respect led to unexpected results. At the southern tip of Dall Island, the southernmost point in the Alexander Archipelago, Vancouver placed the Spanish name Cabo de Muñoz on his chart. In doing so, however, two letters were transposed, and it became Cape Muzon, the name it still bears today.

As was the custom, Vancouver remembered his royal patrons. He named Prince of Wales Island for the oldest son of George III, who became George IV, and the body of water to the east of the island Clarence Strait, after the duke of Clarence, George III's third son, who is more readily remembered in Alaska as Prince William. He also remembered his friends. Vancouver named the long, narrow, and almost circular waterway surrounding Revillagigedo Island Behm Canal after Major Magnus Carl von Behm, who had been the Russian commandant at Kamchatka when Vancouver landed there in 1779 with the remnants of Cook's final expedition. The Spanish had named Revillagigedo Channel the year before for the viceroy of Mexico. Vancouver honored that name on the waterway and added it to the island.

The two ships continued northward, making detailed maps of the intricacies of the Inside Passage. In September off the southern tip of Kuiu Island between Chatham and Sumner (a much later name) Straits, Vancouver faced a decision. Had he explored farther north than his Spanish contemporaries? The voyage of Bodega y Quadra notwithstanding, Vancouver decided that he had. He turned southward, leaving Cape Decision astern.

After wintering for a third year in Hawaii (embracing a tradition that some Alaskans continue to this day), Vancouver sailed due north for Cook Inlet in April 1794. There he began a systematic survey eastward along the Alaskan coast, filling in the blank spots on the chart all the way back to Prince of Wales Island. Vancouver named Hawkins Island in Prince William Sound and mapped the names of Hinchinbrook and Montague Islands that Cook had bestowed in 1778. The two ships skirted Kayak Island and passed the mouth of Malaspina's Puerto del Desengaño. Peter Puget, now commanding the *Chatham*, tried to give the name Digges Sound to this offshoot of Yakutat Bay, but it remains Disenchantment Bay.

On July 10, 1794, *Discovery* and *Chatham* entered Cross Sound and anchored off Port Althorp, a sheltered bay on the north end of Chichagof Island. Vancouver dispatched Lieutenant Joseph Whidbey and a party in longboats to explore eastward. Whidbey mapped Taylor and Dundas Bays and then rowed through the North Passage of Icy Strait between the mainland and Lemesurier Island. "To the north and east of this point," Vancouver later reported, "the shores of the continent form two large open bays, which were terminated by compact solid mountains of ice, rising perpendicularly from the water's edge, and bounded to the north by a continuation of the united lofty frozen mountains that extend eastward from Mount Fairweather."[9] Whidbey was probably at Point Carolus, and the massive glacier front was approximately six miles north of him on a line with Rush Point. Glacier Bay, or what little of it was then free of ice, had been discovered.

Now the names were liberally placed upon the chart: Lynn Canal, named for Vancouver's birthplace; Berners Bay, his mother's maiden name; Douglas Island, for John Douglas, the bishop of Salisbury; Stephens Passage, for Sir Philip Stephens, secretary to the British Admiralty; and Admiralty Island for his naval masters. For this last one most of all, he should have heeded the Tlingit. They called the island Kootznoowoo, meaning "fortress of the bears."

Vancouver named the uncannily straight waterway running south from Icy Strait and Lynn Canal after John Pitt, the second earl of Chatham. The southern reaches of Chatham Strait had already been called Enseñda de Principe (Teacher of the Prince) by Bodega y Quadra. Then there was to be a sound for the older brother of the celebrated Prince William. Vancouver named the waterway between Kupreanof and Admiralty Islands for Prince Frederick, the duke of York, who was George III's second son. Alas, poor Frederick, history dropped his royal title, and he comes off a commoner next to Prince William. Soon Vancouver was back in known territory south of Cape Decision. His ships called at Nootka Sound and Monterey before rounding Cape Horn en route back to England, arriving there in October of 1795.

One might quibble about the appropriateness of some of his place-names, but Vancouver's accomplishment of charting the Alaskan coastline was of immense proportions. Unfortunately, the rigors of life at sea claimed another too young, and Vancouver died just short of his forty-second birthday without seeing his three-volume *A Voyage of Discovery to the North Pacific Ocean and Round the World* in print.[10]

After Vancouver's two-year survey, the English continued to explore Alaska and to assert fur-trading pressure upon it from their Canadian territories. Just where the boundary was between Alaska—whether owned by Russia or later the United States—and Canada would be a source of debate and occasional confrontation for another century. But England's sway over the great land would be limited to major mapping and naming expeditions, in large part because in the press of global colonization England never followed up on the dictates of the Nootka Sound Convention that required settlement to follow the flag. It was a lesson the Russians were determined not to ignore.

Port des Français

Compare maps of North America from the time of Bering's voyages with those after 1763, and it becomes readily apparent just how badly the French lost the North American continental sweepstakes. In 1741, France controlled—or at least claimed—all of North America along and north of roughly the Ohio, Mississippi, and St. Lawrence Rivers. After the 1763 Treaty of Paris ended the Seven Years War (called the French and Indian War in North America), France was reduced to the tiny Caribbean island of Martinique. As the historian Francis Parkman would later term it, half the continent had changed hands at the scratch of a pen. So France's influence on Alaska was fleeting, but it did leave one interesting tale of exploration.

At the center of the tale was the French version of the legendary Captain Cook. Born Jean-François de Galaup in Albi in southwestern France on August 23, 1741, he added "de la Pérouse" to his name when he joined the French navy in 1756—taking it from the name of one of his family's country properties. In time, he was to become known at the French court as the comte de la Pérouse. In between, there were to be many years at sea.

Rising through the ranks, La Pérouse served the Bourbon monarchy ably in ships that roamed the globe. As his early years of service corresponded with France's defeat in the Seven Years War, he no doubt took a certain comfort in seeing the roles reversed when later the French navy successfully came to the aid of Great Britain's American colonies. In 1782 as part of this effort, La Pérouse commanded a flotilla of three ships that surreptitiously entered Hudson Bay and destroyed British posts at Churchill and York. By 1783, La Pérouse was a post-captain, decorated and well respected in French naval circles.

Buoyed by its success in assisting the American colonies, France determined to stride forward once more on the world stage. In that era, few actions signaled that resolve more than expeditions to fill in the blank spots upon the world map. Just as Malaspina would do shortly for the Spanish, La Pérouse was to emulate Captain Cook and take the French flag around the Pacific to discover what lay in between the great captain's landfalls. Of course, it was to be a voyage of science as well as mapping.

La Pérouse took command of the 127-foot *Boussole* and entrusted command of the *Astrolabe* to Paul-Antoine-Marie Fleuriot, chevalier de Langle, who had been with him in Hudson Bay. Indeed, much of La Pérouse's successes during the next two years stem from the fact that he recruited men with whom he was personally acquainted. He also went to great lengths to supervise meticulous preparations to prevent scurvy and ensure a healthy crew.

With a complement of 225 men, the two ships sailed from Brest, France, on August 1, 1785, anticipating a four-year voyage. Following summer into the Southern Hemisphere, they rounded Cape Horn and arrived at Concepcion, Chile, in late February 1786. Here, La Pérouse made a major command decision. To be sure, he had been given great discretion in carrying out his orders, but their focus had definitely been on the South Pacific. Now, La Pérouse decided that he had made such good time in crossing the Atlantic and rounding Cape Horn—turning the latter with "far more ease than I had dared to hope"—that he could afford time to explore the northwest coast of North America. "I knew," La Pérouse rationalized, "that, if I had not been given such an order, it was only because it was feared I might not have enough time to make such a long voyage before the onset of winter."[11] The ghost of Captain Cook was beckoning.

So the *Boussole* and the *Astrolabe* sailed northwest to Easter Island and then north for Cook's Sandwich Islands. La Pérouse landed a well-armed party on Maui, apparently the first Europeans on that island, but was well received by the islanders. Then after sailing westward for two days, the ships passed through the Kauai Channel and headed north.

The blue skies of the islands soon gave way to the overcast gray of the North Pacific. Fog, cold, rain, and more fog became the norm. Early on the morning of June 23, 1786, about three weeks after leaving Hawaii, the fog parted and La Pérouse and his crew glimpsed the towering heights of Mount St. Elias.

No stranger to the Arctic from his days in Hudson Bay, La Pérouse had written his mother afterward that he wished never to return there. He was similarly unimpressed here, describing a "sterile treeless land . . . a black plateau, as though burnt out by some fire, devoid of any greenery, a striking contrast with the whiteness of the snow we could make out through the clouds."[12] The black plateau was the moraine and debris strewn about what would later be called the Malaspina Glacier.

On June 26, La Pérouse dispatched three small boats to reconnoiter the bay to the east of the black plateau. Anne-Georges Augustin, chevalier de Monti, Langle's second in command aboard the *Astrolabe*, led the survey party, and La Pérouse ordered the bay named after him. As Malaspina would later learn, this was not the Northwest Passage but what would come to be called Yakutat, not Monti, Bay.

While one of the naturalists, a man named Dufresne, wrote notes on the fur trade and speculated about France's chances of joining the lucrative rush, the cartographers were soon overwhelmed by the complexities of the fog-enshrouded coastline. They couldn't help but agree with Cook's lieutenant, Charles Clerke, who had written eight years before that "a thick fog and a foul wind are rather disagreeable intruders to people engaged in surveying and tracing a coast."[13]

Sailing gingerly southward along the coast, the ships found clearer weather off Cape Fairweather. As Mount Fairweather towered above them, Tlingit appeared in canoes and on shore. The next day, July 2, a section of smooth shoreline was suddenly interrupted by a wide passage, apparently leading eastward many miles. It was not shown on any of their charts. The Northwest Passage syndrome quickened heartbeats and put a certain look in men's eyes.

La Pérouse ordered the *Boussole* and the *Astrolabe* into the bay, but it was not an easy passage. Two spits of land guarded the entrance, and the opening itself was shallow and blasted by fickle winds and surging tides. Never in his thirty years at sea, La Pérouse later confided, had he seen two ships so near destruction. But once inside, the waters widened and calmed. The ships anchored, and the expedition made camp on a large island almost in the middle of the bay. While the scientific members went about their observations, others probed the bay's upper limits. But they were soon disappointed. Three massive glaciers ringed its head. There was no passage to the Atlantic here.

La Pérouse called the bay Port des Français (Bay of the Frenchmen). Later whalers long called it Frenchmen Bay, but the Russian derivation of a Tlingit name, Lituya, is what remains. Of this uncharted bay, La Pérouse wrote, "If the French government had any plans for a trading post in this part of the American coast, no nation could claim any right to oppose it."[14]

Legend suggests that the Tlingit who cautiously paddled out to the island thought that La Pérouse was Yeahlth, their birdlike creator. Yeahlth, so the story went, was returning to Earth to reward those who had followed his teachings and to turn to stone those who had disobeyed him. This seems unlikely. European vessels along the coast were still rare, but even if the Tlingit in this area had not seen earlier ships—those of Bering, Chirikov, and Cook come to mind—they had probably heard tales of similar visitors from other Tlingit to the south. In any event, these Tlingit offered the island to La Pérouse, and he purchased it with some red cloth and iron tools.

La Pérouse recounts that hundreds of Tlingit visited Lituya Bay during the course of the French stay there. They came in large canoes and camped along the shores, drying salmon and halibut. As some groups left, others came to take their place. Prophetically, La Pérouse wrote that the Tlingit seemed to have considerable dread of the passage into the bay, and that they never ventured through it except during the slack water between flood and ebb tides. Through his spyglass, La Pérouse could watch as canoes approached the passage. The Tlingit leader in the canoe would stand and stretch out his arms to the sun, perhaps both in prayer and as a signal to the others to paddle with all of their strength. When asked about this custom, the Tlingit told the French that seven large canoes had recently been lost negotiating the entrance and that an eighth had barely escaped.[15]

Then the expedition, which to this point had been free of death or major illness, was struck a bitter blow. On July 13, Charles-Gabriel d'Escures, second officer of the *Boussole*, was dispatched with a party of boats to take soundings of the bay so that the chart could be completed. With the Tlingit experience etched indelibly on his brain, La Pérouse specifically warned d'Escures not to venture too closely to the dangerous entrance. Perhaps d'Escures ignored the warning or thought Frenchmen the equal of Tlingit paddlers. More likely, the fury of the seaward-rushing ebb tide surprised him. First his boat was caught up in the maelstrom, then the *Astrolabe*'s pinnace, coming to his rescue, was also ensnared. A third boat barely escaped. Within ten minutes, six officers and fifteen seamen were drowned and washed out to sea and two boats capsized.

La Pérouse was devastated. He erected a small cairn on the southeast corner of the island as a memorial and named the island "Isle de Caenotaphe," or Cenotaph Island, a cenotaph being a monument erected in memory of those whose bodies lie elsewhere. A bottle was placed in the cairn with the names of the dead and an account of the tragedy, captioned with the ominous lines: "At the entrance to this port twenty-one brave sailors perished. Whoever you are, mix your tears with ours."[16]

The tragedy delayed the expedition in Lituya Bay another two weeks, in part because it dictated some necessary reassignments among the crews. Finally on July 30, the two somber ships clawed their way out into the open sea and once again turned south. La Pérouse named Mount Crillon for General Louis des Balbes de Berton, the duke of Crillon, who distinguished himself in the naval battle of Lepanto off Greece in 1571. Later, William Dall would name a major peak south of Mount Crillon after La Pérouse.

Confronted by the jagged array of bays north of Cross Sound, La Pérouse soon realized the gargantuan task of mapping all of the many inlets and islands of the Alexander Archipelago. All of the time that he once thought he would have was beginning to run out. Concentrating on those areas not well mapped by Cook or the Spanish before him, he sailed past Cross Sound, Baranof Island, and Christian Sound, before admitting that the Spanish chart of Bucareli Bay on Prince of Wales Island was of little help. Along the way, he named the southern tip of Baranof Island, now called Cape Ommaney, for Aleksei Chirikov, and Necker Bay and Islands on Baranof's western coast for the French minister of finance, one of the few La Pérouse names to survive.

On August 9, the two ships crossed Dixon Strait and became enshrouded in another fog off the Queen Charlotte Islands. Fog also hid Nootka Sound and the Strait of Juan de Fuca as the winds of autumn chased the expedition south. Finally, the *Boussole* and the *Astrolabe* dropped anchor in Monterey Bay on September 15, 1786, La Pérouse's northern detour having been accomplished. But the voyage was far from over. Pausing at Monterey only ten days, the two ships soon hurried westward across the Pacific. Some of the uncharted lands La Pérouse sought almost wrecked his vessels when what he named French Frigate Shoals suddenly appeared west of Hawaii. Reaching Macao, the ships doubled back to Cavite in the Philippines for repairs and then struck north, mapping Sakhalin Island, the La Pérouse Strait between Sakhalin Island and Hokkaido, and the Kuriles before calling at Petropavlovsk almost fifty years after Bering's first huts were built there.

La Pérouse was indeed well on his way to becoming the French Captain Cook. But the end was to be as tragic as the episode in Lituya Bay. At Petropavlovsk, the captain received dispatches from Paris that directed him to Australia to investigate rumors that the British were establishing a colony of convict labor at Botany Bay. Sailing from Avacha Bay on September 30, 1787, the French ships arrived in Botany Bay four months later, scarcely a week after the British themselves had dropped anchor there. That was one rumor that could be confirmed.

Wary pleasantries were exchanged, and then on March 10, 1788, the *Boussole* and the *Astrolabe* weighed anchor and departed Botany Bay, never to be seen again. For forty years their fates and that of their crews were one of the great mysteries of the seas. In 1827, an Irishman named Peter Dillon finally found relics from the two ships on the tiny island of Vanikoro between the Solomon and New Hebrides Islands. In a cyclone, they had been dashed to bits on deadly reefs. As it turned out, like his hero Captain Cook, Jean-François de Galaup de la Pérouse would remain forever in the Pacific he explored.

Lord of Alaska

The Russians, of course, had not been idle while this ever-increasing parade of foreign flags cruised Alaska's waterways and began to make inroads in its promising fur trade. The rich furs that the survivors of Bering's expedition brought back piqued the interest of the *promyshlenniki.* Beginning in the Komandorskie Islands about 1745 and then island-hopping their way eastward along the Aleutian chain, the *promyshlenniki* undertook a relentless campaign of harvesting furs, particularly those of sea otters, foxes, and migratory northern fur seals. When the supply of animals in one group of islands was depleted, the hunters simply moved on to the next and repeated the slaughter.

The Aleuts watched in horror as the natural cycle upon which their survival depended was first abused and then threatened with extinction. For a time, the Aleuts fought this onslaught, but resistance eventually became futile in the face of Russian guns. Many Aleut men were enslaved and made to hunt, sometimes far away from their villages. Meanwhile, the *promyshlenniki* engaged in drunken revelry with the Aleut women. Warfare, disease, and starvation quickly took their toll on the Aleuts and led to social disintegration.

By some estimates, only 20 percent of the Aleut population of between 15,000 and 18,000 people survived the first quarter of a century of contact with the Russians.[17] Others suggest that it was bad, but not that bad. Regardless, this period of initial interaction between Russian *promyshlenniki* and the Aleuts was the most disastrous and destructive that any group of Alaska Natives was to experience at the hands of outsiders.

To be sure, the marauding *promyshlenniki* had free rein in the beginning, but slowly Empress Catherine II—later to be given the appellation "the Great"—began to take a small interest in Alaska. Catherine, "enlightened despot" that she was, came to power in 1762 and in time expressed concern over the treatment of the Alaska Natives. Unfortunately, no direct action was taken, and by the time the Russian government finally began to exert more control over commercial operations in Alaska after Catherine's death, it was far too late for the Aleuts.

Catherine may have received some of her information about Alaska from the accounts of Stephen Glotov, a Russian fur trader who appears to have traded peacefully with the Aleuts on Umnak and Unalaska between 1758 and 1762. Glotov made a second voyage between 1762 and 1765 that was apparently funded by a group of investors. This was significant because the voyage marked a turning point from the freelance havoc of the *promyshlenniki* to the more organized operations of commercial companies. Glotov sailed east as far as Kodiak. He landed there on September 8, 1763, and was probably the first European to visit the island.[18]

The first government-sponsored expedition to the Aleutians was led by Peter Krentizin in 1768. Leaving the mouth of the Kamchatka River with two ships, Krentizin surveyed the islands of Umnak, Unalaska, Unimak, and the western shores of the Alaska Peninsula. He returned to Kamchatka in 1769 and drowned in the Kamchatka River shortly thereafter. His journals and charts, however, eventually reached St. Petersburg, and they formed the basis for the first reliable map of the Aleutians. Today his name rests on the Krentizin Islands south of Unimak.[19]

By the 1780s, the barbaric chaos of the *promyshlenniki* had given way to the relatively organized exploitation of six trading companies. Named for their founders, all of the companies were privately funded. They controlled individual territories—private hunting grounds almost—stretching across the Aleutians and eastward to Prince William Sound. The largest were the Lebedev-Lastochikin Company operating from the Kenai Peninsula and the Shelikhov-Golikov Company headquartered on Kodiak Island. As each company trapped out its own territory, it looked covetously on its neighbors', and rivalries among these Russian companies were as fierce as any among Alaska Natives.

One of the principals of the Shelikhov-Golikov Company was Grigory Ivanovich Shelikhov, a Siberian fur merchant. On August 3, 1784, Shelikhov arrived in Three Saints Bay on the southeast coast of Kodiak Island with his wife and a party of 192 men and proceeded to construct the first permanent Russian settlement in North America. Shelikhov's wife, Natalya Alexyevna, was likely the first European woman in Alaska. Some of the shores of Three Saints Bay—named after Shelikhov's ship—are quite rocky, but Shelikhov found a small spit of land about a mile and a half south of the Native village of Nunamiut, probably inhabited by Koniag Alutiiq.

At first, relations between the two groups were cordial, but when the Koniag realized Shelikhov's permanent intentions, they made plans to attack. The Russians got wind of this and struck first, subduing the Koniag with the power of Russian cannon. Shelikhov proved a shrewd businessman and was equally forceful with anyone else who got in his way. The Shelikhov-Golikov Company soon stood at the center of the Alaskan fur trade and rewarded its investors with handsome profits.

But competition with rival Russian companies, as well as the British of the Hudson's Bay Company, remained keen. In 1788, Shelikhov went to St. Petersburg with a solution. Shelikhov asked Catherine to grant an exclusive license—effectively a monopoly—to the Shelikhov-Golikov Company for all of the fur trade in North America. Only one powerful company with imperial blessing, he argued, could check the advances beginning to be made on Russian America by both the British and the Americans.

Catherine, admirer of the Enlightenment that she was, embraced its laissez-faire policies of government toward business. Perhaps she declined Shelikhov's request on philosophy alone. Perhaps she had more pragmatic concerns and feared that a Russian trading monolith might heighten, rather than reduce, tensions with Britain, Spain, or others and serve only to involve Russia far more deeply in North American affairs than she cared to be. Whatever Catherine's reasons, Shelikhov left without his monopoly. He and Golikov had to be content with gold medals, silver sabers, and imperial citations, expressing the empress's undying gratitude to them for discovering "new lands and peoples for the benefit of the state." So far, the state had not done much, but the seed of future governmental involvement had been planted.

When Shelikhov left Kodiak Island in 1786, he entrusted the management of his Alaskan operations to two interim managers. Both turned out to be ill-suited to the task, but several years later, Shelikhov found a most unlikely little businessman to run his trading empire. The man's name was Aleksandr Baranov. He was short, slight of build, and had little or no ambition to be either an explorer or a colonizer. What he was good at, it turns out, was the Machiavellian management skills necessary to run a profitable business in a demanding land. Baranov learned early on never to show fear and always to appear stronger than he was. In time, he would come to be called the "Lord of Alaska."

Born in 1746 in Kargopol near the Finnish border, Baranov was a fur merchant in Okhotsk when Shelikhov first approached him about taking on the management of the Shelikhov-Golikov Company's North

American operations. Baranov had reasons for not doing so, but over time Shelikhov persisted, and in 1790, after some financial setbacks, Baranov agreed to take the job. He was forty-three years old and was rapidly running out of chances to make his fortune. Perhaps it lay in Russian America. He may have had second thoughts right from the start. First seasick, then shipwrecked on Unalaska, Baranov finally arrived on Kodiak Island in 1791 in a *baidarka*.[20]

One of Baranov's first moves was to relocate the Three Saints Bay outpost. Baranov chose a new location on the northeast tip of Kodiak Island and called it Pavlovsk Gavan (Paul's Harbor). The site, now the town of Kodiak, was far removed from the tensions with the Koniag neighbors at Three Saints Bay and much drier, but most important, it was surrounded by timber that Baranov intended to use to construct buildings, fortifications, and even sailing ships.

Once established in Paul's Harbor, Baranov turned to an experienced English shipwright, James Shields, and commissioned him to build a ship. When the three-masted, seventy-nine-foot-long vessel was launched on Resurrection Bay in 1794, Baranov christened her the *Phoenix*. It was an appropriate name, and Baranov intended that the ship would give him an increased measure of both self-sufficiency and power. In truth, his very need to build the vessel—under the direction of an Englishman, no less—was evidence of his need for both.

For starters, Baranov's little colony lacked self-sufficiency. Getting supplies from Russia was an almost impossible task. The distance from Okhotsk (not exactly the shopping capital of Siberia) and the harsh weather in the North Pacific conspired time and again to destroy or delay ships and supplies. When ships sailed from the Baltic with a better selection of goods, the trip was even longer. One season a supply ship arrived in Paul's Harbor only to report that the crew, delayed two years en route, had consumed all of the provisions on board.

Far more reliable suppliers, Baranov conceded, were the British and Americans who were beginning to sail these waters. But they were his competitors! Their very presence, however, was indicative of his need for power to repel advances into Russian America. So Baranov bit the bullet and traded with them, doing what was necessary to keep his colony alive. As for needing power, or at least a perception of it, Baranov used the *Phoenix* to intimidate Alaska Natives and to show the Russian flag in the face of foreign ships. The *Phoenix* also made two voyages to Siberia for supplies before she too was lost among the Aleutians.

Competition continued with the other Russian companies. In 1793, the Lebedev-Lastochikin Company built Fort Constantine on Hinchinbrook Island at the entrance to Prince William Sound. When Baranov had occasion to visit in the course of a dispute, he kicked himself for not beating his rivals to this strategic spot. Tensions between the Lebedev-Lastochikin men at the fort and the inhabitants of the nearby Chugach Alutiiq village of Nuchek were high. It seems that the Russians had been stealing furs, attacking women, and burning houses. How, the Chugach asked Baranov, could he condone such behavior when he had told Natives all along the coast that he was the supreme leader of these newcomers?

Realizing that company distinctions meant little and that one Russian was the same as another to the Natives, Baranov knew that he had to take action or see his authority evaporate. Boldly, he ordered eight of the Lebedev-Lastochikin men put into chains and sent back to Siberia to stand trial. They were acquitted, of course, but in the lengthy time that this took, Baranov strengthened the reach of the Shelikhov-Golikov Company and the Lebedev-Lastochikin management decided to curtail its Alaska operations. That was the way you got things done in Russian America!

Although not a religious man himself, Baranov believed that a little spiritual influence by the Russian Orthodox Church would temper some of his men's secular actions. Accordingly, Baranov requested that Shelikhov dispatch one priest to minister to the settlement at Paul's Harbor. Shelikhov had never been bashful about exaggerating or even fabricating the good things the Shelikhov-Golikov Company was doing for the care of Alaska Natives, particularly the now-subjugated Aleuts. Native welfare, after all, seemed to be Empress Catherine's one interest in North America. So Shelikhov promptly went to the Valaam Monastery in Finland and recruited not one but eight priests for the purpose of converting Alaska Natives. When the monks arrived on Kodiak in 1794, no one suspected their lasting influence.

In 1795, Shelikhov died. His successor at the head of the Shelikhov-Golikov Company was his son-in-law, Nikolay Petrovich Rezanov. A year later, Catherine was also dead. Her successor was her son, Paul. This next generation drastically changed the network of competing companies doing business in Russian America. Czar Paul had never been fond of his mother, disdaining her politically for her enthusiasm for European philosophies and personally for her alleged role in the murder of his father. Paul decided that it was time to exert the role of the Russian government in the affairs of Russian America. Rezanov was a minor noble with some access to the court, and he was all too willing to lobby for the same solution that Shelikhov had proposed almost a decade before.

When the dust settled in 1799, the Shelikhov-Golikov Company was left standing and had been granted an exclusive trading monopoly in Russian America. The other companies were given a choice: merge their operations with the new Russian-American Company or liquidate their assets. If it was seen by some as Paul thumbing his nose at his mother's laissez-faire policies, the action readily recognized the economic realities of doing business so far away in North America. Baranov's experiences had made it clear that if the British and Americans were to be bested, or at least kept at bay, it would take a powerhouse with both deep pockets and government backing.

Once the new Russian-American Company had its monopoly, Baranov was free to extend its empire without Russian competition. In 1795, he had established a small settlement on Yakutat Bay, west of a Tlingit village on the site of present-day Yakutat. Reports vary as to whether the settlers were serfs or prisoners—the Russian version of Botany Bay. In reality, there was probably little distinction. Baranov intended that the Yakutat settlers become farmers and grow food to alleviate his perennial supply problem. But the cool, rainy weather and rocky ground were to prove less than conducive to farming.

Leaving the meager outpost at Yakutat, Baranov sailed down the coast past Mount Fairweather and found a much more hospitable location for a settlement along the quiet waters of a sheltered sound. Unfortunately, he also found evidence of "foreigners" carrying on trade there with the Kiksadi Tlingit. Determined to counter what he considered clear encroachments on Russian territory, Baranov sailed back to Kodiak Island. Four years later, he returned to Sitka Sound with a complement of Russians and Aleuts and established Mikhailovsk, or Redoubt St. Michael, about six miles north of present-day Sitka, then a Kiksadi Tlingit village the Tlingit called Shee-Atika. (The fact that it took Baranov four years to undertake the expedition is indicative of the dismally slow speed with which things happened in Russian America, not the lack of a sense of urgency in the matter.)

A mutually beneficial trading partnership might have been possible between the Russians and the Tlingit. But relations soon deteriorated, in large part because the Russians showed little regard for Tlingit sovereignty and the Tlingit were well aware that the price of submission meant allegiance to the czar, free labor for the Russian-American Company, and Russian interference in their affairs.

Trouble came to a head in June of 1802, while Baranov was back on Kodiak Island. Tlingit warriors attacked Redoubt St. Michael and killed or captured most of the 150 Russians and Aleuts living there. A few survivors sought refuge on a British vessel that just happened to be anchored in the harbor. Some thought that the ship's captain, Henry Barber, had actually plotted with the Tlingit to drive out the Russians. Certainly, Barber seems to have acquiesced in the attack, and it didn't help his expressions of innocence any when he took the few survivors to Kodiak and demanded a princely ransom from Baranov for their rescue. Baranov paid a fifth of what was initially demanded and then became obsessed with Redoubt St. Michael's recapture.

In September 1804, the Russians returned in force. Baranov led a flotilla of three small ships and hundreds of *baidarkas* filled with Russians and Aleuts. They all embarked from Paul's Harbor and Yakutat and somehow managed to rendezvous in Sitka Sound largely in one piece. The return of the Russians was not unexpected. Sleek Tlingit war canoes had probably shadowed this motley fleet during its journey along the coast. Baranov found that the Kiksadi Tlingit had abandoned their clan houses atop a promontory that would come to be called Castle Hill, and had withdrawn to a fortified enclosure at the mouth of the Indian River just to the east.

He also found another surprise. Anchored in the sound was the 450-ton Russian frigate *Neva*, a superweapon compared to his tiny ships. On October 1, 1804, the *Neva* and the other ships began a bombardment of the Tlingit fort. Little damage was done, and the Tlingit repelled an attempt by the Russians and Aleuts to storm the fort. A six-day siege followed, and upon advancing to the fort on the seventh day, the Russians found it abandoned. Out of flints and gunpowder, the Tlingit had retreated into the night. What was to be called "the Battle of Sitka" was over. Baranov lost no time in building a fortified town that he called Novo-Arkhangel'sk (New Archangel) on the site.

In time, Castle Hill became the center of the Russian-American Company's operations. Contemporary critics in St. Petersburg and misinformed historians would write of Baranov sitting in Novo-Arkhangel'sk in his "castle," lord, indeed, of an empire. In truth, the initial structures were quite modest, and Baranov spent his years in Novo-Arkhangel'sk first in a small cabin and then in a slightly larger structure that was not upgraded until after his departure. What would come to be called "Baranov's Castle" was not built until long after his death.

Still, that did not stop Baranov from being a gregarious host. Amenities were hard to come by in this land, and entertainment—particularly over the long winter—was at a premium. Baranov is said to have developed a repertoire of songs that he sang or required to be sung for him depending on the occasion. Sometimes there was no occasion at all. Needless to say, such hospitality was fired with vodka and other brews. "They all drink an astonishing quantity, Baranov not excepted," reported American captain John Ebbets. "It is no small tax on the health of a person trying to do business with him."[21]

The following year brought Baranov a rare visitor and an important one at that. In August 1805, the head of the Russian-American Company himself, Nikolay Petrovich Rezanov, arrived in Novo-Arkhangel'sk with big plans. Unlike Baranov, who was essentially a businessman out to make a profit, Rezanov had visions of imperial, in addition to commercial, grandeur. He bemoaned the loss of California to the Spanish, saying that if Peter the Great's instructions to Bering had been carried out, the Russian flag would now be flying over the presidio at San Francisco and other ports to the south. Russia's foothold in Alaska had to be made secure. Novo-Arkhangel'sk was still little more than a collection of huts, and there was a shortage of everything. Rezanov encouraged Baranov to trade with the British and Americans to alleviate his supply problem and even to buy ships from them, but not to become totally dependent on either of them.

In the spring of 1806, Rezanov sailed south for California. Shelikhov's daughter was now dead, leaving him a widower, but he had certainly learned the lesson of marrying well. Seeking to extend Russia's reach well down the Pacific coast, Rezanov called at San Francisco and promptly wooed the daughter of the commandant of the Spanish presidio there. By all accounts, she was striking, a lovely senorita half Rezanov's age. Rezanov traded furs for supplies and then returned with them to Novo-Arkhangel'sk. Then he sailed west across the Pacific for Okhotsk, determined to ask the czar's permission to marry a Catholic and to make plans for a broader colonial empire. While en route to St. Petersburg, he fell from his horse somewhere in Siberia and died from the resulting complications. With him died his plans for furthering Russia's colonial empire, but his admonishment that the nation that controls Alaska is the one that controls the Pacific would reverberate in the years ahead.

In 1808, the capital of Russian America was relocated from Kodiak to Novo-Arkhangel'sk, or Sitka, as even Baranov would begin to refer to the town. Before his death, Rezanov had advocated this move because it fit his political plans to make a strong expansion into southeast Alaska. Politics and business were never very far apart, and now Baranov liked the idea because it furthered his economic interests. For one thing, the sea otters around Kodiak Island were almost gone. Baranov Island was new territory. Sitka Sound offered a deep and sheltered harbor with the surrounding slopes covered with good timber for shipbuilding. Perhaps most important, while Sitka was only a couple of degrees of latitude farther south than Kodiak, its location on the Northwest coast, rather than thrown up against the storm-tossed Aleutians, placed it squarely on the trading routes between the Pacific Northwest and the Orient. And, Baranov conceded, it would be a strategic position from which to guard the southern flanks of Russian America—both politically and economically.

Baranov also took Rezanov's advice about trade, in truth expanding the efforts he had exerted from the beginning to bring required goods to his colony. One might think that if Alaska Natives, particularly the Tlingit, could live so richly off the land and the sea, the Russians should be able to do so as well. Some adaptations were made, of course, but the truth of the matter was that the Russians just couldn't get enough of those cereal grains. Barley, wheat, and corn had long been cultural staples, and the Russians continually sought to supplement their salmon diets with them. Of course, there were other necessities that Baranov had to acquire through trading.

After 1807, Baranov worked particularly closely with the Americans. Yankee ships picked up Russian furs in Sitka and then took them across the Pacific to sell in Canton, where Russian ships were banned. The Americans took a commission on the deal and then returned with trading goods to pick up another load. Baranov had only one problem with the arrangement. The Americans were also trading directly with the Tlingit—not exactly what Baranov had in mind when one talked about the Russian-American Company holding an exclusive trading license.

In 1812, John Jacob Astor, principal of the American Fur Company headquartered at Astoria, proposed a solution that at the time seemed innocuous, but that would ultimately have long-term ramifications for Alaska's boundaries. Astor proposed that the Russians retain exclusive rights to hunt and trade with the Tlingit north of fifty-five degrees north latitude and that the Americans have the same privilege south of that line. Astor agreed to provide all of the goods the Russians needed, and Baranov designated the American Fur Company as its sole agent for the Canton trade.

The British, of course, had strong opinions about the deal and were not about to be cut out of the Northwest coast trade. As it turned out, the War of 1812 upset this little arrangement when Astor suddenly sold out to the Hudson's Bay Company rather than risk his ships and properties being seized by force. Thus, the accord proved short-lived, but though undertaken by commercial companies, it demarcated a boundary line that would later receive the attentions of three governments.

While all of this was going on, the Russians had remained committed to being a player on the world stage. Back in 1803, while American president Thomas Jefferson pondered giving his secretary, Meriwether Lewis, command of an expedition to explore the newly acquired Louisiana Territory, Russia's new czar, Alexander I, determined to show the Russian flag around the world. He turned to an Estonian well tested in the Russian navy, Ivan Fedorovich Kruzenshtern, also known as Adam Johann von Krusenstern. In command of the *Nadezhda*, Kruzenshtern sailed from the Russian naval base at Kronstadt on the Gulf of Finland near St. Petersburg in August 1803. With him sailed the *Neva* under the command of Captain Urey Theodorovich Lisianski.

They first put in at Copenhagen and Plymouth, England, for last-minute supplies and then succeeded in rounding Cape Horn into the Pacific in March of 1804. After a call in Hawaii, the two ships separated as planned. Kruzenshtern steered the *Nadezhda* to Petropavlovsk in Kamchatka and then to Nagasaki. There, an attempt to have the Japanese receive a Russian ambassador and establish diplomatic and trade relations failed, but at least Kruzenshtern managed to spend the winter in warmer Japanese waters. The following spring, he surveyed Sakhalin Island, concluding incorrectly that it was a peninsula jutting out from Asia. Then, with plenty of furs for trading, he sailed for Macao.

Meanwhile, Lisianski had taken the *Neva* north to Kodiak Island with supplies for the Russian-American Company. Arriving at Paul's Harbor on July 14, 1804, he learned that Baranov was off mounting his campaign to recapture Redoubt St. Michael. Lisianski sailed the *Neva* to Sitka Sound and played a decisive role in the ensuing Battle of Sitka. After wintering on Kodiak Island, Lisianski returned to Sitka in the spring, loaded up with furs, and went on to make his own charts of southeast Alaska. In the process, he named Baranov Island for Aleksandr Baranov and mapped what would later be called Lisianski Strait off the northwest corner of Chichagof Island.

This accomplished, he sailed the *Neva* across the expanse of the Pacific and met up with Kruzenshtern in Macao in early December 1805. (On the other side of the Pacific, Meriwether Lewis and William Clark were just about to go into winter quarters at Fort Clatsop on the Oregon coast.) The following February, the *Nadezhda* and the *Neva* sailed from Macao and returned to Kronstadt via the Indian Ocean and the Cape of Good Hope, accomplishing the first circumnavigation of the globe by Russians. Not only had the voyage satisfied Czar Alexander's goals to show that Russia's interests were global, but Lisianski's timely appearance in Sitka Sound may have sealed the fate of the Tlingit.

With this naval success behind them, the Russians, too, were not without the Northwest Passage bug. In 1815, a protégé of Kruzenshtern's, Lieutenant Otto von Kotzebue, captained the brig *Rurik* on a search for the western outlet of the fabled strait. Sailing from Kronstadt on the Gulf of Finland, Kotzebue steered around Cape Horn and into the Pacific, calling at Easter Island and the Marshalls before putting in at Kamchatka. His trip was certainly quite a bit easier than Bering's multiyear, cross-country journey of a century before! From Kamchatka, Kotzebue sailed east and landed on Saint Lawrence Island. Then he passed northward through the Bering Strait and entered a large bay to the northeast. He named the cape on its northern shores after his mentor, Captain Kruzenshtern. While the bay ran eastward for many miles, it closed to end in the deltas of several large rivers. Determined to be remembered for something other than not finding the passage, Kotzebue named the waters after himself—Kotzebue Sound. In truth, Kotzebue should be remembered for much more. On this voyage and another in 1823–26, he crisscrossed the Pacific Ocean as boldly as any Englishman and confirmed that Russia's interests extended far beyond the icy ports of Kamchatka.[22]

Another seeking to extend Russia's influence was to meet with far less success and come to cause Baranov much grief. The man's name was Georg Anton Schaffer, and at stake was the trading relationship that Baranov was nurturing with King Kamehameha in the Hawaiian Islands—the Sandwich Islands as the English still called them. Schaffer was a German surgeon who signed on with a Russian-American Company ship at St. Petersburg in 1813 and wound up in Novo-Arkhangel'sk. Baranov appears to have trusted Schaffer with an errand to Hawaii not because of any exceptional diplomatic skills, but rather because he was the only European available for the job.

Schaffer's assignment was fairly straightforward. He was to sail to Hawaii aboard an American ship and, posing as a disinterested naturalist, gain King Kamehameha's assistance in obtaining compensation for a cargo of furs taken without payment by natives on the island of Kauai. Schaffer did so and first struck up a tenuous relationship with King Kamehameha on the big island of Hawaii. But the king did not hold sway over all of the islands, and when Schaffer arrived on Kauai on May 28, 1816, he was greeted with unexpected pleasantries. Kaumualii was the local ruler of Kauai and Niihau and did not take kindly to Kamehameha's expressions of sovereignty over all of the islands. Kaumualii was looking for an ally against Kamehameha—any ally—and Schaffer fell right into the pot of inter-island politics.

Within a month of his arrival, Schaffer not only had gotten Kaumualii to restore the cargo of furs but also had received Kaumualii's pledge of allegiance to Czar Alexander I and an exclusive trade agreement with the Russian-American Company. But wait, there was more. Schaffer promised Kaumualii the protection of the Russian Empire and agreed to lead an army of 500 men against his adversary, King Kamehameha, provided that Russian forts could be erected on each of the islands, including at Honolulu. The Russians would also receive half of Oahu, plots of land on each of the other islands, and Kaumualii's assurance that as the new master of the Hawaiian Islands he would refuse to trade with the Americans. Here was blatant imperialism the likes of which not even Rezanov had spoken!

Communications between the islands and Baranov in Sitka were tediously slow. It was October before Baranov learned that something had clearly gone awry. An American merchant captain showed up in Sitka and demanded the payment he had been promised by Schaffer for the sale of his ship. Oh yes, Schaffer had thrown in a couple of ships for Kaumualii as part of the deal. Baranov was furious that his instructions had been exceeded, but no more so than King Kamehameha, who did not take kindly to the wholesale sellout of his islands. Baranov immediately distanced himself and the Russian-American Company from the actions of his overly zealous agent and assured Kamehameha of his commitment to continuing trade, and only trade, relations. By the time news of all of this reached St. Petersburg, the Russian government, too, considered Schaffer the ultimate loose cannon. It wanted nothing to do with a deal that had the United States knocking at its door and demanding to know about threats to exclude American ships from the islands.

Meanwhile, King Kamehameha forcefully asserted his claims, and Schaffer was left to flee for his life after being given safe passage to Canton onboard the American brig *Panther*. Incredibly, however, Schaffer wasn't finished. He hurried to St. Petersburg and tried to obtain an audience with the czar. He left only when it became apparent that the policy of the Russian government was to quietly sweep the whole affair under the rug. Schaffer then headed for Brazil, where he passed himself off as "Count von Frankenthal" and proceeded to recruit German colonies for a settlement there.[23]

Interestingly enough, if Schaffer had taken things more slowly, he may have been able to establish an independent trading relationship with Kaumualii that would have complemented what Baranov was already doing with Kamehameha. That might have resulted in a Russian presence in the Hawaiian Islands that would have greatly strengthened a trading triangle between the islands, Sitka, and the California coast, and put Russian America more at the center of the Cantonese trade. Ironically, the United States finally acquired the Hawaiian Islands in 1893 by exploiting native unrest much as Schaffer had done.

The Schaffer affair may have hastened Baranov's departure as chief manager of the Russian-American Company. In truth, Baranov had asked to be replaced a number of times, almost since his first arrival in Three Saints Bay in 1791. Twice a successor was appointed only to die en route. Such events convinced Baranov that he was destined to remain in Alaska. Then, too, there were his children. Like so many Russian men, Baranov had married a Native woman, Anna Grigoryevna, who bore him a son, Antipatr, and a daughter, Irina. Baranov was a doting father and would have found it difficult to return to Russia with his children while his first wife in Russia still lived. And so he stayed, first at Kodiak and then at Novo-Arkhangel'sk.

But the Russian government was taking an ever-increasing interest in the affairs of the Russian-American Company. Even before the full news of Schaffer's misadventures reached St. Petersburg, the company's directors were debating Baranov's future. In 1817, they named Imperial Navy captain Leonty Andreanovich Hagemeister as his replacement. Hagemeister, yet another German in Russia's multiethnic military, had previously cruised the Pacific and was fairly well versed in the affairs of Russian America.

Hagemeister arrived in Sitka in the fall of 1817 and spent several months auditing the company's books—all without informing Baranov of his appointment as chief manager. Suddenly in January 1818, Hagemeister confronted Baranov with a discrepancy showing more goods on hand than reported and summarily dismissed him as chief manager. It was a humiliating and unwarranted end to Baranov's twenty-seven years as the major architect of Russian America.

Now what was Baranov to do? Hagemeister informed him that he had not been voted a pension and that his only course of appeal was to return to Russia and plead his case before the directors. Some have speculated that this was a thinly disguised ploy to get Baranov out of Russian America, where he still held considerable sway despite his ignominious firing. Matters were complicated further by the marriage of Hagemeister's second in command, Lieutenant Semyon Yanovsky, to Baranov's daughter, Irina, after a whirlwind six-week courtship. Yanovsky, who by all accounts was far more gracious than his self-important commander, was put in charge when Hagemeister sailed from Sitka in October 1818. Baranov agonized over his decision but in the end decided to sail with Hagemeister.

It must have been a miserable voyage for Baranov. On April 12, 1819, after contracting a fever, he died as his ship wound her way through the East Indies. He was buried at sea in the warm waters of the Sunda Strait between Sumatra and Java, a very long way from the restless seas of the north. Aleksandr Baranov was seventy-two years old—old for the time and ancient by the rigorous standards of Alaska.

The historian Hector Chevigny called Baranov the "Lord of Alaska," but the title could just as easily have been bestowed corporately on the Russian-American Company. In another sense, the most lasting influence of this period was that of the Russian Orthodox Church, which outlasted both of these secular lords. Commerce led the way; the flag followed; but the church would remain.

God Is in His Heaven

Aleksandr Baranov's body had barely settled to the bottom of the Sunda Strait when the issue of the renewal of the Russian-American Company's exclusive charter was put before the czar. First granted in 1799, the twenty-year license was technically up in 1819, but the bureaucratic wheels turned slowly and it was not officially renewed until 1821. In doing so, Czar Alexander I imposed two new conditions. First, the chief manager of the company in North America must be an officer in the Imperial Russian Navy, and second, the company must increase its financial commitment to the Russian Orthodox Church in Russian America.

The first condition was indicative of Russia flexing its muscles on the world stage. After all, Alexander had recently beaten Napoleon back from the very gates of Moscow, and Russia's Prussian and Austrian allies had helped to finish him at Waterloo. Alexander intended that on this foundation Russia become a world power. Georg Schaffer had been far too bold in attempting to seize Hawaii for Mother Russia a few years before, but the pendulum was continuing to swing toward increased military involvement in Russian America in order to protect the czar's wider interests throughout the Pacific. Increased involvement in Russian America meant increased involvement in the Russian-American Company.

The second condition had its roots in the eight priests that Shelikhov dispatched to Russian America in 1794. From this nucleus, the Russian Orthodox faith was nurtured among the Russians and slowly spread among the Natives. One of the early converts was Baranov's wife, Anna. The Aleuts in particular made for ready converts, perhaps because they had been subjected to such turmoil and were willing to embrace any form of promised salvation. While the efforts of the church were scattered at best during Baranov's tenure as manager, his reports and those of others undoubtedly convinced the czar that converting Natives to the Russian faith was an important ingredient of colonial stability. Thus by 1821, as Alexander asserted a military presence into the affairs of Russian America, he directed the church to expand its role there as well.

The man who would come to stand at the center of the Russian

Orthodox Church in Alaska was Father Ivan Veniaminov. Born Ivan Popov in Anginskoe, Siberia, in 1797, his father was the local sacristan, the caretaker of the vessels used in the church services. When young Ivan enrolled in the seminary in Irkutsk in 1806, there were so many Popovs that the rector made each change his name to the town from which he came. Thus, he became Ivan Aginskii. In 1814 when Irkutsk's Bishop Veniamin died, the rector decreed that in the bishop's honor the seminary's most deserving student would be given the last name of Veniaminov. Ivan Aginskii became Ivan Veniaminov and in time was ordained a priest in the White Clergy. (In the Russian Orthodox Church the White Clergy consists of parish priests who are allowed to marry, whereas the Black Clergy must take monastic vows.)

Fortunately, the new Father Veniaminov does not seem to have suffered from an identity crisis. Indeed, he was tall, athletic, witty, and intelligent, counting among his avocations skills as a craftsman and clockmaker. He married and a few years later had a son. Then in 1823, the bishop of Irkutsk received instructions to dispatch a priest to Unalaska. According to church regulations, no priest could be ordered overseas. Such service was strictly voluntary, but in this case no volunteers came forward. Then Father Veniaminov chanced upon a visitor recently arrived from the Aleutians. Whatever was said, this visitor determined his destiny.

On June 24, 1824, Father Veniaminov and his family arrived on Unalaska. There were no accommodations for the priest, and he immediately set to work with his carpentry skills to build a house for his family. The Aleuts helped him, and it may have been through this sharing of hard labor that Father Veniaminov established the mutual respect and rapport that were to characterize all of his relationships in Alaska. Two years later, the Aleuts helped him complete the construction of the first Russian Orthodox church in the Aleutians.

Father Veniaminov threw himself into his work, readily learning the Aleut language and then conducting services in it. He prepared Aleut textbooks for his school and an Aleut dictionary, and took a keen interest in the ethnography of the Aleuts and the flora and fauna of their islands. He even kept a daily record of temperatures, winds, tides, and barometric readings. Father Veniaminov was clearly there to spread the word of God, but he did so in a way that respected the local traditions of the Aleuts—something no Russian before him had done.

In 1834, Father Veniaminov was sent to Sitka, where he tried the same approach with the Tlingit. Relations between the Tlingit and the

Russians, however, had remained uneasy since their 1804 expulsion from Sitka. In 1821, the Russians invited the Kiksadi Tlingit back to Sitka to profit from their hunting expertise. (The Russians had little choice because that was the year they suddenly banned trading with foreigners.) The Kiksadi lived in the "ranche," an area just outside the stockade walls of the town. Russian cannon were always trained on the site, and the walled town was closely guarded. Father Veniaminov's task was more difficult here, but he met with some success by inoculating many of the Tlingit against smallpox.

Four years later, the priest was summoned to return to Moscow and St. Petersburg for consultation. It was clear to the hierarchies of both the Russian Orthodox Church and the Russian government that Father Veniaminov had single-handedly done more to colonize Russian America through conversions than the Russian-American Company had managed through conquests. There was only one thing to do. Promote him and send him back.

Czar Nicholas I and Metropolitan Philaret both prevailed upon Father Veniaminov to take the monastic vows of the Black Clergy (his wife had recently died) and accept appointment as bishop of the newly created See of Kamchatka, the Kurile and Aleutian Islands. He did so and, taking the name Innokentii (Innocent), was appointed bishop on December 15, 1840.

Russian America was quite pleased, and Sitka prepared to receive its new bishop with a grand new structure. Now it was time for some of that assistance required of the Russian-American Company. Built of Sitka spruce by Finnish shipwrights in the company's employ, the Bishop's House welcomed Bishop Innocent in 1843. The two-story building included living quarters, offices, a seminary, the beautiful Chapel of the Annunciation, and later a parish school. The house served as a bishop's residence until 1969 and was the center of ecclesiastical authority in a diocese that stretched across the northern Pacific.

By 1848, a second grand structure was raised in Sitka. This was St. Michael's Cathedral, for which the bishop himself handcrafted a six-foot clock. St. Michael's was a Sitka landmark and served thousands of parishioners until it tragically burned in January 1966.

But Bishop Innocent was not content to sit in his house or preach only from St. Michael's. Far from it. Between 1842 and 1852, he made three major visitations of some 15,000 miles each to newly created parishes throughout Alaska and Kamchatka. He even visited California, presenting Franciscans there with an organ that he himself had helped to craft. Under his direction, missionary priests traveled vast distances by ship, *baidarka*, troika, dogsled, and foot to spread Christianity. Although they numbered only a handful, their influence was long lasting.

When the Russian-American Company's charter was up for renewal a second time in 1841, it was evident that Bishop Innocent and his church were playing an important role in the continuing transition of Russian America from a commercial venture to a settled colony. Many of the provisions pertaining to Natives in the third charter seem to have come from Bishop Innocent's influence, if not outright from his pen. Force was not to be used upon Natives except as a last resort to keep the peace. Outposts were not to be established among Natives classified as "independent" without their consent. Missionaries were given explicit instructions on how to conduct themselves toward the Natives.

Clauses in the charter provided that Natives who did not profess the Christian faith would be permitted to carry on their devotions according to their own rites. In making converts among the Natives, the Russian clergy was to use only conciliatory and persuasive measures, in no case resorting to coercion. The company was charged with seeing that the Natives were not embarrassed under the pretext of conversion to the Christian faith. And Natives professing the Christian faith who, through ignorance, transgressed ecclesiastical regulations, were to be given remedial training rather than be subjected to fines and punishment. Most important, Bishop Innocent received permission to open a seminary and begin training a local clergy.[24] Perhaps all of this was a long overdue apology for the actions of the *promyshlenniki* a century before.

During the 1840s under Bishop Innocent's direction, four new churches and thirty chapels were built throughout Russian America. By 1848, the clergy's numbers in Alaska had increased to almost fifty. Such influence by the church provided the impetus for other institutions, including a forty-bed hospital in Sitka and a ten-bed hospital in Kodiak. With Bishop Innocent's blessing, the Russian-American Company even supported the construction of a Lutheran church. Long after the company's departure from Russian America, the Russian Orthodox Church continued to maintain its churches, orphanages, and schools in Alaska. Twenty years after the sale of Alaska to the United States, the Russian Orthodox Church was annually spending more for schooling in the territory than was the United States.

In the 1850s, Bishop Innocent was appointed archbishop for all of eastern Siberia, and he divided his time between Sitka and Yakutsk. Then in 1868, he was called back to Russia and appointed metropolitan of Moscow, the ecclesiastical head of the entire Russian Orthodox Church. To this day, long after the transfer of Russian America to the United States and the demise of the Russian-American Company, Russian Orthodoxy continues to thrive in Alaska. It is the most durable legacy of Alaska's Russian past.

Limitations of Empire

Simultaneously with granting the renewal of the Russian-American Company's second charter in 1821, the Russian government determined that there were far too many foreigners and far too much foreign influence in Russian America. Having apparently learned nothing from Baranov's quarter of a century of experience, it decreed that there be no trade with non-Russians in the territory north of the fifty-first parallel—roughly the northern tip of Vancouver Island—and that the Russian-American Company settlements be supplied only by Russian ships. Further, Czar Alexander I issued a ukase (an imperial decree) prohibiting foreign vessels from coming within 100 miles of Russian territory, including all of the islands of the Aleutians.

The results were predictable. Internationally, Great Britain and the United States screamed bloody murder over their exclusion from the fur trade and Russia's arbitrary extension of the internationally recognized three-mile territorial limit. Locally, Sitka and the company's other outposts—deprived of the reliable supply of British and American trade goods that Baranov had labored so long to develop—went hungry. It was still an impossible mission to supply all of the colony's needs from Russia.

In the summer of 1823, as negotiations dragged on over Russia's new claims, Secretary of State John Quincy Adams heatedly informed the Russian ambassador in Washington that the United States would contest the right of Russia to any new territorial claims in North America. Six months later, his words were broadened and incorporated into President James Monroe's annual message of December 2, 1823. "The American continents," Monroe asserted, "by the free and independent condition which they have assumed and maintain, are henceforth not to be considered as subject for future colonization by any European power."[25] This noncolonization clause was one of the two cornerstones of what came to be called the Monroe Doctrine.

The Monroe Doctrine's other cornerstone was its nonintervention clause that essentially told European governments to keep their hands off the affairs of the newly independent governments of their former colonies in Latin America. In later years, this half of the doctrine would receive the most attention and come to be significantly expanded. But in its initial articulation, James Monroe aimed one barrel of the Monroe Doctrine squarely at further Russian advances in North America.

With the American position unequivocal and Great Britain equally adamant, Russia belatedly realized that it had flexed its muscles just a little too much. Unwilling to risk war and divert military resources from his European commitments to enforce his edict, Czar Alexander backed down on his Pacific threats. In 1824, Russia signed an agreement with the United States that recognized the southern boundary of Russian America at 54°40' north latitude and restored the traditional three-mile territorial limit and all trading rights. (Fort Ross in California was conveniently ignored, both because it was an established colony and because the United States correctly assumed that it would soon whither on the vine and drop into its lap.)

A similar agreement with Great Britain was concluded a year later. It was far more complicated because it also attempted to resolve the boundary between Russian America and Canada. *Attempted* is indeed the correct word, because uncertainties over this boundary would fester for almost a century. The Russians, of course, wanted to retain title to all of the major islands along the Inside Passage, particularly Baranof and Prince of Wales. Initially, the long north-south boundary running from near Yakutat to the Arctic Ocean was to be the 139th meridian. From there, the boundary ran southeast in a zigzag atop the crest of the Coast Range all the way to Portland Canal at 54°40' north, but in no event was the boundary to be less than ten marine leagues (about thirty-four miles) from the sea. In other words, Russia was guaranteed a strip of mainland at least ten leagues wide along the coast. A host of surveyors and several international boundary commissions would come to rue that choice of description, but that is the way Alaska acquired its slender southeast panhandle.

As part of this horse-trading, the Russians agreed almost as an afterthought to move the long north-south boundary westward two degrees of longitude to the 141st meridian. It seemed relatively unimportant at the time, but by coincidence, the 139th meridian runs smack through the course of a little stream that would later be called the Klondike. Great Britain also got navigation rights on the rivers flowing through the narrow southeast strip and its desired trading rights, the issue it considered most important at the time.

So they had done it. Knowing that they had their own disagreements to resolve over the Oregon country, the United States and Great Britain had nonetheless acted in reasonable concert to check Russia's advance into North America and finitely define the limits of its territory. Even with its expansion checked, however, Russian America was drawn into increasing interaction with the outside world, and this greatly enhanced the colony's economic stability and level of cultural sophistication.

The 1830s proved to be the glory days of Russian America. Recognizing that Baranov's tenure as chief manager had gone on and on, the directors of the Russian-American Company fixed five-year terms for the navy men who followed him. The only one to rival Baranov in accomplishment was Ferdinand Petrovich von Wrangell, "governor"—as the chief manager was by then being called—from 1830 to 1835. As governor of Russian America, Wrangell oversaw a far-flung realm that extended from the Pribilof Islands to Fort Ross. With a population of more than 1,000, Sitka was the busiest port in the eastern North Pacific. At least one writer waxed eloquently and termed it "the Paris of the Pacific." Such accolades were relative, of course, and even Wrangell did not get to enjoy the comforts of the "castle" many had accused Baranov of maintaining.

After Baranov's tenure, a larger structure had been constructed to serve as the governor's headquarters. The building that inspired the nickname "Baranov's Castle" was begun in 1836 and completed two years later. It was a massive structure of hewn timbers seventy-two feet by forty-two feet with two stories, an attic, and even a glass cupola. The cupola served as the first lighthouse on the Alaskan coast and rose ninty-seven feet above the waters of Sitka Sound. A fine library and a grand piano gave credence to claims of emulating Paris. This building served as the headquarters of the Russian-American Company and the residence of Wrangell's successor governors until the sale of Alaska. The U.S. Army and U.S. Navy then used it during their administrations of Alaska until the stately structure was lost to a fire in 1894.

There was one Russian outpost, of course, well beyond the limits of the 1824 and 1825 treaties. On Rezanov's visit to San Francisco way back in 1806, he had not been so taken with the commandant's daughter that he had failed to notice a sheltered cove that offered a fine anchorage some forty-five miles to the north. On his return to Sitka, Rezanov urged Baranov to make use of this unoccupied stretch of coastline as an agricultural and hunting base from which to supply the settlements in Russian America. Baranov ordered his deputy, Ivan Kuskov, to investigate it further with thoughts toward establishing a permanent post there. Kuskov first visited Bodega Bay in 1808. He reported that it indeed boasted a fine harbor, but in 1811 after several more visits, he selected a gently sloping promontory above a small cove some twenty miles to the north as a satisfactory site for a settlement. While it lacked Bodega's fine anchorage, this location had overall advantages in soil, timber, water, and pasturage.

The following March, Kuskov returned to what came to be called Ross Colony with a complement of men and supplies. The twin goals of the settlement were to grow foodstuffs with which to supply Sitka and the northern posts and to conduct general trading with neighboring Native Americans. The name Ross came from Rossiia (Russia), although the settlement was originally called Ross Colony, Ross Settlement, or even Ross Office by officials of the Russian-American Company. The name Fort Ross was not introduced until the mid-1800s by the Americans.

Despite the commercial purposes of the outpost, a stockade was quickly erected out of hardy redwood. Two blockhouses anchored the northeast and southwest corners, the latter boasting a commanding view of the Pacific and overlooking the anchorage in the cove below. Cannon ports stared ominously from the blockhouses. Accompanying armaments varied over the years, as did the cluster of buildings within the compound. In 1825, a Russian Orthodox chapel was added in the southeast corner of the stockade.

Reportedly, Kuskov paid the Kashaya Native Americans some consideration for the land, but this didn't impress his Spanish neighbors to the south. The Spanish quickly sent an officer to investigate this Russian intrusion into their territory, but while protests were made, they never escalated into open conflict. Each side was content, it seems, to carry on cordial local relations, while blaming any inflexibility on its superiors ensconced back in Sitka and Mexico City.

Unfortunately, Rezanov's long-term goals for Fort Ross were never realized. Decades of hunting had greatly reduced sea otter populations all along the Pacific coast, and revenues from the fur trade dropped steadily after the post was established. When Karl Schmidt took over the management of the post from Kuskov in 1821, he became chiefly interested in shipbuilding, but among his problems was a lack of skilled workers. Schmidt also increased farming operations, but inadequate knowledge of crop rotation and fertilization frequently resulted in less than marginal yields.

Governor Wrangell visited Fort Ross in 1833 in an attempt to jump-start the agricultural program. Wrangell went so far as to try to purchase the San Rafael and Sonoma missions from the Spanish in order to increase agricultural production. The Spanish, of course, turned a deaf ear. Thus, even the farming operations were never successful. What finally sealed the settlement's fate was a contract the Russian-American Company made with the Hudson's Bay Company whereby the latter agreed to deliver all of the food supplies the company needed.

With apparently little thought that he was turning his back on Peter's dream of an empire encircling the North Pacific, Czar Nicholas I gave the order to abandon Fort Ross on April 15, 1839. The post's manager, Alexander Rotchev, was charged with liquidating Russian interests there. Rotchev first approached the Hudson's Bay Company and then the French. When neither expressed interest—perhaps because they recognized that any future operation of the post was likely to put them on a collision course with the United States—Rotchev turned to the Spanish, technically now Californios or Mexicans, Mexico having declared its independence from Spain in 1821. But the Spanish were hardly willing to buy what they still claimed was theirs by right. Besides, the handwriting was on the wall. The Russians were likely to abandon Fort Ross soon whether they received value for it or not.

So in 1841, Rotchev finally negotiated a sale of the operation to John Sutter, a German who had received a nearby land grant from the Spanish. The purchase price was $30,000, payable in produce worth $5,000 for each of years one and two, $10,000 of produce in year three, and the final $10,000 in cash in year four. Significantly, the sale expressly did not include the land. Equally significantly, the record is unclear if these payments were ever made.

The Spanish were ecstatic. Anticipating the Russians' departure, General Mariano G. Vallejo wrote to Governor Juan B. Alvarado: "The news I am going to give you is too good for me not to be persuaded that you will share my rejoicing. The Russians are going at last. They are going and Cape Mendocino will now truly be the northern boundary of the Californias."[26] Six months later, Rotchev and about 100 remaining colonists sailed from Bodega Bay, and thirty years of Russian presence on the California coast came to an end.

The Russians, however, weren't the only ones having trouble with outposts. A decade before it was offered Fort Ross, the Hudson's Bay Company was eager to continue construction of a series of trading posts throughout the Pacific Northwest. The company decided to build a post on the Stikine River within British territory to trade with the Tlingit throughout the Inside Passage. Governor Ferdinand von Wrangell didn't think very much of the idea.

Wrangell knew that his Russian-American Company could not compete directly with the Hudson's Bay Company. The latter's posts were better organized and had both a more reliable and a more diverse supply of items than the Russian-American Company had ever been able to offer. Even if some furs ended up in Hudson's Bay lockers, Wrangell wanted to preserve his company's role as middleman. Clearly, he did not want Tlingit along the coast trading directly with the Hudson's Bay Company.

Wrangell decided to call the British bluff. In the spring of 1834, he sent Lieutenant Dionisy Zarembo in command of the *Chichagof* to build Redoubt St. Dionysius near the mouth of the Stikine River. On June 18, 1834, while the Russian fort was still under construction, the Hudson's Bay supply ship *Dryad* attempted to enter the mouth of the Stikine and proceed up the river to establish the British post. Lieutenant Zarembo informed the ship's captain that he had instructions to prohibit any British vessels from entering the river. The *Dryad*'s captain happened to be none other than Peter Skene Ogden, a legendary figure in the Northwest fur trade. In his earlier rough and-tumble days, it is likely that Ogden would have met such as challenge with force. Instead, he grumbled and withdrew peacefully. A formal protest was then sent through diplomatic channels reminding the Russians that the 1825 boundary treaty with Great Britain specifically granted navigation rights through Russian territory on the Stikine River.

It took five years, but the dispute was finally resolved in 1839 when Russia agreed to a ten-year lease of the entire mainland south of the latitude of Taku Inlet. In exchange for the lease, the Hudson's Bay Company was to make an annual payment of 2,000 river otter (not sea otter) pelts and give the Russian-American Company the right to buy additional furs at discounted prices. It was presumed that these would be harvested east of the Coast Mountains. In addition to the leasehold, the Hudson's Bay Company promised to supply Russian America with foodstuffs and manufactured goods at certain fixed prices.

This agreement in itself was very telling, and it set the tone for the exit of the Russian-American Company from the Alaskan scene. With sea otter populations declining, the Russians were clearly looking to obtain furs from sources farther inland, and they were willing to weaken their control over the mainland in exchange for a remedy to their perennial supply problems. Much of the foodstuffs were to come from farming operations in the Oregon country. In the early 1840s, this agreement worked very well, and operations were so productive that an additional 10,000 bushels of wheat were available. Ironically, Russian America finally had all the food it needed—just as things were winding down.

The Hudson's Bay Company hastened to build Fort Taku at the mouth of Taku Inlet, right at the boundary of its leased area, but abandoned it by 1842. The company also took over Redoubt St. Dionysius, renaming it Fort Stikine. By 1847, this was also abandoned. After all of the effort that had been expended to acquire the posts, the company decreed that the costs of maintaining them were too high and that the fur trade could be serviced more cheaply from a steamer, the *Beaver*. The underlying reason, of course, was that even for the highly organized operations of the Hudson's Bay Company, the fur trade was rapidly declining.

With all of the emphasis on Sitka and relations with the British and Americans, it is sometimes forgotten that during this time the Russians were also exploring inland. One of the most ambitious expeditions was led by Lieutenant Laurenti Zagoskin of the Imperial Navy. During 1843–44, Zagoskin paddled up the Yukon as far as the mouth of the Tanana River and also explored the lower Koyukuk. Perhaps most significant, he directed the construction of a post at Nulato, which was to figure in travels along the Yukon for the remainder of the nineteenth century. He also explored from the Yukon's Innoko tributary east to the lower Kuskokwim, no doubt wondering how two great rivers could come so close, but then empty into the sea so far apart.

Zagoskin was first and foremost a surveyor, but he also made notes on the Native peoples and natural resources of the Yukon Valley. Perhaps his only serious misstatement was that the Yukon was not navigable above the mouth of the Tanana, though in later years, it would sometimes seem that way to riverboat captains who desperately sought the main channel during dry seasons.

A few years after Zagoskin's travels, the panhandle lease to the Hudson's Bay Company expired in 1849. It was renewed for another ten years but with two significant changes in its terms. The Hudson's Bay Company was not required to ship foodstuffs because many of its most productive Oregon farms had been lost to the United States with the 1846 division of the Oregon country. Those that were still in operation above the forty-ninth parallel were suffering from a shortage of workers.

Gold, it seems, had been discovered in California. The other significant change was that there was no requirement that the company sell surplus furs to the Russian-American Company. By then, there was no surplus of anything. Russia realized that times were changing and that it was indeed fortunate to make any deal that gave its colony some measure of income.

The years of cordiality between Russia and Great Britain in Alaska came to an end with the outbreak of the Crimean War in 1854. Precipitated by Russia's designs over the Black Sea, the war was fought in Europe, but it would have lasting ramifications for Alaska. Both the Hudson's Bay Company and the Russian-American Company pleaded with their respective governments to declare Alaska a neutral zone. Both parties agreed to this. Doubtless Great Britain could have easily seized Sitka and other key points during the conflict, but to do so would have raised the ire of the United States at a time when Britain was summarily occupied elsewhere. So an uneasy truce hung over Alaska's islands like a fog, but the bitterness that remained between the two rivals kindled Russian hopes that the United States, and not the more geographically logical choice of Great Britain, would someday acquire Alaska from the exhausted Russian Empire.[27]

The United States was all too glad to step into the void left by the cooling of Russian-British relations. During 1854–55, the U.S. Navy conducted extensive surveying operations throughout the Bering Sea and along both the Siberian and Aleutian coasts. Called the North Pacific Exploring Expedition, it consisted of five ships initially under the command of Captain Cadwalader Ringgold, a veteran of the Charles Wilkes expedition to Antarctica and a widely respected expert on coastal geography. Ringgold was relieved of his command, however, while the flotilla was in Hong Kong, working its way up the western rim of the Pacific. Commodore Matthew Perry took the action because of what he considered Ringgold's declining health. Ringgold protested to no avail, and his ships sailed into Alaskan waters without him. Instead, they were under the command of Captain John Rodgers, who also had extensive surveying experience. Rodgers sailed the *Vincennes* through the Bering Strait and along the coast of the Chukchi Sea, while two of his lieutenants on the schooner *Fenimore Cooper* concentrated on mapping the islands of Attu and Adak. Kennon Island at the entrance to Chichagof Harbor on Attu and nearby Gibson Island are named for these officers.[28]

Of course, the British certainly had continuing reasons to be in the area. The most widely publicized one was the search for the legendary Sir John Franklin. If the disappearance of La Pérouse in the South Seas was the naval mystery of the eighteenth century, the disappearance of the Franklin expedition in the Arctic while seeking the still elusive Northwest Passage was the great question mark of the nineteenth. While the tragedy was very real, principally Great Britain, but the United States as well, mounted numerous expeditions that found humanitarian search-and-rescue operations a noble banner under which to explore new terrain.

For starters, John Franklin was no slouch. In fact, he may have been the preeminent Arctic explorer of his generation. Franklin's first expedition in 1819–22 took him across the heart of northern Canada, from the Great Slave Lake down the Coppermine River to the Arctic Ocean. On his second expedition in 1825, he traveled 5,083 miles in one season from New York City all the way to the mouth of the Mackenzie River delta—no small feat. Then, after what by all accounts was a most contented and well-stocked winter camp at Fort Franklin on Great Bear Lake, he floated back down the Mackenzie and struck west along the coast of the Beaufort Sea. He intended to round Point Barrow and rendezvous with Captain Frederick William Beechey, who was working his way eastward along the Alaskan coast in the *Blossom.*

Franklin made it as far as what he called Beechey Point just west of Prudhoe Bay before turning back to the east in the face of a horrendous storm and rumors of Native unrest. Unknown to him, the *Blossom* was only several hundred miles to the west near what Beechey had named Franklin Point just two days earlier. A party from the *Blossom* had gotten as far as Point Barrow in an open boat. Deprived by only 150 miles of a glorious linkup, Franklin returned to Fort Franklin, having nevertheless covered over 600 miles of unexplored coastline. Meanwhile, his right-hand man, John Richardson, had mapped eastward from the Mackenzie to known ground on the Coppermine. Thus, only that 150 miles remained unexplored along the entire Arctic coast between central Canada and Point Barrow. It was an accomplishment that earned both Franklin and Richardson knighthoods.

Franklin might have done best to rest on his laurels. Instead, years later at age sixty he took off for the Arctic again in an attempt to complete his explorations from the east and once and for all establish a Northwest Passage. His ships, *Erebus* and *Terror*, and 128 men were last seen by a whaling ship on June 25, 1845. Then they vanished into the Arctic mists. By 1848, they had yet to emerge. By one count, during the next decade more than fifty expeditions were launched to search for any sign of Franklin, his men, or their ships. In the labyrinthine passages among the Arctic islands—where a few hours sometimes meant the difference between safe passage and a winter bound tightly in the ice pack—ships searched but to no avail. What did happen, however, was that these searchers inked in the blank spaces on the Arctic map.

In Alaska, Royal Navy lieutenants W. J. Pullen and W. H. Hooper led a party of four small boats and twenty-five men on a traverse of the Alaskan coast from Wainwright Inlet to the Mackenzie Delta. This 1849 trip completed the missing link of Franklin's 1826 trip and provided evidence that wherever Sir John was, it didn't appear that he was west of the Mackenzie. In 1852–54, Commander Rochfort Maguire and the *Plover* wintered in Elson Lagoon east of Point Barrow and made significant contributions to Native ethnology by documenting Inupiat customs and place-names.

After these two expeditions, there was no more "unknown" shown on charts of the Alaskan coasts. But what about the Northwest Passage? That, too, had finally been resolved—sort of. Robert McClure of the Royal Navy was as insatiably ambitious as Franklin had been unassuming. Finally in command of his own ship after twenty-six years, McClure was determined to solve both the riddle of Franklin's disappearance and the Northwest Passage. Turned loose in the North Pacific in 1850, McClure steered the appropriately named *Investigator* north through the Bering Strait and eastward along the Alaskan coast. Forty men in five boats literally towed the ship through the ice pack north of Point Barrow. Clear seas took him east to the mouth of the Mackenzie and then into an ice-bound anchorage at Mercy Bay on Banks Island. That fall, McClure made an overland journey by sledge to an overlook above Melville Sound and confirmed—to his satisfaction at least—that a water route existed from Atlantic to Pacific across the north of Canada.

No trace of Franklin's expedition was found until some years later. Some would say that Franklin—coming from the other direction—had gotten far enough west to almost certainly have known that, save for the ice, a passage to the Pacific was possible. Others insist that Northwest Passage honors must go to the man who finally navigated a ship through its entire length, Norwegian Roald Amundsen in 1905. How appropriate that after centuries of uncertainty about the existence of a Northwest Passage, even its discovery is debated. And how ironic that once found, its ice-bound avenues offered no economic rewards.[29] Such were the limitations of empire.

Postscript to an Era (1728–1865)

By the late 1850s, *Russia's enthusiasm for Alaska was clearly waning. Russian America had become an ever-increasing drain on the czarist government. In large part, this was because Russia's devastation of the fur trade and its inability to diversify into other enterprises had resulted in dwindling economic returns. On its western front, Russia was worn out financially and emotionally by its losing efforts in the Crimean War. The ill-fated "Charge of the Light Brigade" notwithstanding, Great Britain and its allies had managed to keep the Black Sea from becoming a Russian lake. The last thing the czar wanted now was for Alaska to fall into British hands. Victoria and "Rule Britannia" on one border was quite enough.*

Finally, with much of Europe united against it, Russia badly needed an ally, or at the very least a casual friend who would largely ignore it. The United States filled the bill and had already pushed British influence out of the Oregon country. As early as 1853, Nicholas Muraviev, governor general of eastern Siberia, urged the czar to rid himself of his North American possessions and predicted that one way or the other, sooner or later, the Americans would control all of North America. But the United States, for all of its bold talk of Manifest Destiny, was now well down the road that would lead it to civil war. Russia would have to sit in Alaska and patiently bide its time until the conclusion of the American Civil War.

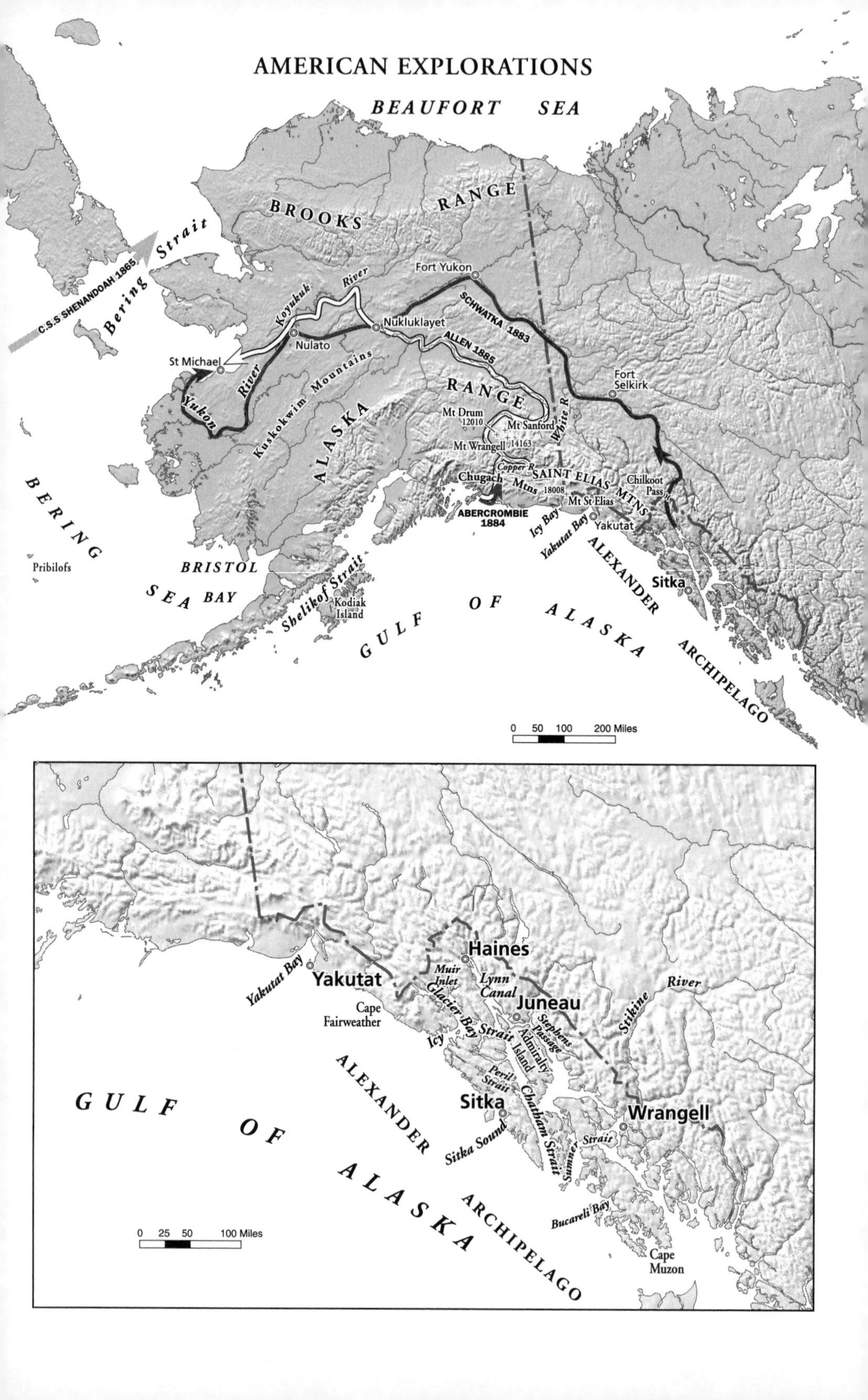

AMERICAN EXPLORATIONS
BEAUFORT SEA
BROOKS RANGE
Bering Strait
C.S.S SHENANDOAH 1865
Fort Yukon
Koyukuk River
Nukluklayet
Nulato
St Michael
SCHWATKA 1883
ALLEN 1885
Yukon River
Kuskokwim Mountains
ALASKA RANGE
Fort Selkirk
Mt Drum 12010
Mt Sanford
Mt Wrangell 14163
White R
Copper R
Chugach Mtns
SAINT ELIAS MTNS
18008
Mt St Elias
Chilkoot Pass
ABERCROMBIE 1884
Icy Bay
Yakutat Bay
Yakutat
BERING SEA
Pribilofs
BRISTOL BAY
Shelikof Strait
Kodiak Island
GULF OF ALASKA
ALEXANDER ARCHIPELAGO
Sitka
0 50 100 200 Miles
Haines
Muir Inlet
Lynn Canal
Yakutat Bay
Yakutat
Glacier Bay
Juneau
Cape Fairweather
Icy Strait
Admiralty Island
Stephens Passage
Stikine River
Peril Strait
Chatham Strait
Sitka
Sitka Sound
Wrangell
Sumner Strait
GULF OF ALASKA
ALEXANDER ARCHIPELAGO
Bucareli Bay
Cape Muzon
0 25 50 100 Miles

BOOK THREE

Seward's Folly: Two Cents an Acre Becomes a Heck of a Deal 1865–1897

The purchase of Alaska,
but it will take the people a generation to find out.

—William H. Seward,
when asked the most important measure of his career

As the American Civil War *came to a conclusion in 1865, the United States looked up from the carnage of its battlefields and determined to make good its earlier boasts of Manifest Destiny. Two years later, despite cries of "Seward's folly" and "Seward's icebox," U.S. secretary of state William H. Seward negotiated the purchase of Alaska from Russia for $7.2 million—roughly two cents an acre. In the three decades that followed, numerous expeditions—some government sponsored, some commercially motivated, and others just individuals out for a look-see—would crisscross much of those 393,750,000 acres to see just what sort of bargain Seward had struck.*

Last Guns of the Civil War

On April 9, 1865, General Robert E. Lee surrendered the Army of Northern Virginia to Ulysses S. Grant—effectively ending the American Civil War. Nine days later, Confederate general Joseph Johnston surrendered his army in North Carolina and purported to do the same for all remaining Confederate forces in the field. But given the communications of the time, there were units that were late in getting the word. Some fought on in Texas until June. And then there were the ships at sea.

The Confederacy had never been able to mount much of a navy, but a few ships—most built abroad—had managed to wreak deadly, if sporadic, havoc on Union shipping. Principal of these was the CSS *Alabama*, responsible for sending fifty-seven Yankee ships to the bottom before being sunk off the coast of France in 1864. But the *Alabama* was not alone.

The linchpin of the Confederacy's naval efforts was Captain James D. Bulloch. Bred of a genteel Georgia family with roots going back to the Revolution, Bulloch joined the U.S. Navy at sixteen. He served on ships of the line for fourteen years and then joined the Merchant Marine, commanding a number of vessels including the mail streamer *Bienville*. This command found him in New York City shortly after Georgia seceded from the Union in early 1861. Some years earlier, Bulloch's half sister, Martha—better known as Mitte—had done the unthinkable and gone and married a Yankee. Mitte now lived in New York City with her husband and growing family, including two-year-old Theodore. Years later, on campaign swings through the South, Theodore Roosevelt would relish telling stories of his Uncle James and the warships of the Confederacy.

Bulloch was soon off to Great Britain, secretly ordered by the Confederacy to secure, equip, and dispatch stealthy vessels capable of raiding Union shipping and then outrunning its heavily armed frigates. It was not an easy task. Great Britain and France were ostensibly neutral, prohibited under international law from aiding either of the belligerents. Whatever Bulloch did, he had to do discreetly, without raising the ire of the Union's British ambassador, the ardent Yankee Charles Francis Adams.

Bulloch set about ordering the construction of speedy merchantmen that could be surreptitiously outfitted as raiders after their launching.

Thus, the CSS *Florida* and the CSS *Alabama* were turned loose on the high seas. Bulloch also kept his eye out for likely ships that could be purchased. In the fall of 1864, with the *Alabama* sunk and Confederate hopes sinking, Bulloch chanced upon the *Sea King*, a long, slender, three-masted vessel of some 220 feet in length and 32 feet in beam, weighing 790 tons empty. She was a combination sailing vessel and steamer and carried an auxiliary steam engine whose propeller could be raised to reduce drag when under sail. Built in Glasgow, the *Sea King* was reported to be one of the fastest ships afloat, making sixteen knots under sail and nine knots under steam in a complete calm. Desperate to stem the downhill slide of the Confederacy, Bulloch and his superiors planned to refit the *Sea King* and use her to destroy the Yankee whaling fleet that cruised the Bering Sea each summer.

When the *Sea King* left England in October 1864, she was equipped with only two twelve-pound cannon, a normal armament for an East India merchantman. After rendezvousing in the Madeira Islands with the supply ship *Laurel* that Bulloch had also secretly dispatched, the *Sea King* took on eight large cannon, a goodly quantity of powder, muskets, and shot, and a new captain and crew. She also took on a new name: CSS *Shenandoah*.

The ship's new commanding officer was James I. Waddell, a lieutenant commander in the Confederate navy. A North Carolinian by birth, Waddell was a seasoned skipper who had graduated from Annapolis and then served in the Mexican War. Like many Southerners, he resigned his Union commission reluctantly, making it clear in his resignation letter that he did not harbor secessionist views, but that he simply could not bear to draw arms against his native state. Waddell received his commission from the Confederacy in March of 1862 and served at Charleston before being dispatched to England to assist Captain Bulloch. Among Waddell's new crew on the *Shenandoah* were veterans of the *Alabama*, including Bulloch's young half brother, Irvine Bulloch. He was Mitte's full brother, and now young Theodore would have two uncles to talk about.

Back in London, there were rumors that the *Sea King* and the *Laurel* were not the innocent merchantmen they appeared. Ambassador Adams protested vigorously that it appeared that a supposedly neutral nation had allowed English subjects to man two vessels whose obvious intentions were to make ready a Confederate man-of-war. But he was too late. While Adams threw a Yankee fit, the *Shenandoah* slipped away into the Atlantic and set sail for Alaska.

The way was still roundabout. The glistening black raider sailed down the Atlantic and took her first prize, the bark *Alina*, out of Searsport, Maine, off Dakar. The *Alina*'s crew was removed to the *Shenandoah* and the ship scuttled. More prizes were sighted and taken as the *Shenandoah* cruised the South Atlantic, rounded the Cape of Good Hope, and sped across the Indian Ocean under sail. Waddell's methods were always similar: give chase to a sail; fly the Union or British flag; hail the ship or fire a shot across her bow to bring her to a stop; board the vessel and ascertain Union registry; and then announce her capture by the Confederate States of America.

By now, the British crews who had first sailed the *Sea King* and the *Laurel* to the Madeiras were back in Liverpool, and Ambassador Adams's suspicions were confirmed. The *Sea King* had become the *Shenandoah*. The American consuls in Liverpool and London spread the alarm to Union naval forces. Slowly, the word reached around the globe: "Stop the *Shenandoah*!"

Waddell's course was for Melbourne, Australia, where the *Shenandoah* would cause almost as much excitement as in London. Waddell's interest in Melbourne was threefold. First, the *Shenandoah* was dismally short on crew. Waddell's attempts to recruit seamen from the original English crews of the *Sea King* and *Laurel* had generally failed, as had his entreaties to the crews of captured prizes. He hoped the blokes of Melbourne would be more interested in a little adventure and profit under the Stars and Bars. Then, too, the *Shenandoah* was having engine problems. The brass band on the coupling of the propeller shaft was cracked and required repair. Finally, Melbourne was a ready port in which to deposit his growing complement of prisoners.

The *Shenandoah* sailed into Melbourne on January 25, 1865. As a British Crown colony, Australia was bound by the same laws of neutrality as Great Britain. Here down under, however, things were a tad more relaxed, and Australian officials were slow to act. Repairs and unlawful recruiting were done—despite loud protests from the American consul. When a British mail steamer left Melbourne, ironically it carried both Waddell's report of successes so far to Bulloch and the American consulate's report to Adams of the raider's whereabouts. On February 18, Waddell sailed out of Melbourne just as he was wearing out his Australian welcome. Then the *Shenandoah* was really on the loose.

The Confederate raider sailed north into the open Pacific. There was only one American warship in the entire western Pacific, the sloop USS *Jamestown,* then laid up for repairs at Shanghai. While the Union navy scrambled to dispatch ships to intercept her—some from as far away as Boston, where the USS *Wachusett* got under way—rumors flew about the gray ghost. The *Shenandoah* was reported back in the Atlantic off Brazil, heading for Peru, inbound to Martinique, and once more off the Cape of Good Hope. But those who gave it some thought knew exactly where she was bound—the Bering Sea—and her target New England's prized whaling fleet.

Indeed, the *Shenandoah* was following her orders and sailing north, pausing only at Ascencion Island in the Carolines (now Pohnpei) to put four Yankee whalers to the torch. After weathering a typhoon and failing to find any of the fast clippers that plied the sea-lanes between the Far East and the American West Coast, she finally sighted the rocky islands of the Kuriles north of Japan. The date was May 21, 1865. General Lee's surrender was now old news throughout much of the world, but Waddell set about the task at hand in ignorance of that fact.

Part of the whaling fleet was supposed to be in the Sea of Okhotsk. The raider sailed into it and a week later stopped the bark *Abigail* and sent her to the bottom. But the shallower waters of the sea's western reaches were still filled with enormous icebergs. Navigating among them made Waddell more than a little nervous, and he turned the *Shenandoah* about and headed toward the open Pacific. Behind him, he left ten New Bedford whalers unseen and unscathed, safely hidden behind the drifting ice.

Waddell steered for the passage between the Komandorskie Islands of Bering fame to the west and Attu Island at the tail of the Aleutians to the east. The *Shenandoah* passed through it and entered the Bering Sea. The weather was typical for the region, but hardly to Waddell's liking. The ship's log recorded sudden changes of weather, heavy fog, and one attempt to chase down a sail that proved to be a rock. But they were getting close. The water was speckled with offal—strips of lean meat from butchered whales. Studying the currents, Waddell calculated that the whalers must be to the southwest and sailed in that direction.

If truth was told, the New England whaling industry was long past its prime. Whale oil was steadily being replaced in lights with kerosene refined from the petroleum being pumped out of Pennsylvania oil fields with a lot less expense. At the highpoint of the whale oil industry in 1853, 238 ships hunted whales in the North Pacific. By 1865, only 85 whalers made up the North Pacific whaling fleet. The *Shenandoah* had already sunk 6 of those in getting to the whaling grounds. In the climactic week following the summer solstice, she was to capture 23 more.[1]

The seasoned New Bedford master of the *William Thompson*, the next whaler to be stopped by the *Shenandoah*, exclaimed, "My God, man, don't you know that the war has ended?" Waddell demanded written proof of the fact. When none was forthcoming, he rapidly commenced seizing and burning whalers from throughout the concentrated fleet. Then, from onboard the brigantine *Susan Abigail*, her captain produced San Francisco newspapers reporting Lee's surrender. Unfortunately for the whalers, they also contained news of President Jefferson Davis's admonishment to fight on to the end. Never mind that Davis himself had been captured only a few days after making the pronouncement, Waddell was now more convinced than ever that his duty was to continue to wreak havoc for the South. His decision was not without major ramifications. If the Confederacy lived, he was a patriot pursuing the South's last gasp under the rules of war. If indeed the war was over, his deeds would be branded acts of piracy.

The *Shenandoah* moved farther north, pausing at St. Lawrence Island just south of the Bering Strait to trade with Eskimos who appeared in kayaks. North of the island the raider took more prizes, almost as quickly as Waddell could dispatch boarding parties, remove the captured crews, and put their vessels to the torch. Smith Lee of the *Shenandoah* led a boarding party onto the deck of the New Bedford bark *Nimrod* and met a familiar face. Her master, James Clark, had been the captain of the *Ocean Rover* when the *Alabama* had captured it two years before. Lee had led that boarding party, too.

The *General Pike*, *Isabella*, *Gipsey*, *Waverly*, and more all fell to the Confederate raider. Then the stage was set for the final hostilities of the Civil War. On June 28, 1865, the *Shenandoah* shadowed another ten whalers among the labyrinthine passages between increasingly numerous ice flows. She struck, and by the time the day was done, eight vessels were burning hulks. The other two were seized to hold the growing complement of prisoners. One grizzled Yankee captain was reluctant to strike his colors, and for a time it seemed that a Rebel broadside would be required, but in the end he too surrendered.

One of the captured Yankees talked of a fleet of sixty more whalers to the north. Whether it was true or a calculated attempt to send the *Shenandoah* chasing into the polar ice is debatable, but Waddell felt he must investigate. The raider sailed north into the Arctic Ocean and crossed the Arctic Circle. No ships were spotted, but the pack ice was building and flowing ominously south toward the strait. Waddell turned his long ship about, and she raced the ice floes through the strait, glancing off them now and again.

Back in the Bering Sea, Waddell took stock and determined that the risks of the ice and of a Yankee frigate appearing on the scene were too great to hunt more whalers. He steered the *Shenandoah* out of the Bering Sea and back to the North Pacific by cutting across the Aleutian chain at Amukta Pass between the Andreanof and Fox Islands.

One of Waddell's men summed up the feelings of the entire crew when he later wrote: "It was a relief to feel we were again bound for the general region of the tropics. I am free to confess that I had enough of Arctic cruising, and if I never look again upon those icy seas and barren shores, fit residences only for Eskimos, seals, and Polar bears, it will not occasion me one moment's regret."[2]

But in San Francisco, Waddell's fears had come true. Newspapers there were branding the *Shenandoah* a pirate ship. As Waddell steered the former raider south across the Pacific, he came up with a new plan. Ignorant of what its newspapers were saying, he thought of sailing into San Francisco Bay, seizing an old Yankee ironclad at Mare Island, and jointly firing on the city unless it agreed to a ransom—bold thoughts for one not altogether certain about his legal status.

While this plan fermented, the cry of "sail ho!" was again raised on August 2, and the raider began her last chase. When the quarry hove to, Waddell dispatched young Bulloch to board her and inspect her registry. The ship proved to be the bark *Barracouta* out of Liverpool. When Bulloch asked how the war was going, the British captain asked incredulously, "What war?"

"Why, the war between the United States and Confederate States," Bulloch replied.

The captain retorted that it had been over since April and, upon learning that the black ship before him was the *Shenandoah*, exclaimed, "Good God almighty! Every navy in the world is after you."

Bulloch returned to the *Shenandoah* in gloomy silence to report to Waddell. This evidence from the friendly British was undeniable. Waddell ordered the *Shenandoah*'s guns spiked and hauled belowdecks.

Now where? And how would they be received? The label "pirate" was a most uncomfortable one for a gentleman of any sort, particularly a Southern one. After flirting with thoughts of Australia, Waddell hurried south for Cape Horn, rounding it well south of the normal shipping lanes, and then beat a cautious path back to Liverpool. On November 6,

1865, the *Shenandoah* sailed into port there under the Stars and Bars and surrendered. In just over a year, she had circled the globe, covered some 58,000 miles, captured thirty-eight ships, sinking thirty-two and releasing six on bond, and taken more than 1,000 prisoners—all without taking or losing a life.

The entire chapter of the Confederate raiders heightened tensions between Great Britain and the United States. In due course, the crew of the *Shenandoah* was freed, and most returned to America. Waddell would command other ships and end up chasing oyster pirates in Chesapeake Bay for the state of Maryland. The claims of the Yankee whalers, amounting to $1,361,983 (a sizable sum in those days), would be addressed as part of the *Alabama* claims, and paid for by Great Britain when a tribunal found the country less than neutral in the actions of the raiders. The *Shenandoah* herself would be sold to the sultan of Zanzibar and later sink after hitting a coral reef in the Indian Ocean.

The *Shenandoah*'s destructive romp through the Yankee whaling fleet in the Bering Sea meant much more to Alaska history than a final footnote in the Civil War. On a grand scale, the incident highlighted the growing controversy of what nation or nations could exploit the riches of the Bering Sea. That issue would take decades to resolve. More immediately, people asked how more than a million dollars of damage could be done in such a distant place. Was there really something of value out there? The questions were particularly profound in the whaling villages of New England and came to weigh upon the mind of one of the states' U.S. senators. In short order, Massachusetts senator Charles Sumner would have occasion to carry on at great length on the floor of the Senate about Alaska, and he would remember the riches plundered there by the *Shenandoah*.

The Telegraph Survey and Mr. Dall

Sometimes, technology moves forward with gargantuan leaps, and what was excitingly new yesterday is outdated at best, ancient history at worst, by the morrow. Examples of this are not limited to airplanes and microchips. In 1860, pioneers plodding across the American West in covered wagons were startled by the blast of a bugle and the pounding of hooves as a solitary rider astride a sweaty buckskin hurried by them in a cloud of dust. The Pony Express! Why, you could send a message from New York to San Francisco in less than ten days! Who would have thought it possible just the day before? Within eighteen months, the Pony Express was bankrupt, busted by a single line of telegraph poles linking the continent. Now, one could answer that message from New York in minutes.

In 1856, an American businessman, Perry McDonough Collins, traveled from St. Petersburg, Russia, into Siberia, and down the Amur River to the Pacific Ocean. Collins had steamship operations on the Mississippi River and in California, but his horizons were global. He saw Russia pushing eastward down the Amur River just as the United States was pushing westward beyond the Mississippi. Collins embodied the commercial component of Manifest Destiny and saw great trading possibilities between these two converging frontiers.

To tie them together, Collins envisioned a telegraph line running from the Pacific Northwest, through western Canada, across Russian America, and under the Bering Strait to Siberia and the Amur Valley. From there, the wire would connect with lines running eastward from Europe, effectively spanning the world and, in a roundabout way, linking Europe and the United States by telegraph. Collins's dream was, as historian Morgan Sherwood suggested, "a vision of the first commercially feasible 'northwest passage.'"[3]

Another determined entrepreneur, Cyrus Field, was also attempting to link Europe and the United States by telegraph. His first attempt in 1857 to lay a transatlantic cable failed, and cheers for success the following year were short-lived when the cable broke again and Field went bankrupt in the aftermath of the Panic of 1857. So Collins's plan garnered support. In 1857, he secured a charter from the Russians and approvals from the British for the international line. William H. Seward, a U.S. Senator from New York, urged congressional support for the project, and Collins was off and running. But if Field's work in the stormy North Atlantic was groping in the dark, Collins's project was even more daunting. The line on the map—such as the map then existed—looked promising, but the truth of the matter was that much of the proposed route lay through unexplored country.

Collins quickly merged his venture with Western Union, but as with many other ventures during the period, his plan's implementation was delayed by the Civil War. When efforts finally got under way again, the project was divided into three divisions—one each in Canada, Russian America, and Asia—and crews were dispatched to the field. To head the Russian America division, Western Union turned to one of the very few Americans who could be said to be an expert on Alaska, Robert Kennicott.

Trained as a naturalist, Kennicott had already made exceptional contributions to cataloging the flora and fauna of Alaska and the upper Yukon watershed. He arrived there via the Porcupine and Mackenzie watersheds—a trip that in itself was enough to make him an "old hand"—and spent the winter of 1860–61 at Fort Yukon. He worked tirelessly under the auspices of the Smithsonian Institution and the Chicago Academy of Sciences to collect some forty boxes of specimens, including a wide array of Native artifacts. Working largely alone on his passion, Kennicott was in his glory. Recruited as an expedition leader, he proved an unhappy failure.

For starters, Kennicott muddied his mission by asking for and receiving permission to include a "Scientific Corps" as part of the main unit of surveyors. Its members included Henry W. Elliott and Charles Pease, who were assigned to birds and small mammals, and William H. Dall, who was initially to focus on invertebrates and fishes. Kennicott was both chief of the Russian American division and head of the offshoot Scientific Corps. His brain would become burdened with the administration of the former, while his heart would continue to lie with the latter.

Kennicott departed San Francisco in 1865 with a hodgepodge of men and equipment that he termed "abominable." They quickly encountered one calamity after another. The *Lizzie Horner*, a small river steamer, was towed to St. Michael to be used on the Yukon, but it broke loose and was wrecked on rocks. Lousy army rations, a lack of sled dogs, and the chaos of a large expedition all wore heavily on Kennicott. No wonder he wrote: "Things don't look at all pleasant and I wish I were quite alone with a small outfit of my own choosing. I could then go in with a lighter heart than I now do when I have the safety and comfort of the others to look after."[4]

Kennicott divided operations into two areas from a base at Nulato on the Yukon just downstream from the mouth of the Koyukuk. One group was to explore westward toward the Bering Strait, and the other was to ascend the Yukon and link up with a survey team working northward through Canada. Little was accomplished in 1865, and just as this plan was to be implemented in earnest in May of 1866, Robert Kennicott died suddenly at Nulato. He was only thirty. Some said it was a heart attack or related malady. Others, including his friend William Dall, suggested that the strain of his position had been too great. There was even speculation that his death was a suicide.

Two weeks after Kennicott's death, Frank Ketchum, whom Kennicott had placed in charge of the Yukon section, Mike Lebarge, a hale and hearty French-Canadian well versed in wilderness travel, and Ivan Lukeen, a Russian Creole fur trader, left Nulato bound for the upper Yukon. The Russian factor at Nulato went with them as far as Nukluklayet at the confluence of the Yukon and Tanana Rivers. Ostensibly, the factor was on his annual summer trading mission, but undoubtedly he also wanted to keep an eye on these newcomers. This was, after all, still Russian America.

Thirty miles above Nukluklayet, the trio overtook an English missionary and traveled upriver with him to Fort Yukon, the Hudson's Bay post where the Porcupine River empties into the Yukon. Despite its location well west of the 141st meridian—the boundary between Russian America and British Canada since 1825—Fort Yukon's remoteness allowed the long-held fiction that it stood on British territory. The Ketchum trip was planned as only a quick reconnaissance to lay the groundwork for a more detailed survey, and at Fort Yukon the party turned around. Floating with the current, they made the return trip from Fort Yukon to Nulato in just eight days. There, they were joined for the winter by William Dall, who had succeeded Kennicott as chief of the Scientific Corps, and Frederick Whymper, a British artist who was recording the landscape and specimens with brush and pencil.

The following spring, Ketchum and Lebarge got an early start on the season and departed Nulato on March 11 to make an overland journey back to Fort Yukon. With them went four sleds, fourteen dogs, and four Natives. Mushing their way up the frozen Yukon, the group was almost out of supplies when it finally reached the trading post two months later. The British factor at Fort Yukon, a man named McDougal, had little to spare, but he gave them some pemmican, and they continued upriver in birch canoes. In due course, they reached Robert Campbell's old Fort Selkirk at the mouth of the Pelly, the extent of their trip.

Returning to Fort Yukon, Ketchum and Lebarge found Dall and Whymper there. They had paddled the 500-some meandering miles from Nulato in a small skin boat after the ice went out. It proved a fast trip for all concerned. The foursome returned to Nulato on July 13 and then were ordered to St. Michael, where they heard the first thunderclap: Field's Atlantic Cable was at last in operation, and the Western Union Telegraph Expedition was ordered to an abrupt end. No telegraph lines would cross the extent of Alaska, but the Scientific Corps that Robert Kennicott had added almost as an afterthought had made the most complete survey of the Yukon River to date. As Morgan Sherwood wrote in his classic study of Alaskan exploration: "It is appropriate and symbolic that the Yukon of Alaska was officially explored by an international party—an American adventurer [Ketchum], a Russian Creole [Lukeen], a French-Canadian [Lebarge], an English artist [Whymper], and a Yankee scientist [Dall]. Their combined background represented the past, the present, and the future of North America."[5]

But Dall was bitterly disappointed. He thought that the investigations had only scratched the tip of the proverbial iceberg. Besides, Alaska had quickly become engrained in his blood. "I felt unwilling," he later wrote, "that the plans of Mr. Kennicott, so far carried on successfully, should be left uncompleted."[6] Dall resolved to stay and asked Western Union to credit part of the salary he was due to the Russian-American Company so that he might outfit himself for a year. On February 3, 1868, he was at Nulato when he heard the second bombshell that would seal his future in the north: Russia had sold Alaska to the United States. Dall took particular delight in communicating the fact upriver to the trader McDougal, addressing him at Fort Yukon, Alaska Territory, *United States of America*.

During the next year, Dall labored diligently to complete the work of the Western Union expedition. Some of the Yukon drainage had previously been charted by Russians and the English, but the work of the

Western Union parties tied these efforts together into a systematic overview of the river. Elijah Smith and J. T. Dyer mapped the wanderings of the lower Yukon below Nulato, and mapping was also done on the Seward Peninsula between Nulato and the Bering Strait.

Perhaps most important, Robert Kennicott's dreams of science were realized in a big way. Landmark work was done in cataloging Alaska's flora and fauna and establishing their relationship to that of Asia and the rest of North America. Dall listed some 77 mammals, 211 bird species, and 52 insect species, including four kinds of mosquitoes, noting that "all are distinguished from the civilized species by the reckless daring of their attack."[7] His associate, J. T. Rothrock, collected 732 botanical species, including many ferns and mosses. And somehow, Dall found time and room to take samples of hundreds of rocks, providing a geologic record of the Yukon's course.

Returning to Washington, D.C., after the 1868 field season, Dall ensconced himself in the bowels of the Smithsonian Institution and wrote the cornerstone of all later Alaskan scientific writing, *Alaska and Its Resources,* which he dedicated to Robert Kennicott. Dall was not right about everything he wrote, but from its publication onward, Dall would be a major Alaskan authority whose shadow would fall over all the scientific surveys of the next quarter of a century. Dall would encourage such exploration, writing that "the field now open to Americans for exploration and discovery is grand. The interior everywhere needs exploration, particularly the great plateau north of the Yukon, the valley of the Kuskoquim [Kuskokwim], and that of the Copper River."[8] Dall's work and these subsequent expeditions—rather than a dreamer's line of telegraph poles drawn on a map—are the true legacy of the Western Union expedition.

Two Cents an Acre

If it was to be called his folly, and indeed characterized by some contemporaries as his greatest blunder, how had William H. Seward come to the point of buying all of Russian America? Seward was born in 1801 at the dawn of the nineteenth century, when the fledgling United States was bent more on its own survival than anything else. By the end of that century, the United States was a world power whose commercial and military interests spanned the globe. Few people championed that progression and laid the foundations for it more than William Henry Seward. Seward's purchase of Alaska was not the senseless pushing of an errant pawn, but rather a major strategic move in a well-thought plan to extend the United States' reach around the global chessboard.

Born in Florida, New York, Seward graduated from Union College in Schenectady and was admitted to the New York bar at Utica in 1822. From the beginning he was involved in politics, but never as an end in itself. He found the emerging Whig Party to his liking, particularly when it came to protective tariffs to stimulate American industry and a national currency system. He ran unsuccessfully as the Whig candidate for governor of New York in 1834, but won in 1838 and again in 1840. Declining a third term, he returned to practicing law in 1843.

By then, the issue of slavery was tearing at the heart of the Whig Party, just as it was at the nation as a whole. No amount of compromise from Whig presidential standard bearer Henry Clay could heal the growing schism or quiet the voices of those who cried out for an absolute and unconditional abolishment of slavery—Seward among them.

The Whigs turned to Mexican War hero Zachary Taylor in the election of 1848, and the New York legislature subsequently elected Seward to the U.S. Senate. Seward was reelected to a second term in 1855, emerging from the ashes of the Whigs to involve himself in the newly founded Republican Party. By 1860, he was a leading contender for the party's presidential nomination. At the Chicago convention, Seward received 173½ votes to Abraham Lincoln's 102 on the first ballot. But when the dust had settled, Lincoln had the party's nomination, and his cabinet came to reflect the deals that had been brokered with favorite sons to achieve it: Salmon P. Chase of Ohio became secretary of the treasury; Simon Cameron of Pennsylvania, secretary of war; Gideon Welles of Connecticut, secretary of the navy; and Caleb B. Smith of Indiana, secretary of the interior. As the defeated front-runner, Seward became secretary of state.

Theirs proved a fruitful partnership. While Lincoln focused on the turmoil of civil war at home and his ultimate goal of preserving the Union at any cost, Seward worked diligently to keep at bay those European powers tempted to take advantage of the situation. Seward also worked tirelessly to build the foundation for later American expansionism. His hand was everywhere. Not surprisingly, he was a supporter of the Western Union Telegraph Expedition. He championed the transcontinental railroad, sought to open trade with China, dreamed of a canal across the Isthmus of Panama, looked to acquire naval bases in the Pacific, and pushed American commerce to all points of the globe.

When Secretary of the Navy Gideon Welles bemoaned the lack of U.S. naval bases in the Caribbean and the detrimental effect that was having on Union blockade efforts, Seward did something about it. He negotiated the purchase of St. Thomas and St. John in the Virgin Islands from Denmark for $7.5 million. As it turned out, Congress was swallowing hard at the time over another bill Seward was presenting. While Congress hesitated, a massive hurricane swept through the islands and wreaked havoc. The end result was that the Virgin Islands treaty was never brought to a vote. (The United States subsequently acquired the islands in 1917 for $25 million.)

The other Seward bill that held the attention of Congress was for the purchase of Alaska. Seward had carefully watched events in Russian America after the third charter of the Russian-American Company expired in 1862. When the company declined an offer of a less favorable charter, the Russian government was faced with taking over its operations and administering a territory that was increasingly unprofitable and indefensible. The other option was to sell it.

Baron Edouard de Stoeckl, the Russian minister to the United States, was called home to St. Petersburg during the winter of 1866–67. When he returned to Washington, he carried with him instructions authorizing the sale of Alaska to the United States for not less than $5 million. When Seward and Stoeckl met in Washington in March, there was no secret about what was on the agenda. Seward was an eager buyer. He promptly made an offer of $5 million. Stoeckl played the reluctant seller very well and held out for $7 million, plus $200,000 to cover the cost of the exchange and to clear the territory of obligations to the Russian-American Company. More cynical minds have read that to include certain bribes necessary to grease the wheels on both sides of the transaction.

Seward and Stoeckl signed the treaty providing for the exchange early on the morning of March 30, 1867, and sent it to the Senate the same day. Neither party knew the exact extent of the land that was changing hands—or, for that matter, what to call it. To Stoeckl, it was Russian America. Somewhere along the line, Seward—possibly with input from Massachusetts senator Charles Sumner, then chairman of the Senate Foreign Relations Committee, and Major General Henry W. Halleck, then commander of the army's Division of the Pacific—applied the name Alaska to the entire area. *Alaska,* the Aleut word meaning "a great country or land," had originally been applied only to the Alaska Peninsula, first by the Russians and later by Captain Cook.

Senator Sumner, also the Republican majority leader, was at first reticent about the purchase, but he warmed to it quickly and delivered an impassioned three-hour speech in support of the treaty's ratification. One of his arguments, as well as a major one used by Seward, was that its ratification was a significant sign of Russian-American friendship—not just to Russia but, perhaps more important, to the European powers. Just as Russia was seeking a friend, so too was the United States. Seward was well aware that much of Europe, particularly Great Britain, had not been judiciously neutral during the recent war. In Great Britain's case the *Alabama* claims would drag on for years, and then there was the little matter of the *Shenandoah.* Let the British worry a little about U.S.-Russian cooperation.

Other arguments advanced by Sumner were from a text approved, if not in fact prepared, by Seward. Wrapped snugly in the cloak of Manifest Destiny, these included extending America's sphere of commerce and settlement on the Pacific coast and widening its base of commerce with China and Japan. It was a good thing that Sumner was not counting on Great Britain as much of a friend, because he went on to argue that "more than the extension of domain is the extension of republican institutions. . . . The present treaty is a visible step in the occupation of the whole North American continent. . . . By it we dismiss one more monarch from this continent. One by one they have retired; first France; then Spain; then

France again; and now Russia; all giving way to that absorbing Unity which is declared in the national motto, *E Pluribus Unum.*"[9]

Sumner was not alone in his pronouncements. Ignatius Donnelly of Minnesota predicted that "the British dominion will be inevitably pressed out of western British America." William Mungen of Ohio declared that "by accepting this treaty we cage the British lion on the Pacific coast," and then went on to predict the decline of the British Empire, the appropriation of its Asiatic possessions by Russia and of its American possessions by the United States, and the coming of a day when "the two great Powers on earth will be Russia and the United States."[10]

As it turned out, the purchase of Alaska was not the prelude to the annexation of Canada, but it left no doubt that the United States would always be a player in whatever events transpired in the Pacific. William Seward's position on the chessboard was shaping up. (If anything, American talk of Canadian annexation fortified the move for Canadian federation and led to the British North America Act of 1867, creating the Dominion of Canada, although two years later Charles Sumner was still suggesting that Great Britain cede Canada in settlement of the *Alabama* claims.) The Senate approved the treaty on April 9 by a vote of 37 to 2.

The deal had been struck and approved, but what status was this new acquisition to enjoy? Previous treaties acquiring new territory, including Louisiana, Florida, the Mexican Cession, and the Gadsden Purchase, had all stipulated that the inhabitants of the newly acquired territory would have U.S. citizenship. They also provided that the territory would, in the words of the Louisiana Purchase, "be incorporated in the Union of the United States." The Alaska treaty gave U.S. citizenship to those inhabitants who chose to remain there, but made no promise of incorporation, meaning a territorial status leading to eventual statehood. Such citizenship did not extend to the Native population or to the Russians, the majority of whom it was presumed would choose to leave the country.

Whether this failure to mention incorporation was intentional because of the sparse Anglo-American population, or whether it was done, as some have suggested, as part of the first steps toward an American imperialism, is difficult—and perhaps pointless—to ascertain. What is clear is that this distinction was to leave Alaska with an uneasy administrative oversight and pose questions over Native claims that were still being debated a century later.

In the broader scope of U.S. foreign policy, the Alaska treaty was a transition between earlier treaties that promised both citizenship and incorporation and the 1898 treaty with Spain, ending the Spanish-American War, that promised neither. In that light, it was indeed an early step toward America's colonial imperialism of the late nineteenth and early twentieth centuries.

The formal transfer of Alaska from Russia to the United States took place at Sitka on October 18, 1867. Brigadier General Lovell H. Rousseau and Imperial Russian Navy captain Aleksei Pestchurov were appointed by their respective governments to serve as commissioners to conduct the ceremony. Brevet Major General Jefferson C. Davis—no relation to the Confederacy's late president—had arrived in Sitka Sound eight days previously with two companies of artillery and one company of infantry totaling about 250 men. So as not to offend the Russians, Davis kept his men aboard the transport *John L. Stevens* until the arrival of Commissioner Rousseau. Also in the harbor were the U.S. warships *Jamestown* and *Resaca* and a host of departing ships belonging to the Russian-American Company.

After Commissioner Rousseau and his party arrived on the USS *Ossipee* on the morning of October 18, they picked their way through the congestion of Sitka Harbor in a launch and arranged to get things under way. At four that afternoon, the various Russian and American contingents assembled around the ninety-foot flagpole in front of the governor's residence, "Baranov's Castle." By all accounts, it was a "bright and beautiful" day—not Sitka's normal fall weather. Prince Dmitrii Maksoutov, the last chief manager of the Russian-American Company and last "governor" of the colony, looked on impassively along with his wife. An era was ending.

To the accompaniment of drums and the thunder of cannon salutes, the double-eagle ensign of imperial Russia began its descent down the flagpole. Then quite inexplicably, the flag caught and refused to budge. Attempts to dislodge it only entangled it more tightly about the halyards. Czar Peter had perhaps turned over in his grave. A marine was sent scampering up the pole. He cut the white, blue, and red ensign free, and it floated downward onto the bayonets of the assembled Russian soldiers. Some reports say that Princess Maksoutov promptly fainted. The Stars and Stripes was raised without incident, and more than a century of Russian presence in Alaska came to an end.[11]

But there was just one problem. The seller had not yet been paid. Under the U.S. Constitution, the Senate must give its advice and consent to all treaties, but all appropriation bills must originate in the House of Representatives. The Radical Republicans who controlled the House, led by Pennsylvania's Thaddeus Stevens, were upset that the president and Senate should present them with a done deal and then expect them to pay for it. They argued that while the Senate might approve a treaty, if its implementation required an expenditure of funds, the House must be involved. Although Seward sweated a little and went about gathering evidence supporting Alaska's value, the issue had less to do with the merits of the purchase than with the intense infighting between President Andrew Johnson and the Radical Republicans. In the end, the House attached a preamble to the appropriation admonishing that any future such purchases required the consent of both houses. The appropriation was finally passed on July 27, 1868. The czar would get his money.

It still remained to define the bargain, but in the end it became clear that those who had cried "Seward's folly" really had no vision of the bigger picture. And in fact those cries against the purchase of Alaska were not nearly as unanimous as historical shorthand frequently suggests. The truth of the matter is that most of the shrill opposition in the press came from one source, Horace Greeley's *New York Tribune*. It was Greeley's acerbic pen that championed the monikers of "Seward's Icebox," "Icebergia," and "Walrussia," and belittled the value of any of Alaska's resources, going so far as to advise other European governments with unwanted wasteland to contact Seward. Evidently, when Greeley had written "go west," he had not meant as far as Alaska!

A great many other newspapers, however, rallied to the cause. Whale oil and fisheries had convinced New England of Alaska's value years before. The region's major papers generally supported the acquisition, and the *Boston Herald* bluntly affirmed that "as to the price, there can be but one opinion—it is dog cheap." Greeley's crosstown rival, the *New York Times,* fired a shot across Greeley's bow by declaring that "while narrow-minded political bigots have been exhausting all their resources in branding him [Seward] as a traitor to his party, he has been quietly pursuing great objects of permanent and paramount interest for his country." And even papers in the Deep South supported the treaty. Out west, support came from the *Portland Oregonian*, the *San Francisco Evening Bulletin*, and the *Sacramento Daily Union.*[12]

So Seward's purchase of Alaska was not universally ridiculed. Nor was it a whimsical or isolated event—far from it. Rather, it was but one piece of Seward's global strategy that took the doctrine of Manifest Destiny to its fullest and extended it from the continental United States to embrace the Pacific rim. Many of Seward's visions did not come to pass for decades—or, in the case of the Panama Canal, half a century. But the foundation for the United States as the world power that emerged from the Spanish-American War was laid by Seward thirty-some years before, and the purchase of Alaska was both a key building block in that foundation and one of Seward's most shining successes. Seward himself would visit Alaska in 1869 after leaving office and come away quite satisfied with his bargain. Many more would soon agree with him.

Boston Men in the Pribilofs

Aleut oral history has its own version of the discovery of the fur seal islands. Igadik was the son of one of the *toyons* (Aleut elder) of Unimak Island. He watched pregnant fur seals swim north through Unimak Pass each spring and then return in the fall with their pups. Caught unexpectedly in his *baidarka* in a strong southerly gale, Igadik was forced to run north with the storm for several days. When the winds died and the seas calmed, Igadik was hopelessly lost in a dense fog. But soon he heard the cacophony of thousands of seals barking. He paddled toward the sounds and came upon the seals' birthing grounds. The Aleuts called the islands Amiq, meaning "land of mother's brother" or "related land."[13]

In June of 1786, Fleet Master Gerassim Pribilof was working for one of the early Russian trading ventures and sailing north of Unalaska in the Bering Sea when he spotted the high cliffs of an island not on his chart. Closer inspection found it teeming with sea otters, fur seals, walruses, and Arctic foxes. He named the island St. George, after the patron saint of the day, and landed a party of hunters there with provisions for the winter. When a return trip was made the following summer, the skies were clear enough that another island was spotted about thirty miles to the northwest. This was originally named St. Peter and St. Paul, but it quickly came to be called just St. Paul. In due course, two smaller islands, each less than a mile in length, were discovered to the southeast of St. Paul. These were later called Otter and Walrus Islands—much more descriptive names. Collectively, these four Islands—St. George, St. Paul, Otter, and Walrus—were frequently referred to as the Fur Seal Islands. Today they are called the Pribilofs.

The Pribilof Islands are the chief breeding grounds for the world's population of northern fur seals. The fur seal's name comes from its thick, waterproof underfur. Incredibly, it can number more than 30,000 hairs per square inch. When sealskins are processed, the longer guard hairs are removed, and this underfur is what makes an extremely soft and warm pelt—the gold of the fur trade.

The annual cycle of the fur seals on the Pribilofs has been going on for thousands of years. Mature bulls, eager to reign over harems of as many as 50 to 100 females, arrive sometime in May and establish their territories. Throughout June, hundreds of thousands of females return to the islands to give birth to a single seal pup and then mate again—all in the flurry of a week or two. The seals usually stay in the Pribilofs until November, when the females and younger males begin migrations far to the south. Some travel as far as southern California and upper Baja. Adult males tend to stay closer to the islands, wintering in the Gulf of Alaska and the southern waters of the Bering Sea. In the spring the migrations are reversed, and the cycle begins all over again.[14]

Russian *promyshlenniki* quickly threatened to disrupt this cycle. They established a skinning operation on St. Paul and used enslaved Aleuts to herd females and their pups along the beaches and then slaughter them with clubs. Later, Aleut slavery gave way to marginal employment, but for a people so in tune with nature's cycles, it must have been a miserable existence.

By the 1830s, both the fur seal and sea otter populations were rapidly declining. Unlike seals that rely on a layer of blubber for insulation, sea otters depend on air trapped in their fur to maintain body temperatures. This very dense underfur of inch-long fibers ranges from brown to almost black. If it becomes soiled or matted, these insulation qualities are lost. That is why sea otters spend so much of their time grooming and preening—their survival depends on it. Older sea otters often develop a silvery head adorned with distinctive, long whiskers.[15]

Unlike fur seals, sea otters generally do not migrate. Thus, it was relatively easy for the *promyshlenniki* to decimate the sea otter population of a particular location. Seeing what unlimited hunting had done to the sea otters and sensing that the fur seals were on a similar road to extinction, Governor Ferdinand von Wrangell set limits on the number of seals that could be harvested in the Pribilofs. It had little impact on overall hunting practices, but Wrangell's action has the distinction of being the first conservation program in Alaska's history. Alaska Natives, of course, had been practicing wise conservation measures for generations.

But the Russians were not the only ones to seek this harvest. American, British, and later Japanese hunters sailed the waters of the Bering Sea in an attempt to get in on the riches. In one Hollywood version based on a Rex Beach novel, *The World in His Arms*, Gregory Peck played an American sea captain out of Boston who was racing Anthony Quinn's character to the Pribilofs. As the two ships tack into a stiff Arctic breeze, Quinn vows that he will beat "the Boston man." Indeed, so prevalent were New Englanders among the fur traders, whalers, and sailors who visited Alaskan waters that for the better part of the nineteenth century Russians and Alaska Natives referred to all Americans collectively as "Boston men"—even adopting the adjective for related things. Thus, John Muir wrote of the Tlingit reference to "Boston food" in his *Travels in Alaska*.

So even before the sale of Alaska to the United States, there were Boston men in the Pribilofs, despite Russian efforts to keep a destructive monopoly to themselves. It has been estimated that the Russians slaughtered several million northern fur seals prior to their sale of Alaska. When the United States acquired the Pribilofs as part of the purchase of Alaska, it was clear that something had to be done to stop the wholesale destruction of the species. In 1869, with the army still in charge of the region, General George H. Thomas visited Sitka, Kenai, Kodiak, Unalaska, and the Pribilofs. Thomas was famous as the Union's "Rock of Chickamauga," and what he knew about fur seals is debatable, but he nonetheless recommended that fur sealing be managed under a leasing system. Congress accepted his recommendations and put out a request for bids from interested companies for a twenty-year, exclusive lease for all fur hunting throughout Alaska.[16] It was a move highly reminiscent of the operations of the Russian-American Company.

Out of the ashes of the Russian-American Company had risen a virtual monopoly that was to have even greater sway—if that was possible—over Alaskan affairs than its Russian predecessor. The Alaska Commercial Company was organized in 1868 with capital stock of $2 million. One of its principal stockholders, Hayward M. Hutchinson of Baltimore, had already bought the assets of the Russian-American Company for $155,000. These included ships, merchandise, and buildings not only in Sitka but also throughout Alaska and the Pribilofs.

In 1870, the Alaska Commercial Company submitted a bid for the exclusive lease of Alaska's fur trade, including the fur seal rookeries on the Pribilof Islands. The lease permitted an annual harvest of 100,000 seals. The presumption, of course, was that the government would accept the highest bid and accordingly receive the highest possible revenue from its resources. Thirteen companies submitted bids, and the lowest price to be paid the government was that of the Alaska Commercial Company, which bid $55,000 in annual fees and a revenue tax of $2.63 per skin. Remarkably, despite submitting the lowest bid, the company was awarded the lease, some said because of the influence of its former attorney, one Cornelius Cole, now a Republican senator from California. In addition to these payments, the lease required the company to hire local Aleuts at the rate of forty cents per seal and also to provide free housing, schooling, and some provisions. That part of the arrangement, at least, was somewhat progressive.[17]

The Alaska Commercial Company continued the Russian practice of driving the seals across the beaches in herds, clubbing them to death, and then skinning them at nearby processing plants. Under the terms of the lease, the 100,000 seals taken annually were not to include females and younger pups, but the reality of these harvesting practices was that many were not separated out of the slaughter. The fur seal trade remained a lucrative business, and the company's dividends offered evidence of why so much political pressure had been brought to bear to ensure that it got the lease. In 1875, the Alaska Commercial Company paid dividends on its capital stock of 37.5 percent; in 1878, the amount was 45 percent, and in 1880 investors got a dividend equal to 100 percent of their capital investment, much of it coming from the fur seal trade in the Pribilofs.[18]

Many have decried the lack of federal attention to Alaska during the first quarter century after its purchase. As the dominant economic power in this broad land, the Alaska Commercial Company liked things the way they were. In the fur business, the more open spaces and the fewer encroachments the better. The less government structure, the more deeply seated the influence of the infrastructure provided by the company. When the federal government did inject itself into Alaskan affairs, it frequently did so in a haphazard way.

As part of the grant of the exclusive license to the Alaska Commercial Company, the United States sought to prohibit the taking of fur-bearing animals in Alaska and its waters by anyone else. It was certainly within the sovereign rights of the United States to prohibit other nations from landing in the Pribilofs and killing seals. Unfortunately, however, this prohibition led to an even more egregious practice.

Russians, Japanese, and especially Canadians began "pelagic sealing," the practice of shooting, harpooning, or otherwise killing seals as they swam in the ocean—outside the territorial waters of the United States. Given the long migrations of fur seals, this opened up quite a chunk of territory. The results were a lose-lose situation. Not only were more females killed in this process, but other females with their own pups will not adopt another pup, so killing a mother doomed her pup as well.

Many of the animals that were slain sank to the bottom before their bodies could be recovered. This uncontrolled slaughter resulted in waste of the worst sort. By one count, ships engaged in pelagic sealing increased from 16 in 1870, to 34 in 1883, and then mushroomed to 115 in 1889. Catches increased from 15,000 seals in 1882 to 60,000 by 1895. Many more seals were killed and sank before they could be recovered.[19]

To counter this new threat to the fur seal population, American revenue cutters were sent to the Bering Sea with orders to seize Canadian and other ships engaged in pelagic sealing anywhere in the Bering Sea. No one questioned the American right to do this within the three-mile territorial limit surrounding each of the Pribilofs. The three-mile limit was well recognized in international law, but proclaiming sovereignty over the entire Bering Sea was an entirely different matter. The United States itself had fought on the other side of the issue in the 1820s when Russia similarly tried to bar other nations from the Bering Sea. Both the United States and Great Britain had made agreements with Russia guaranteeing freedom of navigation and the right to fish in the Bering Sea. Now, the shoe was on the other foot.

Great Britain immediately cried foul at the seizures of the Canadian vessels and protested even more vigorously when the U.S. District Court in Sitka upheld the actions. When President Benjamin Harrison assumed office in March of 1889, one of his first acts was to affirm the prior administration's annual warning that pelagic sealing was banned from all waters of the Bering Sea and that the United States would arrest any violators. Having weathered the storm of the *Alabama* claims with their relations reasonably intact, Great Britain and the United States were eager to resolve this issue short of going to war over seals.

Thus, the matter went through lengthy negotiations and was finally put to international arbitration. A tribunal convened in Paris in February 1893. John W. Foster, who argued the case for the United States, must have felt as if he were being asked to prove that the sky was not blue. To begin with, he relied on Russian documents expressing sovereignty over the Bering Sea and argued that the United States assumed this position upon its purchase. Barely into his case, however, he discovered that these documents were forgeries of the enigmatic Ivan Petroff, a Russian whose nefarious handiwork of historical half-truths and outright lies is spread throughout the history of Russian America. Forced to withdraw those exhibits, Foster next attempted to convince the arbiters that the Bering Sea lay outside the general limits of the Pacific Ocean and as such was not included in the 1824 and 1825 agreements Russia made with the United States and Great Britain concerning international access.

Most incredibly, Foster went on to argue that the fur seals migrating to and from the Pribilof Islands were really domestic animals belonging to the United States. Foster claimed that "the Alaskan fur seal, begotten, born and reared on the Pribilof Islands, within the territory of the United States, is essentially a land animal, which resorts to the water only for food and to avoid the rigor of winter, . . . that it is domestic in its habits and readily controlled by man while on land, . . . that at all times, when in the water, the identity of each individual can be established with certainty, and that at all times, whether during its short excursions from the islands in search of food or its longer winter migration, it has a fixed intention, or instinct, which induces it to return thereto."[20]

There were many snickers in the room when this argument was put forward, and it was facetiously noted that Foster had stopped just short of claiming American citizenship for the seals—something that had not yet been done for Alaska Natives. That Foster himself was evidently sheepish with this argument is best illustrated by the way in which he was very quick to give credit for it to Secretary of the Navy Benjamin F. Tracy.

Not surprisingly, the 1893 court of arbitration ruled against the United States on all issues, denying its claims of exclusive jurisdiction over the Bering Sea and any property rights in the seals when they were outside the three-mile territorial waters surrounding the Pribilofs. Great Britain was awarded almost half a million dollars in damages for the seized Canadian vessels. Great Britain did agree with the United States to ban pelagic sealing east of the 180th meridian (just east of Kiska and the Rat Islands in the Aleutians) and north of thirty-five degrees north latitude (roughly halfway between Los Angeles and San Francisco).

Unfortunately, this agreement didn't stop the Russians and Japanese from continuing pelagic sealing in the western Pacific. In one instance, Japanese sealers were so bold as to land on the Pribilofs. Not until 1911 was a four-party treaty signed by the United States, Great Britain, Russia, and Japan that extended the pelagic sealing ban south to thirty degrees north latitude and widened it across the entire North Pacific. The treaty also made provisions for dividing up those seals taken legally on the Pribilofs between the four nations.

The 1911 Fur Seal Treaty also gave full protection to the decimated sea otter population, but many thought that it came too late. By then, as few as 2,000 animals existed, where a century and a half before they had been prolific across the whole arc of the North Pacific from Kamchatka and the Kurile Islands to southern California. Sea otters had paid the ultimate price for having arguably the finest fur in the world.

Vast areas of the Aleutians were designated a wildlife reserve, but each year the sea otter count decreased. By 1925, a thorough survey failed to detect a single otter, and many thought that the species was extinct. But nature has a way of taking care of its own. Six years later, Frank Dufresne, who was soon to become director of the Alaska Game Commission, was on an inspection tour in the Aleutians when his ship put in at Amchitka Island in about the middle of the chain. Dufresne called on the local Aleut elder, who mysteriously took him aside and told him that there was something he wanted Dufresne to see. Could he be ready at dawn? Intrigued, the next morning Dufresne followed the old man along the northern shores of the island until he ducked down behind a boulder and motioned Dufresne to look carefully among the strands of green and purple seaweed gently rolling in the swells. Almost hidden in the midst of the kelp was a lone otter, drifting on its back with a pup clasped in its forefeet. The sea otter was slowly struggling back from near extinction.[21]

John Muir Visits Glacier Bay

In 1877, army lieutenant C. E. S. Wood and Chicagoan Charles Taylor made a private trip to Alaska, ostensibly to climb Mount St. Elias. At Sitka, they hired a large Tlingit war canoe and a complement of paddlers and started north up Chatham Strait. Supposedly, the Tlingit talked the two out of the attempt on the mountain or at least refused to make the dangerous canoe trip on the open ocean between Cross Sound and Yakutat Bay. Foiled in this attempt, Wood and Taylor returned to Sitka. Taylor had had enough of the trip, but Wood rounded up a couple of local packers and another Tlingit crew and headed back north toward the country on the east side of Mount Fairweather, supposedly having been told by the Tlingit that one mountain was as good as another.

Wood didn't attempt to climb any mountains, but he visited a Hoonah Tlingit seal-hunting camp at what may have been Geikie Inlet and then hunted mountain goats on the higher peaks. He considered traversing into the headwaters of the Chilkat River, but the Tlingit discouraged the idea.[22] Later, when there was some discussion about who was the legitimate "discoverer" of Glacier Bay, Wood freely admitted that he really had had no idea of the significance of where he had been. It was fine with him if the other guy got the credit. The "other guy" turned out to be John Muir.

Acknowledging that he had no definite plan, John Muir departed San Francisco in May 1879 on board the steamer *Dakota*. He was forty-one at the time. How much he had heard about Alaska is questionable, but he admitted to being eager to see new country after eleven years in the Sierra Nevada and the mountain ranges of the Great Basin. The *Dakota* took him to Victoria, British Columbia, from where he meandered around the Puget Sound area for a few weeks before sailing north from Portland. He arrived in Wrangell on July 14, spent a few hours, and then went on to Sitka. By the twentieth, he was back in Wrangell, later describing it as the most inhospitable of towns at first sight. One can only surmise that it was the nearby glaciers that enticed him back and not the town itself.

Muir spent a couple of months exploring the Stikine River valley, or as he wrote it, the "Stickeen." The river was navigable as far as Glenora in British Columbia and sometimes fifteen miles farther to Telegraph Creek. It wound through the Coast Range and gave Muir access to a whole new world of icy glaciers, snow-clad summits, and silt-laden streams. The Stikine had been a route of the gold rush to the Cassiar district of British Columbia, and a few misinformed individuals would try to reach the Klondike via it two decades hence.

After a second major trip up the Stikine and explorations on the Stikine Glacier, Muir was back in Wrangell in October. From a group of prospectors he heard stories about Chilkat Inlet at the head of Lynn Canal, the extreme northern end of the Inside Passage. If it was glaciers he was after, the prospectors told him, that's where he should have a look. Muir made plans to do just that. Samuel Hall Young, a thirty-two-year-old Presbyterian missionary stationed at Wrangell, volunteered to accompany him. Young's wife had just delivered their first child less than a month before, and what she thought of the idea can only be imagined.

Muir, Young, a Tlingit chief named Toyatte, and three other Tlingit left Wrangell on October 14 in Toyatte's canoe. It was extremely late in the year to be paddling northward in search of glaciers. Somewhat offhandedly, Muir acknowledged that fact and later wrote, "On the other hand, though this wilderness was new to me, I was familiar with storms and enjoyed them."[23]

Toyatte's canoe carried the party westward across Sumner Strait, then northward up narrow and rocky Keku Strait. Rolling seas gave them some pause in Frederick Sound, but they traversed it and found easier going in Chatham Strait. The strait led to Icy Strait and the Tlingit enclave of Hoonah. Here, as in other villages encountered, Young gathered as many Tlingit together as would listen and preached the virtues of Christianity.

One of the Tlingit paddlers, Sitka Charley, talked of a place he had gone as a boy with his father to hunt seals. It was beyond Hoonah at a place the Tlingit called Sitakaday, or Icy Bay. Ignoring warnings about making the trip so late in the season, Muir, Young, their four Tlingit companions, and a brave Hoonah, who agreed to act as a guide, crossed Icy Strait and paddled through what is still called Sitakaday Narrows. They passed "smooth marble islands" (today's North and South Marble Islands) and discovered the first of many great glaciers. Scot that he was, Muir named it for James Geikie, a Scot geologist who had written extensively on the Ice Age.

The next day was Sunday, and Young chose to remain in camp. The Tlingit took one look at the cloudy, drizzly weather and decided that a little Christianity suited them just fine if it meant staying in camp. Muir, of course, ventured forth alone into the overcast to see what he could see. In *Travels in Alaska* he recalled: "All the landscape was smothered in clouds and I began to fear that as far as wide views were concerned I had climbed in vain. But at length the clouds lifted a little, and beneath their gray fringes I saw the berg-filled expanse of the bay, and the feet of the mountains that stand about it, and the imposing fronts of five huge glaciers, the nearest being immediately beneath me. This was my first general view of Glacier Bay."[24]

Meanwhile, Young had temporarily put down the Scripture in order to explain to Toyatte and his companions just what it was that Muir was looking for up there: knowledge about glaciers and mountains. Toyatte couldn't quite fathom that and declared that "Muir must be a witch to seek knowledge in such a place as this and in such miserable weather."[25]

Muir made it back to camp, and the following day, with the Tlingit still shaking their heads, the group paddled farther to the head of the bay. They passed a second massive glacier that Muir named for another fellow Scot and geologist, Hugh Miller. Those who travel Glacier Bay today may find it difficult to reconcile the present landscape with Muir's descriptions. The amount of glacial retreat that has occurred in roughly 125 years is mind-boggling. In 1879, Muir found the massive face of the Grand Pacific Glacier draped over Russell Island like a great glob of icing on a tiny cupcake. The glacier extended north and south from the southern tip of the island at least two miles in either direction. Oh yes, said Sitka Charley, this is where he and his father had hunted seals.

Paddling back down the eastern side of the bay, they passed what would later be named the Carroll Glacier and then came to the face of what would be called the Muir Glacier. It covered the entire width of what is now the mouth of Muir Inlet, some five miles wide. The low-slung clouds had hidden it from view when the party first passed. A century and a quarter later, the glacier has retreated almost thirty miles and exposed two major inlets. With the face of every glacier cracking and creaking and groaning, Muir undoubtedly did not have to hasten the Tlingit as they dug their paddles into the frigid waters and hurried out of Sitakaday Narrows and back across Icy Strait to Hoonah.

But Muir's trip was not done. He had, after all, set out to see Chilkat Inlet at the head of Lynn Canal, and Young was intent on preaching to the Chilkat Tlingit there. They paddled east from Hoonah and up Lynn Canal to where a massive tidewater glacier with a two- to three-mile front extended into Chilkat Inlet. This was the Davidson Glacier, named in 1867 by the U.S. Coast and Geodetic Survey for George Davidson, who was in charge of the survey's Pacific coast operations. A century and a quarter later, it too is but a shadow of its former self, shrunk into its outwash plain and no longer extending to tidewater. Just beyond the glacier, the party put in at Pyramid Harbor on the isthmus dividing Chilkat and Chilkoot Inlets. Here Young preached to an assemblage of Chilkat Tlingit and then walked across the isthmus and located what would become the site of the Haines Mission and town several years later.

From here, it was back toward Wrangell via the eastern side of the Inside Passage with a stop to spread Young's word among the Auk Tlingit at Auke Bay below what Muir called the Auke Glacier. When they met a handful of prospectors farther south, Young complained to them about his party's frequent lack of meat. One grizzled miner asked Toyatte why he and his men did not shoot plenty of ducks for the minister. "Because," Toyatte replied, "the duck's friend would not let us; when we want to shoot, Mr. Muir always shakes the canoe."[26]

The Tlingit called this area Sum Dum Bay, a name said to be derived from the booming shots of icebergs calving from the surrounding glaciers. Muir pushed the party to explore the bay's southern arm, but even he was forced to admit—under threats of mutiny from the Tlingit that were only half in jest—that they couldn't make it through the clogged ice to reach the glacier from which it came. Muir had failed to find a glacier and was rather put out. Young termed it only one defeat in an otherwise victorious expedition, but it was to gnaw at Muir until the following summer. Later, Young wrote, "This one defeat . . . so weighed upon Muir's mind that it brought him back from the California coast next year and from the arms of his bride to discover and climb upon that glacier."[27]

With the exploring season now well past, they hurried south, clawed their way through more ice floes, and arrived back at Wrangell on November 21 after a journey of about 800 miles—a rather impressive fall canoe trip by anyone's standards. Young went to work planning his next visits among the various Tlingit clans. Muir spent a month waiting for a steamer headed south and pondering his return to the glacier that had eluded him.

Muir indeed married the following spring and proceeded to arrive on the docks of Wrangell on August 8, 1880—without his wife. Young was there to greet him and eager to join another journey blending glaciers with Presbyterian evangelism. Unfortunately, their loyal companion Toyatte would not be joining them. Toyatte had been shot and killed the previous January while playing the role of peacemaker in a squabble with some Taku Tlingit who had come to Wrangell. In his place went Stikine Tlingit Lot Tyeen and two younger Natives, Joe and Billy. And, of course, there was a sixth companion, Young's lovable mut, Stickeen, who—thanks to Muir's later writings—was destined to have his name enshrined alongside White Fang, Yukon King, and Balto in Arctic dog lore.

The first order of business was to find Muir's "lost" glacier. They left Wrangell in one canoe on August 16, skimming through icebergs at the mouth of the bay the Tlingit called Hutli, or Thunder. Northward then to Sum Dum. Today, this is called Holkham Bay, and its two principal arms, Tracy and Endicott, are named for little-remembered nineteenth-century cabinet officers. This time, Muir and his companions reached the glacier at the head of Endicott Arm. Muir promptly named it after Young. The modest reverend was grateful but later wrote: "Some ambitious young ensign on a surveying vessel, perhaps, stole my glacier, and later charts give it the name of Dawes. I have not found in the Alaskan statute books any penalty attached to the crime of stealing a glacier, but certainly it ought to be ranked as a felony of the first magnitude, the grandest of grand larcenies."[28]

Muir ranked the two days they spent in Holkham Bay among the brightest and best of his days in Alaska, telling Young, "There is your Yosemite; only this one is on much the grander scale."[29] He recalled that he never saw Alaska looking better than when he pushed out of Holham and headed north for Taku Inlet. They found the village of the Taku Tlingit deserted and pressed northward.

According to Young's account, they camped on some gravel flats near a small creek at the head of Gastineau Channel. The next morning, while pushing their canoe through the shallows toward Stephens Passage, they met up with Richard Harris and Joe Juneau, two prospectors whom Young had frequently seen around Wrangell. They were out from Sitka doing some hunting and prospecting. When they heard about the group's last campsite, Harris suggested to Juneau that they too camp there and try panning the gravel along the creek—or so Young's story goes.

The dwindling days and Muir's urge to visit more glaciers pulled them quickly across Lynn Canal and Chatham Strait to Hoonah. It, too, was deserted, all of its inhabitants out hunting, fishing, or picking berries. This time, though, they knew the way and paddled across Icy Strait, stopping as they had the year before at Pleasant Island to harvest wood for fires at their camps among the glaciers.

On this trip they spent a week at Glacier Bay and focused on the Muir Glacier. Muir and Young drove a line of wooden stakes across a portion of its surface and watched with fascination as the line buckled and bent, showing the surprisingly rapid advance of the ice. Young estimated that it was as much as fifty or sixty feet a day. It was the first scientific work in what would become an unparalleled laboratory for the study of glaciation.

But there was still more to see. From the mouth of Sitakaday Narrows, the lone canoe sailed westward, fighting the tides that were running like rivers and dodging the bergs that floated with them down Icy Strait and into Cross Sound. They made landfall in Taylor Bay on the north side of Cross Sound within sight of Cape Spencer and the Pacific. Muir deduced correctly that the gigantic glacier at the head of the bay was advancing, rather than retreating, because of the furrow of plowed-up gravel and dirt at its toe. In fact, a Hoonah Tlingit and his family showed up the next day and complained that the advancing glacier was severely shortening the outflow stream that had once provided a perfect little salmon run. The glacier that Muir and Young called the Taylor was later named after territorial governor John G. Brady. (Taylor Bay at its foot still retains the name of Lieutenant Wood's 1877 traveling companion.)

By now, Stickeen had become an inseparable member of the party. By all accounts, his offhand antics and his matter-of-fact nature endeared him to everyone, particularly Muir. Never mind that the dog had actually been a wedding gift to Young's wife. Stickeen had adopted Muir and vice versa. Early one morning Muir set off to explore the surface of the Brady Glacier, and, despite Muir's attempt to dissuade him, Stickeen insisted on tagging along. What followed, of course, was a horrendous day of crossing wide crevasses and narrow snow bridges amid freezing rain and the flat light of misty clouds. Almost left behind at one nearly impassable snow bridge, Stickeen persevered and made it safely across to the cheers and tears of the thousands who read Muir's story of the faithful dog when he first published it in 1897.

The excursion left both Muir and Stickeen a little less independent, and Muir's thoughts turned to his waiting bride. It was time to head south. Fearing that he had already tarried too long, they set paddles and sails flying and dug for Sitka, where Muir made it in time to take the mail steamer south. Poor Stickeen was left howling after him in the stern of the canoe. Young later reported to Muir that the dog was stolen from him in Wrangell some years afterward.

In the years after Muir's first two visits to Glacier Bay, events moved almost as quickly as the receding glaciers. The same year as Muir's second visit, U.S. Navy captain Lester S. Beardslee took the USS *Favorite* into the bay as far as the Beardslee Islands and charted its lower reaches, including Willoughby Island, named for sourdough Dick Willoughby who happened to be along. Beardslee may have been the first to use the name Glacier Bay. Veteran Inside Passage skipper Captain James Carroll took the first ship, the steamer *Idaho*, into upper Glacier Bay in 1883. Carroll went on to become the best known and most colorful of the steamship captains who brought an increasing number of tourists to the bay during the late 1880s and 1890s. The indomitable Eliza Ruhamah Scidmore was on board the *Idaho* on that first trip and subsequently wrote what may be termed the first "travel" articles about its wonders. Harry Fielding Reid and a host of fellow geologists whose names still adorn many of the bay's glaciers made the first of many scientific trips in 1890. Muir himself returned for a third visit that year.

Did John Muir discover Glacier Bay? No, not really. Tlingit, Vancouver's men, and C. E. S. Wood had all been there before him. But Muir was certainly the first to publicize it and to recognize its incomparable beauty and the scientific value of the glaciology laboratory it remains.

Sheldon Jackson's Missionary Zeal

If ever there was a missionary whose relentless work evoked a chorus of "Onward Christian Soldiers," it was Sheldon Jackson. At five-foot-four, he was far more David than Goliath, making up for what he lacked in height with unbounded energy and a feisty, if at times aggressive, personality. Tact and willingness to compromise were not among Jackson's traits, and he frequently found himself at odds with those opposed to his views.

A New Yorker with a degree in theology from Princeton, Jackson first achieved notoriety as the head of the Rocky Mountain Synod of the Presbyterian Church, establishing churches in a number of Rocky Mountain mining camps. Even when the Presbyterian Church did not share his views, his position as editor of the *Rocky Mountain Presbyterian* and his prominence on the lecture circuit in the Northeast gave him a powerful voice. Jackson also cultivated powerful friends. Perhaps the most important was Benjamin Harrison, a fellow Presbyterian from Indiana. Harrison was destined to be the last American president to be elected without winning the popular vote until the election of 2000, but in the early 1880s, he was a Republican senator from Indiana and the chairman of the Senate Committee on Territories.

Jackson made his first visit to Alaska in 1877 and helped to organize the Wrangell Mission that was run by Samuel Hall Young, John Muir's traveling companion. In 1881, Jackson also assisted Young in establishing the Willard Mission on Chilkoot Inlet near the Tlingit village of Dtehshuh, meaning "end of the trail." Reverend Eugene Willard and his wife did not stay long, and three years later the mission was renamed for Mrs. E. F. Haines, secretary of the Presbyterian Church's Women's Executive Committee for Home Missions. The resulting town of Haines is the only town in southeast Alaska named for a woman.

With the federal government showing little interest in Alaska, this missionary work played a significant role in Alaskan affairs in general and in relations with Alaska Natives in particular. In 1880, it was Sheldon Jackson who suggested that other denominations concentrate their work in a particular area of Alaska rather than compete with one another for souls. Thus, the Presbyterians continued to establish missions in southeast Alaska and remained there with the exception of a commitment for a mission at Point Barrow. In time, the number of Presbyterian churches came to rival that of the Russian Orthodox churches. The Episcopalians continued their work along the Yukon. Baptists took on Kodiak and Cook Inlet; the Methodists were active in the Aleutians; Congregationalists established a mission at Cape Prince of Wales; and the Moravians built missions in the Kuskokwim and Nushagak Valleys.[30]

Jackson returned from his early trips to Alaska convinced that it was his Christian duty to combat the twin evils that he saw rampant among Alaska's population: a decided lack of education and a preponderance of alcohol. These issues inevitably focused on Natives because their numbers accounted for about 90 percent of Alaska's population. First to the Presbyterian Church, then to audiences on speaking tours, and finally to his friend Benjamin Harrison, Jackson championed the twin causes of education and prohibition.

Many embraced Jackson's crusade as noble and in the best of the late-nineteenth-century traditions of carrying "the white man's burden." Others—especially the other 10 percent of Alaska's population—were not so sure. As William Dall reported: "Theoretically, every man is in favor of missionary work; but when, as in the present case, they take up available land for their schools, teach the Indian to work, and to build civilized houses, to ask a good price for his furs and fish, and on no account to sell his young daughters to white men, as was formerly the practice,—such innovations do not meet with universal favor."[31]

During the early 1880s, while Jackson was becoming an increasingly visible advocate for public education in Alaska without regard to race, Benjamin Harrison had been grappling with the deficiencies of Alaska's civil administration. A number of Alaska bills got nowhere in the Congress until Harrison introduced his Senate Bill 153 in December of 1883. Initially, it covered his own administrative concerns, but by the time the bill was passed by both houses the following spring, Harrison had deftly managed amendments to it to also address Jackson's educational concerns.

Harrison's Organic Act of 1884 was the first small step in providing Alaska with some form of government other than the sporadic administration of the U.S. military. (Officially, Alaska was administered by the army from the transfer ceremony of 1867 until 1877, and then by the navy from 1877 until the passage of the Organic Act of 1884.) But the terms of the act were a far cry from representative government. For one thing, Alaska was legally termed a *district,* and not a *territory,* because the latter term would have implied that it was on a track toward statehood. That was certainly not Congress's intent. The act made no provisions for law enforcement and expressly forbade a legislature. In the rush to get things done, the laws of the district were declared to be the laws of Oregon—Oregon being the closest state to Alaska at the time as Washington was still a territory. The president was given the authority with the consent of the Senate to appoint district officials, including district court judges who were charged with the unenviable task of interpreting Oregon statutes in the circumstances of Alaska. If all of this sounds very colonial, it was. At least Congress remembered one cause of the American Revolution and imposed no taxes on Alaska, because Alaskans did not have the right to vote for their representatives.

But Benjamin Harrison had made a start, and Sheldon Jackson had gotten his education plank. Section 13 of the Organic Act of 1884 required the secretary of the interior to make "needful and proper provision for the education of children of school age in the territory of Alaska, without reference to race until such time as permanent provision shall be made for the same."[32] To fund the provision, the act appropriated the less-than-grand sum of $25,000. It also provided that each mission school be assured 640 acres of land that was not to come from lands presently occupied by Natives.

Considering the attitudes toward Native Americans in the American West at the time, the government's view of Alaska Natives was actually quite enlightened. Perhaps with Jackson's input, Commissioner J. D. C. Atkins of the Office of Indian Affairs soon wrote: "The Alaskan Indians, so called, are hardly to be looked upon as Indians in the sense in which the word is applied to the tribes on our western reservations. They are Alaskans, the native people of the land, who know how to support themselves by the resources of the country and the industries naturally arising from them."[33]

But the Office of Indian Affairs also wanted no part in administering programs in Alaska—probably to Alaska's benefit. So the secretary of the interior gave the responsibility for education in Alaska to the Bureau of Education and ordered it to "establish a public school system, not for the whites and not for the Indians, but for the *people* of Alaska."[34] It was a no-brainer, of course, that when the commissioner of education looked around for a knowledgeable and willing individual to develop the program, he quickly settled upon Sheldon Jackson, appointing him the

Bureau of Education's special agent in Alaska. So sure was Jackson in his charge, that he also continued to serve as the director of the Presbyterian Church's mission schools and saw no conflict in imposing his church-centered policies on the entire government program. Whatever cries of separation of church and state might have been raised were largely muted because Jackson was forced to supplement the small government appropriations with major contributions from the various denominations.

By dogsled, *baidarka,* and revenue cutter, Jackson was soon crisscrossing Alaska on inspection tours of what was probably the largest school district in American history. Over the next few years, he established schools at Point Hope, Barrow, and Cape Prince of Wales, nurtured the development of other mission schools with the support of other denominations, and laid the groundwork in Sitka for an industrial boarding school that eventually became Sheldon Jackson College. Whatever else Jackson's critics said about him, they could sometimes question his methods but not his devotion to education. In the ensuing years, this network of mission schools provided many of Alaska's children with a sound educational foundation.

But if the evil of a lack of education was being met straight on, how was Jackson coming with demon rum? The Organic Act of 1884 had included a provision prohibiting "the importation, manufacture, and sale of intoxicating liquors . . . except for medicinal purposes."[35] Jackson and his Presbyterian supporters had also been instrumental in the inclusion of this clause, but more than one Alaska sourdough chuckled at it and said, "Good luck!" One can readily picture a group of grizzled woodsmen huddled around a cabin stove in the dead of the Arctic winter and one of them exclaiming, "Why shucks, everything we do up here is for medicinal purposes!"

One of Jackson's early allies in Alaska in support of both education and prohibition was fellow Presbyterian John G. Brady. Where Jackson divided his time between Alaska and the states, Brady was on every count an Alaskan. He first came to Alaska in 1878 as a missionary and quickly firmed up his views on both the dismal state of military administration and prohibition. "The sending of the soldiers to this country," Brady wrote soon after his arrival, "was the greatest piece of folly of which a government could be guilty. It will require twenty years to wipe out the evils which were brought to the natives. They knew nothing of syphilis, nor did they know how to make an intoxicating liquor from molasses, but now they are dying from these two things."[36]

Truth be told, of course, the Tlingit had been imbibing on home brew at potlatches long before the Americans arrived, and the Russians had done their part to spread the other malady. But Brady joined Jackson's twin crusades and remained a staunch protégé even after hanging up his missionary cloth to become first a merchant and later U.S. commissioner in Sitka.

Given the independent circumstances of Alaska, many argued that a stringent liquor licensing program would be far more effective in controlling the excesses than outright prohibition. When territorial governor Lyman Knapp tried to institute such a system, however, Jackson and his fellow Presbyterians cried foul to Benjamin Harrison, who by then was president of the United States. Harrison chose to remain out of the fray, but in no uncertain terms his secretary of the interior promptly notified the governor that he was to refrain from initiating such a licensing system and enforce the prohibition laws. That, of course, was an impossible task. Even John Brady, who was appointed territorial governor at Jackson's urging in 1897, came to admit that only a licensing system could control the liquor trade.

Meanwhile, Jackson got involved in another campaign destined to be almost as fruitless as prohibition. In 1890, while visiting schools on the Arctic coast on board the revenue cutter *Bear*, Jackson noted that the overhunting of whales and walruses had created a severe food shortage among the Inupiat Eskimos. His solution was to import domestic reindeer from Siberia and raise reindeer herds as part of his overall plans for industrial education. (Reindeer and caribou are technically the same species, but in North America only domestic animals are called reindeer.) When Congress failed to appropriate the necessary initial funds, the good doctor again shook the money tree of religious denominations and raised funds privately.

A motley herd of 16 reindeer landed in Alaska in the summer of 1891 but quickly died. The following year, 171 more were imported, and soon Jackson was visiting reindeer stations around the territory, most built in conjunction with his mission schools. Natives tending reindeer at these stations were about as close as life in Alaska ever got to the cowboy scenes of western cattle ranches. Some sourdoughs even used reindeer for packing supplies and pulling sleds, claiming that they had more stamina than dogs and could forage for their own food along the trail. Jackson's reindeer experiment was never very successful; however, he was still championing it as a way to relieve hunger when starvation threatened the camps along the Yukon during the stampede to the Klondike.

Of Sheldon Jackson's two great crusades, his attempts at prohibition were swept away for good in the flood of gold seekers to the district in 1899, and a liquor-licensing system was finally introduced. But Jackson's education efforts left a lasting legacy on Alaska, in large part because they did not segregate races. Perhaps most significantly, the sensitivity that Jackson and his friend John Brady brought to the Native population helped to preserve a great number of Native, particularly Tlingit, artifacts. Today, these form the core of the collections of the Sheldon Jackson Museum in Sitka. In the end, thanks to his unrelenting missionary zeal, Sheldon Jackson did as much to publicize Alaska to the American public through his educational efforts and lectures as any explorer.

Untangling the Rivers

The U.S. Army in the person of General Nelson A. Miles was feeling more than a little left out. Miles was a doer—definitely not one to be left out. Volunteering for the Union army at the outbreak of the Civil War, Miles rose to a brevet major general of volunteers in command of an army corps. After the war's end, he made the army his career, joining the regular army and working his way back up through its ranks. By 1880, he was a regular army brigadier general in command of the Department of the Columbia. The department included Alaska, but just what the army's role was there, was open to question.

The army's administration of Alaska had come to an end in 1877, when the Treasury Department and navy assumed jurisdiction. Some, including congressmen reluctant to appropriate funds and proponents of civilian rather than military surveys, thought this meant that the army should keep hands off. Miles, however, could not conceive of having military responsibility for an area about which so little was known. In 1882, he made a personal visit to southeast Alaska but ended up having to explain his presence as if he were some interloper. He assured his critics that he was merely inspecting a region that he might be called upon to defend. The trip further convinced Miles that detailed exploration of Alaska was a priority, even if it must be done quietly.[37]

Miles assigned the initial task to Lieutenant Frederick Schwatka. An 1871 graduate of West Point, Schwatka eased the boredom of subsequent army posts by studying both law and medicine. He was admitted to the bar of Nebraska and held a medical degree from New York's Bellevue Hospital Medical College. In 1878–80, he led an expedition sponsored by the American Geographical Society into the Canadian Arctic north of Hudson Bay to search for remains of Sir John Franklin's ill-fated expedition. Suffice to say, Schwatka was qualified and up to the task of exploring the maze of rivers along the upper Yukon.[38]

To counter those who questioned the army's presence in Alaska, Miles couched his instructions to Schwatka as a military reconnaissance to determine relative Native strengths and intentions. Never mind that with the exception of a trading argument or two, relations were peaceful.

Schwatka was to learn "the number, character, and disposition of 'all natives' in the country, their weapons, their relations with each other, their attitude toward the encroachment of whites, and the means of using and sustaining a military force should one be necessary."[39] And, while he was at it, find out just where the heck all those rivers flowed.

With him went a medical officer, a topographical assistant, three enlisted men, and a civilian prospector who was said to be well acquainted with Alaskan travel. Unable, or at least unwilling, to ask his superiors for an appropriation to fund the venture, General Miles covered the expedition's expenses out of what might be termed his department's "petty cash." Consequently, Schwatka later wrote that the expedition "stole away like a thief in the night and with far less money in its hands to conduct it through its long journey than was afterward appropriated by Congress to publish its report."[40]

It's unclear how much discretion Schwatka had from Miles in picking his route. The main stem of the Yukon was relatively well known from the work of the Western Union Telegraph Expedition. Confusion reigned, however, about the Tanana, White, and Copper Rivers and their relationship to the Yukon. Schwatka chose to follow the well-trod path. Unfortunately, it led his quiet little expedition into just the attention that Miles hoped to avoid.

Sailing from Portland, Oregon, on May 22, 1883, on board the *Victoria*, the expedition landed at Pyramid Harbor near the Haines Mission on Chilkoot Inlet. From there, a small steam launch transported them to Dyea at the mouth of the Taiya River. The route up the Taiya and over Chilkoot Pass was well established as a Tlingit trade route to the interior and had been used by a few early prospectors. This was not new ground, but up the river and over the pass Schwatka went, liberally naming features along the way: Nourse River, Baird Glacier, Saussure Glaicer, Lindeman Lake—all after scientists—and Bennett Lake after the publisher of the *New York Herald*. Schwatka even named Chilkoot Pass "Perrier Pass," after a member of the French Geographical Society. "Mon Dieu!" To be sure, it was the convention of the day. One can have only so many "bear" landmarks (Alaska has more than 100 Bear bays, creeks, lakes, mountains, and rivers, plus one Bear Glacier), but thankfully Alaska's fabric is sewn with place-names from all of the cultures that have touched it.

On the shores of Lindeman Lake, the party built a large raft some fifteen by forty feet in size. It was equipped with oars fore, aft, and on the sides and a makeshift sail fashioned from a wall tent. When the wind and the waters were both calm, the men could make a mile an hour at the oars of the cumbersome vessel they christened *Resolute*. From Lindeman they ran the connecting stream to Bennett Lake and then started down the Yukon proper, running the rapids of Miles Canyon with only a little damage.

When not busy sprinkling names, Schwatka provided fascinating descriptions of two of the curses of Alaska river running that continue to this day. "Mosquitoes," he wrote, "now commenced getting very numerous, and from here to the mouth of the river they may be said to have been the worst discomfort the party was called on to endure. [The mosquitoes] . . . will be the great bane to this country should the mineral discoveries or fisheries ever attempt to colonize it. I have never seen their equal for steady and constant irritation in any part of the United States, the swamps of New Jersey and the sand hills of Nebraska not excepted."[41]

The other menace was when the river undercut its banks and caused bristled pine trees to teeter in or just above the water. These trees "form a series of *chevaux de frise* or *abatis*, to which is given the backwoods cognomen of 'sweepers,' and a man on the upper side of a raft plunging through them in a swift current almost wishes himself a beaver or a muskrat so that he can dive out and escape. . . . Such a position is bad enough on any river which has but a single line of trees along its scarp and trending downstream, but on the Yukon it is unfortunately worse, with every branch and twig ferociously standing at 'charge bayonets' to resist anything that floats that way."[42]

Mosquitoes and sweepers aside, Schwatka's third menace that summer was that having crossed Chilkoot Pass, he was no longer in Alaska. He blithely floated on naming features and mapping terrain. The good news was that topographer Charles Homan's survey of the 500 or so miles of the Yukon between Lindeman Lake and Fort Selkirk at the mouth of the Pelly (where Ketchum and Lebarge had turned around in 1867) was the first to be done. Their work held up well under later scrutiny. The bad news was that General Miles was called upon to explain what an American army detachment was doing mapping British territory. Little did anyone know at the time that this very route would soon cause Her Majesty's government much greater consternation when a flood of Americans and others from around the world followed Schwatka's exact route bound for the Klondike gold fields.

The *Resolute* floated on below Fort Selkirk, past the mouth of the

Klondike, and into Alaska. The survey work and Schwatka's naming practices continued. Unfortunately, Schwatka aroused the wrath of prior surveyors, including William H. Dall, by his sharp criticisms of previous mapping work and his disregard for prior usage when assigning names.

In the end, the *Resolute* carried the party some 1,300 miles down the Yukon and arrived at Nukluklayet at the mouth of the Tanana on August 6. (The geography of the upper Tanana was still as confused as ever.) From Nukluklayet, they took a small schooner to Anvik and then the Alaska Commercial Company's *Yukon* to St. Michael. It had been an impressive trip, and the most impressed seemed to be Lieutentant Schwatka himself. Schwatka rushed reports of the trip into the popular press, including *Century* magazine, and a book, *Exploring the Great Yukon*. Miles, evidently, was none too pleased, no doubt in part because he thought the less said about the whole thing the better.

To be sure, Schwatka had surveyed the upper Yukon, descended its entire length, corrected previous mapping work, and added more than enough place-names to the map. His biggest contribution, however, is the fact that his subsequent writings publicized Alaska and the Yukon to the public in general and to a rush of gold seekers in particular a few years later.

Meanwhile, Miles was stuck with the bills and still wanted to know more about the voids in between. The general went to considerable lengths to obtain reimbursement for some of Homan's mapping expenses, and, by one account, Schwatka left a bill outstanding for guiding services. According to one record, a Tlingit guide denounced Schwatka for not paying his guiding fee across Chilkoot Pass. The Tlingit told him, "I will take your name and use it as long as I live," and he was henceforth known as George Schwatka.[43]

Political and financial hassles aside, General Miles was not about to be dissuaded from his original goal. In 1884, he ordered Captain William R. Abercrombie and a civilian scout named Willis E. Everette to make a two-pronged investigation of the vast area between the Yukon River and the Gulf of Alaska. Abercrombie was to ascend the Copper River, while Everette was to follow Schwatka's route down the Yukon and then ascend the White River before crossing over to the headwaters of the Copper. The plan made sense on paper, but once again the magnitude of Alaska's scale would play havoc with its implementation.

Everette's story is simpler to recount because he evidently did little more than follow Schwatka's course on the main Yukon. After the fact, Everette would be full of excuses for why he had not ventured up the White River, including lack of suitable companions. The record seems to indicate, however, that the truth of the matter was that Everette just wasn't up to the role of brave explorer breaking new ground. He spent a month lolling around old Fort Reliance at the mouth of the White River in the company of a female Native interpreter and then floated on down the Yukon.[44]

Abercrombie was to be equally unsuccessful, but at least he gave it a try. Miles ordered him to ascend the Copper River and then cross over into the headwaters of the Tanana. Abercrombie's party, including Schwatka's topographer Charles Homan, landed at Nuchek on Hinchinbrook Island southwest of present-day Cordova on June 16, 1884. Five days later, they reached the Copper River delta and began paddling their way upstream against a vigorous current full of spring runoff. The delta ended abruptly where the Copper River emerged from the Chugach Mountains. Two gigantic glaciers framed the river, each sending large icebergs calving into the river to the accompaniment of thunderous volleys. Abercrombie named the western glacier after George Washington Childs of Philadelphia and the larger eastern one after General Miles. (When all else fails, bring home a name upon the map for your superior.)

Dodging big bergs and a fusillade of smaller ice, Abercrombie and his men rowed through the wide spot in the river at the toe of the Miles Glacier—what would come to be called Miles Lake as the glacier receded over the next century. Beyond, a look upstream was not encouraging. The Copper was a foamy torrent of rapids. Abercrombie saw no plausible way to proceed and called it quits at what would be called Abercrombie Rapids. He reported back to General Miles that any access to the interior up the Copper was impractical. To fellow lieutenant Henry Allen, Abercrombie confided that he would never return to the country.[45] But Abercrombie was not finished with the Copper River, or perhaps it is better to say that the Copper River was not finished with Abercrombie. He was to return some years later and find the route no less daunting.

The Lewis and Clark of Alaska

General Miles was more than a little upset. Schwatka had covered considerable ground, but added little to the map. Abercrombie had failed miserably without even making it through the Copper River's lower canyons. What lay in between the Yukon and the lower Copper River? Who was going to ink in this void on the map? The answers to both questions were forthcoming from the general's twenty-five-year-old aide, Lieutenant Henry Allen.

Henry Tureman Allen was born in Sharpsburg, Kentucky, on April 13, 1859, the thirteenth of fourteen children. These Kentucky roots are still evidenced by the place-names he bestowed across Alaska. By all accounts, Allen was tall, handsome, and fastidious in both manner and dress. His military presence was impressive but not overbearing. In short, he was the gracious, but determined, southern gentleman. Allen graduated from West Point in 1882 and was posted to the Department of the Columbia. He quickly confided to the woman he would later marry, "I am willing to forgo almost any benefit that I might receive by going East for an attempt at exploration in Alaska."[46]

Allen first watched from the sidelines as Schwatka made his trek. The following year he was ordered to support Abercrombie's misadventures by supervising supply shipments from Sitka to Nuchek. Along the way, he quizzed anyone who had even a glimmer of knowledge about the Copper River country and read everything in print about Alaska that he could lay his hands on. By the time he returned from Nuchek in 1884, Allen was confident of his plan. Forget the cumbersome military expedition that Abercrombie had tried to lead. Allen proposed what would become the epitome of a small, alpine mountaineering expedition. He and only two others would make the trip, and they would make it in one season.

General Miles approved the plan but thought that the detachment should include four to ten men with at least a medical officer. Allen won out for a trio when final orders from Commanding General Philip H. Sheridan authorized an expedition of only three men—probably because he was leery of continuing political criticisms of the army's role in exploration in general and the Abercrombie failure in particular. Private

Frederick W. Fickett of the Signal Corps had been at Sitka when Allen passed through and had already volunteered. Sergeant Cady Robertson, an NCO from Allen's own Second Cavalry, was the second volunteer. As it turned out, the party was joined at various points in the journey by two prospectors, Peder Johnson and John Bremner, and several Eyak and Ahtna Natives.

Allen, Robertson, and Fickett were delivered to Nuchek by the U.S. Navy tug *Pinta* in March 1885. A critical piece of Allen's plan was an early season start to facilitate traveling as much of the lower Copper River as possible before the ice broke up. Joined at Nuchek by Peder Johnson, the foursome paddled two canoes east from Nuchek to the tangled delta at the river's mouth—no small feat in itself given the mudflats and fickle currents of that coastline. Then on March 29, with a full complement of mapmaking equipment and linen sailcloth sleeping bags waterproofed with beeswax and linseed oil, the party started up the river. They were none too early.

Within a day they were crisscrossing back and forth, sometimes on sleds and sometimes in canoes, as the ice-choked river demanded. Almost immediately they abandoned about half of their ammunition, food, and clothing—a less than auspicious start that must have left Allen thinking about Abercrombie. But Allen did not pause. He hastened past the Childs and Miles Glaciers, forced his way through Baird Canyon, and found enough snow and ice to make a dash to Taral—some seventy miles upriver. Had he tarried or been somehow delayed, the passage over stable ice and crusty snow would have been reduced to one of roily river ice and snow turning to mud with each step.

At Taral—really nothing more than a small Ahtna village of a couple of huts—they found Johnson's prospecting partner, John Bremner. He had come up the river the previous summer in search of copper and had barely managed to eke out a living and survive the winter. No doubt Allen was not looking his best by now, but he described Bremner in his journal as follows: "Certainly [he was] the most uncouth specimen of manhood that I had, up to this time, ever seen. . . . He was shortening his belt one hole every other day." Allen went on to say, however, "Nowhere did I receive such a warm welcome as at Taral from this naturally heroic specimen of manhood, then so depressed with hunger and destitution."[47]

Just upstream from Taral the Copper River split, and Allen faced a dilemma well known by explorers. Which was the main stem? Postponing a decision on that question, Allen nonetheless determined to explore the eastern branch that the Ahtna called Chittyna, which was said to mean "copper." Copper trinkets and utensils were in evidence among the Natives and, in addition to solving geographic riddles, Allen was eager to document reports of rich mineral deposits.

It made sense to cache most of their provisions at Taral for their return, but the group soon learned a hard lesson. "From this time," Allen wrote, "we began to realize the true meaning of the much-used expression 'living off the country.'"[48] On April 13, Allen's twenty-sixth birthday, the main course was rotten moose meat left from a winter kill by wolves that had been brought into camp by an Ahtna. Even the expedition dogs turned up their noses at it.

For much of its length, the Chitina, as the river is now spelled, runs a course dotted with sinuous braids and shifting gravel bars. As Allen's party followed it eastward, food became even more of a problem. Fickett opined as to how even rotten moose meat would be welcome about now.

Their goal quickly became the village of Nicolai, a *tyone,* or head chief, of the Ahtna of the Copper-Chitina region. Portaging northward from the Chitina toward the Nizina-Chitistone drainage, the famished explorers stumbled exhaustedly into Nicolai's camp on April 20. Here their gracious host allowed them to consume great quantities of moose, beaver, lynx, and rabbit, all cooked whole with their entrails intact. If there were any complaints, Allen did not record them.

Clearly, Allen sought Nicolai's help. His party needed food, transport back to Taral, a guide for the upper river, and information about the surrounding country, including those rumors of copper mines. Nicolai provided all of the requirements, even volunteering himself to serve as a guide. Whether he did so out of genuine hospitality, or simply a notion to have these white men be on their way, is debatable. In any event, under Nicolai's direction a twenty-seven-foot boat was built of untanned moose hides lashed together with rawhide strips and assembled over a frame of willow limbs. Within a few days they were back at Taral.

While his companions rested and replenished their appetites with expedition supplies, Allen again pondered which branch should be called the Copper. Because of the ice along the banks, he still couldn't be certain which was the bigger river. Given the Ahtna meaning of *chittyna,* Allen considered inking in the Copper in that direction. But the western tributary was soon proven to be the larger main stem, and Allen left the name Copper in place upon it, calling the eastern branch by its Ahtna name.

Along the way, Allen mapped the towering monoliths of the

Wrangell volcanoes. The easternmost he named for Joseph Clay Stiles Blackburn, a U.S. congressman and senator from Kentucky. The Ahtna called it K'als'i Tl'aadi, meaning "the one at cold waters." Apparently, Allen knew that Mount Wrangell had already been given that name by the Russians.

Towering over the Copper River at the western end of the range was Hwdaandi-K'elt'aeni. (The first word of the name means "downriver," the second, "the one who controls the weather.") Allen named it for U.S. Army adjutant general Richard C. Drum, who had served in both the Mexican and Civil Wars. As the Copper River bent to the east, the mountain towering above it was Hwniindi-K'elt'aeni. (In this case, the first word of the name means "upriver.") Allen named the mountain for his great-grandfather Reuben Sanford. Years later, Chief Nicolai's thoughts on this nomenclature were recorded at a potlatch speech: "All this time, thousands of years, Indians look up and think that it is K'elt'aeni—hundreds of generations of our forefathers look up and think it Mount K'elt'aeni, but Indians not very smart. First white man come along ten years ago, he say ah hah, Mount Sanford . . . and Mount Sanford it is today, my people."[49]

Chief Nicolai was not the only one disconcerted. Blackburn, Wrangell, Sanford, and Drum are the four great volcanoes of the Wrangell Mountains, but when Allen's map was published it showed five volcanoes, adding a Mount Tillman. Some say that Allen named Mount Wrangell for Samuel Escue Tillman, a professor at West Point, despite the prior Russian name. Others, including an adventurer named Robert Dunn, think Allen simply duplicated Mount Sanford by mistake and named that mountain twice. It would be easy to do given the twisting rivers and the low-slung clouds that frequently obscure the peaks. Dunn wrote about his 1898 sleuthing in search of Mount Tillman as part of a trip to determine if Mount Wrangell was an active volcano. He titled the article "Finding a Volcano and Wiping a 16,000 Feet [*sic*] Mountain from the Map of Alaska."[50] The Tillman confusion was the one glaring error that came out of Allen's mapping.

On a more personal note, Allen named a river flowing south into the Chitina after Dora Johnston, his wife-to-be, to whom he had confided his enthusiasm for this land that had him reduced to eating rotten moose meat. Allen's Dora Creek is probably the Gilahina River. Present-day Dora Creek is a smaller stream flowing south from Gilahina Butte just to the west.

From Taral, Nicolai led the party northward up the Copper River. As the *tyone,* or chief, Nicolai was content to steer the moose-hide boat while all but a bowman pulled on a 150-foot rope to haul it slowly upstream against the muddy current. At the mouth of the Tazlina River, near today's Glenallen, Nicolai turned back, saying that he was at the limits of his territory. Beyond this band of neutral ground, the Natives would be Tanana.

So up the river they continued, as it swung eastward around the flanks of Mount Sanford. On May 27, Robertson and Bremner redesigned the twenty-seven-foot craft and made it smaller and more manageable. Three days later, they decided to abandon the boat and continue on foot. They met Tanana Natives, who shared meager rations with them. Allen wondered how these people managed to survive year-round. His answer soon came from a thrashing in the river. The salmon had arrived, a reliable and life-giving cycle of food rushing upstream. It was none too soon because the boiled meat of the day before had become laced with handfuls of maggots.

But the Copper was ending, turning south and heading toward its source at the glaciers on the north side of Mount Wrangell. Just above the mouth of the Slana River, Allen struck out northward and moved toward a pass in the Mentasta Mountains that he named for Nelson Miles. Now called Suslota Pass after the lake Allen visited at its base, the 4,500-foot pass offered a grand view and showed Allen that he was on the long-sought-after divide between the Copper and Tanana drainages. Allen called it "the most grateful sight it has ever been my fortune to witness," with rivers that ran downstream in the direction he was going. The Tok Cutoff of the Glenn Highway now crosses the range at lower Mentasta Pass just to the west of Allen's route.

Down off the divide they went. On the Tetlin River they built a new boat from three caribou hides and then pushed out into the current. The boat easily ran the Tetlin into the muddy Tanana and made rapid progress down it, sometimes making as many as fifty miles a day. By June 25, the group was in familiar territory on the Yukon after a rapid ten-day descent of the Tanana. At Nukluklayet at the junction of the two rivers, the obvious option was to wait for one of the two small steamers then operating on the river.

But the young West Pointer wasn't quite ready to take the easy way home. Allen instructed Robertson to wait for a steamer and take some of the expedition records down to St. Michael. Johnson and Bremner, still bit by the gold bug despite the hardships of the trip, decided to prospect along the Yukon. Allen, with Fickett still the willing volunteer, turned to fill in the voids on the map of the other great tributary of the Yukon, the Koyukuk.

On July 28, Allen and Fickett turned north and struck out along a roughly defined Native portage route leading toward the Kanuti River, a small tributary of the Koyukuk. From a Koyukon village on the Kanuti, they acquired two birch canoes and paddled swiftly down to the Koyukuk. Once again, the obvious route led downstream, but Allen and Fickett turned north instead and paddled upstream. By now, they must have been feeling at one with the country. They reached the mouth of the Alatna, and an old Native who accompanied them noted that he had more than once been to its head and across the divide to the Kobuk River. Only a couple of unoccupied huts at the confluence marked what is now the village of Allakaket.

Farther upstream, they named a stream coming in from the north the Fickett River. Convinced it was a major tributary, they paddled up it a few miles before deciding on August 9 that the shortening Arctic days suggested that they finally head for St. Michael. Fickett's name did not survive on the tributary, but it ended up being called the John, after none other than their companion John Bremner, who prospected along it during the next two years and was eventually killed by Koyukon Natives.

Allen and Fickett floated down the Koyukuk and reached the Yukon on August 21. Avoiding the 400-mile loop of the Yukon, they took the Unalakleet River portage to Norton Sound and St. Michael. By September 5, they had departed for San Francisco.

General Miles was no longer upset. In fact, he was downright pleased. It was a spectacular achievement: 1,500 miles; three major river basins tied together; valuable observations of the Natives encountered. The voids had been filled in. Miles termed it the greatest act of American exploration since Lewis and Clark crossed the Louisiana Purchase to reach the Pacific. A dozen years after the fact, none other than the dean of Alaskan geologists, Alfred Hulse Brooks, would write: "No man through his own individual explorations has added more to our knowledge of Alaska than has Lieutenant Allen."[51]

So why isn't Henry T. Allen's name inscribed beside those of Alexander Mackenzie, Lewis and Clark, Zebulon Pike, and John C. Frémont in the annals of North American exploration? For starters, Allen was not a gripping writer. He lacked the literary flair of Schwatka and a willing and "expansive" editor such as Frémont found in his wife, Jesse. But most important, Allen was a soldier who saw his duty as reporting the facts without embellishment and doing so in the framework of his official report. Then, too, times were changing rapidly in Alaska, and a host of individual prospectors and other surveyors would rapidly follow his trail and expand upon it. His seminal achievement was a cornerstone, but one which was rapidly built upon.

Allen married the Dora of Dora Creek on July 12, 1887, and remained in the army all of his life. After his Alaskan experiences, he served in the Philippines and in Mexico, proving grounds for the most successful career officers in the early twentieth century. By 1917, he was a brigadier general commanding the Ninetieth Division and later the Eighth and Ninth Army Corps in France during World War I. After the Armistice, he became the commander of the American Army of Occupation in Germany. He finally retired in 1923, forty-one years after his graduation from West Point. In 1928, although almost seventy, his reputation was such that he was considered as a vice-presidential running mate for Democrat Al Smith. Henry Allen died in 1930, a soldier to the end who had always done his duty, no matter the odds against him. He never returned to Alaska.

Juneau, or Whatever Its Name Is

Unlike many prospectors who drifted about Alaska, George Pilz actually knew something about mining. Just about anyone could work a placer claim. Placers yielded the "free" or loose gold that had been eroded long ago from a vein of ore. Placer gold was typically found as grains and tiny nuggets along the sands and gravels of creeks and other watercourses. Lode mining was an entirely different matter. Lode claims mined the gold-bearing vein itself. Extracting gold-bearing ore usually required digging extensive tunnels and shafts. Even then the ore had to be crushed in stamp mills and separated into high-grade concentrates from which the gold could be smelted. Placer claims along watercourses were frequently assigned numbers up- and downstream from the discovery site (hence No. 1 Below, No. 4 Above, etc.), whereas lode claims were usually given names.

Pilz was a German who immigrated to America shortly after the Civil War. He earned his mining experience by working in California in a copper smelter and then a stamp mill. In 1878, at the age of thirty-three, Pilz found himself in San Francisco and in poor health, suffering a particularly bad bout of the inflammatory rheumatism that plagued him all of his life. Quite by accident, Pilz bumped into Nicholas Haley, an acquaintance from his stamp mill days, who was trying to develop a gold lode at Silver Bay east of Sitka. No one, grumbled Haley, seemed to know how to extract the gold from the ore vein. Pilz agreed to have a look and arrived in Sitka early in the spring of 1879. By fall, he had supervised the construction of the first stamp mill in Alaska. The Silver Bay veins proved of low quality, but there were enough rumors of other gold along the Inside Passage to make Pilz think twice. Unable to cover a lot of ground himself, he decided to recruit others to do some prospecting for him so that he could sift the truth from the many rumors.

In 1880, Richard Harris was forty-six. He was born in Cleveland, Ohio, and worked in silver mines in Mexico and camps throughout the Rockies before joining the Cassiar rush in British Columbia in the mid-

1870s. Harris had a jet-black, flaring mustache and a strong-willed temper to match, but he was particularly attached to his friends. One of them was Joe Juneau, a French-Canadian born in Quebec. Harris and Juneau first met in the Cassiar country but quickly found that they had each been around some of the same mining camps throughout the West. At fifty-three, Juneau was old enough to be George Pilz's father. Suffice to say, both Harris and Juneau were rough-cut and hard-core.

Harris and Juneau were at Fort Wrangell in the summer of 1880 when they heard that Pilz was looking for prospectors. Pilz had a number of miners in the field that summer, spread out as far north as Cross Sound and as far south as Prince of Wales Island. Harris and Juneau made the trip to Sitka—getting passage on the steamer *California* with a grubstake—and agreed to explore the mainland for Pilz in the vicinity of Holkham Bay.

With three Tlingit to help paddle, Harris and Juneau left Sitka on July 19 and wound their way through Peril Strait, around Point Gardner, and across Frederick Sound to the mainland just south of the bay. The story of what happened next should be fairly well documented because both Harris and Pilz wrote accounts of the next few months. Pilz, in fact, wrote two accounts. The problem, of course, is that Pilz's accounts differ from each other as well as from Harris's version. Dates, locations, and characterizations of heroes and villains vary in each account.

Apparently, Harris and Juneau proceeded to prospect the mainland northward from Holkham Bay with only minor luck until August 17, 1880, when they investigated a roaring stream pouring into Gastineau Channel. Moseying up the creek, they first found some "color"—traces of free gold—but then stumbled upon gold float—gold embedded in quartz and a strong sign that a rich vein, or even the mother lode, was nearby. How much prospecting the two did along the creek, and how quickly they returned to Sitka, are points that differ in the telling. When Harris and Juneau arrived in Sitka in early September, they told Pilz of their find. He urged them to resupply and return to the stream that they called Gold Creek to stake claims.

Paddling along Gastineau Channel on their return, who should Harris and Juneau meet but John Muir and Samuel Hall Young. The two prospectors must have fidgeted a little when they learned that Muir and Young had just camped near the mouth of what they were calling Gold Creek. Later, Young wrote in his account of the meeting that Harris suggested to Juneau that they try their luck there. Undoubtedly, Harris and Juneau—veterans that they were—had decided to play it very coy until their unexpected visitors left. They knew for sure what was up Gold Creek, and as Muir and Young headed north after glaciers, Harris and Juneau headed up into Silver Bow Basin after gold.

And gold they found. Not scattered placer deposits of a few flakes, but outcrops of a vein of quartz laced with gold. French-Canadian Juneau spoke only halting English and could neither read nor write it. Thus, it fell to Harris to be elected recorder and to draw up the papers forming the Harris Mining District. This he did on October 4, 1880. Then he and Juneau staked claims up and down the basin for themselves, Pilz, and their other grubstakers, including Captain James Carroll and the purser of the steamer *California*. After staking a total of nineteen placer claims and sixteen lode claims in Silver Bow Basin, they returned to the mouth of Gold Creek on October 18 and there staked out 160 aces as the future site of Harrisburgh. So far, it looked as if Joe Juneau was getting the short end of the naming business.

This accomplished, Harris and Juneau returned to Sitka again, but this time they took with them about 1,000 pounds of high-grade ore. Their arrival on November 17 caused quite a stir, and the first major gold rush in Alaska was quickly off and running. Captain Henry Glass of the USS *Jamestown* (the same vessel that had been undergoing repairs in Shanghai when the *Shenandoah* made its Pacific romp) provided one of his ship's steam launches to take Harris, Juneau, Pilz, and some miners back to Harrisburgh. Others swarmed on board the steamer *Favorite* and set off as well. Soon, about three dozen miners were busy staking town lots and nearby claims. The first building in the boomtown appears to have been a prefabricated cabin that Pilz shipped on the *Favorite* and hastily erected.

But there was trouble brewing. Harris's actions in establishing the Harris Mining District had been taken under the provisions of the Mining Act of 1872. This was the cornerstone of mining law, and it gave a tremendous amount of local control and discretion to individual mining districts. Common practice throughout the West demanded, however, that at least five miners be present at the creation of the rules for the district and the election of its recorder. Harris had stretched matters some when he included his three Tlingit companions as "miners." Additionally, there was the question of Joe Juneau's literacy in the whole proceeding.

What seems to have called these issues into question was the fact that Harris and Juneau had been exceedingly generous in the number of claims they had allocated to themselves and others through proxies. As one veteran from the Cassiar who showed up on Gold Creek groused, "We came from the Cassiar, where that didn't go."[52]

Lieutenant Commander Charles H. Rockwell, the executive officer of the USS *Jamestown*, was dispatched to Harrisburgh in a second steam launch to render whatever assistance might be necessary to preserve order. But the miners had matters well in hand. At a meeting on February 9, 1881, thirty-one miners met and declared Harris's rules and election as district recorder null and void. They did, however, ratify the location records of the previously filed claims. The next day, just to show how the tide was running against Harris, thirty-four miners met and voted to rename the town. Lieutenant Commander Rockwell seems not to have done much except administer an oath to each of the meeting participants swearing their American citizenship, but the name Rockwell nonetheless received eighteen votes. Fifteen votes were cast for the name Juneau City and only one for Harrisburgh.[53]

The U.S. Navy quickly embraced the Rockwell name, but the postmaster general, acting upon earlier information, soon gave official status to the name Harrisburgh and appointed a postmaster. Miners who wanted to be certain they got it correctly on their location notices referred to the town as "Harrisburgh, also known as Rockwell," or "Rockwell, also known as Harrisburgh," using or dropping the last h according to personal whim.

By early May the population of "Rockwell, also known as Harrisburgh" had increased to 150 miners. Most were veteran prospectors who had come from the Cassiar diggings via Fort Wrangell or from Sitka. There were also about 450 Tlingit in various camps along Gastineau Channel. They appear to have congregated in order to trade, to hire out as packers, and just to check up on all of the excitement. Some miners expressed concern about being greatly outnumbered, but other than an individual scuffle or two, no conflicts occurred between the two groups.

As for the Tlingit, they had long ago gotten used to these foreigners frequenting first Sitka and then Fort Wrangell, but Harrisburgh was to be the first major white settlement in southeast Alaska except for those two long-established trading posts. The Auk Tlingit from nearby Auke Bay must have been particularly concerned about this very rapid influx of newcomers into their hunting and fishing grounds. Various versions of the story have the Auk Tlingit Cowee playing a role in the gold discoveries, some even suggesting that he almost took Harris and Juneau by the hands and led them into Silver Bow Basin. If so, the subsequent invasion must have had him wondering what changes he had caused.

Before long, steamers were making their way up Gastineau Channel with increasing regularity, including Captain Carroll's *California*. This proximity to easy transportation was one major difference between this first gold rush to Rockwell/Harrisburgh and the later rushes to the Yukon country. The steamers meant that Rockwell/Harrisburgh was comparatively easy to reach, and people and supplies quickly arrived. With a steady supply of building materials, fixtures, and even amenities, the town soon took on the air of permanence. Whereas later Yukon camps would freeze over for the winter, it was easy to leave Rockwell/Harrisburgh at any time of the year. One could take the monthly mail steamer and with the right connections in Portland be in San Francisco in two weeks, or vice versa. More than one miner took advantage of this for a little midwinter spree and to resupply.

Meanwhile, things were also starting to boom on Douglas Island on the western side of Gastineau Channel. Far from being a sideshow, Douglas Island was to become a major attraction in itself. Although not included in Harris's original description of the Harris Mining District, Douglas Island first boasted a number of placer claims, including a discovery made at the mouth of Ready Bullion Creek by William Meehan on December 17, 1880. Meehan is supposed to have picked up a nugget and shouted, "Look at this! Why it's almost ready bullion!"[54] (This stream is now called Bullion Creek, and Ready Bullion Creek is just to the north of it.) These placers reportedly yielded some $15,000 before they played out over the next two years. But the lode claims were a different story.

Beginning with the Ready Bullion and Golden Chariot lodes, claims were staked along the eastern side of Douglas Island during the winter and spring of 1881. Pierre J. Erussard, better known as "French Pete," was a veteran prospector said to be descended from an influential French family. He left France after the Franco-Prussian War and arrived in Seattle in 1872. From then on, Erussard was on the mining trail. On May 1, 1881, Erussard staked what his location notice called the "Parris" lode claim a short distance north of the Golden Chariot. Whether the spelling was intended, a simple mistake in the field, or ignorance on the part of this son of France, the claim and a creek were soon referred to as Paris.

Adjacent to the Paris lode, Henry Borein had trouble staking his lode claim not because of the terrain but because of the large number of bears prowling the underbrush. He finally got the corners in and gave the claim the name Bear's Nest. It turned out to be more of a hornet's nest. Borein thought that he was on an extension of the Paris lode, but the Bear's Nest was later proven to hold only barren rock. This didn't stop several of its later owners from carefully salting the claim with Paris ore and unloading it on some unsuspecting English investors.

The Paris lode soon became important as the cornerstone of the great Treadwell Mine. John Treadwell was a carpenter and builder by training who arrived in Rockwell/Harrisburgh in the summer of 1881. Like George Pilz, Treadwell had gained some lode experience in California. He was sent north by John D. Fry, a San Francisco banker for whom he was building a mansion, to investigate the source of the Alaskan gold that was starting to arrive at San Francisco's mint. Pilz tried to interest Treadwell in some of his claims up Gold Creek, but Treadwell was unimpressed and turned to Douglas Island instead.

French Pete Erussard was also operating a store in town. He was always short of cash, and—depending on who is telling the story—he sold the Paris lode to Treadwell on September 13, 1881, in payment of the freight bill Erussard owed for goods that were arriving on the mail steamer. Supposedly, this bill came to $264, but in later years Erussard screamed bloody murder that he had been taken. It was not the first time—nor would it be the last—that when a toad turned into a prince on the mining frontier, history had a way of being rewritten.

But in the beginning even Treadwell did not recognize what he had in the Paris. He poked around on some other properties and then returned to San Francisco. His opinion of the Paris lode quickly changed when he had some milling tests run on ore samples. John D. Fry found someone else to finish his house, while he and four others hastily organized the Alaska Mill and Mining Company. They hurriedly sent Treadwell back to Douglas Island to buy up adjacent claims.

John Treadwell eventually sold his interests in the operations in 1889 for a reported $1.5 million but the Treadwell Mine complex went on to become the major producer of the area, encompassing a complex system of claims and tunnels along Douglas Island and extending underground beneath Gastineau Channel. By one report, the properties produced $24 million in gold during their first twenty years of operation as compared to only $4 million in the rest of the district. The Treadwell operations came to an abrupt end in April 1917 when the network of tunnels extending under Gastineau Channel collapsed.[55]

Back on the other side of the channel in 1881, Gold Creek and Silver Bow Basin continued to produce. The creek and basin were staked solid from the mouth of the creek far up the neighboring slopes. By the end of the year, 293 placer claims and 131 lode claims had been located in the Harris Mining District, which as previously revised now included Douglas Island. No miners meetings were held over the course of the summer, but when fall brought a slowdown of mining operations, a meeting was called to address numerous cases of overlapping claims, allocations of water (an essential ingredient in placer operations), and lot encroachments on the town waterfront.

One issue to come before the December 12, 1881, general meeting of the town was that nagging question of what it was called. The presence of the U.S. Navy was declining, and while Captain Glass attended the meeting, it is not certain how much of an impetus, if any, his pending decision to abandon the navy post there had on what happened next. It was moved and seconded that the meeting vote on a new name for the town, and out of 72 votes cast, 47 voted for Juneau City, 21 for Harrisburgh (with or without the last *h*), and only 4 for Rockwell.

When the results were announced, Richard Harris quickly moved to call another meeting solely for the purpose of naming the town, presumably to give him some time to lobby in the saloons. But the motion was denied (23 to 43). Word was dispatched on the December mail steamer to the postmaster general, and the change was made on January 10, 1882. Perhaps thinking *city* a little too grand, the department made the name officially Juneau. Harrisburgh, Rockwell, Juneau—take your pick; they were all the same place. Fortunately, the early movement to have it called Pilzburg never got very far.

Joe Juneau went on to make some money but ended up spending most of it in Juneau's saloons. Not content to end up sitting on a bar stool, he followed the mining trail to Circle City in 1895 when he was sixty-nine and then died four years later in Dawson, having lived to see the boom of the Klondike. His last wish was to be buried in his namesake town, a wish that was finally granted in 1903. Richard Harris also made some money but lost most of it in complicated ownership battles, caused in part by overlapping claims. He, too, lived to a ripe old age, dying in Juneau in 1907 at the age of seventy-four. George Pilz, the sickly one, lived in Juneau for a time and then bounced around other Alaska gold camps before finally dying in Eagle in 1926 at the age of eighty-one. The town of Juneau outlived them all.

One for the Duke

With mountains playing so important a role in Alaska's geography, it is not surprising that once the general lay of the land and the avenues of its rivers were drawn upon the map, attention turned to its higher peaks. Because of its close proximity to the coast and its towering height, Mount St. Elias attracted the earliest of these mountaineering forays. St. Elias posed two geographic riddles: Was it the tallest point on the continent, and was it in Alaska or Canada?

Unlike some mythical peaks, this one had an undisputed existence and in fact had been well known since Bering's first sighting in 1741. Its height was another matter. La Pérouse calculated it at 12,672 feet in 1786. Malaspina came remarkably close at 17,851 feet in 1791. Michael Tebenkov's Russian atlas rounded it to 17,000 feet. Dall's calculation of 19,000 feet proved to be a little too bold. But it did set people to wondering about that "tallest" designation.[56] (At the time, few outside of Ohio had yet heard of William McKinley, and the mountain to which his name would be attached had been only vaguely reported.)

The first serious attempt to climb Mount St. Elias was organized by none other than Frederick Schwatka in 1886. Thanks in no small measure to his own self-promotions, Schwatka's Alaskan credentials were by then rather impressive, even if his boat ride down the Yukon in 1883 paled beside Henry Allen's later adventures. Now, Schwatka secured the sponsorship of the *New York Times* for an expedition to St. Elias. In that era of cutthroat competition between major dailies, it was not at all unusual for a newspaper to sponsor an expedition in return for exclusive coverage of its exploits. (Stanley, after all, worked for James Gordon Bennett's *New York Herald* when he first went in search of Dr. Livingstone.) Schwatka was joined on the trip by William Libbey, Jr., a professor of physical geography at the College of New Jersey, and Heywood W. Seton-Karr, an Englishman. Libbey had made the first recorded ascent of Colorado's 14,197-foot Mount Princeton in 1877, but Seton-Karr was the only one of the trio well versed in the use of rope and ice ax.[57]

Schwatka and his party came ashore at Icy Bay on July 17, 1886, from the venerable tug *Pinta*. The bay was then little more than an indentation in the coastline near the mouth of the Yahtse River. The subsequent retreat of the Guyot, Yahtse, Tyndall, and Malaspina Glaciers over the next century greatly increased its size and added two major arms.

The trio and their packers slogged inland, but the enormity of their task soon became apparent. They reached an estimated elevation of 7,300 feet before turning back. In trying to put the *Times*'s investment in the best possible light, Schwatka noted that "nine-tenths of the climb was above the snow level, and which is believed to be the highest climb above the snow limit ever made—a result well worth the expedition." In closing he wrote a line that generations of later Alaska mountaineers would come to identify with: "The lack of continuous fine weather, an absolute necessity in an alpine attempt over unknown paths, was the most formidable obstacle in conquering this king of the continent."[58]

Schwatka, the smooth promoter, got into some hot water when he returned from this relatively minor attempt, having continued his earlier naming habits. He stuck dozens of new names on the map, including that of his benefactor George Jones, the publisher of the *New York Times*. The *Times*'s New York competitors gave him quite a razzing over the "Jones River," and the Tlingit name Yahtse remains on this main stream draining the Malaspina Glacier east of Icy Bay. No one seemed too concerned at the time that the Tlingit called the mountain Yahtse-tah-shah.

Two years later, a more serious mountaineering attempt was made by brothers Harold and Edwin Topham, George Broke, and William Williams. Evidently lacking the political muscle to commandeer the *Pinta*, the party reached Yakutat on July 10 after chartering a local sealing vessel at Sitka. After due negotiations over the price of two Native packers for the trip, they were whisked across Yakutat Bay in Tlingit canoes, reaching Icy Bay in ten hours after a distance of about sixty miles.

The Topham party laid in provisions for a major siege, including 1,400 pounds of bacon, ham, beans, flour, smoked salmon, dried apples, tea, and coffee, and estimated that the climb would take about a month from their base camp on the beach. They headed inland on July 16 and began ascending the Agassiz Glacier as Schwatka had done. But the going was rough. Their plan had been to circle east of the Chaix Hills, but they were forced to retrace their steps and pass the hills on the west. They then climbed a considerable distance up the Tyndall Glacier toward the summit, perhaps even to the 11,461-foot mark that Topham later reported in the *Alpine Journal*. But what remained to be climbed was staggering in scale, as well as swept by thundering avalanches, and the party returned to its base camp on the shore on August 7.

In his account of the trip, Williams wrote that later climbers would be able to follow their route and save valuable time. But he also questioned whether they had indeed found the best climbing route to the summit. "It is not at all unlikely," Williams wrote, "that the true way to the summit is to be found on the northern side, where fewer rocks and better snow would probably be encountered." How to reach the northern side of the mountain was, he admitted, "a problem yet to be solved."[59]

As it turned out, the next expedition to St. Elias did not follow in Williams's footsteps. Instead, a National Geographic Society expedition under the direction of geologist Israel Cook Russell took the field and tried to reach the northern side by attacking the mountain from the east. Russell brought with him Mark Kerr, the expedition's geographer who was charged with solving the riddle of the mountain's elevation, and seven hardy packers who had been recruited in Seattle. Reaching Sitka on board Captain James Carroll's *Queen* on June 24, 1890, the group got passage to Yakutat on the *Pinta*.

Rather than heading west to Icy Bay as the Schwatka and Topham parties had done, Russell went north to the head of Yakutat Bay and started up what he called the Dalton Glacier, well to the east of the mountain. Russell named the glacier and a river draining it for John Dalton, a miner and woodsman who had packed for Schwatka's group. It was a nice gesture, but four years later Dalton was accused of murder, and Russell thought better of it and had the Board of Geographic Names change the name to Turner, after a surveyor for the U.S. Coast and Geodetic Survey.

For three weeks they relayed equipment and provisions northwestward across the Dalton, Lucia, and Hayden Glaciers. It was slow work over rocky moraine and they averaged only about two miles a day. Kerr named the Lucia Glacier, for his mother, who evidently managed to behave herself, as the name survives. The Hayden was just one more honor for the U.S. Geological Survey's Ferdinand Hayden, but their passage up it was soon thwarted. With fog and rain shrouding the avalanches that boomed around them, they retreated southward to a small knob that snuggled up to the edge of the vast Malaspina Glacier. It was covered with flowers and offered a momentary respite of green in a sea of white. Russell named it Blossom Island, and it still appears as such on maps.

From Blossom Island they pondered their next moves. So far, they had covered some ground but not made much progress on the mountain. On August 2, they left the flora of Blossom Island and moved northwestward up the Marvine Glacier, named by Russell for another geologist. At its head, after two days of travel, they reached Pinnacle Pass at about 4,000 feet. North of the pass sprawled a massive glacier, the largest they had yet encountered, save the Malaspina. Russell named it after William H. Seward. If ever Seward was to have his icebox, this was definitely it!

Kerr took measurements of St. Elias and reported it to be only 15,350 feet. Something was clearly amiss, but he seemed quite pleased with his findings. To the north a huge fist of a plateau dominated a jagged jumble of peaks. In keeping with the convention of the day, Russell named it after the former director of the Geological Survey of Canada, William E. Logan.

After waiting for two of the packers who had gone down to bring up more supplies, Russell led the group across the Seward Glacier, through Dome Pass, and onto the Agassiz Glacier, which Schwatka and Seton-Karr had looked up four years before. They crossed the Agassiz and began the long climb up the Newton Glacier to the col between Mount St. Elias and Mount Newton. On August 22, after negotiating one particularly imposing ice cliff upon which they fixed a rope, the party was confronted by heavy clouds and falling snow. Reluctantly, Russell called a halt because "to proceed further would be rash and without promise of success. After twenty days of fatigue and hardship since leaving Blossom Island, with our goal almost reached, we were obliged to turn back."[60]

They were, of course, still a considerable distance from the summit, but Russell's biggest adventure was just beginning. On the morning of August 25, after what Russell called "some consultation"—the limits of which those who have spent time tent-bound debating a climb can well imagine—Russell and Kerr determined to make one more attempt for the summit. The pair reclimbed what they called Rope Cliff and then paused for lunch only to discover that they were running low on coal oil for cooking. Kerr volunteered to descend for more, while Russell took the bulk of the gear and struggled upward to the site of their highest camp. Each went his separate way, but then it started to snow again. Russell managed to reach the high camp and pitch his tent, but he was marooned there for six days as first snow and then the resulting avalanches bombarded the basin. When the coal oil ran out, he was reduced to sticking a wick in some bacon grease in a poor attempt to cook.

Finally, Russell began a cautious descent and was soon met by his anxious companions, who, having endured similar conditions, were climbing upward to look for him. Now Russell called it quits for good, and the party returned to Yakutat Bay. In his account of the expedition, Russell wrote one sage piece of advice. He strongly advised all future parties to position a high camp at the col at the head of the Newton Glacier. Pity he did not take his own advice.

A year later, Russell was back on the mountain. With six packers—three of whom had been with him the year before—he sailed north from Port Townsend, Washington, on the revenue cutter *Bear* and arrived at Yakutat on June 4, 1891. This time, despite Schwatka's warnings of the treacherous surf in Icy Bay, Russell determined to start from there and ascend the Agassiz directly rather than retrace his route of the year before.

On June 5, the *Bear* remained at anchor at Yakutat amid thick fog and stormy seas, but early the following morning she steamed west for the bay. Treacherous surf proved to be an understatement. Three boats capsized during the landings, and Lieutenant L. I. Robinson and four crewmen from the *Bear* and Will C. Moore of Russell's party drowned. It was an inauspicious beginning and the worst disaster to befall an Alaskan mountaineering expedition—albeit not on the mountain—until the 1967 Wilcox Expedition on Denali.

Russell and the remaining five packers struck northward toward the Chaix Hills. He named the range of "hills," as he called them, that towered above the Guoyt Glacier some thirty miles to the west after Lieutenant Robinson and Moore Nunatak at the eastern end of the Chaix Hills after their companion. They followed the Agassiz Glacier around the eastern side of St. Elias and joined their route of the year before above Dome Pass.

Ahead sprawled the stair-step icefalls of the Newton Glacier. With alpenstock and ice ax, they picked their way into the large amphitheater at the head of the Newton Glacier and established what would be their high camp. For twelve days, storms and clouds kept them there. Then on July 24, Russell and packers Neil McCarty and Thomas Stamy set off to reach the col at the head of the amphitheater. Half a day later, slowly surmounting the col, Russell expected to look northward and see at least something green far in the distance. There was nothing, however, but miles and miles and miles of snow-covered peaks and expansive glaciers. Russell named one of the peaks to the northwest after the *Bear*.

They had, however, found the key to the northern reaches of the mountain. High above the col to the south, a long, snow-laced ridge led up the final pyramid of St. Elias. The trio continued up it to the steady tune of ice axes cutting steps. Finally, with readings from two aneroid barometers, they estimated their height at 14,500 feet. If Mark Kerr's summit estimate of 15,350 feet of the year before was indeed correct, they were close. But why was there still so much mountain above them?

It soon became clear that they must turn around if they were to reach the safety of their camp. This they did, with Russell concluding that the only practicable plan would be for them to move their camp up to the col for another attempt—exactly his advice of the previous year. This they attempted to do, but another round of storms quashed their efforts and Russell gave the order to return to Icy Bay.

Repulsed by the mountain, Russell nonetheless went to work solving the riddles. He established a baseline on the beach about three miles long and after repeated measurements for accuracy, calculated the elevation of the summit at 18,100 feet, plus or minus 100 feet. Mark Kerr's lowball figure was discarded, and the currently accepted elevation of 18,008 feet proved to be within Russell's range. St. Elias was lower than Mexico's Orizaba and not the tallest point on the continent. But was the mountain in Alaska or Canada?

Russell computed its location at 60°17'51" north latitude and 140°55'30" west longitude. The Alaska-Canada boundary is the 141st meridian of longitude. Clearly, the mountain was about two and one-half miles east of that, but the convoluted treaty between Great Britain and Russia had set the international boundary east of the 141st meridian at ten marine leagues (about thirty-four miles) from the coastline. Russell calculated that the summit was thirty-three miles from the coast. Mount St. Elias was in Alaska, not Canada, but barely. Later work of the boundary commission would draw the boundary squarely across its summit. The geographic riddles were solved, but an ascent to the high point remained.

Six years later, on January 24, 1897, the *New York Sun* printed a prospector's report of a towering mountain north of Cook Inlet. But that summer it was St. Elias that attracted not one, but two expeditions. In the sporting tradition of those Victorian times, each party discounted that they were rivals, but beneath such gentlemanly denials, the race was clearly on.

Henry Bryant was the leader of the American team. By all accounts, he was in the upper crust of American aristocracy and as close to royalty as Americans were allowed to get. A member of the American Alpine Club, Bryant had financially supported two of Robert Peary's polar expeditions and been on his own trek to the Great Falls of Labrador. He also had experience climbing in Switzerland. In the spring of 1897, Bryant was well into his preparations for a St. Elias expedition when he learned that he would not have the field to himself. In early May, he confided to geologist George Davidson: "On learning of their plans, I felt that I had already gone too far in my preparations to withdraw simply because others had selected the same field of operations. I regret exceedingly the possibility of international rivalry, and a possible duplication of the work."[61]

The real royalty was on the other team. Luigi Amedeo de Savoia-Aosta, the duke of the Abruzzi, was a grandson of Italy's celebrated King Vittorio Emanuele II. His appointed career was the Italian navy, but his avocation was mountaineering. The duke had been planning to attempt Nanga Parbat in the Himalayas and follow in the footsteps—up to a point—of the great British climber A. F. Mummery, who had perished on its slopes two years before. The duke had climbed the Zmutt Ridge on the Matterhorn with Mummery and his partner John Norman Collie, and it was Mummery's good words that helped gain the duke admission to the venerable Alpine Club. When widespread outbreaks of bubonic plague closed India's borders, the duke turned his attention toward St. Elias.

Sporting a thin mustache and looking quite dapper, the duke left Turin in western Italy on May 17, 1897. With him as his aide-de-camp was a close friend from the navy, Lieutenant Umberto Cagni, and his climbing partner from earlier ascents in the Alps, Francesco Gonella. Another navy friend, Filippo de Filippi, was along as expedition doctor and to keep the expedition log. De Filippi's cousin, Vittorio Sella, whose mountain photographs from the Alps and Caucasus had already achieved international renown, had agreed to be the expedition photographer. And then, of course, there were the guides. While previous American efforts on St. Elias had been more homegrown affairs, European mountaineers of that generation would not have thought of going anywhere without them. The duke brought his faithful Joseph Petigax, who had led him on numerous climbs in the Alps, and three others, the best in Italy.

Sailing from Liverpool aboard the *Lucania*, the party arrived in New

York City on May 28 and was met by Professor Charles E. Fay, a founder of the Appalachian Mountain Club and one of America's true proponents of mountaineering for mountaineering's sake. Bryant, Fay informed the duke, was already en route and might already be on the mountain. "I am simply taking this trip for pleasure," the duke told a group of reporters accompanying Fay, and then added, "I shall not remain in New York more than one night."[62]

They crossed the continent by train and spent three days in a San Francisco hotel packing a mountain of canned meats, vegetables, and other foodstuffs into fifty large sacks— 6,600 pounds in all, enough for ten men for fifty days. While in the city, the duke met with Mark Kerr of Russell's 1890 expedition, who drew him a map of their route. Russell himself suggested that the duke hire E. S. Ingraham and a team of packers to help move the expedition's gear. The Alaska Commercial Company hired out its steamer *Bertha* to take the group from Sitka to Yakutat. Finally, the duke chartered a Seattle-based ship, the *Aggie*, to serve as his transport beyond Yakutat. Sella worried that the duke was spending too much money and confided as much in a letter to his own wife. Some expenditures may have been royal prerogative, but by the arrangements made, it seems clear that the duke knew he was in a race.

Upon their arrival, Yakutat was clothed in its normal dress of fog. Dozens of Tlingit villagers paraded along the shore with pine torches in welcome. One of them, the duke learned, had guided Bryant's party safely across the bay a full ten days before. There was no time to lose. Rowboats ferried the expedition's baggage from the *Aggie* to the beach below the Malaspina Glacier, and then a combination of expedition members, Ingraham's packers, and additional Tlingit porters moved it piece by piece onto the glacier.

De Filippi later wrote that the small tents they used weighed only 3.5 pounds each, but that the ten iron bedsteads the duke had brought along for the Italians weighed 14 pounds each. The duke's mentor, Mummery, had eschewed such sleeping arrangements because they created an icy draft underneath the sleeper. The duke took no heed, however, even when Mummery had also warned that "out west, for some mysterious reason, it is considered unmanly to sleep otherwise than on the ground."[63]

Finally, the gear was organized, and on July 1 they struck out north across the Malaspina Glacier bound for Pinnacle Pass. Royalty or not, the duke pulled his weight, scouting for campsites, hauling loads, and frequently being in the lead with compass in hand. Perhaps the man with the hardest task was Sella, who was trying to develop photos in his makeshift darkroom amid the wind and cold.

Whether the duke steered for Pinnacle Pass solely on the basis of Kerr's map, or whether a more direct route up the Agassiz was not advisable, is open to conjecture. But over Pinnacle Pass they went, across the wide, flat Seward, and over Dome Pass to the upper Agassiz, reaching the foot of the Newton on July 16. But where was Bryant? The next day they met four men climbing on the Agassiz Glacier with a sealed letter for the duke from Bryant. Was it news of an ascent? All gathered round as the duke read it aloud. Bryant had thrown in the towel. One of his party was ill, another caring for him, and one look up the Newton had convinced Bryant that he wanted no part of it. In the spirit of the age, he ended by wishing the duke's party every success.

So no one was ahead of them! Up through the icefalls and crevasses of the Newton they went. Bad weather was the norm, once keeping them in tents for three days straight, but on fleeting exceptions St. Elias rose tall and majestic into the blue for Sella's lens. On July 30, the entire group of ten Italians reached the col at the head of the Newton Glacier. The duke named it Russell Col in honor of the man who had asserted that an ascent of St. Elias would require a high camp here. That night was clear, a good omen perhaps. Shortly after midnight, they launched their summit bid.

Perhaps most remarkable, they did it as a team. The duke insisted that the ten climbers on three ropes move together and that all have a chance to reach the summit. The northeast ridge was a straight shot to the summit, long and graced with several rocky steps, and they made slow but steady progress. At 5:00 A.M., the duke called a short breakfast stop at 15,700 feet. De Filippi smoked a cigarette and reported he was able to breathe regularly again. It was hardly the high-altitude medical research that Charles Houston would later do in Alaska, but it seemed to work. At 11:55 A.M. on July 31, with the day still clear, Petigax stepped aside, and the duke, who had been as much "one of the guys" as it was possible to be and still be a prince, put his foot on the summit of Mount St. Elias. They had made the climb of almost 6,000 feet from Russell Col to the summit in ten and one-half hours.

Unfurling a small Italian flag, the duke surveyed the view. To the north, he named a peak beyond Mount Logan after the *Lucania.* To its west, beyond Russell's Mount Bear, he named Mount Bona after his racing yacht. It was a proud day, and all ten climbers returned safely to the col. Amazingly enough, the following day was still clear, and Sella and his assistant Erminio Botta reclimbed a portion of the ridge above Russell Col in order to photograph Mount Logan with a panoramic camera. From then on, it was an uneventful, but welcome, descent to Yakutat Bay.

Russell had found the route, but it was the duke of the Abruzzi who followed it to the summit. Rather than lead a life of idle pursuits, he became one of the great mountaineers of the late nineteenth and early twentieth centuries. Sella's photographs were the first photographic record of a major mountaineering expedition. They served to whet the appetites of many mountaineers of the early twentieth century and illustrate Filippo de Filippi's classic account of the climb, *The Ascent of Mount St. Elias*. Both Sella and de Filippi accompanied the duke on many other expeditions, including his 1909 journey into the Karakoram to attempt K2.

As mountaineering historian Chris Jones noted, the duke of the Abruzzi "belonged to an almost vanished race of explorers who had the private means to organize their own adventures and, more important, the drive to carry them off."[64] St. Elias was not climbed again until 1946, and it remains a very difficult climbing objective.

Postscript to an Era
(1865–1897)

News of the ascent of St. Elias *should have caused great excitement. As it was, all of North America—and soon all of the world—had other thoughts on their minds. On July 17, 1897, while the Duke's party was fighting its way up the Newton Glacier, a reporter for Seattle's* Post-Intelligencer *wrote, "At 3 o'clock this morning the steamer* Portland *from St. Michael for Seattle passed up the Sound with more than a ton of solid gold aboard." Alaska was at another watershed. There was a new frontier to cross. Seward's real estate deal was about to be proven up in a big way.*

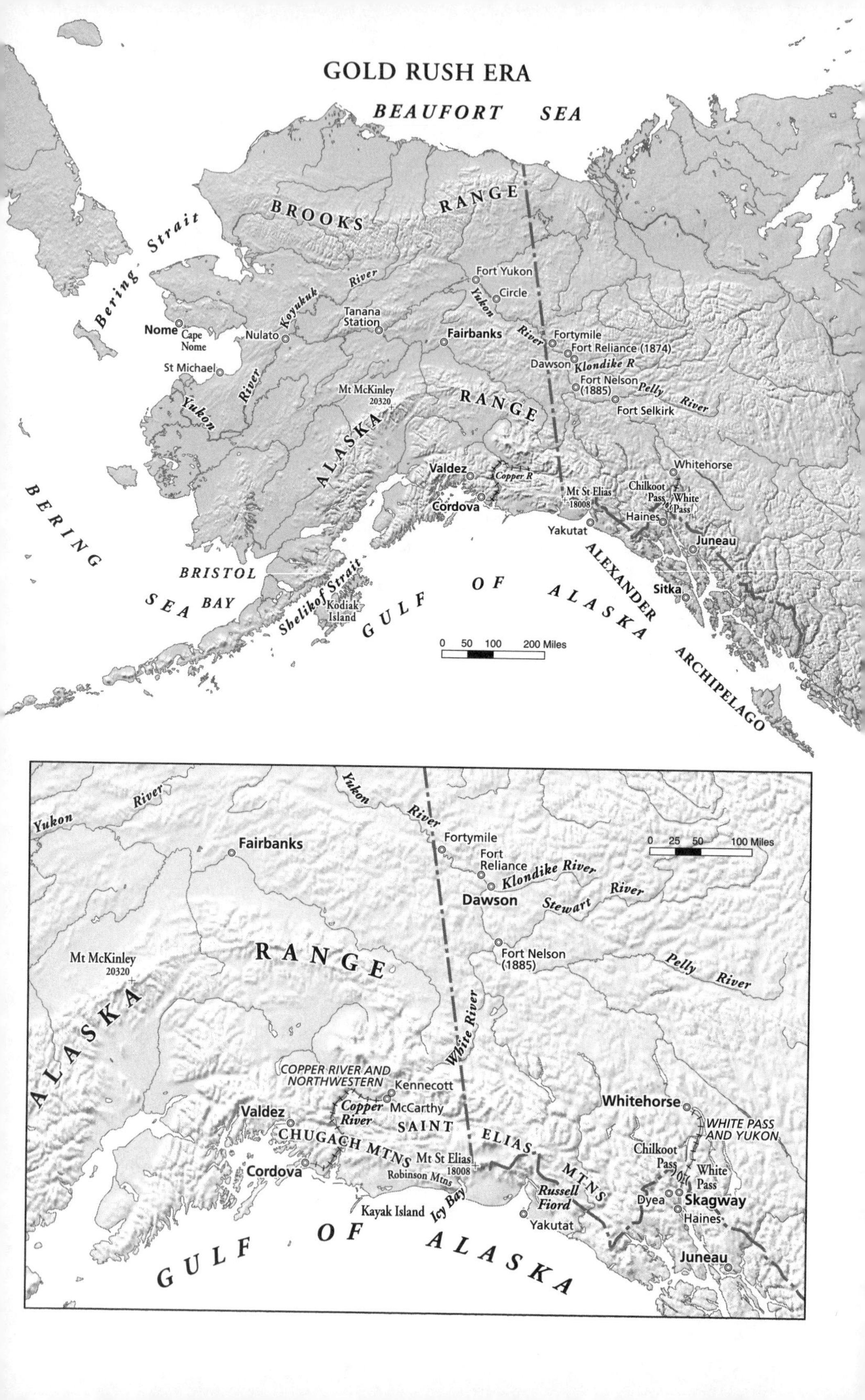

GOLD RUSH ERA
BEAUFORT SEA
BROOKS RANGE
Bering Strait
Koyukuk River
Fort Yukon
Circle
Yukon River
Tanana Station
Nome
Cape Nome
Nulato
Fairbanks
Fortymile
Fort Reliance (1874)
Dawson
Klondike R
St Michael
Fort Nelson (1885)
Pelly River
Fort Selkirk
Mt McKinley 20320
Yukon River
ALASKA RANGE
Whitehorse
Valdez
Copper R
Mt St Elias 18008
Chilkoot Pass
White Pass
Cordova
Haines
Yakutat
Juneau
BERING SEA
BRISTOL BAY
Shelikof Strait
Kodiak Island
GULF OF ALASKA
ALEXANDER ARCHIPELAGO
Sitka
0 50 100 200 Miles
Yukon River
Yukon River
Fairbanks
Fortymile
Fort Reliance
Klondike River
Dawson
Stewart River
0 25 50 100 Miles
Mt McKinley 20320
RANGE
Fort Nelson (1885)
Pelly River
ALASKA
White River
COPPER RIVER AND NORTHWESTERN
Kennecott
McCarthy
Copper River
Valdez
SAINT ELIAS MTNS
CHUGACH MTNS
Whitehorse
WHITE PASS AND YUKON
Chilkoot Pass
White Pass
Cordova
Mt St Elias 18008
Robinson Mtns
Russell Fiord
Dyea
Skagway
Haines
Kayak Island
Icy Bay
Yakutat
Juneau
GULF OF ALASKA

BOOK FOUR

Go North: The Rush Is On

1897–1915

Gold! We leapt from our benches. Gold! We sprang from our stools.
Gold! We wheeled in the furrow, fired with the faith of fools.

—ROBERT SERVICE, "THE TRAIL OF NINETY-EIGHT," 1910

The American mining frontier *began in 1848 with a whisper of "gold" on the banks of the American River near Sutter's Mill. By the following year, that whisper had swelled to a frenzied battle cry that sent tens of thousands of "forty-niners" crossing the North American continent. In the short span of two years, California went from a quiet land of Spanish haciendas to the thirty-first state of the Union.*

The American mining frontier was quite unlike the colonial frontiers of Russia, Great Britain, France, and Spain that had pushed their respective borders in steady geographic advances. The mining frontier was explosive and geographically erratic, characterized by a rapid boomtown growth wherever the cries of "Gold!" and "Silver!" were raised. For fifty years, the American mining frontier raced sporadically about the West: First, gold in California, then the Pikes Peak country in Colorado, the Fraser River valley in British Columbia, gold and silver in Arizona, silver in Nevada's Comstock, back to Colorado again for the silver mines of Leadville, gold and silver in Idaho, gold in the Black Hills of South Dakota, and finally gold at Cripple Creek, Colorado.

Most times, the promises far outweighed the results, and for every miner who struck it rich, there were a hundred who came away disappointed. For every town that boasted of becoming the richest Eldorado on earth, a hundred ghost towns remain. But the American mining frontier was the vanguard of permanent settlement. In the half century between the California gold rush of 1849 and the Klondike gold rush of 1898, the great expanse of the American West was settled with towns, roads, and railroads.

By 1896, silver had been demonetarized and the bimetallism of a generation resolved, with the United States and much of the world firmly committed to the gold standard. In the wake of a national depression, there was to be one last great gold rush where the individual prospector—not the corporate conglomerates of the twentieth century—was still king. In the last great act of the American mining frontier, tens of thousands once again responded to the cry of gold and struck north for the Klondike. There would have to be a geography lesson or two about what country the Klondike was in, but the stampede that followed the arrival of the steamers Excelsior *and* Portland *in San Francisco and Seattle would change Alaska forever and thrust upon it another new frontier.*

Fortymile, Circle, and the Sourdoughs of Rabbit Creek

A lot of folks knew there was gold in Alaska. Russian *promyshlenniki* found traces of golden flecks in creeks and streams for decades, but they were far more interested in the "soft" gold of furs. Russian mining engineer Peter Doroshin discovered small gold deposits on the Kenai Peninsula about 1850, but by then the last thing the czar needed to complicate the administration of the territory was a gold rush by a bunch of outsiders. The news was kept quiet. Crews of the Western Union Telegraph Expedition reported traces of gold out on the Seward Peninsula just after the sale of the region to the United States, but nothing came of it at the time. Then, of course, there was the experience of Joe Juneau and Richard Harris along the Inside Passage. It might not be Eldorado, but they were making out all right. And the upper Yukon had shone a glimmer or two to Hudson's Bay trappers and the few Americans who paddled its many tributaries.

As the Alaska Commercial Company sought to fill the shoes of the Russian-American Company after the sale of Alaska to the United States, it employed two former Hudson's Bay trappers to extend the company's reach along the upper Yukon. Brothers François and Moise Mercier had the blood of generations of hardy French-Canadian *coureurs de bois* coursing through their veins. Both were just the sort of toughened, razor-sharp individuals that one hoped to rely on in a tough and unforgiving land. In 1869, François assumed management of the Alaska Commercial Company's operations out of St. Michael, and Moise took charge of the post at Fort Yukon. While François worked out the logistics of the summer steamboat runs up the Yukon to supply the posts, Moise looked for opportunities to establish additional trading stations.

In 1873, the Mercier brothers met up with three Americans who were lured north by tales of both furs and gold. One of the newcomers had been christened Leroy Napoleon McQuesten. Fortunately for him, he went by the more Alaskan moniker of "Jack." Raised on a farm in New England, Jack McQuesten followed the mining frontier to California and the Fraser River before joining up with the Hudson's Bay Company. One of his partners was a sturdy Irishman named Arthur Harper. Since coming to New York from Ireland in 1850, Harper had also been on the trail of the mining frontier. Pondering the recent discoveries of gold throughout the Rockies, Harper wondered if the gold belt extended to the far north. The third member of the triumvirate was Alfred Mayo, a wiry joker from Kentucky whose work experience included time as a circus acrobat. Over the next two decades, these three would finish what the Mercier brothers started and build the commercial infrastructure that would serve a number of mining rushes.[1]

François Mercier offered McQuesten, Harper, and Mayo jobs with the Alaska Commercial Company in the spring of 1874. Harper chose to continue prospecting along the lower Tanana, but McQuesten and Mayo signed on. Mayo was sent to the old Russian post at Nukluklayet. The company was now calling it Tanana Station. Jack McQuesten drew the assignment to accompany Mercier upriver on the company steamboat, the *Yukon*, to establish another post from which to trade with Han Natives. This they did about six miles below the mouth of the Klondike River—almost fifty miles inside the Canadian boundary. (One has only to look at a map of the current Yukon Territory and see the place-names of McQuesten, Harper, and Mayo spread across it to recognize what little regard was held for the international boundary.) Mercier called the post Fort Reliance and left McQuesten there to see to its completion. Fort Reliance quickly became a landmark on the river, and the Fortymile River downstream and the Sixtymile River upstream took their names from those approximate distances from Fort Reliance.[2]

The following year, the Alaska Commercial Company decided to lease some of its operations to independents, and Harper teamed back up with McQuesten and Mayo to work the upper Yukon posts. A rival company was also on the river that spring, and competition with the Western Fur and Trading Company was frequently fierce. Just as in the days of trading with both the Russian-American and Hudson's Bay companies, the Natives now had more leverage when bartering their pelts for goods. McQuesten, Harper, and Mayo worked these operations but also continued to prospect on the side.

In 1881, during the course of one of his journeys, Harper and several companions crossed from the Yukon near what would later be the site of Eagle to the headwaters of the Tanana. After fording the North Fork of the Fortymile River, Harper took a sample of sand from the riverbank and carefully packed it away. The group reached the upper Tanana, built a moose-hide boat, and descended the river to the Yukon—four years before Lieutenant Allen made a similar trip. When one of Harper's companions went outside that fall, he had the sand assayed in San Francisco and reported to a stunned Harper that it showed $20,000 of gold to the ton. Harper hastened back to the North Fork the next spring but couldn't find the location.[3]

So there was no big strike to report, yet slowly but steadily, by twos and by threes, the number of prospectors on the Yukon was increasing. In part, it was because the firm of McQuesten, Harper & Mayo acted as a sort of chamber of commerce, exhorting the mineral potential of the region and then proceeding to mine the miners who came into the country and bought supplies at their trading posts. In 1883, the Alaska Commercial Company finally bought out those pesky rivals at Western Fur, and the ACC was once more the undisputed commercial power in Alaska. The company controlled what it didn't manage directly through financing arrangements such as it had with McQuesten, Harper & Mayo.

When McQuesten went "outside" on a regular buying trip, he met with the ACC's board of directors in San Francisco. The commercial picture was changing, he reported. The traditional fur trade business with the Natives was declining. The economy of the future was catering to prospectors. Sure, their numbers were small now, but what if . . . The directors agreed, and McQuesten ordered more mining supplies to be shipped north for the following season. Thus, 1885 marked a transition in the economies along the Yukon from the fur trade to the mining supplies trade, and the firm of McQuesten, Harper & Mayo was there to cash in on it.

They were just in time. Coincidentally, in 1885 there was a small strike on the Stewart River, again deep in Canada above Fort Reliance. A year later, in order to accommodate the miners on the Stewart, McQuesten, Harper & Mayo packed up the supplies at Fort Reliance lock, stock, and barrel and moved upriver to establish Fort Nelson at the mouth of the Stewart. But it had barely done so when word came that two miners named Harry Madison and Howard Franklin had struck gold about twenty-five miles up the Fortymile, *below* Fort Reliance and just barely in American territory. This strike wasn't the occasional golden sands of earlier discoveries, but rather genuinely coarse gold. It was the first rich placer discovery on the Yukon, and it set off a rush to the Fortymile.[4]

The first wave of prospectors to Fortymile were locals from other spots along the Yukon, but Harper figured that once word got out, many more from the outside would arrive in the spring. If there wasn't an adequate quantity of supplies along the river, there was apt to be widespread starvation the following winter. Harper sent a hurried letter to McQuesten, who was again buying supplies in San Francisco, that he had best stock up for the boom. Harper's missive was carried over Chilkoot Pass by two volunteers, a steamboat man named Tom Williams and a Native companion. The trip was one of frozen misery. Williams died two days after reaching the trading post at Dyea, but the word reached McQuesten in time for him to be prepared for the spring of 1887.[5]

In short order, a cluster of cabins and assorted buildings sprang up around the McQuesten, Harper & Mayo store at the confluence of the Fortymile and the Yukon. These included the obligatory blacksmith shop, sawmill, and a number of saloons. Glass windows in buildings were almost unheard of, and those wintering over frequently cut clear pieces of river ice to stick in crude openings. The ice let in what precious daylight there was, but was not recommended for solar sites. Most of the diggings were in Alaska, but the growing town of Fortymile was clearly in Canada.

McQuesten located another trading post on the Yukon at a place he called Circle on the south bank of the Yukon about one-third of the way between Fort Yukon and the Fortymile. The name came from the fact that McQuesten thought the post was on the Arctic Circle, although it was actually about fifty miles south of that line. Circle quickly became Circle City and was the mining supply town for a wide region, including the prospectors who were beginning to explore the Birch Creek drainage. Jack McQuesten quickly became the man to know in Circle and built a two-story log building for a store and a fireproof, corrugated iron warehouse.

Someone else who sought to take advantage of the growing trade was a rather dour individual named John J. Healy. Yet another veteran of the mining frontier with stops in Montana, Idaho, and Canada—including the infamous whiskey fort of Whoop-Up—Healy arrived in Alaska in 1886 and set up a trading post at what became the town of Dyea at the foot of Chilkoot Pass. It was at Healy's log post that the brave Tom Williams expired after his Chilkoot crossing. With Fortymile going strong and the town of Circle becoming a supply center, Healy convinced Chicago millionaires Jack and Michael Cudahy of the meatpacking family and P. B. Weare of the Chicago Board of Trade that it was time to once again challenge the monopoly of the Alaska Commercial Company. They formed the North American Transportation and Trading Company, and by the fall of 1892 the company's steamer *Portus B. Weare* was at work on the lower Yukon. The following spring, Healy steamed up the Yukon and set up a rival operation just downstream from Fortymile, calling it Fort Cudahy. Within another year, Healy was also challenging McQuesten's operations in Circle City.[6]

McQuesten and Healy became archrivals with decidedly different ways of doing business. McQuesten was the outgoing, hale-hearty fellow always willing to grubstake a down-and-out prospector. Healy had lower prices, but the rule was cash on the barrelhead—no credit. This difference in styles hinted of a bigger transition in the wind. McQuesten represented the Yukon pioneers who followed a golden rule of "do unto others as you would be done by." Healy was the advance agent of a bottom line–oriented business culture. When the Fortymile miners court found Healy guilty of a minor infraction, its members smugly assessed a stiff fine. Healy decided to teach them a thing or two and call in the real law, the heretofore absentee landlords of Canada.

Healy's complaint to Superintendent Samuel B. Steele of the Northwest Mounted Police coincided with Canadian surveyor William Ogilvie's concerns that Fortymile was a town being run by Americans on Canadian soil. Both Steele and Ogilvie figured that it was time to exercise Canadian sovereignty. Steele dispatched Constable Charles Considine to Fortymile for an inspection tour in 1894. There was no anarchy thanks to the miners court, but Considine reported that thousands of dollars in customs were going uncollected on supplies brought into the town, including a little matter of what he estimated was some 3,000 gallons of whiskey. The following year, Considine returned to Fortymile with twenty constables of the Northwest Mounted Police. A surviving photograph of the group shows a rough-cut image and not the spit and polish of brilliant red uniforms. But the effect was the same: Canada had begun to exercise its sovereignty over the distant Yukon.[7]

Back on the American side in Circle City, McQuesten's way of doing business had paid off in 1893 when two men he grubstaked found gold in the Birch Creek area. The stream ran parallel to the Yukon for much of its 150-mile length before splitting into two different mouths and emptying into the big river some 30 miles below Fort Yukon. But the quickest way to the new diggings was through Circle City, and the town continued to grow, soon boasting between 500 and 700 people. Compared to the Canadian propriety of Fortymile, Circle was a wide-open town where the only law was still the miners committee.

By 1896, Circle City had a music hall, two theaters, eight dance halls, and twenty-eight saloons. The showpiece was the double-decker Grand Opera House run by George Snow, miner by day and Shakespearean actor by night. Snow had his children appear on stage and collect gold nuggets thrown to them by miners hungry for entertainment. The only programming glitch came when a visiting vaudeville troupe was marooned in town for the winter, and seven months of a limited repertoire wore thin on even the least cultured of the audience.

The Episcopal Church bought land for a hospital; the *Chicago Daily Record* dispatched a correspondent to town; the miners established a library; and Anna Fulcomer of the University of Chicago arrived to open a government school. Yes sir, in 1896 Circle City's future looked bright indeed!

Circle's success and all of the neighboring mining activity lured regular steamship navigation farther and farther up the Yukon. The Alaska Commercial Company and both its partners and rivals did more and more business as more and more prospectors found gold in paying quantities. According to one source, the value of mining along the Yukon increased from $30,000 in 1887 to $800,000 in 1896. Thus, while the mining frontier was about to explode on the Alaskan and Yukon scene in a major way, mining had already played a steady and increasing role in the development of the upper Yukon Valley for more than a decade.[8]

Meanwhile, McQuesten's partners—Harper, Mayo, and a more recent arrival named Joe Ladue—had been supplying prospectors on the upper river and continuing to move their trading posts around to accommodate the latest strike. One of those they grubstaked was Robert Henderson. A Nova Scotian of Scottish ancestry, Henderson had followed the mining frontier all over the world and had even prospected in Australia and New Zealand. Now he was trying his luck on the river the Natives called the Thron-diuck—a name meaning "hammer-water." It was so named because the Natives had hammered stakes across its shallow mouth in order to spread nets for catching the fine runs of salmon that ran up it. Prospectors called it the Klondike.

Also in the area was yet another son of the mining frontier, George Washington Carmack. His father had crossed the continent to California as a forty-niner. Carmack himself had sailed to Alaska as a teenager, jumped ship at Juneau, and then pushed north, marrying the daughter of a chief of the Tagish clan of the Tlingit and working for a time as a packer on the trail over Chilkoot Pass.

What happened next is a matter of some conjecture. As with most stories of great mining discoveries, legend, lore, and after-the-fact memories have managed to obscure simpler facts. A century later, the heirs of each participant still champion different versions. In essence, in the summer of 1896, Henderson was taking some color out near Gold Bottom on a tributary of the Klondike River. In the course of his travels he met up with Carmack and suggested to him that he try his luck along Rabbit Creek, another small tributary of the Klondike. The story is told that in the process of several meetings that summer, Henderson made derogatory comments about Carmack's companions, his Native brothers-in-law, Skookum Jim and Tagish Charley. On August 17, 1896, when Carmack, Skookum Jim, and Tagish Charley did indeed make a major discovery on Rabbit Creek, they staked claims and told just about everyone—everyone, that is, except Robert Henderson.

Under Canadian mining law, no more than one claim per person could be staked in any one mining district, except that the discoverer was allowed a double claim. Carmack set off for Fortymile to record his claims and those of Skookum Jim and Tagish Charley. At first there was skepticism. Carmack, after all, had been on the river for years without much success. Some patrons of Bill McPhee's saloon in Fortymile, where Carmack downed a double whiskey, thought it was all a scheme of Harper and Ladue to drum up a little late-season business. Still, the great "what if" of the mining frontier began to gnaw at even the most pessimistic. While Carmack headed to the recorder's office, one by one, with one excuse or another, the miners of Fortymile quietly slipped away and headed upriver for the Klondike.

By morning, Fortymile was as good as deserted. When news reached Circle City and the other camps along the Yukon, they emptied just as quickly. Prospectors who had been content with ten-cent pans of dust now converged on Rabbit Creek, soon to be called Bonanza, and staked claims up and down the creek and its little tributary called Eldorado. The approaching winter be damned. They clawed and dug and burned away at the frozen ground and took out the dreams of many an Arctic night. By the following spring of 1897, the town of Dawson at the junction of the Klondike and Yukon Rivers was a boomtown of about 1,500 people who had congregated there from all over Alaska and northwest Canada. Circle City, "the Paris of Alaska" that had glowed so brightly the previous year, was a ghost town.

In one of history's ironies, Dawson was within a good spit of McQuesten, Harper & Mayo's Fort Reliance. The gold had been almost under their noses all the time. But just like Fort Reliance, those tough old sourdoughs would be left behind by this stampede. Arthur Harper died of tuberculosis in Arizona in 1898—broke and worn out. Joe Ladue also succumbed to tuberculosis shortly after the Klondike strike. McQuesten got lucky on an interest in a share of a Klondike claim and left for California in 1898. Only the wiry little acrobat Al Mayo continued to go with the flow of subsequent mining rushes until he died at Rampart in 1923—outliving his partners by more than twenty years.

Once the ice went out on the Yukon, a motley assortment of river steamers arrived in Dawson to transport several hundred of the lucky ones out to civilization with their winter's take of gold. Down the river they went, as unlikely a crowd of newfound wealth as could be imagined. All had been paupers less than a year before; some had not been "outside" in years; none had been in communication with the "outside" since the previous year. They were not millionaires by today's standards, but $50,000, $75,000, or even $100,000 made them quite wealthy in 1897. Most rode the Alaska Commercial Company's *Alice* or rival John J. Healy's *Portus B. Weare* downriver to St. Michael, where they transferred to the coastal steamers *Excelsior* and *Portland.*

On the evening of July 14, 1897, the steamship *Excelsior* docked in San Francisco and put the word *Klondike* on everyone's lips. Thus, three days later, all of Seattle was on the lookout for the *Portland* as it made its way into Puget Sound. One reporter for the *Post-Intelligencer* wrote, "At 3 o'clock this morning the steamer *Portland* from St. Michael for Seattle passed up the Sound with more than a ton of solid gold aboard." By evening, the phrase "a ton of gold" was being published by newspapers around the world, and the rush to the Klondike was on.

The Trails of '98

First, the obligatory geography lesson: The Klondike was not, is not, and never will be in Alaska. Rather, its richest diggings were a good sixty miles east of the international boundary, well within Canada's Yukon Territory. Somehow, in all the excitement of the rush, that fact was lost upon tens of thousands of would-be tycoons until they came face-to-face with Canadian Mounties manning Maxim machine guns atop the Chilkoot and White Passes. Canada was not about to let the customs duties on goods entering the country escape its coffers, any more than it was going to let its gold slip away down the Yukon without a tax.

Thus, the story of the Klondike and its legends of Charley Anderson, "the Lucky Swede" who bought a million-dollar claim while drunk; Clarence Berry, one of the very few who took out a fortune and managed to die with it intact; "Swiftwater Bill" Gates, who bought up every egg in Dawson because of a woman; and many more—all belong to the Yukon Territory. The Klondike story that belongs to Alaska is that of the amazing routes these argonauts cut across its rugged landscape in order to reach, or at least attempt to reach, this beckoning Eldorado. The more realistic avenues, including the Yukon River and the routes over the Chilkoot and White Passes, would nurture the growth of Skagway, Valdez, and the general commercial development of the territory.

There were many routes to the Klondike, but why were there so many people on these trails? Why this last great rush? What made businessmen in New York and Chicago, schoolteachers in Omaha and Seattle, and budding young Ivy Leaguers up and quit their vocations and join the rush as readily as did newly arrived immigrants, down-on-their-luck farmers, and restless miners who had been on the trail of the mining frontier for decades? By one estimate, between July 1897, when the treasure ships arrived on the West Coast, and the following spring, as many as 50,000 people may have started on the trails to the north.

As the nineteenth century drew to a close, a number of phenomena came together to spark this exodus of biblical proportions. For one thing, there *was* a lot of gold. The phrase "a ton of gold" played well in the yellow journalism of the time, but the truth was that, claim for claim, the

Klondike ground was rich indeed. According to one source, $150 million in gold was taken from its mines.[9] (A ton of gold at the average 1897 price of $20.67 per ounce would have been worth $661,440.) And gold was definitely going to be the currency of the next century. Just the year before, William Jennings Bryan flamboyantly swore not to crucify mankind on a cross of gold, but William McKinley won the presidential election of 1896 and made it clear that gold would be the monetary standard of the future. For many still reeling from the Panic of 1893, which had been triggered by the demonitarization of silver, the golden glitter of the Klondike was a way back.

Then, too, by 1897 the mining frontier had pretty well completed its crisscrossed journey about the American West. From California in 1848 to the most recent strikes at Cripple Creek in the shadow of Colorado's Pikes Peak, the easy ground had been covered. The Klondike might just be the last chance to get rich quick.

Finally, one cannot help but speculate that while the gold was certainly important, this was the last great fling of a generation that sensed that the adventure and excitement of the American West was passing from the scene forever. For many it was, as Robert Service later wrote in "The Spell of the Yukon," "not the gold we were after, so much as just finding the gold."

And so, to paraphrase Service, into the northland they pressed. Clarence Berry warned that it would not be easy. Upon stepping off the *Portland* in Seattle, Berry was quoted as saying: "The country is wild, rough, and full of hardships for those unused to the rigors of Arctic winters. If a man makes a fortune, he is liable to earn it by severe hardship and sufferings. But then, grit, perseverance, and luck will probably reward a hard worker with a comfortable income for life."[10]

Within days of Clarence Berry uttering those words, tens of thousands of gold seekers, their thoughts focused only on Berry's last phrase—"a comfortable income for life"—were bound for the handful of creeks emptying into the Klondike River deep in Canada's Yukon Territory. There was no one route to the Klondike. Indeed, the winter of 1897–98 found Klondike stampeders spread out across the entire northwestern part of the North American continent on a dozen or more routes—all arduous, some downright insane. But from Seattle, Washington, Vancouver, British Columbia, Edmonton, Alberta, and many lesser jumping-off points, the multitudes surged north to the Arctic.

Despite the many routes to the Klondike, the collective experience of this last great gold rush would come to be symbolized by a photograph of a long, single-file line of humanity, struggling upward under incredible weight to reach the summit of Chilkoot Pass. Long before each person in turn reached the top, the reality of the rest of Clarence Berry's words had sunk in.

Chilkoot Pass had been used for centuries by Tlingit as a trading route between the coast and the upper Yukon River basin. After sailing from Seattle or other points to the port of Dyea at the head of Lynn Canal, one headed north up the Taiya River drainage to Chilkoot's 3,525-foot summit and then descended north into Canada to Lakes Lindeman and Bennett. The trip itself was difficult enough, but once Dawson's limited supplies began to run low, Mounties atop the pass required that each person entering Canada have at least a year's supply of food—the equivalent of about a ton, or 2,000 pounds. This meant that travelers were obliged to walk the Chilkoot not once, but ten, twenty, or thirty times, hauling their supplies to the top in stages. Once on the shores of either Lake Lindeman or Lake Bennett, stampeders built boats, waited for the ice to go out, and then sailed and floated down 550 or so miles of the Yukon to Dawson, braving Miles Canyon and Five Finger Rapids in the process.

It was not a route for the faint of heart. One stampeder confided to his diary: "On the road down canyon this A.M. we met a Missouri acquaintance (of boat trip) who was going back to his family, having sold out his share to his partners, and saying to us, 'I've had all I want of this country and am going back to my family and stay there.' There are several more like him, becoming discouraged at the last greatest tug and who sell out for a song."[11]

There were plenty of women on the trail, too. One was Inga Sjolseth, who started north from Seattle on March 9, 1898, on board the steamship *Farallon* in the company of a group of Norwegian immigrants. Arriving in Dyea a week later, Inga's party set up camp on one of the sandbars near the Taiya River while they organized their supplies for the trip ahead. Then began the painstaking work of ferrying loads up the trail. On March 29, Inga wrote in her diary:

> Today we moved from Canyon City to Sheep Camp. The trail in from Canyon is narrow and crowded, lying between two steep mountains. If one looks up one sees only mountains over one's

> head. The snow has turned to water, so progress has been very difficult. Ida and I pulled a loaded sled together through the canyon and heard many comments about it. We heard one man say he wished he had had such a "tram" as we were. Some of the men took off their packs and laughed at us. We have now set up our tent here in Sheep Camp. The snow is thick under our tent.[12]

Spring snows continued to fall. A few days later, Inga was still at Sheep Camp when the deadly Palm Sunday avalanche roared across the trail just north of the camp and buried a long line of stampeders in its path. About sixty (reports of the exact number vary) died in the disaster, and many were buried in what came to be called Slide Cemetery in Dyea. The snows continued, but so did the long line working its way toward the summit.

Finally, after days of poor weather and feeling poorly herself, it was time for Inga to leave Sheep Camp and cross the pass. A little horse, barely alive, hauled a sled with some of her supplies up to the Scales, where goods were weighed and from where she could see the long line leading to the summit. The upward line plodded along ever so slowly, but the downhill route—for those who had cached supplies at the summit and were coming down for another load—was like a toboggan run. Most sat down and slid down on their bottoms, going so fast that Inga claimed she was afraid to watch. When her group's load of supplies finally went over, she was too afraid to accompany it, and so she returned to Sheep Camp.

While waiting for another attempt, Inga washed socks, baked bread, and made pea soup and bread pudding for dinner one evening. One of the single men of the party, Henry Kolloen, delivered a letter to her from a Mr. Sandvig. It was written on a piece of board, and Inga noted in her diary that "this is the kind of letters one receives here in Alaska." Perhaps Sandvig was urging her to make the journey over the pass and catch up so that he could take advantage of some of that cooking. In any event, it was Kolloen that she later married. A week later, her courage up, she made the trip across the pass and arrived at a camp on Lake Lindeman. Even in a howling gale, entrepreneurs were busy at the summit selling tea for ten cents per cup.

Inga's group remained at Lake Lindeman for the better part of two months, while boats were built on Lake Bennett for the journey down the Yukon. During this time, Lindeman was a thriving city of tents, complete with evening entertainment and Sunday church services. Some of the hardships were taken matter-of-factly. Inga wrote at some length in her diary about selling a stove for Sandvig and then succinctly reported that "four men drowned in Lake Bennett today."

By May 28, the ice had broken up on Lake Lindeman and was crashing in great masses down the connecting stream to Lake Bennett. Most boats negotiated this passage safely, but a few smashed into the large rocks and broke apart—the results of weeks of labor gone to pieces within a mile of the start of the water leg. Inga's group was among those who built boats on Bennett to avoid just such an uncertainty.

On June 14, the boats were ready. Inga's group hired a team to transport their gear from Lindeman to Bennett, loaded the boats, and tried to sail north into a ferocious afternoon wind. When headway proved difficult at best, they put back into shore and did what Alaskan adventurers still do in such circumstances: They made a pot of coffee and, in Inga's words, "drank it."

The next day was still blustery, but with a following wind, and they made good time to Tagish House, where they had the luxury of sleeping in bunks. A few days later when they were back in the open, the mosquitoes were so bad that they couldn't sleep. Inga got up at 1:00 A.M. and cooked rice. By 3:30, everyone was up and soon under way. They arrived in Dawson on June 26, just twelve days after departing Lake Bennett. Inga went on to marry Henry Kolloen and later worked several placer claims with him.

If one didn't like the looks of the Chilkoot, the White Pass route took one to almost the same place on Lake Bennett. The White Pass route ran from the port of Skagway, a nearby competitor of Dyea, north up the valley of the Skagway River. It followed the White Pass Fork of the Skagway River to the 2,915-foot summit of White Pass before descending to Lake Bennett and joining the Chilkoot route. At first blush, the White Pass looked deceptively easy, particularly given its lower elevation than the Chilkoot. Perhaps that is why more than 5,000 tried it that first winter of 1897–98.

After a relatively easy beginning out of Skagway, the trail encountered the slippery slate cliffs of Devil's Hill, the boulders of Porcupine Hill, and the liquid mud of the 1,000-foot climb to Summit Hill. Once in Canada—here, too, Mounties imposed a year's-supply-of-food quota per person—the trail skirted a myriad of lakes and crossed two more divides before reaching Lake Bennett. Soon the trail was littered with the carcasses of hundreds of horses brought from Seattle or Vancouver by steamer, but no match for the tough terrain and twenty-four-hour days of packing they were driven to. "The horses died like mosquitoes in the first frost and from Skagway to Bennett they rotted in heaps," wrote Jack London.[13] No wonder the White Pass route was known as the Dead Horse Trail.

Despite such grim scenes, Stewart Campbell chose to try his luck with the White Pass route. He and his party left Kalamazoo, Michigan, on January 26, 1898. His group of businessmen exemplified the fact that stampeders came from all walks of life and were not ruffians at the fringe of society—even if the rigors of the trail might turn some that way in due course. Taking the train to Chicago, Campbell and his cohorts celebrated their first evening at the Chicago Opera House. In Seattle, they nearly got into a scrap at the depot when the freight agent attempted to charge them $448 in excess weight. That they settled the matter for only $64 suggests that there were folks "panning for gold" all along the trail, not just in the diggings.

Seattle was rainy, but bustling with carloads of people and freight arriving all of the time. Campbell's party bought the last of their provisions, including a span of horses, hay and oats, and even some lumber, and then sailed north. By February 15, they were in Skagway, taking meals at the Waldorf Restaurant at one dollar per man per day.

Then they were faced with the prospect of moving twenty-five tons of supplies over what Campbell called "the liveliest gold trail in the world: Horses, mules, jacks, bulls, oxen, cows, goats, dogs, men, women, children, with sleds, packs and every conceivable mode of conveyance zigzagging along a narrow trail (or path) over rocks, between and under rocks, over ice pole and brush, bridges, breaking through now and then and again killing a horse or dog here and there and a continual sound of mush, mush all the time."[14] Two of Campbell's party opted out early, concluding that the trail looked too tough for them.

By March 8, they had only three or four loads left in the lower camp. Campbell was a Scot who was frugal with words both in his diary and on the trail. "Man killed lying on his sleigh a little above Porcupine Hill. Looks like murder. Shot from behind," reads one entry. Then there was the inevitable clash of temperaments in this harsh environment. Campbell wrote of one of his party: "Evers thinks he is only man knows anything about setting up tents. Thinks I don't know anything about it at least."[15]

All along the snowpacked trail were the grisly carcasses of dead horses from the previous fall, as well as new ones that had dropped under the strain. A week later, the weather was warm, but they were still hauling to the summit. On March 16, three of the party went ahead to Log Cabin, a collection of huts over the pass, to prepare a camp. The rest put paraffin sacks over their goods to try to keep out the snow and wet and continued to pack loads toward the summit. Four days later, Campbell got an early start for the summit and made it to the camp at Log Cabin after what he called "a long tramp."

Then the weather changed. March 22 was the stormiest day yet—typical spring storms blowing snow as hard as nails across the pass. Campbell's party kept to camp, where, according to Campbell, "Doyle does nothing but figure how long the oats will last" and "Evers makes bread like grindstone." If it seemed a burden to buy hay and oats in Seattle and pack them this far, the economics of the rush showed the reason. Hay was selling for about $15 per bale, comparable a century later to a staggering $300 in 1998 dollars.

The next job was to fell trees and build a boat. Campbell's party spent more than two months in the process, and it too, like packing on the trail, could strain the nerves of the best of friends. Perhaps the worst team sport was whipsawing logs. The man in the pit was always choking with sawdust and certain that he was pulling harder than the man on top. Campell's diary tells a great deal in a few words:

Mon. Apr. 11, '98
Still cutting logs. Had good feed of beefsteak today. Getting to be an expert at making biscuit.

Tues. May 3, '98
Light from 3:00 A.M. to 9:00 P.M. Weather fine. Doyle exhibits his brains every day.

Sat. May 7, '98
Boiler expert tackles the thing [sawmill] today. No go. Doyle sprained his ankle. Pity it wasn't his neck.

Mon. May 9, '98
A Mrs. Howe buried at end of lake. 72 years old.

Mon. May 16, '98
Working on boat. Doyle comes around once in awhile. Tells us what a fellow told him in regard to building boats.

Thurs. May 26, '98
Working on steamer. Young man by name of Fred Whitcomb of Kerne, New Hampshire accidentally shot himself a short distance from us.

Fri. May 27, '98
Corking steamer today, at least the men are. Greendyke and I are building wheel. Whitcomb was buried today by the Masons.

Sun. May 29, '98
Took several pictures today. One, of the graves at head of lake. One, of the water front with all the boats lined on the shore. Also one of the saw mill. One of line of boats going down lakes for Bennett, belonging to Racine.

Wed. June 1, '98
Fire breaks out at the head of lake and sweeps through the woods like a race horse. We will have to move out whether we want to or not. Pulled our goods all out on beach and sleep for the night on the beach. Took picture of the finest cloud effect I ever saw in my life, caused by the fire.

Fri. June 3, '98
Loaded everything into boats today for an early morning start on unknown waters to the land of gold.[16]

Campbell and his party made it down the Yukon to Dawson and spent the following winter and much of the next spring picking away at the ground without much success. Doyle continued to be a rascal and then an outright thief. By the summer of 1899, Campbell had been away from home for over eighteen months. After cutting some timber and selling it to a local sawmill, Campbell earned $141, which he put toward the passage south. One of his last diary entries from Dawson notes, "A boat load of Anuiser Busch [*sic*] beer just arrived today and it tasted finer than cream."[17] Less than a month later, Campbell was back home in

Kalamazoo, Michigan. A cold beer, a thousand hardships, and maybe a million memories were all he had to show from the trail of '98.

For those who were not inclined to the rigors of Chilkoot or White Passes, there was an easier alternative. The Yukon River was the rich man's route, available to those who could afford the relative luxury of a steamship cruise—even if one might be expected to help cut wood for fuel once on the river. To be sure, it was the long way around. First, 3,000 miles from Seattle to St. Michael near the mouth of the Yukon and then 1,700 miles more up the river to Dawson on board one of a number of hastily constructed little steamers. In theory, the advantage was that one was able to ride all of the way. The disadvantage, of course, was the short navigation season on the river. Once the Yukon froze in the fall, the steamers were stuck until the ice went out sometime in the spring.

The fall of 1897 found a flurry of activity on the Yukon. Those who had struck it rich during the summer, or some who were just calling it quits, were desperately trying to escape downriver. Just as desperate were the crowds they met trying to get upriver and make it to Dawson. Only a handful who started up the Yukon in the fall of 1897 managed to reach Dawson before the freeze-up, and winter found hundreds of stampeders marooned along the length of the river. One group that included the mayor of Seattle ended up in a cluster of shacks that the disgruntled called Suckerville. The mayor had stocked up on goods, which he planned to sell in Dawson at sky-high Dawson prices. In Suckerville that winter, his companions forced him to sell to them—at Seattle prices. One stampeder who made it as far as Circle later wrote, "We were all monomaniacs on the subject of getting to Dawson." After a winter on the river, those who didn't turn back finally made it to Dawson in July 1898, long after the best claims had been staked.[18]

So the promise of the "easy" river route did not hold true, but what about the lure of an "all-American" route that promised to avoid Canadian customs officials? According to some hastily produced and utterly unreliable guidebooks, the solution to dealing with Canadian customs officials was to reach the Klondike through "American" Alaska. That seemed patriotic enough. Never mind that sooner or later one *still* had to enter Canada—that fact was conveniently overlooked by those promoting the route.

One version of the all-American route led north from Valdez on

Prince William Sound, across the Valdez Glacier, down the gorge of the Klutina River to the Copper River, up the Copper and over Mentasta Pass to the Tanana River, on to the headwaters of the Fortymile, and then down the Fortymile to the Yukon, before ascending the Yukon to Dawson. It looked like a fairly straight line when drawn on the crude maps of the guidebooks, but it cut across some of the toughest country in that half of Alaska. Needless to say, few made it, and most got no farther than the Valdez Glacier. Of some 3,000 men and women who landed at Valdez in 1897–98 and attempted to hack their way across the glacier, only 200 or so managed to reach the Copper River. Historian Pierre Berton estimated that only a dozen or so of these reached their goal of Dawson. It would be interesting to know their reaction when finally confronted by Canadian customs officials.[19]

But the other thousands were in even more trouble in Valdez. Port Valdez, the fine ice-free estuary off the Gulf of Alaska, was first mapped and named for a Spanish naval officer in 1790 by Don Salvador Fidalgo. But in 1898, the town of Valdez was nothing more than a cluster of tents, the final stop for the coastal steamers whose captains were cashing in on this so-called All-American route. For those who did not escape the place in 1898, the ensuing winter proved long and supplies meager. The scourge of scurvy was rampant. Most ended up being rescued by a U.S. Army detachment in the spring of 1899.

Remember a certain Lieutenant Abercrombie whose look up the Copper River in 1884 had him swearing never to return to this country? In 1898, William R. Abercrombie, now a captain (promotion in the peacetime army came slowly), landed in Valdez with instructions to explore once again routes to the Yukon in light of the Klondike discoveries. He personally made it as far as Mentasta Pass before descending the Copper and returning to Seattle. When he returned to Valdez with another detachment in the spring of 1899, it was with instructions to construct a military road north to Copper Center. But first he had to lend assistance to the starving stampeders. "My God, Captain," the quartermaster's agent greeted him, "it has been clear hell! I tell you the early days of Montana were not a marker to what I have gone through this winter!"[20]

To Abercrombie's credit, he was far more successful on this Alaskan assignment than he had been years before. After lending assistance, he directed the improvement of a trail out of Valdez, up Keystone Canyon, and over Thompson Pass—roughly the same route the Richardson Highway and the Trans-Alaska Pipeline now follow.

The only thing more ill conceived than the route north from Valdez was the route across the Malaspina Glacier near Yakutat Bay. Obviously, those who tried it had not heard of the exploits of Israel Cook Russell or the duke of the Abruzzi! It was worse than the worst. One party that tried this route was led by Arthur Dietz, who had advertised in the *New York Herald* for partners to form a mining company. From the scores who answered his ad, Deitz recruited seventeen others, including a doctor, two New York City policemen, a mineralogist, a tinsmith, and five clerks. The group met faithfully in New York each Sunday to plan the trip and bone up on Arctic conditions, no doubt strolling around Central Park for exercise or riding a bicycle down to the Battery. With such training, they were put ashore at Yakutat Bay in March 1898.

The Malaspina Glacier is larger than the state of Rhode Island, perhaps 100 times the size of Manhattan, an endless plain of ice, crevasses, and windswept ridges almost devoid of vegetation and substance. Off they went. For three months they struggled to cross the glacier. Dietz's brother-in-law, the physician, became the first to perish when his sled and four dogs disappeared into a crevasse. When winter came, who knew where they were? They tried to winter over in a makeshift cabin, but some went crazy and struck out for Dawson in the dead of winter, never to be seen again. Three others were buried in an avalanche. By spring, the gold fever was gone, and the survivors could think of only one thing: escape.

They thought they were on the headwaters of the Tanana, but somehow they found themselves once again confronted by the ugly Malaspina Glacier. The remaining seven recrossed it to Yakutat Bay, but only four of the original eighteen were still alive when the revenue cutter USS *Wolcott* found them. Dietz was one of them, and he managed to write a book about the ordeal. Most ironic, however, was the reporting of the *Seattle Times* when they were finally brought ashore there. The paper asserted that the four survivors had returned with half a million dollars in gold dust. Obviously, the quartet had nothing but broken bodies, and two were totally blind from the glare on the glacier. With journalism such as that, however, it was no wonder that the rush was on.[21]

There were, of course, all-Canadian routes as well. The Canadians and British promoted them under the guise of their own national pride, but in reality they were attempts to avoid the reciprocal American tariffs of Dyea or Skagway. Some of the publicity and brochures generated by Canadian

Boards of Trade for these routes were so alluring that many Americans ended up trying them, too. The Ashcroft Trail, sometimes called the Spectre Trail, began at the town of Ashcroft, northeast of Vancouver, and snaked its way through the Fraser River country and the Cariboo Mining District. Then it headed north to Telegraph Creek and eventually Teslin Lake and the headwaters of the Yukon.

Others had come this way before. In places there were faint remnants of the swath cut through the spruce forest in 1865 by the Canadian division of the Western Union Telegraph Expedition. At least 1,500 men, including American novelist Hamlin Garland, and some 3,000 horses tried the Ashcroft Trail during the summer of 1898, but it was a never-ending string of black bogs, rushing rivers, and murderous insects. One fellow hung himself from the crosstree of his tent and left behind a hastily scribbled note: "Bury me here, where I failed."[22] Once again, the enormous scale of the land had been overlooked. Only a handful reached their goal.

The merchants of Victoria and Vancouver, British Columbia, also touted the Stikine River Trail. It was advertised as "avoid[ing] the danger and hardships on the passes and the Whitehorse and other rapids." In reality, though, this route delivered one to roughly the same headwaters of the Yukon as did the White Pass and Chilkoot routes, but added at least 300 miles to the trip. The route led up the Stikine River from Wrangell to Telegraph Creek. Wrangell was still a rowdy outpost, and some of Soapy Smith's (see next chapter) henchmen had even managed to get a toehold there. In the early spring of 1898, it was possible to drag sleds up the frozen Stikine River as far as Glenora. But there the trail ended, and as the snow melted, the route on to Telegraph Creek became one interminable sea of deep mud. If one was lucky enough to make it to Telegraph Creek, the route north was essentially the northern part of the Ashcroft Trail, equally nonexistent.[23]

Finally, there were the Edmonton trails, a maze of routes leading north from Edmonton, Alberta, via the complex system of northwest Canada's rivers. The most amazing version ran north down the Athabasca and Mackenzie Rivers almost to the Arctic Ocean and then back south via the Porcupine or Peel River to the Yukon and finally Dawson. More direct routes, somehow piecing together the Peace, Liard, and Pelly Rivers, were almost trackless excursions across Canada's vast north. By the time the handful who survived these routes reached Dawson, it was the summer of 1899, two years after word of the initial discoveries reached the outside. The boom was over, and most sailed down the Yukon again for home without so much as an ounce of gold dust in their pokes. The hardships of these routes are wonderfully told in James Michener's short novel *Journey*, which was originally written to be part of his epic *Alaska*.

Thus, for Alaska the story of the Klondike is one of journeys over rugged geography by rugged individuals—men, women, and even children—who braved the heights and endured the hardships to follow the lure of gold. Most came away empty-handed; some did not come away at all. The trails of '98 resounded with an excitement that captivated a generation and, like a clarion summons, sent its sons and daughters north to Alaska.

Two Towns and a Railroad

In 1887, the Canadian government ordered William Ogilvie, a government surveyor, to locate the 141st meridian on the Yukon River, establishing the Alaskan-Canadian border there, and once and for all determining which flag flew over the mining camp of Fortymile. In Ogilvie's party was "Captain" William Moore, a spry sixty-five-year-old who had earned his sobriquet by sailing throughout the Inside Passage and running a steamer to the Cassiar gold fields. The area at the head of Lynn Canal had long intrigued Moore, and he was well aware of the reports of an ever-increasing number of prospectors using the Chilkoot Trail to reach the upper Yukon. But Moore had also heard rumors that there was another route from the coast to the interior.

While Ogilvie's main party headed north up the Taiya River valley bound for Chilkoot Pass, Moore and a Tlingit named Skookum Jim Mason (by no small coincidence, George Carmack's brother-in-law) paddled about five miles east from the Healy trading post at Dyea to the mouth of another river. Taiya Inlet was much deeper here than it was along the wide tidal flats near Dyea. The Tlingit called the area Skagua, meaning "home of the north wind." Moore and Mason made their way up the river valley, and in between fighting alder thickets, devils club foliage, and rocky cliffs, Moore enthusiastically made notes for the route of a wagon road. The two reached the crest of the Coast Range and descended northward to meet up with Ogilvie and the main party on the Chilkoot route at Lake Lindeman.

Moore convinced Ogilvie that he had found another viable route up from the coast. Whether it was easier than the Chilkoot would be vehemently debated a decade hence, but what was without doubt was that this new pass was a good 600 feet lower in elevation. Ogilvie incorporated Moore's information into his survey and named the pass after Canada's minister of the interior, Thomas White, who had authorized his expedition. In subsequent years, more than one traveler crossing its heights would be certain that White Pass had been named for the color of the snowy landscape and not for the largely forgotten Thomas.

Crotchety old coot that he was, Moore was also a big promoter. He quickly pronounced White Pass suitable not only as a wagon route but also as a railroad route to the headwaters of the Yukon. The captain had been saying for years that someday a big strike was going to be made up that way. So, with the patience of Job, Moore staked a 160-acre homestead at the mouth of the Skagway River and waited for the rush he was certain would come.

When it finally came in the fall of 1897, Moore suffered the indignation of being evicted from a portion of his homestead as hundreds of crazed gold seekers pitched tents all over his property. Sure, Moore had a valid homestead claim, but try telling that to these hordes on the trail of Eldorado. Besides, where could Moore turn for assistance? In the time required for any appeal to reach Sitka, half a town was built. Within a few short months, streets were laid out, substantial wooden buildings erected, and dock facilities constructed. By January 1898, Skagway had four long wharves extending into Taiya Inlet that bustled with the comings and goings of assorted ships.

E. B. Wishaar arrived in town on board one of them and found the trails north to be impassable. Unseasonably warm weather had melted the snow and ice into a quagmire of bottomless mud. Wishaar wrote in his diary for January 3, 1898, that "the weather continues abnormally warm, the thermometer hanging between 32 and 36°, with a regular thaw making pedestrianism very unpleasant. . . . Shooting scrap in town last night. Some lot jumpers attempted to jump Moore's 5 acres. No one hurt and Moore won the day."[24] Poor Captain Moore! He almost lost the last parcel he had.

A month later, Skagway's first newspaper, the *Daily Alaskan*, hit the streets. It soon reported on the glories of Skagway's new hotels, among them the St. James, Dewey, Fifth Avenue, Seattle, and Golden North. Originally constructed as the two-story Sylvester Building at Third and State Streets, the Golden North was later moved a block east to Third and Broadway and a third story added. Its recognizable golden dome remains the dominant building on Broadway.

Skagway's most famous citizen in those heady days of 1898 was Jefferson Randolph "Soapy" Smith. Like most dictators, he was both revered and feared. While his gang of henchmen duped newcomers, fleeced returning sourdoughs, and even resorted to murder out on the trails, Smith stood innocently and supported everything from stray dogs to destitute stampeders—the latter frequently made so by the work of his gang. Already run out of other towns on the mining frontier, notably

Denver and Creede, Colorado, Smith was destined to meet his end in Skagway. He got his nickname, "Soapy," from running a confidence game where twenty- or even hundred-dollar bills were wrapped around bars of soap. The "lucky" ones were then placed in a barrel with similarly wrapped bars. Five dollars a shot was all it took to try to win, but of course the only winners of the big bucks were Smith's hired shills.

Fact frequently blurs with fiction when recounting tales of Soapy's reign in Skagway, but one story is too good to pass up even if its sources are dubious. One of Smith's operations, it seems, was a telegraph station that purported to send messages to loved ones back home. No one seemed to notice that the telegraph lines ran a few miles out of town and then stopped. But that didn't stop Smith from receiving replies from the outside—always sent collect.

Finally, Smith's gang pushed the locals too far. When a miner who had just returned from the Yukon was robbed, duped, or otherwise conned—the versions of the story vary—out of some $3,000 in gold dust, a vigilante committee determined to confront Smith's iron rule. On July 4, 1898, Soapy grandly rode through town on a white horse at the head of the Fourth of July parade. Four days later, he was dead, shot down on Juneau Wharf in a shoot-out with the town surveyor, Frank Reid. Smith died instantly. Reid lingered for days before being given a hero's burial that the *Skagway News* said was to Skagway and Alaska what General Grant's had been to New York and the nation. Perhaps.

There were also reports, long buried by grateful adoration, that Reid himself was not free of taint. No matter; the legend has become fact with both men. Smith the villain and Reid the hero still lie in Skagway cemetery, while their characters come alive in summer melodramas. Soapy's grave marker is simple. Reid's is a tall stone spire that bears the inscription "He gave his life for the Honor of Skagway."

One who claimed to have witnessed Soapy's shoot-out was Harriett Pullen, a woman also destined to leave her mark on Skagway. Compared to earlier mining rushes, the Klondike stampede included a higher percentage of women. Some were wives, others were daintily captioned in one photograph as "entertainers," and still others were early-day feminists bound and determined to cast off the era's suffocating Victorian constraints.

Within days of the *Portland*'s landing in Seattle, Harriet and her husband left a bankrupt farm and four children in Washington State and arrived in Skagway on September 8, 1897, with the vanguard of the rush. Starting a restaurant in a tent, Harriet cooked meals, while her husband ran a pack string across White Pass. Harriet soon earned enough money to move the restaurant into a log building and send for her three sons. Somewhere along the line, husband and wife sold the packing business and split the sheets as well. Later, after her husband left town, she would tell guests that she was widowed.

Harriet briefly joined the gold rush to Atlin in nearby British Columbia but quickly decided that there was more money to be made by feeding the miners than by joining them in the creeks. Breads, cakes, and pies became Ma Pullen's specialties, not gold pans and sluice boxes. Her next project was a boardinghouse. She first rented and later purchased a large frame house that Captain Moore built in 1899. Harriet named it the Pullen House. With numerous additions and two additional buildings moved onto the property, the Pullen House became widely recognized among Alaska's finest hotels. Because Ma Pullen served a lavish fare, including fresh vegetables that she grew on land near Dyea and fresh milk from cows, many travelers called it the best in the territory.

Harriet Pullen always claimed that she witnessed Soapy Smith's demise. More likely it was a story that made for good telling to her guests. What does appear to be true, however, is that for years she received money from an anonymous out-of town source that paid her to take flowers to Soapy's grave. When she finally died in 1947 at the age of eighty-seven, she was an Alaskan icon and the grande dame of Skagway.[25]

Meanwhile, there was another town in the running for the title of "Gateway to the Klondike." Just as the boom hit Captain Moore's quiet homestead in the fall of 1897, so too did the tempo increase dramatically around the Healy trading post at Dyea. Within days of the arrival of the first stampeders in late July 1897, a townsite was laid out near the trading post and the neighboring Chilkoot village. Dyea got its name from a mispronunciation of the Taiya River. In the beginning, Dyea was better known than Skagway, and the trail over Chilkoot Pass was far better established than the suggestion of a "trail" over White Pass. Tlingit came from as far away as Sitka to join local Chilkoot Tlingit in packing supplies across the pass for the stampeders. Packers included Tlingit women and small girls, and wages ran between ten and twenty dollars per day. Good money then.

The established route over the Chilkoot and this ready source of local labor should have boded well for the future of Dyea, but the town had one major problem. It was located on the tidal flats at the mouth of the Taiya River. Building construction was easy, but lacking Skagway's deep port, Dyea had to contend with the constant fluctuations of the tides over miles of beach. Stampeder Wilson Mizner described the chaos of landing in Dyea in December 1897 when men and their outfits were "almost pushed off into waiting scows, rafts, and rowboats." Mizner reported: "With people wrangling and fighting over freight, with confusion, great avalanches booming down the mountain sides all about us, and absolutely no one able to give us anything but abuse, my first view of Dyea was accompanied by one long and two thousand short blasts of profanity."[26]

A large tidal range of twenty feet or greater over a flat, sandy tidal zone meant that there was more than a mile of beach exposed between high and low tides. "Hurry, the tide is turning" became the watchword of the beach. Goods were tossed into anything that would float, and then the steamers backed off for deeper water. The smaller barges, scows, and assorted watercraft were soon left high and dry. Wagons came out to them to unload their goods and transfer them to higher ground. Some, like William E. Crain of Illinois, bristled at the arrangements. On March 12, Crain wrote in his diary: "A.M. and we are reaching our jumping off place, Dyea . . . and we are stranded one mile from Dyea (on a lighter) on a sand bar, and we have either to pay enormous rates else stay here until high tide." A day later, safely ensconced at the Palace Hotel, Crain reported that a man had been shot and killed the night before, but that all was quiet otherwise. "This place last fall had one cabin [but] now is a thriving city of 150 or more frame buildings."[27]

The solution to the mire of the tidal flats was to build a lengthy wharf, but it would have to be a long one. Deep water was 1.2 miles south of the high-tide line. The Skaguay Wharf and Improvement Company claimed a site on the western side of Taiya Inlet, but it required that a rocky wagon road be built along some cliffs. The Dyea-Klondike Transportation Company also staked out a site due south of town. The company opened a short dock in February 1898 and then three months later completed the Long Wharf, a narrow strand of wooden planks plunked down on pilings that marched an amazing 7,500 feet from downtown to deep water. Despite these efforts, many ship captains opted to land at Skagway instead.

But Dyea was booming nonetheless. Over 150 businesses soon offered services, including hotels, restaurants, bathhouses, attorneys, a dentist, an eye doctor, seamstresses, meat markets, a candy factory, two newspapers, and even a reading room. There were also about forty saloons, variety theaters, and bordellos. Ed Lung passed through town in May 1898 and reported that there was plenty of entertainment to be had with dance halls and saloons going full blast. Lung had planned on getting up early the following morning to start up the Chilkoot, but he did not get a very restful sleep because of all of the business one gravel-voiced woman was getting beyond the thin partition. From a population of about 1,000 late in the summer of 1897, Dyea grew to 8,000 by May 1898. Meanwhile, by some estimates, 30,000 to 40,000 may have passed through the town that first year. But who was counting as long as the boom lasted![28]

Bailey's Hotel, the Glacier Hotel, and the Pacific Hotel did a brisk business. The Pacific's manager, J. J. Bell late of San Francisco, boasted that his establishment's cuisine was the best in Alaska, bar none. The first issue of the *Dyea Trail* quoted Bell as asserting that "all the essentials and delicacies of the season will be served, and the house will be a credit to Alaska's greatest city."[29] Supposedly, the Pacific could accommodate 300 guests—obviously, private rooms were not an option. But even the Pacific could not rival the three-story Olympic, built at Third and River Streets—conveniently close to the river, it turned out. The Olympic was Dyea's finest hotel, with 112 rooms each renting for three dollars per night.

Dyea even experienced its own little sideshow gold rush. Fed by rumors that gold was where you found it, a number of prospectors staked claims on the hillsides around town during the winter of 1897–98. Whether this local gold fever was merely a winter diversion while storms blanketed Chilkoot Pass, or hype stoked by local merchants to keep people in town, is debatable. No gold was found, however, and soon the throng was moving on.

Not surprisingly, a battle of community boosterism, the likes of which was seen in few places on the mining frontier, erupted between Skagway and Dyea over which town would control the lion's share of trade bound for the Klondike. For a time, Dyea seemed to be winning, thanks more to the tramways over Chilkoot Pass and Skagway's reputation for lawlessness than Dyea's second-class port facilities. But Skagway grew calmer after Soapy Smith's passing, and soon it had another ace up its sleeve. Skagway was about to become a railroad town.

Captain Moore had long boasted that a railroad could be built across the White Pass, but it was British dollars and Irish determination that made it happen. The London banking firm of Close Brothers & Company supplied the capital. Close Brothers bought out the franchise of another company and incorporated the White Pass and Yukon Railway Company. In Alaska, this British-owned company operated legally as the Pacific and Arctic Railway and Navigation Company—a mouthful that never had to be said except in a room full of lawyers. Samuel H. Graves, Close Brothers' main representative in North America, became president of the White Pass and Yukon Railway and turned the ground operations over to two civil engineers, Erastus Hawkins and his assistant, John Hislop.

One night in April 1898, Hawkins and Hislop sat sipping scotch in Skagway's St. James Hotel with Sir Thomas Tancred, a civil engineer of international reputation with whom Close Brothers had contracted to conduct a feasibility survey of the White Pass route. Tancred was skeptical. The 2,900-foot elevation gain from sea level to the summit in such a short distance was daunting, and the route would require major blasting and bridging. Tancred hadn't even been able to travel the route from the summit down to Bennett given the winter weather, and who knew what problems really lay between there and the Yukon for a railroad? Nevertheless, Hawkins pressed Tancred to give the project his endorsement. Sir Thomas refused.

At an impasse, the three men downed the last of their scotch, flicked the ashes off their cigars, and approached the desk for the keys to their rooms. As they did so, a bone-weary traveler splattered with mud trudged across the hotel lobby in front of them and asked for a room. "Who was that?" Hawkins asked the desk clerk, as he picked up his key. "Name's Heney," was the reply, "been on a scouting trip for a railway."

Hawkins's eyes widened. He asked the desk clerk to invite the newcomer to join his group for a nightcap. The clock was approaching midnight, and Heney had left Bennett at three that morning and made the long journey over White Pass, but he was always ready to talk about railroads. Over more scotch and cigars, the three engineers listened to Heney recount the terrain between Bennett and the summit and how he proposed to solve the problems of the grade between there and Skagway. It would be tough work, Heney assured them, and it would take considerable capital, but quite matter-of-factly Heney told them that he would like to be the man to build it—not to *try* it but to *build* it. Tancred and Hawkins exchanged glances. Sir Thomas gave an almost imperceptible nod. The project had his approval. Close Brothers would supply the capital, and Michael J. Heney would supply the determination.

Born in Ontario, Canada, to Irish immigrants, Heney got his first taste of railroading when he ran away from home in 1879 at the age of fourteen to work on a construction crew grading the line of the Canadian Pacific Railroad. His older brother soon arrived to retrieve him, but from then on railroading was in his blood. Four years later, Heney left home for good and headed for the Canadian Pacific's railhead in Manitoba. He started out leveling roadbed with a mule and a scraper; his hard work soon had him laying rail and then working on a survey crew in British Columbia's Selkirk Mountains. By the time this Irishman arrived in Skagway in the spring of 1898, he had acquired extensive mountain railroading experience and a knack for making a crew of unruly men do the impossible for him. Affectionately, they called him "the Irish Prince" or simply "M. J." [30]

So Heney went to work. The Canadian Pacific was a standard-gauge line, built with four feet, eight and one-half inches between the rails. This measurement had been adopted in North America during the Civil War when the need to move equipment over many different lines proved the folly of each railroad having its own gauge rather than a standard width between the rails. Many mining railroads throughout the mountains of the American West, however, adopted a three-foot "narrow gauge" because of its utility in climbing steeper grades, turning tighter curves, and generally costing less to construct than standard gauge. Clearly, the geography of the White Pass dictated that the White Pass and Yukon Railway be a narrow-gauge line.

While Heney arranged for men and equipment to be shipped north from Seattle, a roadbed was graded right down the middle of Broadway in downtown Skagway. Work proceeded quickly up the lower reaches of the Skagway River. By coincidence, the first locomotive left Skagway's wharf and went steaming up Broadway pulling two flatcars loaded with ties and iron on July 20, 1898, the same day that Frank Reid finally died of his wounds. One era ended; another began. During one frantic three-day period that month, over one mile of rail was laid, and by late July, there were seven miles in place.

Then it was past the many braids of Pitchfork Falls and on to White Pass City, ten miles north of Skagway. Here, both the river and the railroad grade made a sharp turn to the east. White Pass City and what the railroad called Heney Station became the temporary railhead. After some convoluted negotiations, the railroad finally purchased George Brackett's toll road that led north from Skagway and over the pass. Very little of the toll road was actually used for right-of-way, but as the grade was blasted out of the cliffs high above the North Fork, debris frequently obliterated large portions of the wagon road. Brackett was going to get his money one way or the other.

Despite Heney's rapport with his crews, it was difficult to keep men on the job digging rock when the lure of digging gold was so strong. Not only was the Klondike calling, but reports of the Atlin gold strike and the hardships of tough construction work constantly cut into the workforce. Many workers signed on with the railroad in Seattle, got free passage to Skagway, worked for a week or two, and then slipped away to the gold fields. By one estimate, more than 35,000 men worked on the White Pass and Yukon during the two-year course of its construction between Skagway and Whitehorse. Out of this number, no more than 2,000 were ever in the field at one time.[31]

Under Heney's determined drive, however, they got the job done. On February 20, 1899, under clear skies, but with the temperature a bitterly cold minus-four degrees Fahrenheit, the rails reached the White Pass summit. Heney jumped crews ahead to work on the grade between Bennett and Log Cabin, while the main force continued north from the summit. When the first train rolled into Bennett on July 6, 1899, the town was convinced that it was here to stay.

Bennett, of course, was already booming. Since the first flood of stampeders emptied off the Chilkoot in the fall of 1897, the shores of Lakes Lindeman and Bennett had resounded with the whir of sawmills and the banging of hammers as men built a fleet of hundreds of barges, scows, and assorted craft. By one Northwest Mounted Police count, in May 1898 there were 778 boats under construction at Lake Lindeman, 850 at Lake Bennett, and 198 at Caribou Crossing and Tagish. Inspector Samuel Steele estimated that at least 1,200 more boats were built in the following three weeks.[32]

That spring, the Upper Yukon Company packed two fifty-foot steamers over White Pass in pieces while the ground was still frozen and then assembled them on the shores of Lake Bennett. They were then launched in May when the ice went out. The Bennett Lake and Klondike Navigation Company did likewise and ran a regularly scheduled steamer from the mouth of One Mile River connecting Lakes Lindeman and Bennett to just above Miles Canyon. On the other end of the spectrum were the single-occupant craft like the one Harley Tuck saw constructed. "Our neighbor who is going alone," wrote Tuck, "built a boat that was more like a floating coffin." When the fellow loaded his stuff and pushed the vessel out to see how it would float, it turned bottom up and spilled his goods into about four feet of water.[33]

So Bennett boomed, boasting restaurants, merchants, banks, a drugstore, bakery, Northwest Mounted Police post, and at least six hotels, including the New Arctic Hotel, which stood near the railroad station. Of course, there was also the Presbyterian church, destined to be the town's lone survivor. With the arrival of the White Pass and Yukon, there were an estimated 2,000 people who remained in town, even as the transient hordes continued northward. But the railroad, too, was soon moving northward. Bennett enjoyed the boom of being a railhead for about a year. Just as he had done coming off White Pass, Heney threw crews ahead to Whitehorse and had them grade southward, while the main force moved along Lake Bennett. The two met at Caribou Crossing, later to be called Carcross, and trains began running to Whitehorse on July 29, 1900, after twenty-six months of construction. The project cost $10 million and would never have been completed without Mike Heney's iron will.

With steamboat navigation on the Yukon between Whitehorse and Dawson, the railroad was never extended any farther, but it quickly left its mark on the Chilkoot and White Pass routes. The tramways shut down over the Chilkoot within days of the railroad's arrival in Bennett. Sheep Camp was deserted. By the end of the summer of 1900, Bennett itself was down to fewer than 100 people. Left behind were slopes devoid of timber from the frenzied boat building. The White Pass and Yukon Railway managed to survive the rush to the Klondike, and though operations were sporadic at times—even shutting down for a period in the 1980s—a new gold rush of tourists keeps it operating during the summer more than a century after its construction.

Like most mountain railroads, the White Pass and Yukon Railway endured the inevitable play on its initials. Locals said they stood for "Wait Patiently and You'll Ride," but that's only part of the story. On the broader scale, British capital and Michael J. Heney's Irish determination teamed up to overcome bottomless mud flats, raging rivers, and towering cliffs to build one of the most fabled of the mountain railroads that responded to the economic pull of the mining frontier.

The completion of the White Pass and Yukon Railway out of Skagway sealed Dyea's fate and that of the Chilkoot Pass route. Dyea died as quickly as it had boomed. Characteristic of the mining frontier, it was an overnight boomtown that lasted as a viable community for less than two years. Most of its buildings were torn down, and the lumber was used elsewhere. The short-lived glory of the Olympic Hotel survived one year longer than most of the town, but it too closed in 1900. The next year, C. W. Young, a Juneau businessman, cut the building into sections and barged them south, to be reassembled as the Hotel Northern in downtown Douglas, standing there until it was destroyed by fire in 1911.

Today some scattered foundation ruins, a few rotting stubs of the piers from the Long Wharf, and the Slide Cemetery are all that remain of Dyea. In the end, the White Pass and Yukon Railway proved that Skagway was the town to bet on in the future. Even after the Klondike boom passed, the railroad and the town's deepwater port continued to make Skagway a gateway to interior Alaska. Thanks to the railroad, Skagway boomed, while Dyea went bust.

One Man's Summer Vacation

In the summer of 1899 as the nineteenth century drew to a close, one of America's up-and-coming powers, soon-to-be rail baron E. H. Harriman, took a two-month vacation. He went to Alaska. His wife, Mary, and their five children, including seven-year-old Averell, went with him. There, the similarities with other family vacations ended. Also along were 119 others: three relatives and a friend; a retinue of Harriman staff and servants, including two stenographers, a doctor, a nurse, and a chaplain; 65 officers and crew of the chartered steamer *George W. Elder*; 5 budding artists and photographers; and 25 of the era's most distinguished geologists, botanists, ornithologists, and ethnologists. It was, to paraphrase John F. Kennedy, the most distinguished gathering of scientists, writers, and naturalists since Thomas Jefferson dined alone. And if a single event can be called the crucible of Alaskan conservation efforts, the Harriman Alaska Expedition is it.

Harriman's doctor had told him to take a break. His rescue of the Union Pacific was in full swing. His acquisition of the Southern Pacific and his fight with Jim Hill for the Northern Pacific were two years in the future. Harriman may have gone purely for the diversion. He may have underwritten the venture as a grand philanthropic gesture to East Coast society, which still saw him as somewhat of an upstart outsider. Some later speculated—without much basis in fact—that the coterie of scientists was mere window dressing to obscure his real motive: a quiet investigation of the feasibility of a railroad between New York and Paris via a tunnel under the Bering Strait. Given Harriman's nature, it was probably a combination of motives, including one purely personal one. The Wall Street financier was determined to bring home an Alaskan brown bear. The summer's itinerary was left quite open, except that a definite hunting stop was scheduled at Kodiak Island.

The *George W. Elder* was not your average yacht. Originally built in 1877 for the Oregon Steamship Company, the ship first saw service as a regular on the run between San Francisco and Portland. She was 250 feet long, had a beam of 38.5 feet, and was registered at 1,709 tons. Two slender masts towered fore and aft of a large smokestack that sat squarely amidships. Her coal-fired boiler and large three-bladed propellers were capable of making twelve knots. The iron steamship was plying the waters of the Inside Passage as a mail streamer before Harriman ordered it refitted for his cruise. When the *Elder*'s makeover was complete, the vessel had become a luxury liner, furnished with staterooms, scientific labs, and a 500-volume library boasting almost everything that had ever been written about Alaska. In her holds, there were livestock pens crowded with cows, pigs, and chickens—just in case the hunters and fishermen on board were not successful.

Harriman left the invitations and details of organizing the scientific participants to C. Hart Merriam, then chief of the U.S. Biological Survey. The inner circles of exploration and science were particularly highbrow in the later decades of the nineteenth century, and Merriam recruited heavily from among his fellow members of the venerable Cosmos Club of Washington, D.C., and the National Academy of Sciences. The Harriman party included ornithologist and author John Burroughs, glaciologist Grove Karl Gilbert, artists R. Swain Gifford, Frederick S. Dellenbaugh, and Louis Agassiz Fuertes, and photographer Edward S. Curtis. William H. Dall was along as sort of the "grand old man" of Alaska, as was John Muir, by now well established as America's foremost wilderness sage. Henry Gannett, chief geographer of the U.S. Geological Survey, joined the group and would pen several of the expedition's most memorable lines.

But the expedition's most influential writer was probably George Bird Grinnell. He was an ethnologist, author of books about Native Americans, and publisher of *Forest and Stream*, one of the most widely read outdoor magazines of the day. As early as an 1887 editorial, Grinnell had proposed an organization to curb the excesses of sport hunting and promote renewable wildlife and natural resources management. Shortly thereafter, Theodore Roosevelt—then known largely to *Forest and Stream* readers only as the author of *Hunting Trips of a Ranchman*—hosted a dinner party for a group of gentlemen sportsmen. Out of this gathering grew the Boone and Crockett Club, America's first national conservation organization. (Five years later, John Muir founded the Sierra Club.) Doubtless TR himself would have delighted in the Harriman adventure, but by 1899 he was up to his eyebrows in work in his first year as governor of New York.

After a trip by special train across country from New York to Seattle, the assembled party left Seattle on board the *Elder* on May 31, 1899. As the ship steamed north, the scientific luminaries engaged in impromptu discussions on a host of topics and took turns giving formal nightly lectures. To be sure, there was a collegiate atmosphere on board. The Harriman Alaska Expedition even had its own cheer: "Who are we? Who are we? We are, we are, H.A.E.!"[34] It was the ultimate semester at sea.

The *Elder*'s first stops were at Annette Island, where the Reverend William Duncan had established a reservation for Tsimshian Native Americans from British Columbia in 1887, and at Wrangell, still a relatively rough-and-tumble outpost. Then it was on to Juneau with a tour of the great Treadwell Mine and Skagway with a ride on the newly constructed White Pass and Yukon Railway. The railroad was complete to Lake Bennett, but officially operating only between Skagway and the White Pass summit. Harriman paid close attention to Mike Heney's handiwork and got so engrossed in the area that the train departing the summit for Skagway left without him—a mistake that was quickly discovered and rectified. It simply would not do to run off without the man who was paying the bills!

From Skagway, the *Elder* called briefly at Juneau to retrieve a scientific party and then headed for John Muir's old stomping grounds at Glacier Bay. The party spent five days there, the longest stop of any on the trip. Muir held court on the glacial retreats that had occurred since his first visits and no doubt delighted in telling the Harriman children the story of the faithful Stickeen. From the bay, it was south down Peril Strait to Sitka, where the party was feted by territorial governor John Brady and given a tour of Sheldon Jackson's budding museum. Farther up the coast at Yakutat Bay, they visited a Tlingit sealing camp, and Harriman intrigued local Tlingit by playing a phonograph recording he had made of Tlingit chants at Sitka only a few days earlier. Obviously, he had all of the latest "toys."

The *Elder* sailed west from Yakutat and put in to the harbor at Orca (soon to be overshadowed by Cordova) before steaming amid the mixed puzzle pieces of Prince William Sound. Scientists and laymen alike marveled at the huge glacier west of Port Valdez that rivaled anything in Muir's Glacier Bay. Collectively, they gave it the name Columbia, after Columbia University. Farther west, the *Elder* poked its bow into two largely unexplored fjords. In 1794, Vancouver gave the name Port Wells to the wide northwestern arm of Prince William Sound. Just the year before the *Elder*'s arrival, army captain Edwin F. Glenn camped there with orders to probe the towering Chugach Mountains for routes to the

Copper and Susitna Rivers. Glenn found none, and the area became the one location of the *Elder*'s entire cruise where expedition members liberally sprinkled place-names.

At the head of the main northerly extension of Port Wells, the expedition continued the university tradition and named twin glaciers after Harvard and Yale. To four large glaciers draping the western side of the fjord, they gave the names Smith, Bryn Mawr, Vassar, and Wellesley, and then decided that the entire location should be called College Fiord. A decade later, geologist U. S. Grant (no relation to the famous general and president) added the names of the remaining Seven Sisters, Baltimore, Barnard, and Holyoke, to three smaller glaciers also along the western side of College Fiord.

The westernmost of the two fjords was a surprise. It was not marked on the charts, but the *Elder* gingerly picked her way past the Barry Glacier at its mouth and steamed into a large fjord rimmed by glaciers and snow-clad peaks. The fjord and the massive glacier at its head were named after their host, but around Harriman Fiord, the group also bestowed a host of descriptive names, including the Cascade, Serpentine, Stairway, Surprise, and Toboggan Glaciers. Later, U. S. Grant and his associates also added names here to commemorate the expedition: Mounts Gilbert, Curtis, Muir, Gannett, and Emerson for the scientists and Mount Doran for Captain Peter Doran of the *Elder*.

Now it was time to keep the appointment on Kodiak Island. They put in at Kodiak, and while Mrs. Harriman organized picnic excursions and rehearsed the upcoming Fourth of July program, Mr. Harriman and Merriam, backed up by a properly armed retinue, went tramping across Kodiak Island. The Wall Street baron got his forest bruin, and Merriam was delighted to confirm that it was indeed a Kodiak brown bear. The fact that it was on the small size, and a younger female with a small cub at that, was downplayed, and only John Muir winced at the sight of the two skins being brought on board the *Elder*.

Supposedly, it was Mrs. Harriman who insisted on seeing the Bering Strait. Perhaps, but the *Elder* would not have moved one inch in that direction if EHH had not approved. Maybe he was thinking about a rail tunnel under its waters after all. Captain Doran and local pilot Omar J. Humphrey steered the ship to Dutch Harbor and then the Pribilofs with only minor mishap, it being not at all uncommon in those days for a coastal steamer to run aground or hit a rock or two here and there. In the Pribilofs, Merriam gave a well-received lecture on seal rookeries before the steamer turned toward Siberia. After a stop on the Siberian coast, Mrs. Harriman got her wish, and the *Elder* steamed east across the Bering Strait to Port Clarence, an Inupiat settlement just south of Cape Prince of Wales. This was the northernmost point visited, and on July 13, the *Elder* turned south.

Given the methods of the day, it had been a whirlwind trip. How much relaxing Harriman had done is open to question. As with everything he did, Harriman had participated to the fullest in all of the shipboard activities and shore explorations, including piloting small boats through ice-choked bays and raging surf. By the time the ship was again back in southeast Alaska, it was the looming railroad wars that again weighed heavily on his mind. Supposedly, even Harriman soon growled, "I don't give a damn if I never see any more scenery."[35]

But there was one more major stop to be made. On the trip north, Harriman had heard rumors of an abandoned Tlingit village south of Wrangell. If time permitted, he intended to stop. It did and the *Elder* dropped anchor off the Tlingit village of Cape Fox southeast of Ketchikan. Cape Fox was not inhabited, its residents having moved north to the village of Saxman closer to Ketchikan. Nineteen totem poles, burial grounds, and houses containing belongings remained. Whether these were *abandoned* in the Anglo-Saxon sense is debatable. Nonetheless, the expedition members swarmed ashore and collected a number of artifacts, including several totems. To the ethnologists' credit, these artifacts ended up at the California Academy of Science, Chicago's Field Museum, and the University of Michigan, but it would be some decades before this collecting mentality was replaced by the preservation mentality. Muir, who had already looked the other way at some of his companions' actions, was horrified.

The *George W. Elder* dropped anchor in Seattle on July 30. Nineteenth-century anthropological collecting methods aside, the trip was a remarkable documentation of Alaska on the verge of a new frontier. C. Hart Merriam was charged with organizing the expedition's papers and photographs. They appeared in eleven volumes over the next fifteen years and became the most complete collection of Alaskan data since William H. Dall's *Alaska and Its Resources* was published in 1870. Two other volumes, with descriptions of Alaska's mammals to be written by Merriam himself, never appeared, perhaps because Merriam became too bogged down in his editorial duties to write them.

E. H. Harriman immediately again immersed himself in railroading, but the camaraderie and collaboration of the scientific members of his expedition would have a decided impact on the conservation consciousness of Alaska and the nation. Edward Curtis would go on to become a world-renowned photographer, Gilbert a foremost authority on glaciation. Grinnell, Merriam, Burroughs, Muir, and others would all work together in the conservation battles of the next three decades.

If there is a fitting valedictory for the Harriman Alaska Expedition, perhaps it was the words penned by Henry Gannett in his section of the report on the general geography: "There is one word of advice and caution to be given to those intending to visit Alaska for pleasure, for sightseeing. If you are old, go by all means; but if you are young, wait. The scenery of Alaska is much grander than anything else of the kind in the world, and it is not well to dull one's capacity for enjoyment by seeing the finest first."[36]

Then again, perhaps the words were more fittingly a prelude, because the Harriman Alaska Expedition highlighted the dichotomy that would grip Alaska throughout the next century: Was it a land of rare beauty to be preserved as wilderness at all costs? Or was it truly the "last frontier" to be tamed with roads, railroads, airfields, pipelines, and more. A century after E. H. Harriman's summer vacation, the questions still remain.

And what of the venerable *George W. Elder*? Released from the Harriman charter, the *Elder* returned to regular service and then hit a rock in the Columbia River in 1905 and sank. It was raised a year later and repaired, but its final resting place is unknown. It was last reported bound for Puget Sound from South America in 1918 with a load of fertilizer nitrates on board.

Last Stops of the Mining Frontier

In all the tumult and excitement of the rush to the Klondike, it is easy to forget that it was but one chapter in a much broader saga that draped the mining frontier across the entire length and breadth of Alaska. Many of the dreamers who converged on the Klondike arrived too late to stake claims. Those who didn't turn around and rush back outside were eager to follow the rumors of other strikes. Other stampeders hadn't reached Dawson before they were diverted elsewhere by even grander promises. Then, too, there were those who struck out for Alaska with no intent whatsoever of reaching the Klondike. They headed for other locations where a prospector or two had once turned a few shovels of gravel and caught sight of a golden glimmer. One such place was the Seward Peninsula.

The Seward Peninsula is that long thumb of land jutting westward from Alaska toward the Bering Strait and Siberia—the remaining high ground of the massive land bridge connecting Asia and North America. To the north are Kotzebue Sound and the deltas of the Kobuk and Noatak Rivers. To the south are Norton Sound and the delta of the Yukon. Those who thought that the Klondike was the end of the world should have seen the Seward Peninsula. The Klondike had forests of trees for fuel and buildings. Sure, it could get cold in Dawson, but it was generally a dry cold, unlike the bone-numbing stabs of the icy daggers that blow off the Chukchi and Bering Seas. The Seward Peninsula had, well, not much of anything—except maybe gold.

There was one person who knew for certain that there was gold on the Seward Peninsula. His name was Daniel B. Libby, and he knew because he had found traces of it there while surveying for the Western Union Telegraph Expedition way back in 1866. When news of the rich strike in the Klondike hit the West Coast in the summer of 1897, the proverbial lightbulb seems to have turned on in Libby's brain. Persuading four wealthy San Francisco businessmen to back his venture, Libby chartered a vessel, filled it with tools and supplies to last four years, and sailed north from the Golden Gate on August 18, 1897. With him went A. P.

Mordaunt, L. F. Melsing, and H. L. Blake. A month later, the *North Fork* landed the quartet in Golovin Bay on Norton Sound, about sixty miles east of Cape Nome. Golovin Bay was the site of a Native mission operated by the Swedish Evangelical Mission Covenant of America. Cape Nome had been noted on Russian charts as Mys Sredniy (Cape Middle) but seems to have gotten its English name when a draftsman in the British Admiralty took the query "? name" on a draft chart to mean "C. Nome" or Cape Nome. Strange, but apparently true.

Libby, Mordaunt, Melsing, and Blake headed north up the Fish and Niukluk Rivers. Throughout the fall they prospected from a camp that came to be called Council City and found color in a number of creeks flowing into the Niukluk. Around Christmas, an Inupiat named Too rig Luck showed Blake ore samples that reportedly came from farther west near Cape Nome. Guided by Too rig Luck, Blake and Nels O. Hultberg, the Swedish missionary at the Golovin Bay mission, set out to have a look. Traveling with sleds pulled by reindeer, they made the trip in the dead of winter with only about four hours of daylight each day. Breaking through the ice, Blake and Hultberg panned gold from a frozen creek somewhere north of Cape Nome. Whether this was Anvil Creek or some other tributary of the Snake River is in dispute. Blake and Hultberg could never agree exactly where they had been, and this disagreement was only a precursor to the infighting and complicated legal maneuvers that were about to characterize the entire Nome rush.

Hultberg appears to have been more fired by the prospects of finding gold than saving souls, because over the course of the winter, he made two additional prospecting trips. One was in February east toward Norton Bay in the company of Daniel Libby, and the other was in March up the headwaters of the Niukluk. Then in April, Hultberg arrived back in Council City with three other Swedes. With Libby and his partners, the Swedes proceeded to stake claims along the neighboring creeks and to organize a mining district.

By the time the ice went out on Norton Sound that June, a steady trickle of men began arriving in Council City from St. Michael. Council City was, of course, nothing more than a handful of tents and crude shacks, but its proximity to St. Michael gave it a boost. St. Michael was—as it had been for decades—the gateway to the Yukon. Would-be tycoons heading upriver or dejected turnarounds just come down changed between river steamers and ocean vessels at St. Michael. The Council City diggings were within 100 miles of St. Michael and offered a ready alternative to the 1,700-mile ride upriver or returning home empty-handed. Some made the trip while waiting for ships.

Meanwhile, Hultberg and Blake had been planning to return to the streams north of Cape Nome. Accounts differ as to who invited whom and why, but in the end two Swedes, John Brynteson and John L. Hagelin, and two Americans, Chris Kimber and Henry L. Porter, joined the trip. The six men arrived at the Snake River west of Cape Nome on August 4, 1898. Working their way upstream along Anvil Creek, they found some color, but nothing to cause a stir. Hultberg then went farther upstream alone. A couple of hours later, Blake joined him and asked, "Well, what luck?" "None," replied the Swede somewhat nervously, "let's head back."

The group soon parted and went their separate ways. Later that month, however, Hultberg returned to Council City and quietly urged Brynteson to take Eric O. Lindblom and Jafet Lindeberg and return to Anvil Creek. Significantly, they went armed with both prospecting tools and numerous powers of attorney. On September 22, 1898, the three staked a discovery claim along Anvil Creek and then proceeded to stake adjacent claims up and down the creek. As Blake later grumbled, these three "lucky Swedes" took the power of attorney of every Swedish missionary in the vicinity and located the whole country.[37]

Word of the discoveries on Anvil Creek reached Council City in mid-November; the place emptied almost overnight. From all over the Seward Peninsula, St. Michael, Unalakleet, and points along the Yukon, men descended on Cape Nome. By the spring of 1899, many passengers disembarking at St. Michael bound for the Klondike changed their destination to Nome instead. What they quickly found was that not only had the best claims already been staked—that was perhaps understandable—but also that almost every conceivable claim had been staked. How could this be?

The Swedes may have been lucky, but they were also shrewd, if not downright devious. Organizing the Cape Nome Mining District, they made certain that its rules permitted an individual to file multiple claims and to file through powers of attorney. By the time the lucky Swedes were done, they had filed forty-three claims in their own names and forty-seven others through powers of attorney.[38] Normally in the West, their fellow miners would have given short shrift to such nonsense, but that's where the devious part came in. Among the many claims the Swedes filed were those in the names L. B. Shephard, the U.S. commissioner at St. Michael, and Captain E. S. Walker, in command of the troops stationed there. When newcomers, as well as old-timers like Blake who had been left out of the staking, attempted to challenge matters legally, Commissioner Shephard upheld the local rules. When vigilantes threatened stronger action, Captain Walker was quick to intervene with his troops.

The Swedes were not alone in their overreaching. Both the Alaska Commercial Company and John J. Healy's North American Transportation and Trading Company were accused of hiring Laplanders working in the area with Sheldon Jackson's reindeer herds to stake claims at the rate of two dollars a claim. Once again, this disregarded one of the well-established tenets of mining law: that a claim must be proven up before it can be recorded. All of this meant that the gold fields of Nome became some of the most heavily litigated in the history of the mining frontier. Litigation in one case that involved whether a claim had been staked under a power of attorney or for the benefit of two Inupiat boys lasted twenty years, involved eleven different courts, and went to the U.S. Supreme Court four times.[39] Reading it even today is enough to give third-year law students hives.

What saved the Nome district from total anarchy was the discovery in the summer of 1899 that there was gold on the beaches. Not gold in the rivers and creeks that had to be sluiced, or gold held tightly by the permafrost that had to be chipped and thawed, or gold entwined in ore that had to be milled, but free gold in the black sandy beaches—gold for the taking. All one needed was a shovel and something with which to sift. One might just as well be picking up coins. Placer gold like that was just as good as currency and could be taken right from the ground to the bank. No millionaires came off of Nome's beaches, but the sands there provided a needed safety valve and left thousands able to say, "Yup, I was in on the rush at Nome."

George Edward Adams, a correspondent for *Harper's Weekly*, didn't put it quite that succinctly. Adams wrote that "two years ago, if any one had even suggested digging for gold on the ocean beach, he would immediately have been classed as a fit subject for an insane asylum, but within four months after the discovery of gold in paying quantity at Cape Nome, the sea beach for upwards of fifty miles was prospected, and in most instances with highly satisfactory results."[40] Then Adams really got carried away: "With the garish sun shedding its kindly beams nearly every minute of each twenty-four hours, the 'land of the midnight sun' is a pleasing and satisfying sight to the seeker after its hidden wealth, and it is no wonder that the Cape Nome miner accomplishes as much in four months as does his brother miner in double that time in a more temperate zone."[41]

One of the reasons the rush to Nome attracted such crowds was that compared to the Klondike and other parts of Alaska, it was relatively easy to reach—just as Juneau had been during its first boom. One could leave Seattle or San Francisco and take the rich man's route to the Klondike, but make only half the trip. One stayed on board ship for the entire journey to Nome, and the hardest part was the jostling ashore in lighters once the ship anchored off its beaches. There was no Chilkoot or White Pass to surmount, no endless maze of glaciers north of Valdez, no heated debate about whether or not one would be iced in on the Yukon. Nome was easy to get to and, equally important for some, easy to get out of. By the early fall of 1899, the new boomtown had at least 3,000 inhabitants. Some reports say that 5,000 people spent the following winter in the district.

But even the most vocal of community boosters was forced to admit that the sanitary conditions in Nome that first summer were deplorable. Malaria was common, and nearly every day brought a death from typhoid fever. No one wanted to dig cesspools when he could be digging for gold. But help was on the way, in the form of both civic-minded town fathers and a high number of women. "There are a thousand women in the district," George Adams wrote, "and they have, as usual, taken vigorous action with the sanitary questions; brave, noble, energetic, self-sacrificing American women—God Bless them!—mothers, wives, daughters, sisters, and sweethearts who will urge upon those near and dear to them the pressing need of better sanitation."[42]

Those who came the following year were almost inevitably out of luck, even for the gold of the beaches. But that did not stop them from coming. A case in point was Eleanor Caldwell, who left Washington, D.C., with her daughter on May 11, 1900, "[their] destination Cape Nome, [their] object a fortune." Needing to buy supplies in Seattle, she found its stores plastered with signs proclaiming, "No more orders taken for Alaska." Caldwell desperately needed a stove for her outfit. Recognizing that its procurement was "a special and difficult problem of itself," she later confessed, "The process by which I became the owner of the coveted article [stove] was a lengthy one; feminine arts and cajolery were not entirely left out of consideration; failure seemed imminent, when, lo! my pressing needs above all other Nome travelers were recognized."[43]

Caldwell and her daughter sailed on the steamer *Tacoma*, whose passengers included about 100 women. One admitted that she was a professional gambler, but many were nurses; some were intent on opening restaurants and hotels; and a few, Eleanor noted, were accompanying their husbands "just for fun." Obviously, they did not know much about Nome!

Making its way into Norton Sound, the *Tacoma* ran aground in the fog while trying to steer past the Yukon flats. The crew and passengers tossed some 300 tons of coal (worth upwards of $100 per ton in treeless Nome) overboard in order to lighten the ship for the high tide. Three days later, the steamer dropped anchor a mile and a half off the beaches of Nome, and Caldwell was at the mercy of the opportunistic lighter operators who ferried goods and passengers between ship and shore. The beach was a chaotic jumble of people and supplies, every square foot covered for miles in either direction. "Possession," wrote Eleanor Caldwell with due melodramatic afterthought, "was nine points of the law, the tenth having resolved itself into a question of skill in pistol practice." [44]

Mother and daughter went to a hotel to wait while their trunks and gear, including the precious stove, were ferried to shore. The thought of a room seemed to offer a pleasant respite, but for a dollar a head they were each given a "shake-down." Not quite as bad as it first sounds, a shake-down was a skin thrown—shook down—upon the floor. One rolled up in it and tried to sleep while the business of Nome continued unabated all around in the glare of the Alaskan summer night. But in the morning, they got a bargain: three buckets of water for twenty-five cents.

For a week, the Caldwells patiently watched for their trunks and gear to appear on the beach, all the while searching for an overlooked corner upon which to stake a claim. There were none, of course, and by the time their gear was finally unloaded, it never made it off the beach. Rumors of a smallpox scare on one of the ships, lack of any ground upon which to stake a claim, and probably a few too many nights in a shake-down convinced Eleanor that it was time to sail for home. She and her daughter reboarded the *Tacoma* just as it weighed anchor to steam south, having sold her outfit and the highly valued stove at what she confessed was "a slight loss." Like so many who were lured to Alaska by golden dreams, Eleanor Caldwell left empty-handed, but having done her part for the local economy.

But for all it was lacking locally, Nome became a bustling city of amenities quite quickly. The following winter (1900–1901) at a charity ball, Alfred H. Dunham, soon to be Alaska's chief game warden, wore "the first dress-suit ever exposed to public view in Nome. Needless to add," Dunham later admitted, "I was the uncomfortable center of all eyes." Judge Clark broke the ice by shaking hands with the "man who had nerve enough to wear a dress-suit to Nome." But, as with so many things, a trend had been born in the most innocent of circumstances, and Dunham recalled that "it was a most amusing thing to see the dress-suits appear after that."[45]

One of the most remarkable things about Nome was the speed with which it grew, both in population and in the accoutrements of civilization. By one account, Nome had 25,000 inhabitants by 1905, and its streets included paved, electrically lighted thoroughfares lined with banks, schools, churches, and theaters. There were telegraph and telephone systems and three separate railroad lines. Three daily newspapers bannered the news about the diggings, the latest legal maneuverings, and the numerous social clubs, including the Arctic Brotherhood. There were even substantial greenhouses for growing vegetables.[46]

One of Nome's early social highlights was G. L. "Tex" Rickard's Christmas dinner in 1900. Tex was a city marshal in Henrietta, Texas, when he took a leave of absence and visited Juneau in 1895. He chanced to meet up with sourdough Alfred Mayo, who filled him so full of tales of the upper Yukon that Tex mailed his resignation back to Henrietta and headed north over Chilkoot Pass a full year and a half before the route got crowded. He landed in Circle City and then joined the exodus from there to Dawson when news of the Klondike broke.

Not much of a miner, Rickard nevertheless showed his business and promotion skills and turned a tidy profit buying and selling claims. Taking his profits down the Yukon, he detoured to what was first called Anvil City to see what the excitement was all about. Believing that it was Dawson all over again, he made some investments and opened a saloon. That Christmas, partly through generosity and partly for publicity's sake, Tex cornered the market on turkeys, hired an orchestra, and invited everyone from near and far to a sumptuous feast. Meals were served in shifts until both hunger and turkeys were effectively dispatched.

The following spring, Nome was officially incorporated, and Tex Rickard was elected to the first city council. Like so many on the mining frontier, Tex didn't stay long, of course. By 1906, he was in Goldfield, Nevada, operating the famed Northern Saloon with its stable of eighty barkeepers. There, he turned to boxing instead of turkey dinners to promote his saloon, eventually becoming one of the country's leading fight promoters.[47]

Fairbanks came about a little differently. About 200 miles upstream from its confluence with the Yukon, the Tanana River sweeps close to a range of hills. These are the foothills of the White Mountains that divide the Yukon and Tanana drainages. The southern bank of the river blends into wide, flat wetlands. In the late summer of 1901, the steamboat *Lavelle Young* was doing her best to huff and sputter her way upstream through this meandering maze of braided channels and shifting sandbars. Her captain, C. W. Adams, had agreed to take one Elbridge Truman Barnette, always known as E. T., to the head of navigation on the Tanana. The *Lavelle Young* was loaded down with goods and supplies. Barnette was intent on opening a trading post and general store where the new road that Captain Abercrombie was surveying between Valdez and Eagle would cross the Tanana just west of present-day Tok.

It should be noted that Barnette's record to date had not been flawless. Born in Akron, Ohio, about 1863, he spent time in prison in Oregon for stealing his partner's share of the proceeds from a horse-trading venture in Canada. Barnette claimed that he had been robbed of the money. That failed to explain, however, his high-roller lifestyle and a record of him exchanging Canadian money for U.S. currency shortly following the "robbery." Denied an explanation, the jury gave him four years. Family connections from Ohio appear to have prevailed upon Oregon's governor to commute Barnette's sentence after little more than a year with the proviso that he never return to Oregon.

In the summer of 1897, E. T. Barnette was in Helena, Montana, when news of the Klondike hit. He rushed to Seattle and boarded the steamer *Cleveland*, one of the many aged, overloaded hulks that departed West Coast ports for Alaska that summer. The *Cleveland* survived a near grounding and a fire and managed to chug into the harbor at St. Michael just five hours after the departure of the river steamer that was supposed to take the ship's passengers up the Yukon. Undeterred, the Klondikers all chipped in, bought another little steamer outright, and set off up the Yukon after electing Barnette the captain, a title he was to wear for the rest of his life.

"Captain" Barnette managed to baby the steamer up the Yukon as far as Circle City before freeze-up marooned her for the winter. Reaching Dawson the following spring, Barnette spent some time there but quickly became enamored by the prospects of a boom along the proposed Valdez-to-Eagle road. He returned to Montana, married Isabelle Cleary, and then bought $20,000 worth of goods that he shipped to St. Michael. The captain never had much luck with steamers, and when the one he bought to take him again up the Yukon ran aground and sank off St. Michael, he presented himself to Captain Adams.

Head of navigation was a fairly vague term, especially since neither Captain Adams nor Barnette had ever been up the Tanana. Added to that, it was late in the year, and the heavily laden little steamer was not flat bottomed. When it became clear that the Tanana was impassable a couple of hundred miles below his intended destination, Barnette convinced Adams to try a detour via the Chena Slough, a backwater on the Chena River just above where it empties into the Tanana. A few miles up the Chena, Adams met with a similar impasse and told Barnette that this was as far as he was going. The captain turned the *Lavelle Young* downstream and put in at a low bluff that had a good stand of spruce. Here, Barnette was going to have to be content to build his trading post.

Barnette thought that things looked pretty bleak, but before they could even finish unloading the steamer, along came his first customer. The man's name was Felix Pedro. He was an Italian immigrant who had worked in the states as both a coal miner and a hard-rock miner. Pedro had been knocking around the Yukon since 1895. His base of operations was Circle City, but only because it was the closest place around. From there he had worked Birch Creek, taken a stab at the Klondike, and then prospected throughout the mountains between the Yukon and the Tanana. A store right here would be a lot more convenient. He hadn't struck it rich yet, but Pedro assured Barnette that not only would he become Barnette's first customer, but also that the good color he was finding in the hills north of the river would someday mean a big stampede. Barnette, with his goods setting on the riverbank, his new wife in tears, and the *Lavelle Young* disappearing downriver, hoped that he was right.

Barnette spent part of the winter of 1901–2 on Chena Slough and then went outside to Seattle to sell furs and obtain more supplies. He left his brother-in-law, Frank J. Cleary, in charge of the post with orders not to extend credit to anyone. While Barnette was gone, Felix Pedro came in for more supplies, and Cleary evidently grubstaked him out of Cleary's personal account. That summer, Pedro's boast came true, and he duly reported word of rich diggings to Cleary. By then, Barnette had returned. Ever the promoter, Barnette sent word upriver to Dawson that a big strike had been made. Things had quieted down some in Dawson, and the first wave to descend on Barnette's little camp comprised not prospectors but the overflow of Dawson's gamblers, saloon keepers, and ladies of the night—all more accustomed to mining the miners than mining the creeks. Barnette's place was beginning to boom, all right, but what it seemed to lack were miners in the creeks.

Judge James Wickersham, that pillar of Arctic jurisprudence, passed through the following spring and offered Barnette some sage political advice. Wickersham suggested that Barnette name the growing cluster of cabins after Charles W. Fairbanks, the senior senator from Indiana. That way, the judge reasoned, if ever there was anything that the town needed in Washington, there would be at least one friendly ear. Barnette did so, but Senator Fairbanks was elected vice president under Theodore Roosevelt in 1904 and, in the proverbial manner of the times, was never heard from again. (Roosevelt supposedly said of the choice, "Who in the name of heaven else is there?") Indeed, Judge Wickersham and Captain and Mrs. Barnette later enjoyed a memorable dinner with the vice president and his wife in Washington before both the judge and the veep sought to distance themselves from Barnette's nefarious schemes.

But the budding town of Fairbanks was not without a rival. Others started the town of Chena right at the confluence of the two rivers. Chena was where all of the big steamboats docked. For a while it gave Fairbanks a run for its money, but Chena's founders were overly proud of their prices for city lots. Many newcomers headed upriver a few miles to Fairbanks, where Barnette was busily attracting businesses with the offer of free lots. As will be seen, Barnette never gave anything away, but so it seemed in the summer of 1903.

Later that summer, gold was found on Cleary Creek, and then there were strikes on Fairbanks and Ester Creeks. More miners arrived, the Dawson malcontents were appeased, and the boom finally appeared to be gathering steam. One of the reasons things got off to a slower start in Fairbanks than at many of the other strikes was that the gold in this region took more work to find. Much of it was placer gold, but instead of lying about on the beaches as at Nome or glistening up from Bonanza Creek, the gold around Fairbanks was frequently buried under many feet of overburden. Felix Pedro understood this and dug deeper. The surface gave a promise of riches, but the really rich deposits were not found until substantial work was done and the heavier equipment required for major hydraulic mining operations was put in place. The Fairbanks rush was like slowly opening a water tap; there wasn't much of a cleanup the first year or two, but after 1904 things really started to flow.

In fact by 1904, those bound upriver for Dawson who detoured to Fairbanks were convinced that Fairbanks was the more booming town. Judge Wickersham helped out the cause by moving the headquarters of the Third Judicial District, including courthouse, post office, and jail, from Eagle to Fairbanks. The town voted to incorporate in November 1903 and elected its first mayor and city council. Whoever got the most votes was to have been mayor and the runners-up on the city council. Barnette received 67 votes for mayor to John L. Long's 73 votes. That should have made Long mayor and put Barnette on the city council, but the captain was going to be mayor of his own town no matter what. He badgered the newly elected city council until it finally made him mayor despite the prior rules, an outcome that did not sit well with the independent miners. One disgruntled voter groused, "Mr. Long is certainly the choice of the majority of the voters in Fairbanks, and the people's voice at the polls should have been considered by the council."[48]

These politics aside, the town went about building a school and sanitation system and organizing a volunteer fire department. Soon there were churches to challenge the numerous saloons and the long line of cribs in the red-light district. By 1905, there were an estimated 5,000 people in town, and even a raging flood that year and a disastrous fire the next couldn't keep Fairbanks from soon rivaling Nome as the largest town in Alaska.

What kept Fairbanks growing was that it quickly became the supply center for a number of mining camps that sprang up in the hills north of town. Pedro Camp, Livengood, Fox, Gilmore, Olnes, and Chatanika at one time or another each had hundreds of inhabitants. Teamsters charged anywhere from one dollar to ten dollars per ton to haul supplies to these mining camps. Soon the valley was to have what every successful mining center needed: a railroad.

First named the Tanana Mines Railroad, the Tanana Valley Railroad was incorporated in 1904. As with almost every other western railroad, its financial backers were from the East, in this case New York and Chicago businessmen. But the man on the scene and the key cog in the wheel was Falcon Joslin, an attorney who had made money at Dawson and who was now determined to see Chena and Fairbanks boom. One of Joslin's supporters would later term him "the Harriman of the North," no doubt with more local pride than anything else, but certainly no one could argue that Joslin was not sworn to do his utmost for the cause of Alaska railroads.

The Tanana Valley Railroad was built in two main stages. In 1905, construction crews laid twenty-six miles of track from Chena to Gilmore on Pedro Creek and then a 4.7-mile spur to Fairbanks. Judge Wickersham was the principal speaker at the gala golden spike ceremony on July 17. Mrs. Isabelle Barnette graciously accepted the golden spike as a souvenir. Two years later, the line was extended another twenty miles to Chatanika. In its relatively short length, it encountered all of the obstacles known to mountain railroads, from mucky bogs along the Chena, up narrow winding valleys, to the steep grades around Pedro Dome. Spruce for ties and timber was readily available, but everything else—including rails and rolling stock—had to come up the Yukon by steamer. That trip was not without risk, and six flatcars once wound up on the bottom of the river instead of on the tracks.

Outside rules sometimes had to be bent a little to fit Alaskan reality. The Interstate Commerce Commission had a rule that no dynamite or explosives could be carried closer than ten car lengths from the locomotive. Fine, but the Tanana Valley line rarely put more than four cars on any of its trains, and such explosive cargo was a requisite in a mining district.

Joslin spoke enthusiastically of extensions east to Circle City and west to Nome, and a line built directly to Haines at the head of Lynn Canal. Many of the transcontinental lines in the states lived off land grants, but in Alaska Joslin had another idea. He confidently predicted that "a station could be established every twenty miles along the route and that gold or copper mines would be opened near every station."[49] The Tanana Valley Railroad was built without federal subsidies, but Joslin regularly called for federal aid to Alaska's railroads—a call that went unheeded at the time.

While it was late in getting started, the boom at Fairbanks may have been the most profitable of all of the gold rushes that swept Alaska. Between 1902 and 1914, $63 million was produced from area mines. Significantly, almost all of the gold came from placer deposits. The peak year was 1909, with production of some $9.5 million, but from then on it was downhill fast as even the deeper placer deposits played out.[50]

As the boom crested, E. T. Barnette figured that it was time to cash in his chips and leave town. No doubt he could have done it in an honorable way, but that was not the way Barnette had ever done things. He was always involved in one type of litigation or another and never seemed to be able to come clean until forced to do so. Even his partner in the original charter of the *Lavelle Young* was forced to sue him for his portion of the profits before finally being awarded in excess of $40,000 in 1908. It was in the midst of this prolonged litigation that word of Captain Barnette's prior stay in the Oregon penitentiary became known on the streets of Fairbanks. The *Fairbanks Daily Times*, never a fan of Barnette's, told the news in a banner headline that screamed "EX-CONVICT" in bold letters across the entire top of its front page.

Such revelations and lawsuits aside, Barnette had been as busy collecting banks as some men had been collecting mining claims. By 1909, two of the three banks in town had been absorbed one way or the other into Barnette's Washington-Alaska Bank. Most folks still held that it was the cornerstone of the town's economy, but in January 1911, shortly after Barnette resigned as its president and left town, the bank failed. Fairbanks depositors were out roughly $1 million. Recriminations flew left and right, but all ended up targeting E. T. Barnette.

In Barnette's defense, it should be noted that the captain's only conviction in the entire bank matter was a misdemeanor of making a false report of the condition of the bank with intent to deceive and defraud. But that did not stop the *Fairbanks Daily Times* and a host of others from branding him a common thief and an outright scoundrel. One popular theory held that Barnette had gotten his first bank, the Fairbanks Banking Company, to invest in shares of the Gold Bar Lumber Company at an inflated price. By no small coincidence, the lumber company was run by one of Barnette's ubiquitous brothers-in-law. Each time a merger was completed and Barnette's personal stake in the matter valued, these inflated shares allowed him to skim considerable cream off the deal.

Then there was the accusation that through some fanciful bookkeeping Barnette had actually used the cash of the Washington-Alaska Bank to acquire it—effectively buying the bank with its own assets. The bank got worthless collateral and promptly went broke. It took Fairbanks years to recover. However it happened, the bottom line was that Barnette got all the money and promptly left town. Whatever one chose to believe, what was clear was that somehow in a little over a decade Barnette had managed to extract enough money from his Alaska ventures to buy millions of dollars of property in California and Mexico.

After a lengthy receivership, the depositors of the Washington-Alaska Bank got back about fifty cents on the dollar for their deposits, but for years the oldest joke in Fairbanks among the old-timers was how appropriate it was that the most crooked street in town was named Barnette Street. The *Fairbanks News-Miner* used his name as a verb for any illegal shenanigans, reporting "Barnetting Going on Still." Ironically, all of this seems to have been lost to history when in 1960 the town dedicated the E. T. Barnette Elementary School. Doubtless the captain smiled in his grave a long time at that one.

After Nome and Fairbanks, there would be other gold discoveries in Alaska, including those at Kantishna, Chisana, and Iditarod. And there would be rushes for other metals, such as copper and molybdenum, and of course the black gold of oil, but after Nome and Fairbanks no longer would the mining frontier be the domain of the individual. Companies with heavy equipment, outside investors, and bottom-line orientations would take the place of grizzled prospectors begging a grubstake to make it through another year. Nome and Fairbanks were the last stops of the American mining frontier.

Crest of the Continent

The mountain's existence was certainly no secret. Captain George Vancouver made reference to it in his log in 1794 while charting Cook Inlet. The Tanaina Natives around the inlet called it Traleyka, meaning "the high one." Governor Wrangell placed the massif on his 1839 map. The Russians called it Bulshaia Gora, or "the great mountain." William Dall looked at the entire snow-clad range from the Tanana Valley in the late 1860s and proposed calling it the Alaskan Range. Lieutenant Henry Allen crossed the eastern part of that range in 1885 and then reported very high snow-clad peaks to the south as he floated down the Tanana. The Athabaskans of interior Alaska called it Denali, meaning "the high one." Only Anglo prospectors along the Yukon River bestowed a nondescript name. They called it Densmore's Mountain because in 1889 Frank Densmore had led a party north of the mountain and enthusiastically described its grandeur.[51]

Then in January 1897, the *New York Sun* published a dispatch from a prospector named William A. Dickey, who had spent the latter half of 1896 prospecting in the Susitna River valley at the head of Cook Inlet. More than a little presumptuously—this was, after all, only a private prospecting party—Dickey wrote, "We named our great peak Mount McKinley, after William McKinley of Ohio, who had been nominated for the Presidency, and that fact was the first news we received on our way out of that wonderful wilderness."[52] By the time Dickey's dispatch was published, McKinley was president-elect. Later, Dickey's rationalization for the name was that he had grown tired of guff from the free-silver prospectors he had encountered and had chosen to honor McKinley, a solid supporter of the gold standard.

But how high was the mountain? Dickey, a Princeton graduate, had spent about a decade in the Seattle area beneath the hulk of Mount Rainier. One of his party, a Mr. Monks, had actually climbed Mount Rainier the previous summer. Dickey reported in his dispatch that Monks was of the opinion that the front peaks in the range (including 14,573-foot Mount Hunter) were about the same altitude as 14,411-foot Mount Rainier and that their "Mount McKinley" was at least 6,000 feet higher. Either Monks had a very sharp eye, or it was a very lucky guess!

Once the duke of the Abruzzi and his party reached the summit of Mount St. Elias later that year, attention turned to Dickey's "Mount McKinley" and the intriguing question of its altitude. Was Monks correct? Was the mountain really 20,000 feet in height? In 1898, a U.S. Geological Survey party led by George Eldridge and Robert Muldrow ran a survey line up the Susitna River and from six locations along it triangulated the summit of McKinley's South Peak. Muldrow's calculations indeed showed it to be above 20,000 feet.

The next question was, could it be climbed? The first to attempt to answer it was Alfred H. Brooks. It is dangerous to give out superlatives in a land filled with them, but Brooks is the recognized dean of Alaskan geologists. An 1894 graduate of Harvard, Brooks first joined the U.S. Geological Survey's Alaska field teams in 1898. After four successive seasons in eastern Alaska, in 1902 he was assigned a traverse of the Alaska Range from Cook Inlet to the Yukon River, with specific instructions to "bend close to the base of Mount McKinley."[53] Perhaps there were unknown others before him, but in the course of this exploration, Alfred H. Brooks became the first person to document his own footsteps on the great mountain and then propose a plan for reaching its summit.

With six men and twenty packhorses, Brooks led his party up the Susitna, Yentna, and Kichatna Rivers and across the Alaska Range at Rainy Pass, just southwest of the present-day boundary of Denali National Park and Preserve. Then he swung northeast, paralleling the giants of the Alaska Range until he camped on the headwaters of Slippery Creek, which flows north from beneath Peters Dome just northwest of Denali. In the process, Brooks got a good view of Denali's giant western neighbor. Three years before, army lieutenant J. S. Herron had crossed the range at Simpson Pass, about twelve miles northeast of Rainy and much more difficult, and proceeded north to the Yukon. From the vicinity of Lake Minchumina, Herron compounded the Ohio politician problem by naming this mountain after Joseph B. Foraker, a U.S. senator from the Buckeye State. The Tenaina Natives around the lake called it Sultana, meaning "the woman," or Menlale, meaning "Denali's wife."

Although surveying, not mountaineering, was his charge, camped so close Brooks could not resist at least one day's scramble on the mountain itself. On August 6, 1902, he climbed alone above the camp, intent on reaching a shoulder northeast of Peters Dome at 10,000 feet. When the difficulty of the task and the magnitude of the mountain became apparent, Brooks built a cairn, placed a cartridge shell from his pistol inside it, and estimated his elevation at 7,500 feet. In 1954, a USGS party found the Brooks cairn at 6,300 feet almost exactly three miles northeast of Peters Dome.[54]

Brooks and his party continued northeast, roughly following the course of the future Denali Park road, and descended to the Nenana River near Riley Creek. They followed the Nenana north and eventually reached the Yukon. Half a century later, the Brooks maps and geologic data were still recognized as the most complete and reliable on the Denali region. The first publication from the expedition, however, was an article Brooks wrote titled "Plan for Climbing Mount McKinley." It appeared in the January 1903 issue of *National Geographic Magazine* and suggested that the best way to approach the mountain for a climb was from the north, even though it would mean wintering over on the north side of the range. His recommendation would be remarkably prophetic.

The following year, 1903, Judge James Wickersham, U.S. district judge for Alaska's Third District, took Brooks's advice and reached the mountain from the north via a steamer on the Kantishna River. Wickersham made camp near the terminus of the Peters Glacier, which Brooks had named the year before for a USGS topographic chief. Wickersham christened the glacier after yet another Ohio politician—McKinley's patron, Mark Hanna—but fortunately, the name did not survive. What has survived is Wickersham's claim to the first serious mountaineering attempt on the mountain. Wickersham and his party climbed the Jeffery Glacier, just south of the Peters, and reached 8,100 feet before Wickersham reported, "We recognize that we are inviting destruction by staying here." The massive wall above them, now known as Wickersham Wall, would not be climbed until 1963.

As they retreated from the mountain, Wickersham, too, wrote a prophetic line: "The glacier rising from the northeast angle is clearly the one that the Munkhotena Indians referred to as leading to the summit. . . . It seems to afford the road we have been looking for."[55] That glacier is now called the Muldrow.

Considering the vastness of the terrain, it is surprising that two months after his departure Wickersham's camp at the base of the Peters Glacier received another expedition. Its leader was Dr. Frederick A. Cook, destined to become one of the most controversial explorers in history. What was undisputed at the time was that Cook was well versed in

Arctic travel, having participated in polar explorations with both Peary and Amundsen during the 1890s. But Cook, like so many others, failed to grasp the enormity of Alaska's scale.

Cook left New York City in mid-May 1903—incredibly late in the season. It was late August before he and four others reached Wickersham's abandoned camp after approximating the Brooks route of 1902 from Cook Inlet, but crossing rugged Simpson Pass instead of Rainy. With only a few days to climb—they were not prepared to winter over and had to reach the coast before freeze-up—the party made a valiant effort and ascended the Peters Glacier to between 10,000 and 11,000 feet on the West Buttress of the North Peak. From their high camp, chief packer Fred Printz, a veteran of the Brooks expedition, remarked to Cook: "It aint that we can't find a way that's possible, takin' chances. There aint *no* way."[56] On September 4, they began a frantic march to escape the approaching winter.

Some, including mountaineer Bradford Washburn, have suggested that Cook was now intent on proving something by the trip after failing to climb the mountain. Instead of retracing their route, the party struck eastward, as had Brooks. But Brooks was bound for the Yukon. Cook's route to safety led to the south. Somehow, they had to find a pass through the Alaska Range suitable for packhorses. Above the East Fork of the Toklat River, Sable Pass blocked their route to the east. They turned south up the East Fork and by lucky happenstance found the only pass in the 140 miles between Simpson Pass and the head of the Sanctuary River that could be crossed by a pack train. The party crossed the 6,000-foot pass and descended the valley of the Bull River to the Chulitna and Susitna Rivers, arriving back on Cook Inlet on September 26.

Had Cook's Alaska work ended here, he would have had the distinction of making a serious mountaineering attempt on McKinley's summit and the first circumnavigation of the core of the Alaska Range. But Cook was intent on reaching the summit of Mount McKinley.

In 1906, Cook returned to Alaska with a party that included Professor Herschel C. Parker of the department of physics at Columbia University and Belmore Browne, a respected artist and outdoorsman. Repeating his mistake of 1903, and ignoring Alfred Brooks's advice of a northern route after wintering over, Cook again approached the mountain from the south, via the Susitna River and its tributaries. But now he made an even bigger mistake. Instead of the Brooks route over Rainy Pass, or even the more difficult but known route he had taken in 1903 across Simpson Pass, Cook insisted on forging a new route a few miles to the north. It was supposed to cut across the Alaska Range closer to Mount McKinley and save them a few miles, but instead it got the party entangled in the labyrinthine valleys beneath Mount Dall. Fred Printz, who was once more along as chief packer, must have wondered what was going through Cook's head.

Retreating from this debacle, Cook turned his attention to a more direct route to the glaciers on the mountain's south side. The expedition hacked its way up the boulder-strewn moraine of the Tokositna Glacier, a Tanaina name meaning "the river that comes from the land where there are no trees," and then tried to find a way up its neighbor to the east. Cook had named this glacier for his stepdaughter, Ruth, as he passed down the Chulitna in 1903. In the end, after a summer spent battling swollen rivers, alder thickets, and insufferable mosquitoes, the expedition returned to Cook Inlet convinced that a southern route to the mountain was not feasible.

Then, after separating from Parker and Browne, and in the company only of a packer named Ed Barrill, Cook returned to the vicinity of the Ruth Glacier. A few weeks later, he emerged with the claim that would tarnish his reputation and deprive him of what may have been a valid claim to being first to reach the North Pole. Cook claimed that he had reached the summit of Mount McKinley. From the south. With only one companion. In less than two weeks. In September conditions.

The controversy would fly about for years. The exposure of Cook's infamous "fake peak" photograph would be the most damning evidence of a bald-faced lie. But Belmore Browne would say it most simply and most graphically: "I knew the character of the country that guarded the southern face of the big mountain, had travelled in that country, and knew that the time that Dr. Cook had been absent was too short to allow of his even reaching the mountain. I knew that Dr. Cook had not climbed Mount McKinley, in the same way that a New Yorker would know that no man could walk from the Brooklyn Bridge to Grant's Tomb in ten minutes."[57]

Frederick Cook returned from his polar expedition in September 1909, announcing to the world that he had reached the North Pole on April 21, 1908, almost a full year before Peary. While most of the world greeted him triumphantly, the Explorers Club in New York had grave questions to ask him about his 1906 Mount McKinley claims, particularly in light of his 1908 book, *To the Top of the Continent*, published while he was in the Arctic. Having lied once, Cook compounded his misdeed by seeking to embellish it with details. The most damning embellishment was the photograph of the "fake peak." Parker and Browne, among others, seized on this piece of evidence and determined to use it to prove the truth.

The result was that 1910 saw three expeditions take the field to attempt to reach the summit of Mount McKinley and prove Cook's claim one way or the other. Two were organized by veteran exploring organizations, the Mazama Mountaineering Club of Oregon and the Explorers Club–American Geographic Society Expedition from New York, which included Herschel Parker and Belmore Browne. The third was a group of sourdoughs from Fairbanks. The impetus for the latter was a barroom bet and a matter of local pride.

The organizer and leader of the sourdough party was Tom Lloyd, a veteran prospector who had spent time in the Kantishna and Toklat Mining Districts. With him were fellow prospectors Pete Anderson, Billy Taylor, and Charlie McGonagall, as well as Charles E. Davidson, a qualified surveyor and photographer who would later become surveyor-general of Alaska, and two of Davidson's friends, Robert Horne and William Lloyd—no relation to Tom. Three days before Christmas 1909, the *Fairbanks Daily Times* reported that "Cheers Are Given as Climbers Leave." Here was a party getting a head start on the mountain.

By early March, the party had established a winter base camp at Cache Creek, south of Wonder Lake, and stockpiled moose and caribou for food. But the strain of two wintry months on the trail caused tension between Tom Lloyd and Davidson. Abruptly, Davidson and his two friends packed up and returned to Fairbanks. Lloyd, Anderson, Taylor, and McGonagall continued up the mountain via the Muldrow Glacier, establishing a high camp at 11,000 feet. Here, Lloyd stopped.

After a week of hacking out an icy staircase atop what came to be called Karstens Ridge, the three younger men set out for the top on April 3, 1910. Incredibly, they hauled along a fourteen-foot-long spruce pole to erect as a summit marker and six doughnuts each for food. They reached the northeast ridge of the North Peak and erected the pole at a point some distance below the summit, certain that the flag flying from it could be seen by their friends in the Kantishna district. McGonagall stayed at the pole, but Anderson and Taylor continued the remaining yards to the summit.

With weather variously described as "30 below" and "colder than hell," Anderson and Taylor nonetheless spent some time on the summit enjoying the view. If the peak two miles to the south appeared higher, neither made mention of the fact. They rejoined McGonagall, who was very cold from waiting, and then descended to their 11,000-foot camp after an eighteen-hour day with some 8,500 feet of elevation gain over new ground—an extraordinary achievement! Subsequent events, however, would draw a cloud over it.

The first problem for the sourdoughs was that while they had indeed reached the highest point visible from Kantishna and Fairbanks, this was only the North Peak. It was a staggering 19,470 feet high, to be sure, but still two miles north of and 850 feet below the 20,320-foot South Peak, the true summit of Mount McKinley. Now the magnitude of Davidson's departure was brought home. Quite possibly, his surveying skills would have recognized the error. Whether the sourdoughs could have traversed to the South Peak and still returned safely to the Muldrow camp is grist for countless campfire debates, but Davidson's photographs might have at least verified their North Peak ascent.

The second problem was of Tom Lloyd's doing. While Anderson, Taylor, and McGonagall returned to the Kantishna diggings, Lloyd returned alone to Fairbanks and, having learned nothing from Dr. Cook, promptly reported that all four men had been to the tops of both peaks.

When Charles Sheldon and others, including Herschel Parker and Belmore Browne, expressed some skepticism at the claim, Lloyd sent word to Kantishna that Anderson, Taylor, and McGonagall should make another climb of the peak and photograph the flagpole. Not waiting for word as to whether or not the three were even willing to do so, Lloyd sent a report of the success of this "second climb" to newspapers in Fairbanks, New York, and London. Another case of embellishing a lie with details! As Billy Taylor later put it, "We three didn't get out [from the Kantishna] until June, and by then they didn't believe any of us had climbed McKinley."[58]

Meanwhile, the Mazama and Explorer Club expeditions were probing the mountain's southern defenses around the Ruth Glacier and Dr. Cook's reported ascent route. Neither party made a serious climbing bid, but Parker and Browne found and photographed Cook's "fake peak"—not the top of the continent but a small point just above the Ruth Glacier, almost twenty air miles from McKinley's summit. Determined to resolve unequivocally the Cook and sourdough claims, Parker and Browne planned a 1912 expedition to the mountain.

For this attempt, Parker and Browne decided on a completely new approach to the northern slopes of the mountain—by dogsled, in winter, from the south. By now, the possibilities of the northeast route via the Muldrow Glacier were well known to them—as were the difficulties of southern approaches. When ice blocked the head of Cook Inlet, Parker and Browne landed in Seward on the opposite side of the Kenai Peninsula. Leaving Seward on February 1, 1912, they mushed north across the Kenai Peninsula along the incomplete Alaska Railroad line, around Turnagain Arm, and past what would become the site of Anchorage. They joined up with photographer Merl La Voy and Arthur Aten, veterans from the 1910 Explorers Club expedition, at the tiny village of Knik. The foursome then continued north up the Susitna and Chulitna rivers and crossed the Alaska Range near Anderson Pass, about thirty miles east of the mountain. April 17, 1912, found them camped near the big timber on the McKinley River at the terminus of the Muldrow Glacier.

Like the sourdoughs before them, they first spent time stocking up on supplies of moose and caribou and then began the slow process of relaying supplies to successively higher camps. Also like the sourdoughs, they found the key to avoiding the great icefall at the head of the Muldrow Glacier and climbed south to atop Karstens Ridge. Unlike the sourdoughs, they patiently established and stocked a series of higher camps above the ridge.

Finally, on June 29, 1912, Parker, Browne, and La Voy were camped at about 16,600 feet in the basin between the North and South Peaks. Leaving camp at 6:00 A.M., they made slow but steady progress contouring across the basin and onto the upper northeast ridge of the South Peak. Carefully recording their progress with La Voy's camera and an altimeter, the trio stopped briefly to congratulate one another when they realized that they had bested the duke of the Abruzzi's North American altitude record of 18,008 feet on Mount St. Elias. (That in itself was evidence of what they thought of the sourdoughs' claim.)

Then the mountain turned against them. The temperature plummeted, and an icy gale ensued. As Browne later wrote, "As I again stepped ahead to take La Voy's place in the lead, I realized for the first time that we were fighting a blizzard." Onward they battled. Parker's altimeter registered 20,000 feet, but they were all beginning to freeze—literally. Browne turned to Parker and yelled above the wind, "The game's up; we've got to get down!"[59] So close. So close. But down through the blizzard they stumbled until they finally reached their camp at 16,600 feet.

There, they rested the next day and tried to consume a miserable diet of pemmican. The dried meat had served them well at lower altitudes but had become indigestible fat at 16,000 feet. Then, at 3:00 A.M. the next morning, they began a second assault on the summit. Somewhere about 19,300 feet, they again encountered blizzard conditions and were forced to descend. Three days later, they rejoined Arthur Aten in the base camp on Cache Creek.

But the mountain had one more point to make. On July 6, as the party recovered in base camp, an enormous earthquake swept the area, sending avalanches roaring off the peaks and boulders plunging down the slopes. Browne wrote, "My strongest impression immediately after the quake was surprise at the elasticity of the earth. One felt as if the earth's crust was a quivering mass of jelly."[60]

Thanks to La Voy's photographs, Parker and Browne brought back indisputable proof of their climb. Had they seen the sourdoughs' pole on the North Peak? they were asked. No, they replied, they had not.

It remained for Alaskans to complete the ascent of Denali. The foursome that set off from Fairbanks for the mountain in the winter of 1913 was Alaskan to the core. Dr. Hudson Stuck was the Episcopal archdeacon of the Yukon and a veteran of climbs in the Colorado and Canadian Rockies and of Mount Rainier, as well as a decade of missionary work in Alaska by dogsled. Harry Karstens, later to be the first superintendent of Mount McKinley National Park, had joined the 1898 Klondike gold rush to Dawson at nineteen and crisscrossed Alaska on dogsled in the years since then. Walter Harper was the son of an Alaskan Native mother and the indomitable Arthur Harper of Yukon River fame. Walter was just twenty-one, but he had spent three years with Archdeacon Stuck as interpreter and dogsled driver. Robert Tatum was also twenty-one and a postulant for holy orders at the mission at Nenana, but he, too, had just spent the past winter on the trail in a determined attempt to get supplies to two women missionaries stranded at Tanana

Crossing. Two younger Native boys accompanied the expedition, looking after base camp and the dogs.

Stuck and Karstens followed the now established route up the Muldrow Glacier and prepared to climb into the upper basin. The landscape, however, bore little resemblance to the photographs from the Parker-Browne climb. Instead of a smooth ridge, there was a sharp cleavage, a pinnacle, and a great gap, all strewn with a jumbled mass of blocks of ice and rock. What could have caused such chaos? Then they remembered the earthquake Parker and Browne had reported the previous year. What the sourdoughs had scampered up in a day, and what Parker, Browne, and La Voy had also easily surmounted, was now to be a treacherous undertaking. The 1912 climbers were barely to safety when the mountain had snatched their relatively easy ascent ladder and dashed it to pieces. Karstens and Harper spent an exhausting three weeks chopping a three-mile-long staircase in the shattered ice of the ridge. Finally, on May 27, 1913, they were above what Stuck named Karstens Ridge. The mountain had almost thwarted another attempt.

With the ridge behind them, the party painstakingly established five high camps in the upper basin, determined to have adequate supplies to weather any storm. One day while resting and talking about the sourdoughs, Walter Harper suddenly exclaimed, "I see the flagstaff!" Indeed he did. The sourdoughs were vindicated.

On June 7, 1913, after more than forty-eight days above the snowline, the party rose at 3:00 A.M. from a high camp at about 18,000 feet and began the final summit attempt. Stuck, Karstens, and Tatum were feeling poorly, but the indefatigable Harper led the way and remained in the lead all day. Finally, at about 1:30 P.M., this son of Alaska reached the summit. There was no more uphill. Stuck, Karstens, and Tatum soon followed. The false claims of Cook, the half-truths of Lloyd, and the near-miss of Parker, Browne, and La Voy were all put in perspective. This was not a mountain to be taken lightly. It would be almost two decades before other footsteps marked the crest of the continent.

Copper, Kennecott, and One Heck of a Railroad

Ahtna Natives had long traded copper artifacts with the Tlingit and other Natives of the coast. Some pieces found their way into the hands of Russian traders. On his 1885 trip through the Copper River drainage, Lieutenant Henry Allen tried to locate Chief Nicolai's copper mine. Then in 1899, while charged with building a road north from Valdez over Thompson Pass, Captain William Abercrombie dispatched Lieutenant Oscar Rohn to follow in Allen's footsteps and seek an alternative route. Rohn moved up the Chitina River but soon suffered from a shortage of rations, much as Allen had.

Luckily, Rohn stumbled across a prospector named James McCarthy who loaned him some surplus horses and supplies. In return for this favor, Rohn named a creek after McCarthy and then became the first to describe in detail the mineralization of the area. He also described the Kennicott Glacier and named it for Robert Kennicott of the Western Union Telegraph Expedition. Perhaps most significant, Rohn reported finding "copper float," slivers of copper ore indicative of nearby veins, among the gravel of the creeks.

Every great mining boom has a special story about its initial discovery. Usually, legend has managed to obscure fact, and "Tarantula Jack" Smith and Clarence Warner's discovery of the Bonanza copper mine is no exception. Late in July or early August 1900, prospectors Smith and Warner stopped for lunch on the banks of National Creek beside the Kennicott Glacier. Looking upward several thousand feet, they saw a green spot at the crest of a sawtooth ridge that looked like grass. Warner speculated that it might be a good place for sheep. Smith wanted to investigate, but Warner was nursing a sprained ankle and declined to make the climb. Then Smith found some copper float in the creek. Quickly putting two and two together, Smith made a beeline for the green spot on the ridge—with Warner limping along close behind. The spot of green turned out to be an enormous outcropping of chalcocite, an immensely rich copper ore. Warner proposed a host of names for the claim, but Smith stuck with the obvious. He said it was a bonanza, and Bonanza it became.

The greatness of the Bonanza discovery—and the extraordinary efforts men would expend to exploit it—is evident when one considers that copper mines in Utah and Arizona were then mining ore concentrations of 2 percent copper; the Bonanza ore initially assayed at up to 70 percent copper. Smith, Warner, and their partners soon sold their claim to a young mining engineer, Stephen Birch, who proceeded to buy up all of the surrounding claims. With the backing of J. P. Morgan, the Guggenheims, and other mining capitalists, Birch formed the Alaska Syndicate. This led to the formation of the Kennecott Mines Company and, eventually, to the incorporation of the Kennecott Copper Corporation in 1915.

Don't get confused about the spellings. The rule of *i* before *e* holds true here. Oscar Rohn named the Kennicott Glacier and Kennicott River for Robert Kennicott. For some reason, when the incorporation papers were filed for the Kennecott Mines Company, the *i* became an *e* and the mining company, the mines, and the company town were spelled Kennecott. Recent usage has promoted the Kennicott spelling for the town as well as the glacier and the river—apparently a belated expression of local independence from the historic power of the company—but this is historically incorrect. The mining company and mines continue to be spelled Kennecott.

Once the early assays were completed, there was no question that the Kennecott copper deposit was among the richest in the world. Rather, the question was how to get ore concentrate from a mill deep in the heart of the Wrangell Mountains to the Guggenheim smelter in Tacoma. The answer was a railroad. The history of mining railroads in the American West is one of rising to meet seemingly insurmountable challenges—providing, of course, that the promise of economic riches waited at the end of the line. In few instances did the physical challenges loom so large and yet the economic prize gleam so brightly as in the construction of the Copper River and Northwestern Railway from the Gulf of Alaska to the Kennecott mines.

The Copper River and Northwestern was born amid a tangled mess of intrigue involving Alaskan coal deposits, East Coast politics, and half-hearted starts on other railroads in southcentral Alaska. Not surprisingly, the man who made it happen was none other than Michael J. Heney. On April 1, 1906, the vanguard of Heney's construction crews unloaded at Eyak, the site of an Alaska Packers Association salmon cannery near the mouth of the Copper River. Attempts to survey lines north from Valdez had not gotten very far, and Heney, with his hot-blooded Irish determination, was convinced that a route directly up the Copper River was feasible. Tiny Eyak was quickly overshadowed by Heney's new town of Cordova, which he named after the original Spanish name for nearby Orca Bay.

Within a few months, there were about 500 men at work grading alongside Eyak Lake. Buildings popped up almost overnight. Freight and materials were quickly unloaded from ships and moved off the docks thanks to a tram system salvaged from the cannery. There didn't seem to be any doubt that the new railroad would soon be hauling both copper from Kennecott and coal from nearby Katalla well ahead of any of its paper rivals. The reason for the optimism, the fledgling *Cordova Alaskan* reported in its very first issue, was "because M. J. Heney who has the contract to build this railroad is known all over Alaska as having built the White Pass & Yukon Railroad." [61]

But Heney knew that there was one major challenge to the route. The twin glaciers that had intimidated Abercrombie on his first visit to the Copper River were still there. Visiting Miles and Childs Glaciers today, one has a hard time imagining the scene in 1908 when Childs Glacier extended well into the Copper River and Miles Glacier presented an imposing 300-foot face along the *western* edge of Miles Lake. (Today, it is several miles east of there.) There was not much room between the glaciers for the river, let alone a railroad.

The challenge was met by the construction of a 5-million-pound steel bridge that stretched 1,550 feet over a river that was ornery by any standard. Called the Million Dollar Bridge, it cost half again that much and consisted of four sections: two 400-foot spans, the critical number-three span of 450 feet, and a final span of 300 feet. The longest span crossed the swiftest part of the river and was subjected to the greatest concentration of ice flews. Huge concrete prows were constructed to protect the piers from the grinding assaults of enormous icebergs. Construction required sinking caissons into the river bottom for the piers and then building the spans on falsework—temporary piers and scaffolding erected on the ice after the river froze during the winter. The trick was to get the work completed and the span self-supporting before the ice went out.

In May 1910, as the number-three span was being put in place, the Miles Glacier unexpectedly began a rapid advance that pushed a concentration of icebergs toward the bridge site. The congestion was within hours of crushing the entire structure when the ice went out in the river and relieved the pressure. On June 19, 1910, the fourth and final span was put in place with cranes, and trains carrying construction material were soon running across it.

But the elements were not quite finished. That summer, the Childs Glacier just downstream of the bridge also began a rapid advance. It surged forward almost 200 feet, moving from within 1,775 feet of the bridge on June 3 to only 1,571 feet on October 5. This was as close as the bridge was long, and with the glacier's icy front towering some 200 feet above the river, one could only wonder what would happen next. The glacier advanced another 100 feet the following year but then began a gradual retreat. Records show that the annual snowfall at the bridge site during the construction years exceeded 400 inches. This was about twice the normal amount and proved that this herculean task had indeed been performed under almost impossible conditions.[62]

Besides icebergs, glaciers, and shifting sandbars, there was the notorious Copper River wind. It howled out of the canyon, and it swept with a fury across the delta. Railroad legend has it that a heavy chain was fastened at the entrance to the Flag Point Bridge at Mile 27. If the chain stuck straight out in the wind, engineers didn't dare to take their trains across the bridge.

The Copper River and Northwestern Railway reached Chitina at the confluence of the Copper and Chitina Rivers on September 12, 1910. But there were still sixty-five miles of rough country to traverse to reach Kennecott. The line required two more large bridges, less daunting than the Million Dollar Bridge, to be sure, but still major challenges. The steel Kuskulana Bridge was 525 feet long and consisted of three spans of 150, 225, and 150 feet. The center span hung 238 feet above the Kuskulana River and bridged the Kuskulana Gorge at Mile 146. The wooden Gilahina Trestle at Mile 160 was built on a grade and ranged from 80 to 90 feet high. It was 880 feet long and required half a million board feet of lumber. Both the Kuskulana Bridge and the Gilahina Trestle were built in the dead of the winter of 1910–11.

On March 29, 1911, the rails finally reached Kennecott. Unfortunately, tough Mike Heney did not live to see it. He had died the previous fall of pulmonary tuberculosis. The entire line from Cordova to Kennecott cost $23 million, then a staggering figure. In between Cordova and Kennecott were 196 miles of track laid over raging rivers, through narrow canyons, and across deep gorges. Excavations exceeded 5 million cubic yards, more than half of which was solid rock. A full 15 percent of the line was built on bridges or trestles, much of it across the Copper River delta. There were 129 bridges between Cordova and Chitina alone—an average of one per mile. The copper had better be worth it![63]

And it was. A latticework of tunnels was driven into the ore bodies of the Bonanza and Jumbo mines. Two long tramways were constructed up the mountainside, each capable of carrying about 400 tons of ore every twenty-four hours. The tramway to the Jumbo extended some 16,000 feet and began operations in 1913. Increased demand for copper as the United States became electrified, and inflated prices during World War I, spurred production ever higher. Soon the mines and mill were operating around the clock and employing about 500 men.

During the course of the Kennecott operations, workers mined 4.6 million tons of ore averaging an incredibly rich 13 percent copper. Nine million ounces of silver were smelted as a nice side product. According to one source, this netted the Kennecott Copper Corporation a $100 million profit on between $200 million and $300 million in ore sales. By volume, Kennecott operations rank eleventh among the great copper mines of the United States, but no other property could match Kennecott for the high percentage of its mineral content. If the ore body at Kennecott had been larger, it might have become the world's richest producer.[64]

Kennecott, the company, was synonymous with copper from the beginning, but Kennecott, the town, boomed right along with them. With the Copper River and Northwestern Railway its only link to the outside world, Kennecott was the classic company town. Everything that happened there was company-controlled—subject only to the whims of nature and the price of copper. The company built a hospital, company store, grade school, dental office, dairy, and bunkhouses along with the massive buildings needed for the operation of the mine. A recreation hall hosted town dances, Christmas festivities, winter basketball games, and those newfangled moving picture shows. There was an ice-skating rink, the obligatory mining-camp ball field, and even a tennis court. Where could one go for some real excitement? The only escape from such civility and order was in the bars and brothels of McCarthy, four miles back down the road.

In July of 1906, a prospector named John Barrett had staked a homestead of 296 acres at the mouth of McCarthy Creek where it enters the Kennicott River. The following year, when surveyors for the Copper River and Northwestern Railway came through, they quickly found that Barrett had staked the land most suitable for a turntable and switching operations. These had to be located there because of limited room between the glacier and hillside four miles farther up the valley at Kennecott. By the time the railroad was completed in 1911, the business opportunities of being on a railroad near a great mining center were too obvious to miss. As a company town, Kennecott was closed to all but its employees, so all of the other traffic—miners from the surrounding hillsides and in time even a tourist or two—ended up at the railroad station on John Barrett's land.

Copper, as it turned out, was not the only mineral that quickened the pulse of the area. In mid-July 1913, four men walked into the McCarthy Creek railroad station and whispered that clarion summons, "Gold!" They had come from the Chisana country (then called Shushanna) eighty miles to the east on the other side of the Wrangell Mountains. The quickest way to reach the new placer finds was to take the Copper River and Northwestern Railway from Cordova to McCarthy and then start hiking. Even though the last eighty miles involved crossing wild rivers, crevassed glaciers, and Skolai Pass, it was the most direct route to the new diggings.

The Chisana rush convinced John Barrett that things would never be the same again. Around the part of his homestead that contained the railroad station, he platted the town of McCarthy and started selling lots. Stores quickly opened to serve the Chisana miners, who frequently had to tramp back to town for supplies since it was the closest source around. Even with a ready supply of goods arriving on the railroad, merchants were still readily mining the miners. One grizzled prospector growled, "McCarthy prices are like our sheep, when they see us coming along, they go up."[65]

The Chisana rush rapidly peaked, but it was the impetus that spawned McCarthy's development. McCarthy became everything that Kennecott wasn't—everything the company wouldn't let Kennecott become. Both towns were legally dry, but where the company enforced prohibition laws at Kennecott, McCarthyites just laughed at them. Restaurants, pool halls, hotels, saloons, a dress shop, photography studio, garage and auto repair shop, shoe shop, hardware store, and more came to grace the town's streets. And, of course, there was the infamous "row" along McCarthy Creek. More than one Kennecott miner got so inebriated that he had to be taken home the next morning strapped to a dogsled. Then, after another month's wages, he would do it all over again. Kennecott managers frequently looked the other way and came to view McCarthy as a necessary safety valve. Besides, in this land where labor was sometimes hard to come by, miners could leave only via McCarthy. They seldom got past the town without going broke and then having to report for work again.

During the early 1930s, the Great Depression sent the price of copper plummeting to rock bottom; McCarthy's economy plummeted right along with it. Then an ominous blow fell. The railroad bridge across the Copper River at Chitina washed out in a 1932 flood. Kennecott Copper Corporation almost breathed a sigh of relief and used it as an excuse to shut down mining operations until the price of copper rose. Mrs. Roy Snyder, proprietor of the Mecca Cafe and Pool Hall in McCarthy, summed up the situation in a 1933 letter: "I don't know what we are going to do here if things don't pick up. It is terrible here and no prospect for better times until next year, if then."[66]

When President Franklin Roosevelt's New Deal legislation guaranteed the price of copper at nine cents a pound, the mine reopened—briefly. The end was sudden. In 1937, the company was painting its buildings a brand-new red; the following year, operations were closing down for good. The dismal price of copper was only part of the reason. Since the 1920s, it had become increasingly apparent that while Kennecott ore was the highest grade in the world, the ore body was comparatively small. Kennecott was running out of copper.

Long before this realization, the Kennecott Copper Corporation had used its Alaskan profits to diversify its holdings to include mines, railroads, and related operations all over the world, becoming the world's largest copper conglomerate. Kennecott, the company town, was abandoned almost overnight. McCarthy withered on the vine.

Kennecott's demise also spelled the death knell of the Copper River and Northwestern Railway. As with so many western railroads, the Copper River and Northwestern's final terminus at Kennecott was originally thought to be only a branch line before the railroad continued its drive north from Chitina up the Copper and down the Tanana River to Fairbanks, connecting all of southcentral Alaska. This never happened, in part because the harshness of the terrain on this initial leg gave the line more than its share of problems, but also because there was no economic pull nearly as strong as Kennecott had been to lure it farther into the interior of Alaska. The line had been built to serve but one master—the mines of Kennecott—and any thoughts of further expansion were quashed after the Alaska Railroad arrived in Fairbanks via Anchorage in 1923.

Kennecott Copper Corporation filed a notice of termination of service on the line with the Interstate Commerce Commission on September 13, 1938. With winter on the way and without waiting for a decision from the commission, the company ran the last train on the line on November 11. It pulled into the McCarthy depot with instructions for the station agent that he had an hour and a half to collect his papers and belongings and get on board.[67]

As railroads go, the Copper River and Northwestern's twenty-seven years of operation were brief. In its later years, locals came to say that the railroad's initials stood for "Can't Run and Never Will." Maybe. In retrospect, the Copper River and Northwestern stands as a most impressive achievement in the annals of western railroading. Such was the lure of copper.

Preserving the Bounty

As armies of gold seekers and others sought to take from the land, there were also stirrings that parts of it should be preserved inviolate at all costs. The first decade of the twentieth century saw the establishment of Alaska's two national forests and its first national park, initial lobbying for a second national park, and the first major confrontation over the development of its natural resources. While most Alaskans bristled at what they considered federal government attempts to impede the territory's growth, others cheered the reservation of these enormous tracts of lands.

In 1891, Congress passed the Forest Reserve Act, authorizing the president to place federally owned public domain lands into "forest reserves," the equivalent of what would soon be called national forests. Within a decade, three presidents transferred some 50 million acres nationwide into these reserves. These included Afognak Island north of Kodiak that was set aside by Benjamin Harrison in 1892 to protect its salmon-spawning streams, timber, and animals. The Afognak Island Forest Reserve, which later became part of Chugach National Forest, was the first federal conservation designation in Alaska.

After 1901, Theodore Roosevelt enlarged the system of forest reserves with a passion nurtured by his early days in the Dakotas, as well as his close association with the Boone and Crockett Club. Roosevelt named Gifford Pinchot, a well-connected advocate of scientific forest management, as chief of the Bureau of Forestry. Then, with Pinchot's wholehearted support, Rossevelt proceeded to set aside another 150 million acres of reserves over the next seven years. When development interests finally lobbied Congress to pass a revision to the 1891 law limiting such widespread presidential discretion, Roosevelt quickly huddled with Pinchot and reserved another 16 million acres of the best of the remaining parcels before the law took effect. In 1907, at Roosevelt's urging, this system of forest reserves became national forests under the administration of the U.S. Forest Service, and Pinchot was appointed its first chief forester.[68]

For his first Alaskan act, Roosevelt designated 4.5 million acres scattered among the islands of southeast Alaska as the Alexander Archipelago

Forest Reserve on August 20, 1902. Five years later, he signed an executive order creating Tongass National Forest with another 2.25-million acres. This first Tongass National Forest encompassed the mainland from Portland Canal north to the Unuk River on Behm Canal, essentially the area now designated as Misty Fiords National Monument. In February 1909, during Roosevelt's last full month in office, he combined the Alexander Archipelago Forest Reserve with these Tongass lands to make one national forest with almost 7 million acres. Over the ensuing years, more additions were made, and today the 17-million-acre Tongass National Forest is the largest in the United States.

Meanwhile in southcentral Alaska, Roosevelt created Chugach National Forest on July 23, 1907. Its name (CHEW-gatch) evolved from the Russian understanding of what the Natives around Prince William Sound called themselves, Chugatz or Tchougatskio. Originally, the Chugach encompassed about 5 million acres stretching across the Kenai Peninsula and Prince William Sound, but six major boundary adjustments, including adding and then deleting Afognak Island and later transferring areas to national-wildlife-refuge and national-park status, were made by 1980.

The purpose of both the Tongass and Chugach National Forests—indeed the initial purpose of all national forests—was to prevent the selective clear-cutting of the best and most accessible timber while leaving behind denuded and quickly eroded landscapes. The forest service's charge was the prudent management of renewable resources to ensure their longevity. But as was to be the case many times in Alaska during the next century, the creation of these first national forests left unsatisfied parties at both ends of the spectrum.

A majority of Alaskans were dismayed and some more than a little confused by the reservations. Their response was, "Hey, wait a minute, for a generation the federal government has been stifling growth up here through benign neglect, and now that it's finally paying some attention, it's telling us that we can't use or develop much of the best land." At the other end of the spectrum there were the diehard, "preserve the wilderness at any cost" advocates. One of them was John Muir, who crossed swords early on with Gifford Pinchot over the concept of multiple-use management, a dilemma—or reality, depending on one's viewpoint—that continues to confront land managers to the present day.

In southeast Alaska, where much of the territory's population then lived, the combined Tongass National Forest encompassed a high percentage of the land most suitable for homesteading and agricultural development. But at least there were trees in the southeast. Much of the land set aside for the Chugach was not tree-covered, and many thought that it was "at the very best too marginal an area to be reserved as a national forest."[69] But southcentral Alaska had coal, and there Theodore Roosevelt really threw the fat into the fire.

As the gold rush waned, many Alaskans were banking on coal to be the economic savior for the territory. As part of his Progressive thinking, however, Roosevelt issued an executive order in 1906 prohibiting additional coal mining on the public domain in Alaska. Coal reserves were further locked up by the initial creation of Chugach National Forest and its subsequent expansions to include the bulk of the Bering River coal fields around Katalla. Through a strawman named Clarence Cunningham, the Morgan-Guggenheim syndicate held many unpatented claims in this area. Pinchot encouraged Roosevelt to make the withdrawals to counter what Pinchot thought was the syndicate's plot to corner the market on Alaska's coal. For his part, Roosevelt believed that such withdrawals would encourage Congress to create a coal-leasing system that would both be open to competition and also provide a source of revenue.

After Roosevelt left office, his handpicked successor, William Howard Taft, continued these policies, but Congress failed to create a leasing system and the coal remained untouchable. Finally, Pinchot and Taft's pro-development secretary of the interior, Richard Ballinger, who at one point had represented Cunningham in his attempts to patent the claims, had a severe falling-out. The whole story is long and complicated, but the gist of this first Alaskan battle of conservation versus development is that it led directly to Pinchot's firing by Ballinger and Ballinger's subsequent resignation. At least indirectly, it fueled the political schism that saw Roosevelt and the Progressives challenge Taft as a third party in 1912.

For Alaska, the coal reservations had the effect of not only impeding mining operations but also discouraging railroad development. No one was very interested in financing a railroad that had to operate on imported coal. The result was a double whammy. First, railroads, steamships, and locals were paying $11.00 to $12.00 per ton for imported coal when high-grade coal was readily available from the Bering fields for $2.50 to $3.50 per ton. Second, coal tonnage that was historically a ready and reliable source of freight revenue for the West's railroads was nonexistent.[70]

Nowhere was the resentment higher than in Cordova, where the completion of the Copper River and Northwestern Railway to Kennecott was supposed to be just the first step in ensuring the town's economic future. Kennecott, after all, was initially only to be a branch line on a route that led all the way to Fairbanks. On May 4, 1911, with visions of tea being dumped into Boston Harbor, 300 Cordova businessmen and citizens armed themselves with shovels and marched to the wharf of the Alaska Steamship Company and proceeded to shovel several hundred tons of imported British Columbia coal into Orca Bay. When Richard J. Barry, the company's general agent, demanded that this Cordova Coal Party cease, he was met with continued shoveling and shouts of "Give us Alaska coal."

If there were indeed parallels to be drawn with the Boston Tea Party, some, including the *Seattle Times*, were quick to point out the decidedly Alaskan character of the Cordova affair. The Bostonians of 1773 had gone "in dead of night . . . and disguised as Indians," the paper noted, whereas the Alaskans marched unabashedly in broad daylight. Even the far-off *Philadelphia Bulletin* called the event "a demonstraton of a not unreasonable impatience with the dilatory federal policy relating to the development of Alaskan resources."[71]

Meanwhile in nearby Katalla, which was still hoping to become the gateway to the Bering coal fields, Gifford Pinchot was burned in effigy amid a flurry of posters plastered around town that proclaimed:

> PINCHOT, MY POLICY
> No patents to coal lands!
> All timber in forest reserves!
> Bottle up Alaska!
> Save Alaska for all time to come![72]

Those words turned out to be mild (and representative) compared to what journalist George E. Baldwin later wrote about Pinchot after the forester had visited Alaska: "When the high priest of conservation, the prince of shadow dancers, recently visited Alaska to gloat over his handiwork of empty houses, deserted villages, dying towns, arrested development, bankrupt pioneers, and the blasted hopes of sturdy, self-reliant American citizens, it is a striking comment on the law-abiding character of our people that he came back at all."[73] It's probably safe to say that Baldwin was not on John Muir's mailing list. Congress finally passed an Alaska coal-leasing bill in 1914.

The U.S. Forest Service was just one aspect of the federal government's presence in Alaska. Another was the still-neophyte national park system. The National Park Service was not created as a separate agency until 1916, but Congress began the national park tradition by setting aside Yellowstone as early as 1872. In addition to such acclaimed treasures as Yosemite, Glacier, Mount Rainier, and Sequoia National Parks, Congress created a number of Civil War–era military historical parks in the late 1800s before designating a national park to commemorate the Battle of Sitka. Established in 1910, Sitka National Historical Park was the first national park in Alaska. Over the years, its mission has expanded to tell the larger story of Tlingit and Russian cultural influences in southeast Alaska.

Interest in the cultural heritage of Sitka and all of Alaska had been growing thanks in part to the efforts of Governor John Brady and Sheldon Jackson's museum. In 1903, Brady collected about twenty totem poles from the Tlingit villages of Tuxekan and Klawock and a number of Haida villages on Prince of Wales Island. These totem poles were shipped to St. Louis and assembled as part of the Alaska Exhibit at the 1904 Louisiana Purchase Centennial Exposition. While bands played "Meet Me in St. Louis," hundreds of thousands of fairgoers got their first look at totem poles, other Alaska Native artifacts and crafts, and mounted Dall sheep and brown bears. To many, Alaska became much more than just exaggerated tales of the gold rush.

After the exposition, some of the totem poles were in such poor condition that they were sold, but others were displayed at the Lewis and Clark Exposition in Portland the following year before being shipped to Sitka. While hindsight might cast Governor Brady's collecting in much the same light as the Harriman Alaska Expedition's visit to Cape Fox, the end result was that the remnants of these poles form the basis for the totem park that still lines Sitka Sound adjacent to the 1804 battleground site. Because of decay and weathering over the years, only a few of the poles along the park trail still contain fragments of the original poles that Brady shipped to St. Louis, but his efforts were a critical step toward recognizing and preserving the multicultural heritage of southeast Alaska.

The creation of the new national park at Sitka was relatively noncontroversial because it involved a small area of land around a specific site. Those who dreamed of national-park status for vast sections of Alaska would face tougher challenges. No one dreamed a bigger dream—nor worked so hard to see it come true—than railroader, hunter, and naturalist Charles Sheldon. Born in Rutland, Vermont, in 1867, Sheldon was a graduate of Yale and a product of the eastern elite. Over the years, however, his Alaskan associates came to admire him as a man to be trusted on the trail no matter whether it led across frozen tundra or through the halls of Congress. After a law degree, Sheldon went to work in railroading and served as general manager of a railroad in Mexico from 1898 to 1902. During that time, his investments in Mexican mining operations allowed him to retire from active business at the early age of thirty-five.

A member of the Boone and Crockett Club, Sheldon counted C. Hart Merriam and Theodore Roosevelt among his many friends, and he was on a first-name basis with many of the participants of the Harriman Alaska Expedition. Bolstered by these connections, Sheldon became a tireless champion of the cause to preserve North American game animals, particularly mountain sheep.

Sheldon first acquired his interest in mountain sheep while working in Mexico. He hunted both desert and mountain bighorns. Then in the summer of 1906, Sheldon heard the call of Alaska. He hired Harry Karstens as a guide and packer and headed for the north side of Denali in search of Dall sheep. The pair began their quest along the moraine of the Peters Glacier, but finding no sheep there, they moved progressively eastward until they reached the headwaters of the Toklat River. "Sheep at last! I thought," wrote Sheldon, "but the field glasses revealed a grizzly bear walking along smelling the ground for squirrels or pawing a moment for a mouse. Under the bright sun its body color appeared to be pure white, its legs brown."[74]

But the sheep were close, and before Sheldon left that summer to the honking of sandhill cranes winging south, he observed hundreds of sheep in the rocky crags between the Toklat and Teklanika Rivers, including seven rams that he shot as specimens. Sheldon quickly determined, however, that his study of Dall sheep would not be complete unless he observed them for an entire year. "With this in view I planned to revisit the region, build a substantial cabin just by an old camp on the Toklat, and remain there through the winter, summer, and early fall."[75]

Sheldon did just that, returning to the shadow of Denali the following summer, again in the company of Harry Karstens. Sheldon's year in residence on the Toklat, from about August 1, 1907, to June 11, 1908, allowed a systematic, season-by-season observation of the Dall sheep, and his meticulous field notes and keen powers of observation set a high standard for all of the naturalists who followed him to Denali. Somewhat surprisingly, when Sheldon left Denali and Alaska in the summer of 1908, he was never to return. Perhaps he knew in his heart that what he had experienced would never be the same again and that he could not bear to see the change. But the passion he took with him from that year on the Toklat fanned the spark that slowly built into the creation of Alaska's second national park.

Using the bully pulpit of the Boone and Crockett Club, and enlisting the aid of a host of fellow conservationists, including Theodore Roosevelt himself, Charles Sheldon became a tireless advocate for the creation of a national park around Denali. It was the only way, he believed, to preserve the complete ecosystem of Denali and protect its sheep, bears, moose, and wolves from market hunters. Sheldon first put the park idea into words in his journal entry of January 12, 1908, penned during the winter that he spent in his little cabin on the banks of the Toklat. In that entry, he called it Denali National Park, just as he had always called the mountain by that name.

After years of lobbying, the authorization bill for the park finally passed. Sheldon's pleas for the name Denali, however, fell on political ears, and Congress chose the political, not the historical, name in creating Mount McKinley National Park. But it has remained Sheldon's Denali in spirit. In recognition of his efforts to bring it to fruition, Charles Sheldon was delegated to deliver the bill personally to President Woodrow Wilson, who signed it on February 26, 1917, and gave the pen to Sheldon.[76]

The establishment of Mount McKinley National Park was to exemplify the frequent gulf between congressional designation and funding. Congress did not appropriate funds for the park in 1917. Nor did it do so in 1918, when the report of the secretary of the interior noted that "the failure of Congress to provide funds for the administration of Mount McKinley National Park has again prevented the National Park Service from taking possession of that area."[77]

Finally in 1921, again at the urging of Charles Sheldon, the first appropriation of $8,000 was made, and a superintendent was appointed to organize a small ranger force to monitor poaching. Thanks to Charles Sheldon's unqualified recommendation, National Park Service director Stephen Mather was able to write, "There is no question in my mind about [Harry] Karstens being the man for the place."[78]

Years later, while studying the wilderness that Charles Sheldon labored so long to save, naturalist Adolph Murie wrote in his journal: "During various trips down the Toklat in the course of my field work, I usually stopped at [Sheldon's] cabin, lingered to examine the walls, the shelves, the wooden pegs used for nails. I would stand before the cabin and look across the gravel bars to the mountains, a scene Sheldon must have enjoyed. Although the cabin is deteriorating, and a swing of the river may destroy it suddenly, I have a feeling it should be left alone. I think that Sheldon, with his love for wild places, would like to have his cabin crumble to earth with age."[79] So it has.

The Day the Sky Turned Black

A row of volcanoes lines the Alaska Peninsula above the waters of Shelikof Strait as it separates the mainland from Kodiak Island. They lend a heightened air of uncertainty in an unpredictable land. Extinct volcanoes are one thing, but the distinction between dormant and active vents can be a fine one—undisturbed for millennia or suddenly and sharply awakened by unseen forces just below the earth's crust.

In 1799, the Russian-American Company established a trading post at Katmai just northeast of the mouth of the Katmai River. Koniag Natives traded at the post and soon established a little village at the site. A trading route connected the village on the eastern, or Pacific, side of the Alaska Peninsula with villages on the western, or Bristol Bay, side. This trail led west up the Katmai River and Mageik Creek, across a wide pass between 7,250-foot Mount Mageik and 6,010-foot Trident Volcano, and then down the valley of the Ukak River to Naknek Lake. Mount Katmai rose to the north of Trident Volcano and at some 7,500 feet was estimated to be one of the tallest points in the region. It usually took four days to travel between Katmai Village and Naknek on Bristol Bay. Although the 2,600-foot pass was relatively low, it was a tough route, notorious for fog, rain or snow, and a seemingly endless wind that roared through the gap.

After a sojourn in the Lake Clark and Iliamna country, John Clark and Alfred Schanz crossed the pass with dogsleds from west to east in February 1891 while returning to the coast. Schanz was in the field writing for *Frank Leslie's Illustrated Newspaper*, and his subsequent article was the first print reference to "Katmai Pass," even though Mount Katmai and the headwaters of the Katmai River were some distance to the north.

Seven years later, Josiah E. Spurr was traipsing around the Alaska Peninsula for the U.S. Geological Survey when his party also crossed the pass. Spurr characterized the two volcanoes towering above Katmai Pass as extinct, despite his report that Natives told him that one of them—which one Spurr did not say—occasionally smoked. Spurr reported extensive hot springs on the Katmai side of the pass and frequent earthquakes and "other evidences of volcanic activity." In fact, Spurr's party experienced a minor quake just after crossing. Dormant, or even active, might have been a better characterization.

Spurr left what was to be the only written description of the pass from this period, characterizing it even late in the season as "high, snowy, and rocky, and has no definite trail." Spurr went on to report, "The wind is often so cold and violent here, even in summer, that the natives do not dare to cross except in calm weather, for the gusts are so powerful that stones of considerable size are carried along by them."[80]

Meanwhile, Katmai Village had become enough of a settlement that it boasted 218 people in the 1880 census and 132 people in the 1890 census, mostly Natives. Shortly after Spurr's visit in 1898, there was excitement in the area when prospectors bound to and from the Nome gold fields began to use Katmai Pass as a shortcut, rather than sail the rocky waters of the Aleutians. That plan worked, assuming one could get a steamer between Katmai and Kodiak. The lone European trader in Katmai built a bunkhouse to accommodate travelers who were forced to wait for such connections.

But by and large, life was pretty quiet in Katmai. On the morning of June 4, 1912, most of the inhabitants of the village were on an extended fishing trip far away. Two families who remained grew increasingly nervous about the rumblings that had been coming from the mountains to the west. Deciding to leave the village, they got in a dory and rowed south toward Puale Bay.

Two days later, on the morning of June 6, the steamer *Dora* left Uyak on the western shore of Kodiak Island almost directly across Shelikof Strait from the now-deserted Katmai Village. Captain C. B. McMullen noted in his log that there was a strong westerly breeze blowing (no surprise there) and that the sky was clear (now that *was* a surprise). By 1:00 P.M., the *Dora* steered eastward to enter Kupreanof Strait at the northern end of Kodiak Island. Suddenly, a deep and powerful rumbling of tremendous volume was heard from the Alaska Peninsula. Captain McMullen looked astern and soon saw a thick, dark cloud rising into the air. He calculated the source of the cloud as Katmai Volcano, some fifty-five miles away. The cloud grew in size, and within two hours it had overtaken the vessel and engulfed it in ash.

Meanwhile, in the Ukak Valley on the western side of Katmai Pass, the only people around were a Native named American Pete and his party. The initial rumblings did not faze them, but then came the loud explosion. American Pete later reported: "The Katmai Mountain blew up with lots of fire, and fire came down trail from Katmai with lots of smoke. We go fast Savonoski. Everybody get in baidarka. Helluva job. We come Naknek one day, dark, no could see. Hot ash fall. Work like hell."[81]

The blast was heard as far as Fairbanks and Juneau, 500 and 750 miles away, respectively. The enormous cloud of ash that was immediately seen from Lake Clark and Cook Inlet settled over the Alaska Peninsula and swept eastward to blanket much of Kodiak and Afognak Islands. Unable to make a safe anchorage at Kodiak, the *Dora* turned back toward the open sea and headed for Seldovia with her decks covered with slippery ash. The town of Kodiak, 100 miles from the eruption, was plunged into a total darkness that lasted sixty hours. The more optimistic said that it was impossible to see beyond fifty feet. Others swore that one could not see a lantern held at arm's length. One ship's searchlight could not penetrate the thick haze beyond the ship's bow. Ash was two feet deep in places and so heavy that roofs caved in and buildings collapsed. One photograph from the aftermath of the catastrophe shows a stately upright piano sitting forlornly in a parlor piled high with drifts of white ash.

The revenue cutter *Manning* was in the harbor at Kodiak taking on coal when the suffocating veil dropped over the town. As ash clogged the streams and wells, the *Manning*'s captain, K. W. Perry, provided drinking water to the dazed townspeople. It was an eerie time. Lightning struck the wireless station on Woody Island and cut off communications with the outside world. Perry himself was unsure of the magnitude of the disaster. By noon on June 7, the cloud of ash had reached the eastern limits of Prince William Sound, some 350 miles away.

Two more major eruptions occurred that day. By now, the *Manning* was a refuge for hundreds of evacuees, frightened all the more by the reverberating sounds of ash-laden landslides roaring down the surrounding hillsides.

Ash fell as far away as Juneau, Ketchikan, and the Yukon. Fumes were reported in Vancouver. Dust falling there and throughout the Puget Sound region blotted out the sun and dissolved linen on the line with its acidity. Brass fittings on ships passing the Cape Spencer lighthouse 700 miles to the east were tarnished minutes after being polished. Closer to the epicenter, there were reports of vast fields of pumice floating in Shelikof Strait. It was like a sea of black polar ice. One witness swore that the accumulation of lava rock in the water was thick enough to support the weight of a man, although no one appears to have volunteered to test the claim. Miraculously for all of this, no one was killed. Had such an explosion and accompanying chaos occurred in a more populated area, it is almost impossible to imagine the resulting death and destruction.

The dust had barely settled when expeditions were organized to survey the extent of the disaster. Robert F. Griggs, a botanist at Ohio State University, led four expeditions to Katmai under the auspices of the National Geographic Society beginning in 1913. Returning to Kodiak on his second trip, Griggs was amazed at how quickly the vegetation on the island was recovering. Ground that was buried under barren ash a short time before was now lush and rich in grass as high as one's head. The salmon berries in particular seemed to be thriving.

On the mainland beneath the volcano, however, things were quite different. What was once the deep flowing mouth of the Katmai River was now a braided maze of quicksand and intertwining streams. Griggs couldn't even see the far shore and described the river here as "five miles wide and five inches deep." What was confusing, as well as somewhat disconcerting, was that in addition to the volcanic destruction, there was evidence everywhere of a very recent flood. High-water marks were all around, and the Russian Orthodox church in Katmai Village had been floated from its foundation and left to sink into the mud flats.

The deserted site of Katmai—never to be rebuilt—showed some signs of new vegetation in 1915, but the higher mountainsides were sparse. Alder seemed the most decimated of the trees and shrubs, but Griggs noted that this had its advantages, because it was much easier to pick one's way through their entanglements. With Griggs on this trip were B. B. Fulton, an entomologist, and Lucius Folsom, a Kodiak resident well versed in the currents of Shelikof Strait.

As the trio climbed toward the higher peaks, Griggs noticed that a tremendous column of steam 1,000 feet in diameter and more than a mile high was rising from a point where no volcano was indicated on his maps. "Even from our position," Griggs later wrote, "it was evident that this was at present the most active volcano of the district." Briefly, Griggs flirted with the truth and doubted the conventional wisdom about the source of the eruption. "And it was not at all certain but that this," he went on, "rather than Katmai, had been the seat of the great eruption whose effects we were studying; for, curiously enough, there has never been any very positive evidence, beyond the statements of a few natives who saw the beginnings of the eruption."[82]

Close as he was to the truth, Griggs went on to err in concluding that the ash deposits were deeper nearer Mount Katmai. He also surmised that given the size of the blast, the much larger crater atop Mount Katmai must have been the source of the eruption rather than the smaller vent. Griggs and his companions named Mount Martin to the south for George C. Martin, National Geographic's first man on the scene in 1912, and then climbed Trident Volcano before returning to the coast. An attempt to reach Katmai's broken summit would have to wait for another year.

The photographs that Griggs brought back from this trip caused great excitement, and National Geographic sent Griggs back to the mountain in 1916. This time, he was again joined by Folsom and also by D. B. Church, a photographer. Because of his thankless experience thrashing through trackless mudflows and crossing wild streams the year before, Griggs also employed a packer. Natives on Kodiak Island had been less than enthusiastic about venturing too close to the smoking mountain, so Griggs recruited Walter Matruken, a hard-bitten woodsman known as the "one-handed bear hunter of Kodiak Island."

The party made its way up the main stem of the Katmai River, past what Griggs named Fultons Falls after his companion of the previous year. Above the falls, the valley narrowed into a steep canyon of colorful rock strata. Here, too, there were marks of the powerful flood that was still so evident along the river's lower course. Finally, they discovered the source. During the eruptions, a dam of ash and debris had built at the head of the canyon. Melting snow and subsequent rains formed a lake that continued to rise behind the dam. The dam finally burst only a few days before Griggs's 1915 arrival, scouring the valley with its resulting torrent.

On this 1916 visit, Griggs and his party climbed to the summit of Mount Katmai—or at least what was left of it. The upper 1,500 feet of the mountain were gone, replaced by a huge caldera almost two miles square where the summit had once been. Peering over the rim of the caldera, Griggs looked into a massive crater that was filling with a green lake surrounded by numerous steam vents. All four explorers admitted to being a little nervous, even the one-handed bear hunter. They later made a second ascent of the peak to take more observations and photographs, and then turned their attention to finding the source of the clouds of steam continually rising to the west.

The group moved toward Katmai Pass. On July 31, 1916, Griggs was out ahead of the others and just about to turn back when he climbed a small hill atop the pass for one more look to the west. He was met with an extraordinary sight. "I can never forget my sensations," Griggs later wrote, "at the sight which met my eyes as I surmounted the hillock and looked down the valley; for there, stretching as far as the eye could reach, till the valley turned behind a blue mountain in the distance, were hundreds—no, thousands—of little volcanoes like those we had just examined. . . . If all of these vents were to be counted, their numbers would undoubtedly reach into tens of thousands. . . . It was as though all the steam-engines in the world, assembled together, had popped their safety-valves at once and were letting off surplus steam in concert."[83]

Time was too short in 1916 to explore the valley adequately, so Griggs returned with a larger party in 1917. He spent the intervening winter worrying that the smoking fumaroles would be gone by his return. Others in the group worried about just the opposite—that the valley was ready to erupt again. In time, the fumaroles in the Valley of Ten Thousand Smokes would close and cease to steam, but as remarkable as it is today, the valley was even more so in 1917. The Valley of Ten Thousand Smokes became the principal reason that President Woodrow Wilson created Katmai National Monument on September 24, 1918.

For years everyone, including Griggs, was certain that it was Mount Katmai that had erupted. Its huge crater and greatly reduced elevation seemed to offer irrefutable proof of such an event. Actually, the culprit that made so much noise and caused the resulting ash flows that turned the valley of the Ukak River into the Valley of Ten Thousand Smokes was Novarupta, a small vent situated just north of Katmai Pass that may not have erupted previously. Not until 1953 did geologist Garniss Curtis show that despite the Griggs report, the ash flow indeed thickened the closer one got to Novarupta. Novarupta, and not Mount Katmai, was the source. Underground, Novarupta, Katmai, and Trident all share one massive magma source. When magma burst through the overlying sedimentary layers and Novarupta exploded, the magma that was supporting Mount Katmai was drained away, causing its summit to collapse inward. Thus, the top of Mount Katmai did not blow off but rather sank inward, leaving its huge summit crater. When the last bit of lava oozed out of Novarupta's vent, it solidified and formed a lava dome, the tenuous cap that remains.

So how does the Katmai eruption compare on the scale of the world's volcanoes? In its sixty-hour eruption cycle, Novarupta disgorged an estimated 4.8 cubic miles, yes *miles*, of dense rock and another 3.6 cubic miles of ash. Only one eruption in historic times—that of Greece's Santorini Volcano in 1500 B.C.—displaced more material. Indonesia's infamous Krakatoa spewed out little more than half as much in 1883, but killed 35,000 people. A more recent eruption along the Pacific's rim of fire, the 1980 eruption of Mount St. Helens, packed only one-tenth the punch of Katmai. Core samples in the Valley of the Ten Thousand Smokes show evidence of at least ten previous eruptions during the past 7,000 years. Trident Volcano has erupted four times since the major 1912 event, most recently in 1968. Someday, and it is only a matter of geologic time, the sky will turn black again.

Postscript to an Era (1897–1915)

By 1915, *the American mining frontier had run its course, and the curtain fell on its grand Alaskan finale. Some people had even managed to be in on most of it. Dick Irwin was a case in point. Born in Canada too late to join the first great rush to California, Irwin was lured to Colorado in the 1860s by the discoveries at Central City and Black Hawk. In 1865, along with Fletcher Kelso and John Baker, Irwin located the Baker Mine on Kelso Mountain in the shadow of Irwin's Peak. What does that say about self-promotion? Irwin's Peak was later renamed after botanist John Torrey, but Dick Irwin built the first horse trail up neighboring Grays Peak and charged some of Colorado's first tourists a fee to ride up it.*

Next, Irwin was in Colorado's San Juan Mountains, then Custer County, where he helped to found the town of Rosita. When the cries of "Silver!" were raised on Colorado's Western Slope, Dick Irwin was there to incorporate the town of Irwin on August 21, 1879. Naturally, he boasted that it would become one of the grandest in the West. But the mining frontier was moving on, and Irwin stayed only about long enough to donate a bell for the first school. By 1884, he was in Coeur d'Alene, Idaho, before prospecting around Wagon Wheel Gap with a fellow named Nicholas Creede in the early 1890s. Dick Irwin never struck it rich, but neither could he ever cure the itch. Is it any wonder that he wound up in Nome in 1901?

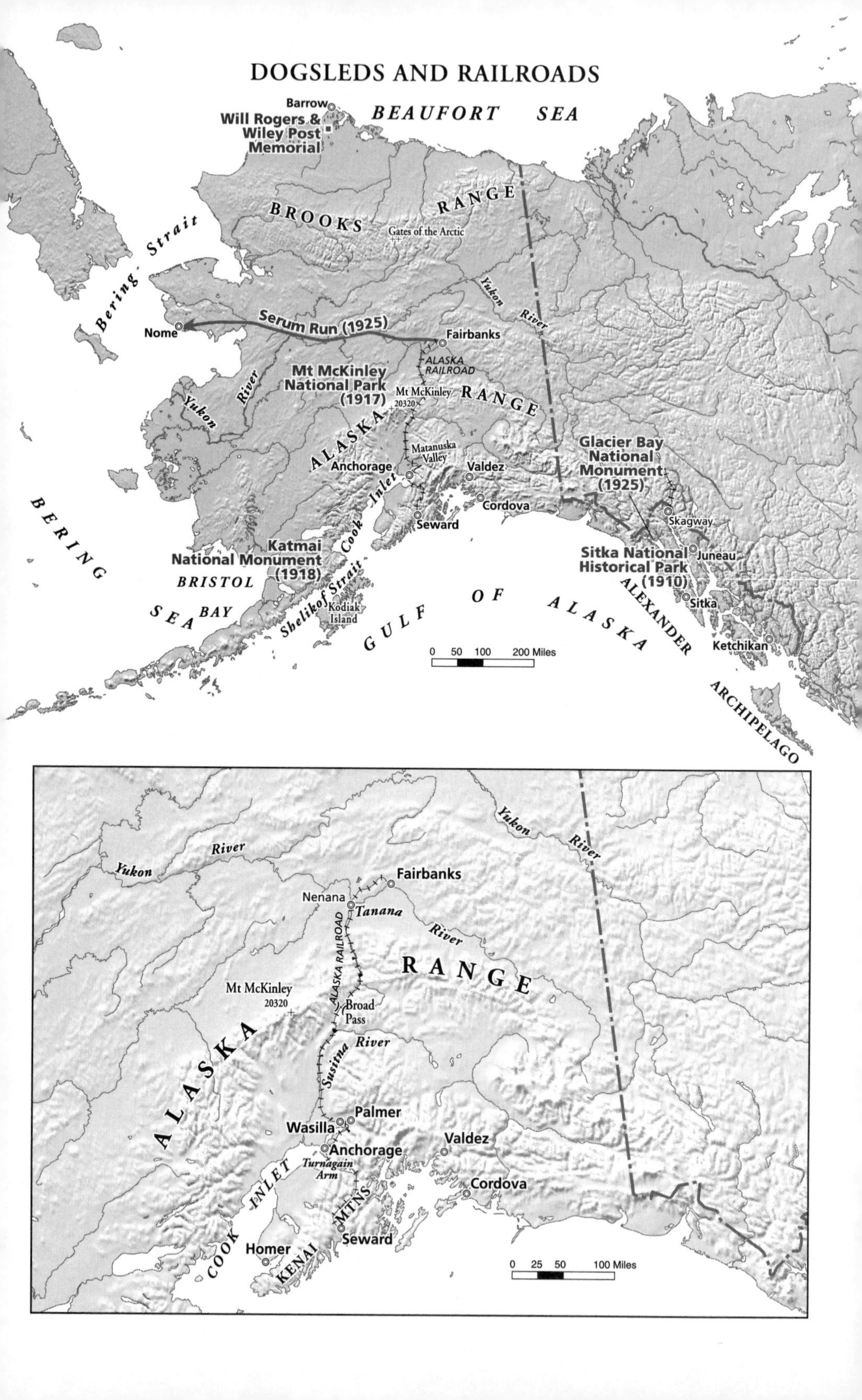
DOGSLEDS AND RAILROADS
Barrow
Will Rogers & Wiley Post Memorial
BEAUFORT SEA
BROOKS RANGE
Gates of the Arctic
Bering Strait
Yukon River
Serum Run (1925)
Nome
Fairbanks
ALASKA RAILROAD
Mt McKinley National Park (1917)
Mt McKinley 20320
ALASKA RANGE
Matanuska Valley
Anchorage
Valdez
Cordova
Seward
Cook Inlet
Glacier Bay National Monument (1925)
Skagway
Juneau
Sitka National Historical Park (1910)
Sitka
Ketchikan
BERING SEA
Katmai National Monument (1918)
BRISTOL BAY
Shelikof Strait
Kodiak Island
GULF OF ALASKA
ALEXANDER ARCHIPELAGO
0 50 100 200 Miles
Yukon River
Fairbanks
Nenana
Tanana River
ALASKA RAILROAD
ALASKA RANGE
Mt McKinley 20320
Broad Pass
Susitna River
Palmer
Wasilla
Anchorage
Valdez
Turnagain Arm
Cordova
COOK INLET
KENAI MTNS
Seward
Homer
0 25 50 100 Miles

BOOK FIVE

Interlude: The Calm between the Storms

1915–1941

Always, after any stampede, it's not the successes who build up the country. They go home with the stakes they made. It's the failures who stay on, decade after decade, and establish homes.

—Albert Ness to Bob Marshall, in *Arctic Village*, 1931

Alaska's Railroad

The enormity of Alaska's rugged terrain has always placed a premium on transportation. In a land where transport by dogsled across frozen creeks and hard-packed trails or by summer steamboat on the larger rivers was long the standard—and a highly preferred alternative to boggy, bottomless ruts—railroads were initially viewed as halfhearted dreams, and for good reason. Who, after all, has heard of the Alaska Pacific, the Valdez and Yukon, or the Alaska Home Railroad? These were just three of the dozens of "paper" railroads that incorporated with big Alaskan plans, but never laid a single rail. The two exceptions, the White Pass and Yukon and the Copper River and Northwestern, were only qualified successes that were nurtured and prodded along by the economic pull of the mining frontier. But even they couldn't complete a contemplated link between Alaska's southcentral coast and the interior waterways of the Yukon.

The first serious attempt to do so was born of the mind-set of an "all-American" route to the Klondike. As early as 1900, John E. Ballaine, a Seattle newspaperman and real estate developer, scouted the coasts of western Prince William Sound and the Kenai Peninsula for possible port sites from which to build a railroad inland. Ballaine rejected a site at the mouth of Ship Creek on Cook Inlet because of frequent winter ice floes, major silt problems, and tidal fluctuations that ranged as much as thirty-five feet. He also rejected a site on Passage Canal at the extreme western end of Prince William Sound. The waters there were deep and ice-free in winter, but a railroad route out of the mountain-ringed enclave would require extensive and expensive tunneling. The best site was at the head of Resurrection Bay, where Baranov's men had built the *Phoenix* over a century before.

In 1902, Ballaine organized the Alaska Central Railroad and dispatched a crew of surveyors to Resurrection Bay. Ever the real estate man, Ballaine also sent along his brother, Frank, to establish the townsite of Seward and to gain control of as much of the surrounding land as he could. In August 1903, Alaska Central construction crews began work on the Seward terminal, wharf, and dock facilities. By the end of 1904, they had completed 18.3 miles of standard-gauge track running north from town across the Resurrection River and over the low divide to the head of Kenai Lake.

Ballaine sold out his interest the following year, but with Fairbanks finally beginning to boom, the railroad's new owners pushed construction forward and reached Mile 47 just beyond Grandview. North of Grandview, construction difficulties through the heart of the Kenai Mountains and the financial Panic of 1907 soon conspired to slow the railroad's advance. By the fall of 1909, the Alaska Central was bankrupt, and its bondholders foreclosed.

The bondholders quickly reorganized the railroad as the Alaska Northern Railway and then vainly continued its push northward. The railroad skirted the toe of the Spencer Glacier, bridged the waters of Portage Creek and the Twentymile River, and then reached Kern Creek. The line's new name could not, however, change the harsh economic realities of Alaskan railroad operations. At Kern Creek, still in sight of the vast mud flats at the head of Turnagain Arm and only seventy-two miles from Seward, the Alaska Northern sputtered to a halt

Not only was the railroad far short of its goals of Fairbanks and the Yukon, but also it was without a ready source of revenue. It wasn't long before the entire route was a real mess, a would-be avenue of commerce whose light rail and untreated ties linked together a succession of tight curves, steep grades, narrow cuts and fills, and high spindly bridges. For a time, it appeared as if the Alaska Northern was going to wither on the vine.[1]

Meanwhile, Falcon Joslin, that "Harriman of the North," had been busy building the Tanana Valley Railroad out of Fairbanks. Joslin had done all right locally, but he had come up far short of his dream of a rail line across all of Alaska, linking Fairbanks with Nome to the west and Haines to the southeast. Joslin's voice, however, continued to be as shrill and incessant as the whistles of his locomotives in urging railroad construction. "Not a foot of railroad construction has been done in Alaska in two years," bemoaned Joslin. "None is planned. Railroad building has ceased."[2] It was a situation that Joslin found abominable. James Wickersham, the early champion of Fairbanks and by now Alaska's lone and nonvoting territorial delegate to the U.S. Congress, agreed and determined to do something about it.

In August 1912, after no small amount of arm-twisting by

Wickersham, Congress passed Alaska's Second Organic Act, the first significant change in Alaskan administration since the Organic Act of 1884. The 1912 act formally established Alaska as a territory and created an elected territorial legislature. While many hailed this "home rule" act as a huge step toward shedding Alaska's colonial status, Wickersham's critics charged him with a "sellout" because Congress placed numerous restrictions upon the powers of the new legislature. Those political arguments would continue until—some would say, "well past"—statehood, but Section 18 of the act was to have an immediate and lasting impact on Alaska's future. That provision granted the president the authority to study and recommend railroad routes that would best "develop the country and the resources thereof for the use of the people of the United States."[3]

President William Howard Taft, who had strongly supported the provision, wasted no time. On August 31, 1912, one week after signing the act, he appointed the Alaska Railroad Commission. It was statutorily composed of an army engineer, a navy engineer, and a civil engineer—who were all to be unconnected with any existing Alaskan railroad venture—and the unimpeachable Alfred H. Brooks. The president charged the commission with making its study and recommendations within an almost impossible ninety-day timetable, a schedule that was undoubtedly dictated more by Taft's pending departure from office than by his knowledge of the rigors of Alaska's far-flung geography.

After a hurried field visit and some frantic writing, the commission delivered its report on January 20, 1913, almost two months late. It was padded with prior government documents and transcripts of hearings and interviews, but it proved prescient in two major recommendations and one salient requirement: First, that a line be built to link the Copper River and Northwestern at Chitina with Fairbanks; and second, that the Alaska Northern Railway be extended into the Matanuska-Susitna Valley with a branch line to the Matanuska coal fields and then continued over the Alaska Range to the Innoko-Iditarod mining country along the Kuskokwim. But even with what turned out to be wildly optimistic construction estimates, the commissioners could not report that either route would sustain a profitable private business venture. Thus, the overriding requirement was that of government involvement, either by way of federal subsidies of a private line or a railroad operated outright by the federal government.

Taft, the conservative Republican, was strongly opposed to government operation of railroads. Given the restrictive economics of the

Alaskan situation, he did, however, support government ownership of the line if that was the only practical way that it was going to be built, so long as its operations were leased to a private company. But Woodrow Wilson, a Democrat with decidedly different views of the role of government, succeeded Taft as president in March 1913. The following December, Wilson, the bespectacled Princeton orator, stood before Congress and established the modern tradition of personally delivering his State of the Union message. In his first such address Wilson declared: "Alaska, as a storehouse, should be unlocked. One key to it is a system of railways. These the Government should itself build and administer, and the ports and terminals it should itself control. . . . But the construction of railways," the president continued, "is only the first step; is only thrusting in the key to the storehouse and throwing back the lock and opening the door. . . . We must use the resources of the country, not lock them up."[4]

With that clarion summons, the Alaskan railroad bills—introduced in Congress in the wake of the Alaska Railroad Commission's report—began to pick up steam. Alfred H. Brooks spoke in favor of one, noting that people would soon pour into Alaska because of the population pressures in what—with the admission of New Mexico and Arizona in 1912—had just become the "lower forty-eight." Delegate Wickersham had his own bill before Congress and urged that a government railroad would loosen the tight grip over Alaska held by the Morgan-Guggenheim syndicate through its ownership of the Copper River and Northwestern.

In the end, the Alaska Railroad bill that President Wilson signed on March 12, 1914, gave the president wide-reaching powers to choose a route and construct a railroad from the Gulf of Alaska to the Yukon. Where he did it, and how he did it, was largely up to him, so long as the aggregate mileage did not exceed 1,000 miles and the total expenditures over the life of the project—which were subject to continuing appropriations by Congress—did not exceed $35 million. An initial appropriation of $1 million was made as part of this enabling legislation.

To implement the project, Wilson created the Alaska Engineering Commission in May 1914 and turned to his secretary of the interior, Franklin K. Lane, to staff it. On Lane's recommendation, Wilson appointed William C. Edes of California, who had worked for the real Harriman and had extensive mountain-railroading construction experience in the Sierra and Siskiyou Mountains, commission chairman. Lieutenant Frederick Mears of the U.S. Army, who was a veteran of locating and operating the Panama Railroad in conjunction with the construction of the canal, and Thomas C. Riggs, Jr. of Utah, who was an old Alaskan hand and former chief of the Alaskan Boundary Survey, were also appointed.

The new commissioners lost little time. By midsummer, they had fourteen field parties surveying possible railroad routes across the breadth of southcentral Alaska. Crews investigated the routes north from Cordova and Valdez—with and without the existing Copper River and Northwestern—via some combination of the Copper, Tonsina, and Delta Rivers to the Tanana Valley and Fairbanks. Other parties worked over the route from the Kern Creek railhead of the Alaska Northern on around Turnagain and Knik Arms, up the wide Susitna Valley, across the Alaska Range at Broad Pass, and then down the Nenana to reach Fairbanks from the west. Throughout all of the preliminary discussions, tapping the Matanuska coal fields had loomed large as an immediate goal, and a team surveyed routes to them from the Susitna, as well as via the "backdoor" from Chitina. Finally, the Iditarod component of the 1912 report was not overlooked as a crew ran a line west from the Susitna to the Kuskokwim country. The result of each party was detailed maps at one inch to 400 feet.

With the reports of the 1914 field season in hand, the commissioners went to work to determine a final route and which existing lines, if any, the government would buy. In the end, the commission's recommendation and President Wilson's decision bespoke the obvious. Not only was J. P. Morgan's asking price for the Copper River and Northwestern a whopping $17.7 million, but also any deal with the Morgan-Guggenheim syndicate was apt to revive the past battles of the Ballinger-Pinchot feud over coal lands. No, the final answer was both politically and financially easy. On April 10, 1915, Wilson announced that the government would build its new rail line via the western route from Seward to Fairbanks and in the process acquire the Alaska Northern Railway.

The final price for the Alaska Northern's seventy-two miles of haphazard track was $1.15 million, well below the commission's appraised value of $1.5 million. The price included the right-of-way, yards, office buildings, dock, and waterfront at Seward. Hindsight would question the decision to buy the Alaska Northern rather than building a new port on Passage Canal and blasting a tunnel west through the mountains to the head of Turnagain Arm. Such a decision would have eliminated sixty-four miles of mostly tough, mountainous terrain, and even if more costly in its initial construction would likely have made up for that expense in reduced operating costs. Nonetheless, the Alaska Northern's bondholders got paid, and Seward got the nod to be the southern terminus of the Alaska Railroad. The Alaska Engineering Commission's initial construction estimate for the 471 miles from Seward to Fairbanks, not including the purchase of the Alaska Northern, was $22.6 million.

Congress's 1915 appropriation for the railroad was $2 million, with which to make a final location survey of the entire route, rehabilitate part of the Alaska Northern, and construct as much new line as possible. Less than three weeks after President Wilson's announcement, Frederick Mears arrived at the motley collection of tents and crude buildings at the mouth of Ship Creek, charged with building the 230 miles of new main line between the terminus of the Alaska Northern and the summit of Broad Pass. The tides and winter ice of Cook Inlet were still as disagreeable as when John Ballaine viewed them, but those problems be damned. Ship Creek was about to become a railhead.

The obvious first goal was to reach the Matanuska coal fields. The Alaska Northern had, after all, learned what it was like to be a railroad without a solid source of revenue. With large deposits around Chickaloon some thirty-five miles up the Matanuska River from its mouth on Knik Arm, Matanuska coal could both fire the railroad's locomotives on the construction drive north and also be shipped south to Seward for export to points in southeast Alaska. By the summer of 1915, 1,500 men were working along Cook Inlet "making dirt fly," and more than thirty miles of new roadbed had been cleared and graded by teams of contractors responsible for piecework sections. Day laborers were paid three dollars per day and then were charged one dollar per day for board. Much of the required heavy construction equipment was transferred to Alaska from the recently completed Panama Canal.

Ship Creek changed its name to Anchorage—taken from the Knik Anchorage—and quickly experienced an overnight boom of its own. It wasn't exactly gold fever, but the next best thing to gold to encourage a boom was the hoot of a steam whistle and the signs of railroad construction. By the end of the summer, Anchorage boasted about 2,000 inhabitants, and 749 town lots had been auctioned off by the government for a total sale of $164,210, or about $220 per lot. Bets were that the thirty-eight-mile spur from Matanuska Junction, where the main line crossed the Matanuska River and curved west, to the Matanuska coal fields would be opened in 1916. "From that point," one source reported, "the question of how soon the first locomotive whistle blows in Fairbanks will depend very largely upon how fast Congress makes available the money necessary for construction work."[5] The report proved correct, and the first shipment of Matanuska coal arrived in Anchorage on August 17, 1916.

Two months later, the full thirty-eight-mile Matanuska branch line was completed to Chickaloon, but southeast of Anchorage there was still a large gap to be bridged before Matanuska coal could reach the docks at Seward. In the summer of 1917, Chairman Edes confessed to Secretary Lane that the seventy-two miles of the Alaska Northern were in such a sorry state that it would take an estimated $4.5 million to make grade reductions and strengthen trestles along the route in order to handle heavy coal trains. Maybe the line's purchase price had not been such a bargain after all.

If rehabilitating the dilapidated Alaska Northern was a challenge, the majority of the forty-four miles between Kern Creek and Anchorage that bordered Turnagain Arm was a nightmare. The twists and turns of the new railroad had to be blasted and picked out of crumbling cliffs. The roadbed was exposed to frequent rockslides and snowslides thundering down from above and relentless tidal surges driving in from the sea. And then there was the wind. Sometimes it blew so strongly that four-cylinder gasoline rail cars trying to move against it were stopped dead on the rails.

Meanwhile, Thomas Riggs had been given charge of the northern division between Broad Pass and Fairbanks. The task here was no less daunting. In many locations the challenge was to find the bottom of—or at least adequately secure footers and pilings into—the quagmire of the summer tundra. Everyone except Fairbanks merchants agreed that it didn't make much sense to start construction from Fairbanks, because the town was located at the end of the proposed line and on the far side of the Tanana River. Any construction equipment and materials that reached Fairbanks had to come up the Tanana only to backtrack along its northern banks to its confluence with the more rambunctious Nenana. Consequently, the commissioners chose to establish a construction camp at the mouth of the Nenana, where the main line would cross the bigger river, and then build south from there. Lots were laid out in the new town of Nenana, and the right-of-way was graded south toward what would be another source of revenue, the Nenana coal fields.

But the railroad fortunes of Fairbanks were not completely on hold. Now it was Falcon Joslin's turn to cash out. The Tanana Valley Railroad was in a sorry state, and Joslin's dreams of an east-west line spanning Alaska had long since been dashed, when the Alaska Engineering Commission decided in 1917 to purchase the line for $300,000. This price was $90,000 below the commission's appraisal and far below the Tanana Valley's own valuation of its property and equipment at $2.1 million. The action gave the government railroad undisputed control of the Fairbanks yards but was done mostly to provide yet another point of construction. A roadbed was eventually graded westward from the Tanana Valley tracks to the north side of the Tanana River at Nenana, and narrow-gauge equipment operated over it until there was a hookup with the main line after completion of the Tanana Bridge.

By the end of 1917, Alaska's new railroad had made great progress in only three years of construction work, but it had also managed to spend or commit almost $27 million out of the $35 million aggregate expenditures cap spelled out in the 1914 legislation. Even more ominous, it had failed to complete the link between Seward and Anchorage. But there were even bigger problems "outside." The United States was finally embroiled in Europe's Great War, and Woodrow Wilson, who had once spoken so adamantly about opening up Alaska's resources, was now preoccupied with making the world safe for democracy.

Alaska was far removed from the battlefields of France, but the war's impacts were significant nonetheless. Frederick Mears resigned from the railroad in January 1918 to accept an army colonelcy and command of the Thirty-first Engineer Regiment. Mears led the regiment overseas to work on the French railroad system. Thomas Riggs stayed closer to home, but he also left his position on the commission that April to accept appointment as Alaska's territorial governor. Many others joined the armed forces or became committed elsewhere. The war sped the decline of Alaska's population, which had already been decreasing since the waning of the gold rush. In 1918, the railroad's workforce was half that of the prior year. Labor unrest, frequently fired by socialist rhetoric, kept officials uneasy and on the lookout for sabotage.

Despite the slower pace that summer, Chairman Edes was finally able to drive the last spike between Seward and Anchorage at Mile 78.5 on Turnagain Arm on September 10. But the connection to the Kenai proved fragile at best. From November of that year until the following July, a massive snowslide north of the Grandview summit blocked the line. Another slide at Mile 80 shut down more track and increased the length of the bottleneck to almost forty miles. Passengers, mail, and freight were shuttled between trains waiting on either side of the gap by dogsled. It was a tried-and-true Alaskan solution.

Frederick Browne replaced Riggs as boss of the Fairbanks division and took advantage of the frozen ground that winter to work south from Nenana to the Burns coal mine at Mile 363 just north of Healy. Rails reached the mine on April 4, 1919, and coal was hurriedly shipped north to Nenana and fuel-hungry Fairbanks before a temporary trestle across the ice of the Nenana River at Mile 370 was swept away by the spring thaw. War and weather weren't the only enemies. A deadly influenza epidemic hit Alaska hard, and in April 1920 at Nenana alone forty-three people, more than one-tenth of the town's population, died.

The light at the end of the tunnel was dim at times, but it was there. Chairman Edes, in poor health and worn out from five years of nonstop railroading, resigned from the commission, but Colonel Mears was soon back from the war and ready to lead the completion efforts as chief engineer. By the end of the 1920 construction season, the remaining gap in the line was closing, but it still stretched more than eighty miles between Healy at the mouth of the twisting Nenana River canyon and the long crossing of the Susitna River at Gold Creek. Centered on the Broad Pass area, these miles also contained tough crossings at Hurricane Gulch and Riley Creek, both deep incisions that required great steel bridges.

Difficulties and delays were the norm. At the Susitna crossing at Mile 264, one pier collapsed into the river because of improperly mixed cement. It was repoured, and crews hurriedly erected the steel for the 504-foot main span even before the new concrete was completely cured. The first train rolled across the Susitna River on February 6, 1921. Twenty miles north at Hurricane Gulch, it took until August to finish the bridge that carried the steel rails 296 feet above a cascading torrent. The centerpiece of the Hurricane Gulch Bridge was a massive 384-foot arch span, flanked by 120-foot truss spans and then another 240 feet of steel viaduct on the north end. At Riley Creek just south of the McKinley Park station, concrete foundations were poured in the dark of December with little more than six hours of daylight per day. A 570-foot cantilever bridge of box girders supported by five four-legged towers was anchored to the south side, and another 330 feet of wooden trestle were built to link it with the high ground on the north. With the line hacked and blasted through the Nenana Canyon, that left only the mighty Tanana to be crossed.

Originally, the plan was to cross the Tanana via a 1,600-foot system of piers and multispan bridging built out from the flats along the low riverbanks. A 200-foot center lift span would be raised and lowered to accommodate steamship traffic bound for Fairbanks. But Frederick Mears was never keen on the idea. For one thing, he was concerned about the continuing operational costs associated with the lift span, but mostly he was worried about the massive ice sheet that formed along the broad flats each winter. The ice sheet was regularly three to four or more feet thick. Each spring the river would flow under and around it until rising waters eventually floated the whole thing and sent it churning and grating downriver. Concrete piers driven to bedrock were impossible in the deep silt and gravel of the riverbed, and Mears could easily envision the entire line of piers being picked up with the ice and swept away downriver. No, there had to be a better solution.

Moving about a mile upriver to above where the Tanana spread out, Mears chose a site where the main river could be bridged in a single 700-foot span. Work began on two gargantuan pier foundations in the spring of 1922. After the steel erection was complete, the bridge deck hung forty-seven feet above the average high-water mark of the river to permit easy steamship passage without a lift span. At the time of its construction, the only simple truss span in the world that was longer was a 720-foot span in a crossing of the Ohio River. The Tanana Bridge and its approaches, including hundreds of feet of wooden trestle on each side, alone cost about $1.3 million. No wonder that the original $35-million cap was a thing of the past! The first train chugged across the structure on November 23, 1922, and it was hailed as complete on February 27, 1923. With the Tanana breached, all that remained was to widen the three-foot rails to standard gauge along the fifty-five miles between Nenana and Fairbanks. Work started on this project in early April and was completed by June 15.

After nine years of construction and by now some $60 million, the timing of the completion was crucial because important visitors—President Warren G. Harding and party—had been planning all spring to visit. In fact, Warren and Florence Harding had been excited—*really* excited—by the prospect of visiting Alaska for more than twenty years, ever since hearing about a friend's adventures there during the Klondike gold rush. But Florence's ill health and Warren's political commitments always postponed any definite travel plans. Now in the summer of 1923, Florence was adamant that they finally make the trip as part of a grand western tour. She was momentarily well, although the president was increasingly ill from the twin strains of an undiagnosed heart condition and a rapidly disintegrating presidency.

The presidential party, including Secretary of Commerce Herbert Hoover and his wife, sailed from Tacoma on board the SS *Henderson*, escorted by the destroyers *Cory* and *Bull.* Florence Harding first got her long-held wish to see Alaska off the shores of Annette Island on July 8, 1923. Alaska territorial governor Scott Bone, whose appointment had been a reward for having worked on Harding's campaign, met the party there and escorted it northward, calling at Ketchikan, Wrangell, Juneau, and Skagway. Along the way there were gifts of salmon, berries, and wildflowers, and local seafood began to be routinely served in the *Henderson*'s mess. Then, while the president played endless rounds of shuffleboard and poker, the *Henderson* and its escorts crossed the Gulf of Alaska en route to Seward. As the ships passed between Rugged and Cheval Islands and into Resurrection Bay, Governor Bone grandly christened the entrance to the bay the Harding Gateway. Later, the U.S. Geological Survey also gave the president's name to the massive icefield atop the Kenai Peninsula.

At Seward, the presidential party transferred to a special train, including the presidential car, *Denali*, for the journey north to Fairbanks. James G. Steese had replaced Frederick Mears as head of the Alaska Engineering Commission and was now president of the Alaska Railroad. Steese ordered a Dodge roadster specially equipped to run on the rails to follow the train, and Florence Harding rode a ways in it with Steese to look at the scenery. At Chickaloon, the president climbed 190 wooden stairs to take in the view despite looking exhausted and breathless halfway up. Few seemed to think much of it at the time. Later, Harding sat in the engineer's seat of the locomotive as the presidential special made the run between Wasilla and Willow. The presidential special stopped for twenty minutes at McKinley Park station and Superintendent Harry Karstens joined the president's party for the trip to Nenana.

As it turned out, Alaska, that land of extremes, was in the midst of a summer heat wave. On July 15, 1923, in temperatures near ninety degrees Fahrenheit, Harding tapped on a golden spike at the northern end of the Tanana Bridge to commemorate the Alaska Railroad's long-anticipated completion. The president kept his suit coat on despite the heat and then made a couple of errant blows at the iron spike that was to replace the ceremonial one. From Nenana, the party moved on to Fairbanks with plans to return to the coast via the Richardson Highway and the Copper River and Northwestern. This itinerary was canceled, however, when Florence Harding suddenly became ill. Two days later, after a purposely slow train trip to lessen the jerks and rocking motions, the party was back in Seward. There they reboarded the *Henderson* and sailed eastward, calling at Valdez, Cordova, and Sitka. At the last stop, Harding watched seamen load heavy boxes of crabs on board.

Sometime between the time the *Henderson* left Sitka and its arrival in Vancouver, British Columbia, Harding ate some shellfish, almost certainly some of those recently delivered crabs, and became sick. Others in the party, including Mrs. Lou Hoover, were similarly affected. Where the crabs came from, when they were served, and by whom were hotly debated questions in the aftermath, but it seems clear that rather than a sinister plot to kill the president, the culprit was simply shellfish poisoning. A little over a week later, on August 2, 1923, Warren G. Harding lay dead in a San Francisco hotel room, the victim of a probable heart attack after his condition was exacerbated by poor medical advice and further weakened by the bout of food poisoning.

A few weeks after the Hardings' visit, passenger and freight traffic began in earnest along the entire route. Standard-gauge rails now linked Seward with Fairbanks, but the Alaska Railroad was far from a line in tip-top condition. Realigning grades, strengthening bridges and trestles, and reworking the ballast along frost-heaved portions of roadbed became continuing rituals—far more so in Alaska than the routine maintenance of other roads. But the route was open, and two significant events quickly followed: the expansion of the railroad into the steamboat business and the opening of the floodgates of tourism.

Riverboat service with railroad connections was begun almost immediately from Nenana down the Tanana and Yukon Rivers to Holy Cross. The White Pass and Yukon, whose directors had grumbled plenty over the years about the prospect of their privately financed railroad competing with a government-run operation, now howled even more as their transportation lock on the upper Yukon was challenged. A threatened steamboat war along the Yukon was averted through an uneasy peace that saw the government boats of the Alaska Railroad's River Boat Service operate on the length of the Tanana and on the Yukon between the mouth of the Tanana and Holy Cross. The White Pass–controlled American Yukon Navigation Company ruled the remainder of the Yukon from the mouth of the Tanana upstream to its rail terminus at Whitehorse. White Pass boats were allowed to call at Nenana and make connections there with the Alaska Railroad, but they were to chug up the Tanana without pause and offer no competition on it.

Such improvements in the speed and relative reliability of travel inevitably encouraged visitors to come and see the sights of this grand land. In particular, the completion of the Alaska Railroad had a profound impact on Mount McKinley National Park because it offered relatively easy access to a heretofore remote area. The railroad had barely been completed to Riley Creek when Harry Karstens welcomed the first tourists to Mount McKinley National Park. Accommodations were leftover construction cabins, but that didn't deter the determined.

For years, the plush facilities were at Curry, a depot and hotel built in 1922 some twenty miles north of Talkeetna and about 100 miles south of the park entrance. Passenger trains routinely stopped here overnight. Known during the railroad's construction days as Dead Horse Camp, its name was changed in 1922 to honor Charles Forrest Curry, a California congressman. A touch of luxury in an otherwise rugged land, Curry's hotel boasted a swimming pool, tennis court, and a three-hole golf course. Subsequent competition from the McKinley Park Hotel and faster through-train service led to its demise, and the entire structure burned in 1957. Prior to the Second World War, however, Curry set the tone of things to come.

Back in 1922, when the railroad's completion had at last seemed secure, the name of the Alaska Engineering Commission was replaced with the designation "The Alaska Railroad." The capitalized *T* in the article was no mistake. Other railroads laid tracks across Alaska's wild terrain, but none came to embody more of its spirit and form a more important artery of its commerce than the Alaska Railroad. The White Pass and Yukon and the Copper River and Northwestern became legends; the Alaska Railroad became Alaska. It was, and remains, Alaska's railroad.

The First Iditarod

Nome was still a very isolated place in the winter of 1925. The ships that had crowded its beaches during the gold rush were far fewer now and came only during the summer months when the ice was gone from Norton Sound. The first airplanes to visit Nome arrived in 1920, but air travel this far north was still a decided novelty. Even the most daring pilots were smart enough not to challenge the numbing Arctic cold in open cockpits. The only reliable way in or out of town in the winter was by dogsled. It was a grueling journey that frequently took thirty days between Nome and Nenana on the recently completed Alaska Railroad. Nome's only other link to the outside was by radiotelegraph, or wireless as it was called—a lonely series of dots and dashes that worked as long as there was not too much atmospheric disturbance from the northern lights. Nome was used to this isolation, but in January 1925 there was trouble in town.

The lone doctor in town that winter was Curtis Welch, a graduate of Yale who had turned down a lucrative practice in southern California to come north to Nome with his wife, Lula, in 1906. There were a lot fewer people in town now, but Dr. Welch still found great reward in fulfilling the role of country doctor, ministering to a population of about 1,500, two-thirds of them white and one-third Inupiat Eskimo. In addition to the routine of new babies, old deaths, and inevitable injuries, sick calls in the dark of winter were the norm. Children down with sniffles, sore throats, and coughs were not uncommon, particularly given the fact that many played and attended school together.

This routine was rudely interrupted in mid-January when two Native children died from what appeared to be severe coldlike symptoms. Dr. Welch was both disturbed and puzzled. He could not pinpoint the killer. Then, on January 21, the doctor was summoned to the home of a six-year-old boy who had similar symptoms as well as a high temperature and very shallow and labored breathing. Examining the boy's mouth, the doctor found the telltale membrane indicative of diphtheria. Caused by bacteria, diphtheria was a highly contagious, communicable disease that blocked air passages and if untreated could cause death. Because of lower immunities, Nome's Native population was particularly at risk, just as it had been during prior outbreaks of measles and flu. It was difficult for Dr. Welch to imagine how this killer had raised its ugly head in the isolation of wintry Nome, but it had.

Both the treatment of those infected by the bacteria and the inoculation for those exposed or likely to be exposed depended on an antitoxin serum. Dr. Welch quickly checked his supply and found grim news. Despite his request for an increased quantity the summer before, he had only 75,000 units of antitoxin on hand. One thousand to 3,000 units were enough to prevent the spread of the disease in those exposed, but the average victim might require 30,000 units to combat the disease. Within ten days, Dr. Welch was to record five deaths, twenty-two diagnosed cases, thirty suspected cases, and another fifty persons who had definitely been exposed.

Dr. Welch and George Maynard, Nome's mayor and the publisher of the *Nome Nugget*, immediately formed the Board of Health and imposed a strict quarantine across the town. Welch sent frantic telegrams to Seward, Juneau, and Fairbanks, reporting Nome's plight and requesting all available serum. The largest supply was thousands of miles away in Seattle, but Dr. J. B. Beeson, the surgeon at the Alaska Railroad hospital in Anchorage, located 300,000 units there. If delivered to Nome quickly, it might be enough to stem the tide until a greater quantity could arrive from the lower forty-eight.

Among those who responded to the distress call was the town of Fairbanks. Fairbanks was quite proud of its budding air operations, and several hardy souls volunteered to uncrate their machines that had been disassembled and stored for the winter and try to make the trip. Territorial governor Scott Bone pondered the offer seriously, but in the end vetoed it. Bush pilots would soon change the face of Alaska, but airplanes were still a little new in 1925 for the race to Nome. If a plane went down, the loss would be twofold: not only a brave pilot but also more victims in Nome because quite likely the essential serum would be lost as well.

No, Bone thought, this was a job for the toughest of the tough. Bone turned to the sled drivers of the Northern Commercial Company, which operated the 674-mile route between Nenana and Nome. The company sent out a flurry of instructions and asked its sled drivers to attempt to make the normal thirty-day run in as little as fifteen days. Dogsledding, of course, was a tried-and-true form of winter transportation in the Arctic. Alaska Natives had long used dogs to pull sleds and haul supplies.

Russian trappers, Hudson's Bay Company traders, and early prospectors all came to rely on sled dogs for winter transportation across rugged terrain. Not surprisingly, friendly competitions sprinting up the final hill of a journey soon led to organized racing. In 1908, the Nome Kennel Club sponsored the first All-Alaska Sweepstakes Race, a demanding 408-mile round-trip run across the Seward Peninsula between Nome and Candle. Of the trails that led to Nome, one had been blazed in 1910 and 1911 all of the way from Seward via the gold camp of Iditarod.

In Anchorage, Dr. Beeson packed the glass vials of serum into a cylinder container, wrapped it in padded quilts, and delivered the twenty-pound package to the Anchorage train depot on January 26, 1925, for the rail leg to Nenana. The temperature was a nose-tingling minus-forty degrees Fahrenheit at 11:00 P.M. the following evening when the first driver in the relay, "Wild Bill" Shannon, secured the package to his sled and hurried west out of Nenana. A succession of drivers received the precious cargo and sped it down the Tanana and Yukon, averaging about thirty miles a leg. Drivers included Natives Johnny Folger, Sam Joseph, and Titus Nickoli. When the twelfth driver, Charlie Evans, left Bishop Mountain early on the morning of January 30 for the final push into Nulato, the temperature was minus-sixty-four degrees Fahrenheit, and the serum was fifty-five hours out of Nenana.

Meanwhile in Nome, local mushers were preparing to send out their best drivers to intercept the serum delivery and carry it the remaining miles into town. The town's top sled driver was Leonhard Seppala, a forty-eight-year-old Norwegian and three-time winner of the All-Alaska Sweepstakes Race. Seppala selected twenty of his best dogs and sped out of Nome heading eastward, intending to stage some of the dogs at posts along the route as fresh replacements for the return trip. In the lead position was his favorite dog, a twelve-year-old gray husky named Togo—after the Japanese admiral who defeated the Russian fleet during the Russo-Japanese War. Seppala left behind a large black dog named Balto, whom he considered more of a solid freight hauler than a team leader.

Seppala intended to meet the serum carriers at Nulato, but the west-bound teams made such good time that he almost missed them. As Seppala mushed past Henry Ivanoff just west of Shaktoolik—some 140 miles west of Nulato—Ivanoff frantically flagged him down. The serum was only 169 miles out of Nome. The exchange was made, and Seppala started back westward. Earlier that day, he had crossed the ice of Norton Bay, a north coast inlet of Norton Sound, in order to avoid the lengthier shore route. Now, however, the wind was roaring from the northeast, creating a dangerous condition that tended to push the ice out of the bay and away from shore. What if he encountered open leads in the darkness or, worse yet, found open water between him and shore after he had battled the twenty miles across the bay? Like everything else on the run, it was a high-stakes gamble. Seppala chose the shorter route across the ice and won. By Saturday evening, January 31, Togo guided Seppala's sled up the north bank of Norton Bay and to the safety of a Native cabin.

Now, for the first time in over 500 miles, the delivery effort paused as Seppala rested briefly, hoping that the roaring blizzard would subside. Early the next morning, when the storm showed no signs of slackening, Seppala knew that he could wait no longer. He started westward again, finally arriving at Golovin. Seppala, Togo, and their team had carried the serum 91 miles, almost twice the distance of any other driver and through certainly the most severe of the horrendous conditions along the route. Togo settled down to a well-deserved rest, but up ahead on the trail the freight hauler Balto was about to go to work.

Two other Nome sled drivers, a big Norwegian named Gunnar Kaasen and a veteran racer named Ed Rohn, were dispatched eastward in support of Seppala, figuring to relieve him over the last fifty-three miles between Bluff and Nome. At Bluff, Charlie Olson, the proprietor of the Bluff roadhouse, was sent eastward across Golovin Bay to wait for Seppala at Golovin. After Seppala and his beleaguered team pulled into Golovin, Olson left there at 3:15 P.M. on Sunday, February 1, and made the return trip to Bluff. The temperature continued to hover around minus-thirty degrees Fahrenheit, and the wind was still blowing about forty miles per hour. Olson made it back to his roadhouse—barely—and handed the serum package to Gunnar Kaasen.

Perhaps because of his own size, Kaasen knew that strength and endurance were every bit as critical as speed in a race. Kaasen worked for the same mining company as Seppala, and when he was asked to head east to assist in the relay, Kaasen without hesitation picked Seppala's Balto to lead his team. Seppala had named the dog after Lapp Baltow, another fellow Norwegian who had been with Fridtjof Nansen on his explorations of Greenland.

It was a thirty-four-mile run between Bluff and Port Safety, where Ed Rohn was waiting to make the final twenty-mile sprint into Nome. Rohn had a crack team and had won every race he had entered during the preceding year. When Kaasen left Bluff, the blizzard was still howling. Olson urged Kaasen to delay his trip, but like others before him, Kaasen pushed off into the teeth of the gale.

Before long, drifting snow obliterated the route, and Kaasen was forced to grab Balto's harness and help break trail. Somehow, each time they lost the trail, Balto managed to find it. In time, they were back on hard pack and flying across the Topkok River. Suddenly, Balto stopped short and refused to budge despite Kaasen's commands. When the driver walked forward to investigate, he found his leader standing in water from a crack in the ice. Had Balto not stopped, all of the team would have been drenched and quickly frozen.

As Kaasen drove on, unbeknownst to him, the Board of Health in Nome had telephoned a message to Solomon, at the eastern end of the Nome telephone system, that because the weather was so abominable Kaasen should wait at Solomon for conditions to improve. Kaasen and Balto never got the message. They continued past Solomon and drove on until a hurricane gust flipped their sled. By the time Kaasen untangled the team and righted the sled, he felt for the serum package and was horrified to find it missing. Frantically, he groped through the blowing drifts in the Arctic darkness until he found it. After more than 600 miles, the precious cargo had almost been lost.

At 2:00 A.M. on Monday, Kaasen reached Port Safety. The wind was finally dying down, and what remained was now at his back. Balto and his team had covered the last twelve miles of trail in just eighty minutes. The cabin at Port Safety was dark; no dog team stood ready to relieve him; Ed Rohn was inside asleep, having assumed that Kaasen had received the message to hold up at Solomon. Now Kaasen faced a decision that would lead to considerable controversy. He could wake Rohn and wait for him to harness his team, or with Balto in the lead, he could finish the trip to Nome himself and be there in a few hours. He chose the latter, and at 5:30 A.M. on Monday, February 2, Kaasen's team trotted down Nome's deserted Front Street and up to Dr. Welch's front door. All were exhausted, but they had covered the final fifty-three miles in seven and one-half hours. What normally required thirty days, what the Northern Commercial Company had promised to try and deliver in fifteen days, had taken a relay of twenty determined drivers and hundreds of dogs just five days, six and one-half hours.

Dr. Welch began dispensing the serum, and within days Nome's epidemic eased. Another 1.1 million units rushed from Seattle aboard the SS *Alameda* to Seward were again carried along the Nenana-to-Nome trail.

There were no more diphtheria deaths in Nome, and by February 21, the quarantine was lifted. Two questions seemed to remain. How had diphtheria spread to Nome in the first place, and who were the heroes of the race? Lula Welch answered the first question when she traced the outbreak to a Native family that had arrived in Nome from Holy Cross the previous November. The second question was more complicated.

The *Nome Nugget* said it succinctly in its banner headline: "Nome Mushers Figure in Race Against Time," but that wasn't the whole story. Gunnar Kaasen and Balto were hailed by some as the heroes of the run. The serum manufacturer gave the driver $1,000, and offers arrived from Hollywood for a film. The other drivers received $18.66 each from a public donation fund and $25.00 per day from the territory for their time on the trail. And don't forget Governor Bone's parchment citations praising their efforts. Seppala, in particular, felt slighted. He and Togo had carried the serum almost twice the distance of Kaasen and Balto, and had traveled a much farther total distance from Nome. Ed Rohn was miffed that he had been left out of the last leg, and even accused Kaasen of bypassing the Port Safety exchange point in order to get the final glory in Nome. The unsung heroes, of course, were the eighteen other drivers spread across the Yukon. But it just goes to show that in Alaska, even an errand of mercy is not without controversy.

In the end, there was enough glory to go around. When the Dog Mushers Hall of Fame opened in 1966, Leonhard Seppala and Togo were among the first inductees. Togo now graces the headquarters of the Iditarod Trail Committee in Wasilla. Schoolchildren in Cleveland, Ohio, bought Balto so that he could spend his later years at the Cleveland Zoo. Later, he was preserved in the Cleveland Museum of Natural History, where he has remained save for a brief reprieve to visit the Anchorage Museum of History and Art on loan in 1998.

A half century after this heroic dogsled run, mushers again took to the winter trail toward Nome. The modern Iditarod began in 1967 with a 50-mile race organized by Dorothy Page of Wasilla as part of the Alaska Purchase centennial celebration. Leonhard Seppala was to have been honorary race marshal, but he died just before the event. His wife took his place and scattered his ashes along the trail. Afterward, Page, who was destined to become "the Mother of the Iditarod," suggested extending the race from Wasilla to Iditarod, a distance of some 500 miles. Joe

Redington, Sr., who would soon be called "the Father of the Iditarod," agreed, but he asked the obvious questions. Why stop in Iditarod? Why not race all the way to Nome?

Such a competition required a very different mental and logistical effort than the shorter sprint races, but since its first run in 1973, the annual Iditarod race from downtown Anchorage for over 1,000 miles to Nome has become an Alaskan institution. Twenty-two men finished the race the first year, but it was the only year that the Iditarod would be a men-only sport. Beginning with Mary Shields and Lolly Medley in 1974, women raced and began to move up in the field. When Libby Riddles became the first woman to win the Iditarod in 1985 and Susan Butcher captured the first of her four titles in 1986, T-shirts said it all: "Alaska—Where men are men and women win the Iditarod!"

The impact of the very first Iditarod in 1925 was threefold. First, newspapers across the United States and the world published the story of Nome's plight and the heroic response. As a result, millions not only focused on Alaskan heroics but also were encouraged to get diphtheria inoculations. In the early 1920s there were 210,000 cases of diphtheria and 20,000 deaths annually. By the end of the decade, the disease had been largely eradicated. Second, happy though the ending was for most, the episode provided an impetus to the Post Office Department to speed plans for airmail service in Alaska. Finally, the stories of Togo and Balto grew into legends and numerous children's books. A bronze statue of Balto was even erected in New York City's Central Park, far, far away from the trails of Nome.

As a representative of hundreds, Balto's inscription reads: "Dedicated to the indomitable spirit of the sled dogs that relayed antitoxin six hundred miles over rough ice, across treacherous waters, through arctic blizzards from Nenana to the relief of a stricken Nome in the winter of 1925. Endurance—fidelity—intelligence." Endurance—fidelity—intelligence. Who could ask for anything more?

Conflicting Views, Continuing Battles

When President Woodrow Wilson signed the 1917 legislation creating Mount McKinley National Park and handed the pen to Charles Sheldon, it signaled much more than the creation of Alaska's second national park. The act was a major milestone along a long path littered with conflicting views about how to manage Alaska's resources that dated back to the days of the Russian *promyshlenniki* and that would continue well into the next century.

The key impetus that got the Mount McKinley legislation passed was the impending doom facing big-game animals, but the legislative debate raised much broader issues. Shortly before Congress passed the Mount McKinley legislation, geologist Stephen R. Capps, who spent almost three decades in Alaska for the U.S. Geological Survey, wrote of the immediacy of the threat to wildlife in an article intended as an integral part of a final lobbying volley fired by park proponents. "How necessary is it that this park should be reserved immediately, rather than at some indefinite date in the future?" Capps asked. "Is there any danger that the park will not keep, even if not reserved? The answer is plain and admits of no argument. The scenery will keep indefinitely, but the game will not, and it must be protected soon or it will have been destroyed."

Capps had already alluded to the much broader concept of wilderness preservation on a much grander scale than that of merely saving Sheldon's sheep and bears. His premise was that "the Mount McKinley region now offers a last chance for the people of the United States to preserve, untouched by civilization, a great primeval park in its natural beauty."

Capps went on to describe the geography and concomitant resources of the ubiquitous river valleys and surrounding hills along this part of the Alaska Range, and then—to counter those who had long railed against the economic impact of such federal withdrawals—he posed the question, "Will it pay?" Capps had no doubt about the answer: "Considered as a purely business measure, without taking account of the esthetic value of such a permanent national reserve in its influence on the development of the American people, the Mount McKinley National Park will be a tremendous financial asset to the territory of Alaska and to the United States as a whole."[6] The immediacy of the threat, the preservation of an entire wilderness ecosystem, and the positive or negative economic impact—presciently or unwittingly, Capps had enunciated the three themes that would be central to all future national-park debates in Alaska.

Things move more slowly in Alaska, however, than in most places. Although Mount McKinley National Park was created in 1917, it was 1921 before the fledgling park had its first superintendent, Harry P. Karstens, a man well acquainted with both the mountain and its animals. Karstens was Alaskan to the core—the sourdough's sourdough. He had been a Klondike stampeder, Seventymile River miner, sled driver for the Kantishna mining camps, and hunting pal of Charles Sheldon's. When Archdeacon Hudson Stuck asked Karstens to join the 1913 attempt on Mount McKinley, Karstens had already explored its lower slopes and mused with Charles Sheldon that they should be the ones to climb it together.

On April 12, 1921, National Park Service director Stephen Mather sent Karstens a ten-page letter of instructions for the new park. By all accounts, Karstens had his work cut out for him. The first priority was to protect the wildlife. Keeping outside market hunters at bay was one thing, but his fellow Alaskans were accustomed to using the whole spread of the public domain at their pleasure. Now someone—albeit one of their own—was telling them to cease and desist hunting there, and arresting them if they didn't. It was a major change of both habit and mental attitude that Alaskans would have to suffer through with each national park reservation.

During his seven years as park superintendent at Mount McKinley, Karstens excelled at the pioneering aspects of the wild and undeveloped park, but frequently found his no-nonsense, direct-action approach at odds with his Washington superiors. Among his achievements were the establishment of ranger patrols, including with dog teams during winter, to reduce poaching; supervision of the construction of the Riley Creek operations base and park headquarters; construction of a wagon road west to the Savage River; and welcoming the first tourists. No doubt inspired by his association with Charles Sheldon, Karstens also supported the first detailed natural history research in the park, including Olaus Murie's caribou studies.

It was the bureaucratic battles that finally wore down Karstens. He resigned his position in October 1928 and went back to Fairbanks to resume his transportation business. But Karstens was definitely sold on the national park concept. Once in Anchorage, he had addressed the Women's Club on the park's progress and concluded: "A natural park is being preserved in its naturalness for you and for me and for our children—unspoiled, unmarred. To enjoy this pleasuring ground for the benefit and enjoyment of the people, its sublimity of beauty and grandeur, one must be in tune with these things. There is little to offer visitors who need attendants to make them comfortable, . . . but there is much to offer those who understand the language of the great silent places."[7]

The majority of 535 members of Congress acting to establish a national park was bad enough for some Alaskans, but when the president undertook to establish a national monument under the 1906 Antiquities Act by executive order alone, that really caused a stir. Such action was one time that things could move fast in Alaska! President Woodrow Wilson did just that in creating Katmai National Monument in 1918 at the urging of those who had seen the Valley of the Ten Thousand Smokes. Exuberant *National Geographic* articles aside, not everyone agreed.

Thomas Riggs, Jr., late of his work on the Alaska Railroad and now governor of the territory, reacted vehemently to Wilson's action—despite the fact that he was a Wilson appointee. Riggs wrote in his 1918 report to the secretary of the interior: "For the sake of the future of Alaska, let there at least be no more reservations without a thorough investigation on the ground by practical men and not simply on the recommendation of men whose interest in the Territory is merely academic or sentimental." Two years later, Riggs was even more succinct, declaring that Katmai Monument served no purpose and should be abolished.[8]

Meanwhile on the other side of Alaska similar battles were raging over the area around Glacier Bay. Central to the story here was a man whom Governor Riggs no doubt included in the category of "merely academic or sentimental." But William S. Cooper was no armchair advocate. During three summers in the Colorado Rockies—beginning in 1904 when he was only twenty—Cooper covered a wide range of territory and made first ascents of a number of peaks, including Pigeon, Vestal, and Arrow, three jagged San Juan summits. Upon his return to Detroit after his 1908 Colorado trip, Cooper complained of chest pains and dizziness. A specialist told him that he had strained his heart on his climbs and should avoid such exertion in the future. The diagnosis was probably wrong, but Cooper never climbed seriously again. Instead, he threw himself into the next best thing: traveling among the mountains as a botanist and ecologist.[9]

In 1916, as a professor at the University of Minnesota, Cooper made the first of four scientific expeditions to Glacier Bay. Out of these trips eventually came nine publications, but in the beginning he was looking for an area where he could study a full ecological cycle. Given the rapid retreat of its glaciers, the Glacier Bay area was the perfect laboratory in which to compress time.

Cooper began on the glacial moraine beneath the faces of the receding glaciers and investigated the black crust of mostly algae that formed there. As he moved farther from the glacier—down the bay but backward in plant development time—Cooper documented what became the classic stages of plant succession, more recently called terrestrial succession. These led from the black crust, to moss, to species that add nitrogen, to willows and cottonwoods, and finally to the spruce and hemlock forests around Bartlett Cove that were several hundred years old. Subsequent research would show this succession process to be more complicated than Cooper's model, but his pioneering work at Glacier Bay over two decades laid the groundwork for more advanced botanical and ecological research.

On Cooper's first trip, he established nine permanent research quadrates that he would visit and monitor on succeeding visits. On all of Cooper's expeditions, Captain Tom Smith shuttled him around the bay in first the *Lue* and later the *Yakobi*. Smith got a good dose of "glacier sense" on their first trip. He anchored the *Lue* at what he thought was a safe distance from the face of the Rendu Glacier after putting Cooper and his party ashore. A sizable chunk of the glacier's face calved off in due course, and the resulting surge sent the *Lue* heaving and pitching violently. According to Cooper, Smith was always reluctant to venture very close to a face again despite Cooper's urgings.

World War I postponed Cooper's second visit until 1921. By all accounts, including their own, he and Tom Smith became and remained good friends, but they also came to disagree strongly on the future of the area. Cooper wanted it preserved. Smith thought it was ripe for mining.

At the Ecological Society of America's 1922 annual meeting in Boston, a suggestion was made that the organization should investigate preserving the entire Glacier Bay area. William Cooper agreed to chair the committee, which included that apostle of Katmai, Robert F. Griggs. When the committee presented its conclusions at the society's 1923 annual meeting, it was no surprise that it enumerated five key reasons why the Glacier Bay area should be afforded some level of protection: its tidewater glaciers, its old-growth coastal forests, the opportunities to study plant succession, the historical association with Vancouver, and the area's relative accessibility. Whether its accessibility was seen as an asset that would encourage tourist visitation, or as a detriment that might destroy the resource without such protection, is debatable. In any event, that point was left out when the society incorporated the other four reasons into a resolution urging the president to withdraw the region.[10]

Next came a demonstration of the power of the presidency. Scarcely five months after the Ecological Society of America's annual meeting, President Calvin Coolidge by executive order in April 1924 temporarily withdrew more than 2.5 million acres (about 4,000 square miles) around Glacier Bay from any kind of activity—mining, homesteading, timbering, or otherwise. Essentially, this covered everything from the Pacific coast eastward across the Fairweather Range all the way to Lynn Canal, including the homesteads around Gustavus.

Most Alaskans went ballistic. The *Juneau Daily Empire* labeled the proposal to establish a national monument there "A Monstrous Proposition." There were more than 30,000 acres of surveyed agricultural lands and three or four times more than that capable of agricultural development, the paper argued. What about the canneries, operating mines, patented mineral claims, and settlers on homesteads within the area, not to mention the vast opportunities for mineral prospectors, water power developments, lumbering, and more?

But the *Daily Empire* was just getting warmed up. "It is said the proposed National Monument is intended to protect Muir Glacier," the paper noted, "and to permit of the study of plant and insect life in its neighborhood. It tempts patience to try to discuss such nonsensical performances. The suggestion that a reserve be established to protect a glacier that none could disturb if he wanted and none would want to disturb if he could or to permit the study of plant and insect life is the quintessence of silliness. . . . It leads one to wonder if Washington has gone crazy through catering to conservation faddists!"[11]

Barbs continued to fly from both sides. When Coolidge's temporary withdrawal was made permanent the following year, however, the new Glacier Bay National Monument had been pared down to less than half of its original size. Most of the spruce forests of the lower bay and the agricultural lands, including those around Gustavus, had been excluded from the new monument, and the eastern boundary stopped at the crest of the Chilkat Range rather than extending all the way to Lynn Canal.

William Cooper returned to Glacier Bay for a third time in 1929, again in the company of Tom Smith. On this trip, they made their first visit to Johns Hopkins Inlet. The glaciers there were in rapid retreat during this period, and Cooper and Smith ventured into Johns Hopkins Inlet again in 1935. Smith was more excited than ever about the area's mining potential, and on this visit—in the middle of the depression—there were plenty of people who shared his views. With glimmers of the gold rush not too far removed from recent memory, many folks thought that the way out of the depression was a resurgence of mining just about anywhere that it looked the slightest bit promising.

While Tom Smith sailed Glacier Bay's inlets with William Cooper looking at glaciers, he was also a partner in a mining venture within the boundaries of the new monument. Whether or not the claim was legal depended on who was telling the story. Smith's partner and the real impetus in the venture was Joe Ibach, a likable enough sort who was also somewhat of a character. The home that Joe shared for more than thirty years with his wife, Muz, at Willoughby Cove on Lemesurier Island was a welcome port to many a traveler. And if anyone truly loved the Glacier Bay country, it was Joe Ibach.

In the summer of 1924, *after* Coolidge's temporary withdrawal, but *before* the final order establishing the monument, Ibach went ashore just northwest of the Reid Glacier, found what he took to be gold-bearing ore, and staked three claims. When the monument was declared, the land office warned Ibach not to work the claims, but he proceeded to do so surreptitiously nonetheless. With Muz's help, he sorted the ore and hauled the best stuff out a couple of sacks at a time, frequently on board Tom Smith's boat. One year, the Ibachs and Smith actually owed the Juneau smelter money for freight after the run, while another year they split the smelter proceeds and got a grand total of thirteen dollars each. It was hardly enough to start a rush, but Ibach was as optimistic as any who had ever tramped the trails of '98.

Enter novelist Rex Beach. Ibach and Beach went way back, having been on hunting trips together decades before. Beach showed up on Lemesurier Island in 1935 and quickly convinced himself—undoubtedly with a little encouragement from Joe—that opening Glacier Bay

National Monument to mining was exactly what was needed to lift Alaska out of the Great Depression. Beach soon wrote a magazine article and letters to President Franklin Roosevelt and others asserting that the monument was "absolutely barren and the only timber, such as there is, lies along the southern edge. It is not a good game refuge, nor are there any fishing streams or lakes in which salmon spawn. Presumably there are some sheep and goats in the St. Elias Range but it is the last place anybody would go for bear, moose or caribou. In fact the whole area is like a haunted house and I doubt if ten white men have visited it in the last ten years."[12]

Not much of that was true, but Beach, whose reputation rests on fictionalized accounts of some of Alaska's more colorful history, was never one to let facts stand in the way of the particular story he was telling. Beach found an ally in Alaska's territorial delegate, Anthony Dimond, and together they exchanged a flurry of letters with both FDR and Secretary of the Interior Harold Ickes. Over the course of the spring, Roosevelt changed his view of the situation at least twice, going from being inclined toward permitting mining, to against it, and then once again supportive of it.

Somehow in June 1936, in scarcely more than a week before Congress adjourned for the political conventions, a bill that opened Glacier Bay National Monument to mining was rushed through both houses of Congress and signed by the president. William Cooper summed up the feeling of those who had worked so hard to preserve the region when he wrote: "The entire procedure was thus carried through in one week. No one likely to attempt opposition knew of the affair until it was a *fait accompli.*" Beach only grinned. "It is absurd to assert, as those scientists did," he himself later wrote, "that in helping Tony Dimond establish a precedent for which Alaskans have long fought, we threatened the integrity of the national park system."[13]

As soon as the legislation opening the monument to mining passed, Beach wired Ibach a coded message that he should go out and stake more claims. Joe did so, but results were much the same as his earlier efforts. There was no Eldorado to be found in Glacier Bay. Interestingly enough, however, the following year, Roosevelt transferred more than 1 million acres from the Tongass National Forest to the monument. The action angered Ernest Gruening, soon to be Alaska's territorial governor, because it put several dairy farmers out of business and in Gruening's view was another example of the federal government locking up the land. Other battles would soon be fought on Alaska's soil, and skirmishes in the Glacier Bay land-use war would continue until the area was finally designated a national park in 1980.

From the conservationists' standpoint, they had been lucky. Glacier Bay apparently held no great gold riches, but on a larger scale, the action of a publicist who obviously had the presidential ear set a dangerous precedent when the gates to what had once been thought safe from development were suddenly thrown open. In the big picture, that is the significance of the activities at Glacier Bay in the 1930s. On the smaller, more personal side, there are two footnotes to the story.

Joe and Muz Ibach lived on Lemesurier Island until 1956. Devoted to each other, they had a long-standing pact that if something happened to one of them, the other would not be far behind. In 1959, Muz died in a Juneau hospital. Joe was beside himself, and he took her body back to Lemesurier Island for burial. The following spring, he made plans to return to the mines at Reid Inlet, but the morning he was to depart he left a handwritten will on some brown wrapping paper and then shot himself. At the bottom of the paper he had written, "There's a time to live and a time to die. This is the time."[14]

William Cooper returned to Glacier Bay in 1956 and again in 1966, the fiftieth anniversary of his first visit. He was hailed as the "Father of Glacier Bay National Monument" but seemed to delight most in crawling around on his hands and knees at Blue Mouse Cove pointing out plant species. William S. Cooper, the would-be mountaineer who was told to take it easy in 1908, died in 1979 at the age of ninety-five.

Salmon on the Run

Alaska's territorial governor and future U.S. senator Ernest Gruening once wrote that "salmon and Alaska have been as closely intertwined as cotton and the South."[15] Whatever else may have been exaggerated about Alaska over the years, that much was indisputable fact. Alaska Natives, particularly the Tlingit, knew this well and counted salmon as a principal source of food. The Russians should have survived nicely on salmon, but the meat, grains, and potatoes of their culture were too deeply ingrained.

To the uninitiated, salmon nomenclature can be confusing. For starters, there are Atlantic salmon and Pacific salmon. All the species of Pacific salmon are anadromous, meaning that they hatch in fresh water, spend part of their lives in salt water, and then return to fresh water to spawn and die. Conversely, Atlantic salmon are catadromous; they migrate to salt water to spawn and, unlike their Pacific cousins, do not die shortly afterward. While there are only five species of Pacific salmon, each has both a common name and a nickname in addition to its Latin name. To add to the confusion, some species have different local nicknames outside of Alaska.

The king of Pacific salmon is the chinook, nicknamed appropriately enough "kings." The chinook salmon has always been one of the most important fish along the entire north Pacific coast. It is also the largest salmon species, averaging sixteen to twenty pounds along the Inside Passage, but as much as forty pounds on the Kenai Peninsula. Chinooks are easily identified because their teeth emerge from a very dark gray or black gum line, and they have large black spots on both upper and lower tail lobes. Chinook salmon run (head upstream to spawn) from May through July. Not only do they run small streams and rivers, but they also travel the length of the Yukon, a herculean odyssey that takes about sixty days and covers more than 2,000 river miles.

Coho salmon are also known as "silvers" because of their color. These salmon quickly became favorites among sport fishermen because they are spectacular fighters. Cohos spawn a little later than chinooks, and their runs peak in August, usually associated with periods of high runoff. The normal weight range is eight to twelve pounds, but the Alaska state record for a coho is twenty-six pounds.

Chum salmon are smaller than either chinooks or cohos. Fifteen pounds is trophy weight for these salmon, which are nicknamed "dogs" because of their hooked snout and large teeth. Chums are easily confused with cohos, especially in salt water, but one reliable way to tell the difference is to look them in the eye. Chums have much larger pupils than cohos. Chums spawn from mid-July into the fall. Because of their wide range—throughout Alaskan waters, including the Arctic coast east to the Mackenzie Delta—chums have always been a major source of dried fish for winter subsistence. Like a good dog returning home, chums may be the most territorial of salmon varieties and seem to return to their birthing streams with geographic precision after two to five years.

Pink salmon, also called "humpies" because of the elongated hump forward of their dorsal fin, are small in comparison to the other species, but have long been called "the bread and butter" fish in many Alaskan fishing communities because of their commercial importance. Pinks average three to five pounds in weight and have small scales and large dark oval spots on their backs and both lobes of the tail. Pinks run between July and mid-October and seem to spend only one winter at sea, which accounts for their smaller size.

Finally, there are the salmon that make Alaskan streams run red. Sockeyes are silver-blue in salt water but turn bright red with a green head in fresh water prior to spawning, hence the nickname "reds." (Along the Columbia, sockeyes are called "bluebacks.") Sockeyes are difficult to catch with sport tackle. The Kenai Peninsula is one of the few places where sockeyes seem interested in lures and flies—a fact that has made certain areas on the Russian and Kenai Rivers run as full of fishermen as the reds themselves.

The first commercial salmon cannery in Alaska was built at Klawock on Prince of Wales Island in 1878. From an initial shipment of 15,000 cases valued at $60,000, Alaska salmon production rose quickly. By 1889, almost 700,000 cases worth just under $2.8 million were processed at thirty-seven Alaskan canneries spread throughout the southeast, as well as Cook Inlet, Kodiak Island, and Bristol Bay. In addition to the canneries, there were a few salteries that packed and exported salmon in barrels.[16]

Most cannery operations had sporadic populations. Crews arrived and the place boomed for the short processing period—whenever the salmon were running in local streams—and then the cannery would be boarded up, save for a watchman or two, until the next season. Other canneries became the nuclei of small communities such as Hoonah, Tenakee Springs, Chatham, Kake, Kasaan, and Hydaburg. Still others spawned larger towns, including Ketchikan. A fire there in 1889 destroyed a cannery along Fish Creek, but it was rebuilt and the town grew up along with it. Eliza Scidmore, that grande dame of early Alaskan travel writers, reported in 1890 that ten canneries clogged the mouth of the Karluk River on Kodiak Island alone. For a time, it seemed that every salmon in Shelikof Strait was bound for Karluk Lake. And, for a time, many thought that there was no limit to Alaska's salmon bounty.

One observer who did not share the common view and had concerns about overfishing was Tarleton H. Bean. As early as 1889, Bean warned in a report for the U.S. Fish Commission that "under judicious regulation and restraint these fisheries may be made a continuing source of wealth to the inhabitants of the Territory and an important food resource to the nation; without such regulation and restraint we shall have repeated in Alaskan rivers the story of the Sacramento and Columbia; and the destruction in Alaska will be more rapid because of the small size of the rivers and the ease with which salmon can be prevented from ascending them."[17]

The only immediate action that was taken in response to Bean's plea was the passage of an 1889 federal law that purported to prohibit the erection of dams, barricades, or similar obstructions that had the effect of capturing all or most of the salmon trying to ascend a particular stream. Such traps severely limited the number of salmon that successfully completed the run to their spawning grounds, and it took only a few seasons to reduce dramatically the populations of those streams. The problem, of course, was that despite enacting the law, Congress allowed three years to pass before it appropriated funds to enforce it. One inspector and an assistant were hired for all of Alaska. To add insult to injury, since no commercial transportation existed, the inspector usually had to depend on the vessels of the canneries he was charged with regulating for transportation.[18]

A school of thought quickly developed that the way to ensure a continuing supply of salmon was to raise them in hatcheries. In 1900, the federal government issued regulations requiring every company taking salmon in Alaskan waters to establish hatcheries and to return red salmon to the spawning grounds at the rate of at least four times the number of fish taken the preceding season. That sounded noble enough, but once again the regulation was not enforced. Perhaps most important, however, the science behind the regulation was never questioned or tested. Could salmon really be raised in hatcheries?

In 1903, supervision of Alaska's fisheries was transferred to the Bureau of Fisheries in Theodore Roosevelt's newly created Department of Commerce and Labor. Over the next several years, bureau officials managed to obtain appropriations to construct two hatcheries in Alaska, despite widespread opposition from the salmon-packing companies. Indicative of the divergent views of resource protection versus unlimited commercial exploitation was the fact that the Bureau of Fisheries wanted to regulate fishing within three miles of the mouths of all streams and rivers. The packers successfully lobbied Congress to have that figure set at 500 yards instead. The salmon lobby was certainly powerful in Congress, but there were also those who thought things a little too cozy between the would-be regulators and the industry. When the Bureau of Fisheries established a branch office in Seattle, it was located in the building where twenty of the major salmon companies had their offices.[19]

Alaskans took none too kindly to what they saw as unbridled interference with the territory's resources by both the federal government and the packing companies. Even in granting Alaska a territorial legislature and some measure of home rule, Congress had bowed to the salmon-packers lobby and reserved the right of fishery regulation in the territory to the Bureau of Fisheries—not the legislature. This was just one of the many compromises in the Second Organic Act of 1912 for which territorial delegate James Wickersham was roundly criticized by independent Alaskans.

Beginning in 1911, the annual reports of the territorial governors began to sound the alarm about reduced salmon runs. In 1914, Governor John F. Strong reported to the secretary of the interior that "the waters of Alaska have been exploited for their wealth for many years and they have yielded many millions of dollars, and large individual fortunes have been accumulated therefrom. At no time, however, have the exploiters contributed anything like an adequate return for the privileges they have enjoyed." Two years later, Strong asked the Bureau of Fisheries for a status report on Alaskan salmon runs. No problem, was the official reply, but Strong vehemently disagreed and asserted that all one had to do was to look at the streams to be aware of the ever-decreasing runs.[20]

Dr. Hugh Smith, a biologist, was appointed commissioner of the Bureau of Fisheries in 1913. Smith readily assured everyone that science could manage the salmon decline. Once again, hatcheries were promoted as a cure-all, despite a lack of data to support such leaps of faith. By the early 1920s, thanks in part to the boom associated with feeding canned salmon to thousands upon thousands of troops in World War I, fishing had replaced mining as Alaska's major industry. But by then, even the packers were starting to think that there could be a problem.

As annual harvests declined, Commissioner Smith became somewhat of a fall guy and was forced to resign. His successor, Henry O'Malley, finally admitted to Congress that artificial propagation in hatcheries was simply not compatible with the long-ingrained natural migrations and spawning habits of Pacific salmon. Break the cycle, and it was not readily fixed.

Secretary of Commerce Herbert Hoover became so concerned by the salmon decline that he took the extraordinary step of recommending to President Harding the establishment of the Alaska Peninsula Reserve and the Southwestern Fishery Reserve, the latter of which included Kodiak Island and Bristol Bay. Harding created the reserves by executive order in the spring of 1922, and fishing was allowed there only by permit.

Two problems arose immediately. If Alaskans had been miffed by their inability to hunt on national park reservations, "no fishing" signs in Alaskan waters were the final regulatory insult. These reserves were the first places in American history—save for a few areas set aside to Native American tribes—that U.S. citizens had been denied the right to fish. Some thought it was a God-given right; others desperately had to do it to feed their families. The second problem was that what permits there were seemed to be going en masse to employees of the large packing companies. The whole situation smacked of outright favoritism, and charges that a salmon version of the Teapot Dome scandal was in the offing got so shrill that Hoover was forced to hold hearings on the fishing permit system when he came north with Harding in 1923 for the Alaska Railroad dedication.[21]

The short-term result of all of this was the White Act of 1924. Introduced into the U.S. House of Representatives by Congressman Wallace H. White of Maine, the law established mandatory "escapement limits" and gave the Bureau of Fisheries the authority to arrest violators. This meant that at least 50 percent of the salmon run in any given stream had to be permitted to pass upstream to spawning grounds. At some levels—mostly those of the federal government and the packers—the White

Act was hailed as a major milestone in the effort to preserve Alaska's renewable salmon resources. At Alaskan levels, folks weren't too sure what to expect. For one thing, an attempt by territorial delegate Dan Sutherland to amend the White Act to outlaw certain trap locations ultimately went down to defeat. It was quite clear to most Alaskans just who was calling the shots on salmon management. Regulation of coastal and inland fisheries would not be transferred to Alaska until it became a state.[22]

Politics and bureaucratic wrangling aside, during the 1920s fishing in Alaska, principally for salmon and halibut, became the economic bedrock that mining had never been able to find. Homer Pennock might have been a prospector when he first frequented a long, slender spit of land jutting into Kachemak Bay, but to the town that grew up there, Homer was synonymous with halibut. Ketchikan soon boasted that it was the "salmon capital of the world." And then there was Petersburg—rock-steady, reliable Petersburg.

Peter Buschmann was a Norwegian immigrant living in Seattle when he ventured north in 1895 looking for prime fishing waters and a new base of operations. At the northern entrance to Wrangell Narrows on Mitkof Island, Buschmann found the perfect harbor. It was close to both fishing grounds and the Seattle-to-Alaska steamship runs, and close as well to nearby supplies of ice from the Baird Glacier. Peter's fishing camp was not planned as a "Little Norway," but as Norwegian friends and acquaintances in the Seattle area heard about it, they drifted north to join him. In time, a town grew to celebrate Norwegian holidays and blend its culture with that of the Tlingit, sharing recognition of the importance of maintaining the balance of nature's bounty. Unlike most Alaskan towns that rose and fell—adapted or died—to a succession of frontiers, Petersburg has been anchored in only one. Fishing was its founding purpose, and fishing remains its lifeblood.

Thrill 'em, Spill 'em, but Never Kill 'em

There is an old pilot joke in the continental United States that distinguishes between the two types of flying conditions, VFR (visual flight rules) when visibility is good and IFR (instrumental flight rules) when visibility is limited or nil. To the question "Do you fly IFR?" the response was "Of course, I fly IFR—I Follow Railroads." In Alaska, as with so many other things, the joke had to have a little different twist. The Alaskan answer: "Of course, I fly IFR—I Follow Rivers," since railroads, or any kind of roads, were in limited supply.

During the 1930s, before such distinctions were made, some people contended that there were actually three sets of flight rules for Alaska's differing weather conditions. First, there was "Pan American weather," because Pan American Airways flew only when the skies were clear. Then there was "flying weather," the extremes of which varied greatly depending on who was doing the talking and, more important, who was doing the flying. Finally, there was "Gillam weather," weather so bad—usually dense clouds that "socked in" a field along with strong, gusty winds—that only veteran pilot Harold Gillam would fly in it. Seasoned Alaskan bush pilots were always content to sit out "Gillam weather" on the ground, reasoning that "God's plenty busy taking care of Harold."[23]

Harold Gillam was an Alaskan legend, but he was neither the first nor the best known of Alaska's bush pilots. The first airplane in Alaska was optimistically christened *Tingmayuk,* an Eskimo word for bird. Weighing about 500 pounds and built of light wood, muslin, and piano wire, it was the creation of "Professor" Henry Peterson, a Nome music teacher. Interestingly enough, it was also probably the first airplane anywhere to be outfitted with skis.

On May 9, 1911, outside of Nome, Peterson fired up his single engine and eased the biplane forward. He opened the throttle, but the contraption still only managed to "ease" forward. Even with a push down a hill, the *Tingmayuk* never got up enough speed to take to the air. Probably, it was just as well. The headline in the *Nome Nugget* summed it up best: "Peterson Unable to Defy the Law of Gravity."[24]

Alaska's first successful flight occurred two years later, but only after the aircraft had gotten there by steamship and railroad. The town fathers of Fairbanks wanted something extra special to celebrate the Fourth of July of 1913. They contracted with James V. Martin, a New Englander temporarily in Seattle, to bring his aircraft north and fly over the Fairbanks ballpark on the big day. Martin shipped his disassembled aircraft from Seattle to Skagway by steamship, put it on a flatcar of the White Pass and Yukon for the trip to Whitehorse, and then barged it down the Yukon and up the Tanana to Fairbanks. It was a big effort, but the novel airplane made a huge hit with the Fairbanks crowd. Martin flew for eleven minutes and then touched down safely, ending Alaska's first flight but presumably setting the more visionary minds in the crowd to thinking.[25]

World War I stifled domestic aviation for a time, but when it was over, the skills and techniques honed in the skies above France catapulted America into the air age. On July 15, 1920, General Billy Mitchell of the Air Service of the U.S. Army dispatched eight men and four airplanes of the Black Wolf Squadron from New York to Nome on the nation's first cross-country flight. The exercise was one of Mitchell's many demonstrations of the potentials of air power. Mitchell, by the way, was no stranger to Alaska, having worked as a young lieutenant on the military telegraph line between Valdez and Eagle in 1901–2.

Commanded by Captain St. Clair Streett, a combat pilot in the recent war, the De Havilland biplanes of the Black Wolf Squadron became the first aircraft to be flown from the lower forty-eight to Alaska. They entered the territory at Wrangell, flew up the Inside Passage and across White Pass—with 100 feet to spare, according to Captain Streett's report—and then made stops at Whitehorse, Dawson, Fairbanks, Ruby, and Nome. There were many days of delay along the route caused by bad weather and broken parts, but all in all, the squadron covered the 4,500 miles between New York and Nome in fifty-three hours and thirty minutes of flying time. "Some day," Captain Streett speculated in an article recounting the adventure, "this trip may be made overnight—who knows?"[26]

New York to Nome! Now that really got people thinking. The first to do more than just dream about Alaskan air service was a New Yorker from Buffalo named Clarence Prest. He tried to take off from a beach at Juneau in 1922 and fly over the Coast Mountains to the Yukon. He survived four engine failures and made it as far as Dawson before one final failure left him to walk out of the wilds near Eagle—without his airplane. Mosquito-bitten and discouraged, Prest returned to New York.[27]

One pilot who went to Alaska and stayed was Carl Ben Eielson, a flight trainee during the war who had earned his wings too late for combat. Afterward, Eielson tried both barnstorming and law school—perhaps not so disparate preoccupations as they might seem at first glance—and watched the Black Wolf Squadron take off for Alaska after a stop in North Dakota. Determined to follow, he soon ended up in Fairbanks—not as a pilot but as a high school science teacher. In 1923, however, Eielson persuaded several Fairbanks businessmen to back him in the Farthest North Airplane Company. Eielson ordered a plane, assembled its pieces, and promptly took off from the Fairbanks ballpark.

Eielson flew passengers and supplies between Fairbanks and some of the outlying mines and soon was carrying both passengers and deliveries between Fairbanks and Nenana. This was the first commercial flight service along the Yukon and only the second in Alaska. (A few months earlier, C. O. Hammondtree had started taking sightseers up at Anchorage and flying supplies to nearby canneries.) But what Eielson really wanted was to land a mail contract between Fairbanks and McGrath, a mining town about 300 miles to the southwest along the Kuskokwim. Given the traditional uncertainties of the region's economy, a government mail contract was like money in the bank and as much of a guarantee of a rock-solid job as one was likely to get in Alaska.

Eielson continued to badger the U.S. Post Office Department until it finally awarded him a Fairbanks-to-McGrath run. On February 21, 1924, with the temperature minus-five degrees Fahrenheit at Fairbanks, he climbed into the open cockpit of his De Havilland and took off for McGrath. Flying at more than eighty miles an hour in the open cockpit, Eielson was subjected to mind-numbing wind chill, but three hours later he landed in McGrath, successfully completing the first air mail delivery in Alaska. Along the way, he flew over one of the dogsled mail carriers mushing along the winter trail. The sled driver looked up as Eielson waved and then roared overhead. The handwriting was on the wall—or in this case, in the sky.

But as the famous serum run to Nome the following winter was to prove, for a few more years dogs were still more reliable than airplanes. In May, on his eighth trip between Fairbanks and McGrath, Eielson survived a crash landing. Given the new technology, it was perhaps to be expected, but the Post Office Department took a dim view of the episode and abruptly canceled Eielson's contract. Dogs would carry the mail a little longer.

Eielson tried to do battle with the Post Office Department, but when that didn't get him very far, he met up with British explorer George Hubert Wilkins and soon shared Wilkins's dream of flying over the North Pole. After a series of crashes, they were beaten to that goal, but on April 15, 1928, the two left Barrow in a Lockheed Vega intent on crossing the polar basin. They flew eastward across the tops of Canada and Greenland and then after more than twenty hours in the air were forced down by a storm near Spitsbergen. For five days they sat marooned and waited for the sky to clear. When it did, they tramped out a crude runway in the newly fallen snow, and Wilkins had to push the plane through the snow to get it going. Miraculously, Eielson got the Vega into the air, but Wilkins could not make it into the plane before it took off. Now that must have given Wilkins just a bit of a knot in his stomach. But Eielson landed again, and on the third try, the plane made it into the air with both men, and they continued on to Spitsbergen.

The flight with Wilkins brought Eielson considerable recognition, and he started his own flying service. Whatever else may be said about Eielson—he certainly undertook heroic feats—it does not appear that he was a natural flier, the kind who flew by the seat of his or her pants and knew up from down in even the most threatening of circumstances. In fact, Eielson never seemed comfortable in a plane. Contemporaries recalled that he had an "uncanny ability of getting lost no matter how many times he traveled a route."[28] Eielson made his final mistake when he and his mechanic took off from Teller on the Seward Peninsula on November 9, 1929, on a mission to salvage a shipment of furs from an icebound ship off the Siberian coast. It was Gillam weather, but the impatient Eielson flew straight into the teeth of a blizzard. The bodies were recovered two months later by Eielson's protégé, young Joe Crosson, and—Harold Gillam.

If Carl Ben Eielson was one extreme of Alaskan bush pilot, Noel Wien was the other. Born in Lake Nebagamon, Wisconsin, in 1899, one of four brothers, Wien began flying in 1921 and barnstormed for two years before heading north for Alaska in 1924. From the start, he was at home flying across the bush. Wien made the first flights between Fairbanks and Anchorage and Fairbanks and Nome and flew all over the Yukon Basin. He had an uncanny sense of direction, even over new territory, and by 1927 he had set himself up with a flying service in Nome. Two years later, he opened a branch in Fairbanks but soon sold out to Eielson and returned to the States. By 1932, however, Wien was back in Fairbanks. Despite contracting polio in 1935 and the loss of an eye in

1939, Noel Wien flew until 1955, establishing one of Alaska's premier airlines, nurturing his sons' own flying careers, and, in the process, becoming a legend.

There is no question that early Alaskan flying presented its own special problems: freezing cold that was almost as hard on equipment as it was on the pilots, fickle weather that was given to change in the time frame of a few minutes or the distance of a few miles, and marginal landing facilities that routinely lacked lights, beacons, and other navigation aids. Navigation itself posed a whole different set of challenges in the far north.

Harold Gillam got started flying cargo from Cordova to mining operations scattered throughout the nearby mountains. It was risky business. The landing sites were tenuous at best, always ringed by mountains and subjected to low-hung clouds and those same erratic winds that had once played havoc with even the locomotives of the Copper River and Northwestern. It was out of Cordova that Gillam got his reputation for an uncanny ability to fly no matter how low the ceiling. And he got his nickname there, too. Once when some schoolchildren in Cordova were asked to write a poem about their favorite person, one third-grader picked the daring pilot and wrote:

He'll thrill 'em,
Chill 'em,
Spill 'em,
But no kill 'em,
Gillam.[29]

After three years flying out of Cordova, "Thrill 'em, spill 'em, but never kill 'em" Gillam went north to Fairbanks. Admirers said that he had the eyes of a cat. Critics called him suicidal. The truth was that Gillam was a meticulous, natural navigator who was able to perform his feats of bad-weather and night flying because he had a healthy respect for his instruments—crude though they still were. He was among the first in Alaska to fly IFR, the real IFR.

Gillam got his first mail contract between Cordova and Eagle in 1931. By 1938, he had mail contracts between Fairbanks and twenty outlying communities. He developed such a perfect record of on-time deliveries that the Post Office Department declared him the best in the country. But such success was not without local uproar. The government money spent on mail contracts had long had a trickle-down effect, benefiting sled drivers, roadhouses, and strings of tiny supply towns along the mail routes. Airplane mail deliveries distributed the wealth differently and benefited mechanics in the bigger towns and fostered airport development. By the end of the 1930s, airplanes carried 500,000 pounds of mail, and the Post Office Department ceased dogsled operations in most areas for good. The lifeline of the overland mail trail declined, as did the towns along it.[30]

Cordova was also the proving ground for another legendary pilot. Merle K. Smith came along after Harold Gillam left for Fairbanks. Among his contracts was one with the Bremner Mining Company to fly supplies to a mountain-ringed strip that was about 300 feet long and 20 feet wide with a ditch on either side. That's the length of a football field, but decidedly less than the width between the hash marks. After watching one of Smith's landings, the local miners advised him to check the runway before he took off to make certain that his Stearman had not stirred up any rocks or created any new potholes upon touching down. Nah, said Smith, it would be OK. Down the runway Smith went until one of the wheels dropped into a new pothole and the plane nosed over into the mud. Like Gillam, the sobriquet bestowed on him that day remained with him for the rest of his life, and the airport at Cordova is still proudly known as the Merle K. "Mudhole" Smith Airport.

Federal promotion of aviation was late in coming to Alaska. Both the Air Mail Act of 1925 and the Air Commerce Act of 1926 provided incentives to aviation, including contract air routes and navigation aids such as emergency landing fields, radio range stations, and beacons, but neither act applied to Alaska because it was not a state. Not to be deterred, the territorial legislature authorized the expenditure of road moneys for landing fields. By 1927, forty-four landing fields had been built, and three flying services were in operation. Some of these fields were little more than a graded stretch of ground, and, in fact, when Anchorage built its first airstrip in May 1923, the whole town turned out to help smooth over what was really nothing more than a dirt field.

More often than not, however, and what really set Alaska apart from the lower forty-eight, was that its landing fields quickly turned out to be anywhere and everywhere, from sandbars in rivers to the rivers themselves, beaches, small clearings, lakes and other waterways for floats, and frozen tundra for skis. The roadless vastness of Alaska encouraged the new innovations of aviation as nowhere else and made Alaskans, as Jean Potter wrote in *The Flying North*, "the flyingest people under the American flag and probably . . . in the world."[31]

The airplane changed the face of Alaska and provided outlying communities such as Nome and Barrow with year-round connections. There was still a very high degree of self-sufficiency, but supplies and medical care were now hours away instead of weeks or even months. Mapping, photography, and weather observations were just a few of the things that now began to be done from the air. By the end of the 1930s, one report showed ninety-seven established civilian airfields, although only the bigger fields at Anchorage and Fairbanks had lighted runways.[32]

All of these efforts were by necessity homegrown—staffed and financed by Alaskans—and all operations that went on were strictly intrastate. There were no connecting outside flights. In 1934, Lieutenant Colonel Henry H. "Hap" Arnold flew into the territory in command of the first nonstop flight of army planes direct from the lower forty-eight to Alaska. Later testifying before Congress, the future father of the U.S. Air Force remarked, "I thought I had seen and talked to air-minded people . . . but I had to come to Alaska to really find a place where air transportation is taken as a matter of course and has become a necessary adjunct to the economic life of the country."[33]

And, of course, that necessary adjunct quickly came to demand interstate connections. Pan American Airways started service overseas from the States in 1927 and had routes to Latin America by 1931. But not until June 1940 did the airline start clipper service between Seattle and Ketchikan and Juneau. From there, connections were available to Fairbanks, Nome, and Bethel via Pan American's subsidiary, Pacific-Alaska Airways, which had purchased the equipment and routes of a number of Alaskan operations. This was a major event and Alaska's first transportation link to the continental United States other than by ship.

In the summer of 1935, another major event occurred in Alaska that set the aviation world buzzing and left all of the country mourning. Wiley Post was thought by some to be America's best pilot—or at least its most famous—after Charles Lindbergh. Will Rogers was thought to be, well, America's best friend. Post, a diminutive five-foot-five roustabout who had lost his left eye in an Oklahoma drilling-rig accident, flew into fame by setting two round-the-world flight records.

On June 23, 1931, Post and Australian Harold Gatty left venerable Roosevelt Field in New York in a Lockheed Vega, a high-wing, single-engine monoplane that was painted white with blue trim and christened the *Winnie Mae*. Eight days, fifteen hours, fifty-one minutes later they were back. The only major mishap of their global flight was a bent propeller suffered while attempting to take off from a beach near Solomon, Alaska, just east of Cape Nome. Temporary repairs got them as far as Fairbanks, where local pilot Joe Crosson, who was a good friend of Post's, came to their aid with a new prop. Post and Gatty flew on to New York, where the world record and a ticker-tape parade were theirs.

But Post thought that he could make the trip faster—and alone. On July 15, 1933, again in the *Winnie Mae*, he left Brooklyn's Floyd Bennett Field, flew nonstop to Berlin, and then across Russia to Alaska. He was ahead of his record time when he got lost in bad weather and landed at Flat, one of a cluster of old mining camps around Iditarod. Flat might have been the town's name, but the field wasn't, and upon touching down the *Winnie Mae* broke a landing gear strut. Once again, Joe Crosson came to Post's aid and helped him make the necessary repairs. Even with this second Alaskan mishap, by the time Post landed back in New York, he had shaved twenty-one hours off his old record.

As the old saying goes, however, all glory is fleeting, and Post needed something to keep his name in the headlines. In February 1935, he retired the trusty *Winnie Mae* and bought a new airplane. By all accounts, it was a hodgepodge of mixed components. Joe Crosson was more succinct. He later termed it "the Bastard." The main airframe was a Lockheed Orion modified with a stouter low wing from a defunct Lockheed Explorer. Post outfitted it with an overpowered, 600-horsepower Pratt & Whitney engine—in later automobile terms, something akin to putting a Ford 390 V-8 into a '64 Falcon. Then he changed out the landing gear for pontoons. At best, the plane was nose-heavy. At worst—well, who was going to argue with Wiley Post? In later years, he would never have gotten the certification to take it off the ground.

Will Rogers, actor, author, humorist, and in the depths of the Great Depression America's diversion as well as its conscience, was always intrigued by aviation. A 1925 flight with General Billy Mitchell made him an advocate. Later, Rogers reported on Wiley Post's record-setting flights and tried to boost his career as best he could—they were, after all, fellow

Oklahomans. Will Rogers was also interested in seeing what he termed "that Alaska." Post and his wife, Mae, showed up at the Rogers home in California in the summer of 1935, and a few days later Rogers flew with the Posts in the new Orion on a quick trip to New Mexico. The die was cast.

Both Post and Rogers were more than a little coy about their intentions. Rumors circulated that Post had something up his sleeve that involved Siberia. In any event, on August 7, 1935, the Associated Press reported, "Ignoring reports of storms on their route, Wiley Post and Will Rogers took off here [Lake Washington near Seattle] today in Mr. Post's new plane for Alaska."[34]

The weather was indeed stormy, and Ketchikan was fogged in, so Post flew on to Juneau, landing there in the rain. Territorial governor John Troy invited the two celebrities to dinner at the governor's mansion and then twisted Will's arm to make an evening appearance on KINY radio. From Juneau they flew to Fairbanks via the long route, first to Dawson, then to Aklavik near the Mackenzie Delta, Herschel Island on the Arctic coast, and south again to Fairbanks, landing on Chena Slough in the afternoon of August 12.

There, they met up with Joe Crosson. On August 14, Crosson flew Post and Rogers on a tour south to Anchorage in a Pan American Airways Lockheed Electra, the same type of aircraft that Amelia Earhart would try to fly around the world two years later. That Crosson chose to fly himself and pilot one of his company's planes is understandable, even without his profane characterization of Post's plane, but the move put Post in the unlikely role of passenger.

The Electra landed briefly at Savage Camp in Mount McKinley National Park—about long enough to say that Will Rogers had been there—and then continued south to Palmer. Rogers wanted to see the new Matanuska relief colony there, but upon circling Crosson ruled the field too rough for the big Electra. He flew on to Anchorage and landed at Merrill Field instead. A local pilot took Rogers and Post on the short hop to Palmer and back, and then with Crosson again at the controls, the twin-engine Electra got everyone back to Fairbanks in time for a round of socializing.

The next morning, August 15, the weather report from Barrow was Gillam weather: a temperature of forty degrees Fahrenheit, dense fog, no ceiling, no visibility. Rogers spent the morning working on his columns, while Post looked at houses with Crosson. Post liked it here and was serious about moving to Fairbanks and settling down. With that interest, one wonders what his rush was to get out of town, except that Wiley Post was always rushing somewhere. Despite his friend Crosson's cautions, and apparently showing little concern for the safety of his famous passenger, Post taxied the Orion into Chena Slough and took off for nearby Harding Lake, where Crosson had arranged to top off the plane's fuel tanks before it flew north to Barrow. If ever the ghost of Carl Ben Eielson should have cried out, it was now.

If Will Rogers had misgivings about the flight, no one heard him voice them. He appears to have placed great, if somewhat naive, trust in Post and taken a lot about Alaskan flying for granted. The fact that Rogers was photographed along the route wearing a topcoat and looking very much as he would have deplaning at Floyd Bennett Field suggests that he didn't fully understand the potential hazards of Alaskan air travel. Then again, perhaps the very fact that he was dressed that way shows how commonplace air travel had become in just a couple of decades. Whatever else *Alaskan* travel may be, however, it is not commonplace.

Five and one-half hours after leaving Harding Lake, Post was clearly lost. The weather report had been very accurate, and the Orion circled and circled above the clouds near what Post's dead reckoning told him was close to Barrow. Gingerly descending through the thick overcast, Post finally caught sight of a person walking along a stretch of beach and set the Orion down on what turned out to be Walakpa Lagoon about a dozen miles southwest of Barrow. Nearby was an Inupiat Eskimo summer fishing camp, now occupied only by the Clair Okpeaha family.

Pilot and passenger climbed out of the plane and asked Clair Okpeaha and his fourteen-year-old son, Patrick, for directions to Barrow. Supposedly, Will's first words to the two were "Anyone here from Paducah?" but that may be more Rogers legend than fact. Certainly, the Okpeahas were surprised to see an airplane flying in this sort of weather. Indeed, no one was flying in or out of Barrow at all that day. Learning that Barrow was just down the coast, and clearly trusting the luck that had taken him twice around the world, Post climbed back into the plane and took off with Rogers.

According to Clair and Patrick Okpeaha, the Orion took off quickly and steeply and then made a sharp bank to the right—the direction of Barrow. Several hundred feet above the water, the engine was heard to sputter. For an instant, the plane seemed to hang in the sky, and then it plummeted into the lagoon, apparently killing pilot and passenger instantly. Once again, someone of fame, someone by all accounts well qualified, had miscalculated.

To be sure, the modified Orion had its mechanical and aerodynamic faults, but tragically at the core of the crash appeared to be that cause for which no equipment can compensate: pilot error. For starters, there was the weather. No one else was flying in it. Then there was the pilot's experience with this particular airplane. Post had been flying the Orion with its floats for only a few weeks. Despite his experience at the controls of the *Winnie Mae* and other land-based planes, Post had probably made no more than a dozen takeoffs and landings with floats. For some reason, Post, who had always babied the *Winnie Mae* smoothly into the air, quickly developed a tendency with the Orion to—in pilot's terminology—"horse" the aircraft off the water, pulling it quickly into the air and then climbing steeply. One Fairbanks pilot watched Post do this on his departure from Chena Slough and shook his head in disbelief. It may have had something to do with the aircraft's tendency to be nose-heavy, but it is a dubious technique with land-based planes, and one that is even deadlier in a floatplane, where the floats create considerable drag.

Post took off quickly and steeply and then committed the third strike when he banked the plane sharply, perhaps for a last glimpse of the lagoon to get his bearings. All of this reduced the aircraft's lift. That was the end. Probably, the aircraft stalled aerodynamically, which in aviation terms means that it stopped flying through the air. Others have suggested that the steep climb drained the low fuel in the tanks away from the carburetor feed, and the engine experienced a mechanical stall.

America mourned two heroes, and once again tragedy fixed the nation's eyes on Alaska. And once again tragedy raised cries for government supervision. Most important to Alaska, the Civil Aeronautics Act of 1938 introduced a whole slew of rules and regulations to Alaska's heretofore fast-and-loose skies. Suddenly, the Civil Aeronautics Administration was assigning routes on the basis of "convenience and necessity," telling pilots who could fly which routes, checking logbooks, and monitoring passenger loads.

Most pilots grumbled, but they complied with the new orders and kept right on flying, Harold Gillam among them. At the height of the frenzy of World War II, "Thrill 'em, spill 'em, but never kill 'em" Gillam crash-landed a twin-engine Lockheed Electra a few miles outside of Ketchikan after being sucked into a tornado of a downdraft. Gillam died while trying to walk out and get help for his injured passengers. The truth was, of course, that they always thrilled 'em, sometimes they spilled 'em, and once in a while there was a tragedy that killed 'em.

Knocking around the Gates

If there was an area of the United States that in the 1930s could still be called a largely unexplored wilderness, it was at the heart of the massive Brooks Range of northern Alaska. It wasn't so much that people hadn't tried to figure out its maze of rugged mountains and meandering rivers, but that there was just so much of it to figure out. Nunamiut Inupiats had long followed the caribou herds through its infrequent passes and hunted Dall sheep and brown bears on its rocky slopes. The indomitable Captain Cook reached Cape Lisburne in 1778 and saw its far-flung western foothills trail off into the Chukchi Sea. Royal Navy captain Frederick William Beechey sailed beyond Cape Lisburne in 1826, hoping to meet up with Sir John Franklin. Seal hunters and whalers visited the Arctic coast for decades, and even Sheldon Jackson arrived in Barrow to help found a Native school. But all of these journeys bespeak the obvious: They had all been via the sea. What lay at the heart of the country between the Arctic coast and the great Yukon River to the south—an area encompassing almost half of Alaska?

The first serious reconnaissance of the Brooks Range was undertaken in 1884—from the sea, simultaneously by the coast guard *and* the navy. Lieutenant J. C. Cantwell of the U.S. Revenue Cutter Service surveyed the lower reaches of the Kobuk River. Incredibly, Lieutenant George M. Stoney of the U.S. Navy led a similar but separate team over the same ground only days behind him. Cantwell heard Inupiat reports of a large lake at the headwaters of the Kobuk from which a portage could be made to the upper Colville on the northern slope of the range. The following year, Cantwell took a steam launch up the Kobuk as far as she would go and then switched to skin boats in order to reach Walker Lake beneath Mount Igikpak and the Arrigetch Peaks. Cantwell marveled at the spectacular surroundings and the lake's huge trout and wrote, "We would have called it Utopia had not the mosquitoes nearly driven us wild."[35]

Meanwhile, one of Cantwell's assistants, S. B. McLenegan, made a similar reconnaissance up the Noatak River, reaching its Grand Canyon. That same year (1885), the determined Lieutenant Henry Allen made his detour up the Koyukuk after a tour of the Copper, Tanana, and Yukon

Rivers. With the steadfast Private Fickett, Allen pushed up the Koyukuk as far as the John River. Early the following February, navy lieutenant George M. Stoney mushed up the frozen Kobuk by dogsled and crossed the divide over to the upper Alatna River. Stoney recorded the first written description of the Arrigetch Peaks: "They appear in every conceivable way and shape: there are rugged, weather-scarred peaks, lofty minarets, cathedral spires, high towers and rounded domes; with circular knobs, flat tops, sharp edges, serrated ridges and smooth backbones. These fantastic shapes form the summits of bare, perpendicular mountains."[36]

Then, of course, there were the miners. John Bremner, who had been with Lieutenant Allen on most of his journey, mined Tramway Bar on the Middle Fork of the Koyukuk before meeting his death in a misunderstanding with Koyukons in 1891. The backwaters of the Klondike tidal wave filled up the lower Koyukuk. Freeze-up trapped steamers, and about 350 prospectors spent the winter of 1898–99 at scattered collections of ramshackle shacks along the river. The names given to these assorted hovels—Rapid City on the Alatna, Union City at the junction of the Koyukuk and its South Fork, and Jimtown up the South Fork at the mouth of the Jim River—bespoke grander things than ever came from them. Gordon Bettles opened a trading post on the Koyukuk about a mile downstream from the mouth of the John, and there were even a few cabins built up the river's Middle Fork at a place called Slate Creek. Whether it was the placer mining in the frigid waters, or the fact that many turned around here and called its quits, is debatable, but Slate Creek was soon called Coldfoot.

News of the Nome discoveries sent most of these prospectors packing, but there continued to be little flurries of mining activity along the upper Koyukuk. When gold was found along Nolan Creek, most of the town of Coldfoot began to drift north ten or so miles to what was soon a little settlement called Wiseman. The miners numbered in the dozens and the finds in the cents per pan, but that and some hunting were enough to keep a man going.

So by the 1930s, the high Arctic Divide in the heart of the Brooks Range at the heads of these rivers was hardly terra incognita, but there were still plenty of blank spaces on the map. Enter Bob Marshall. As his brother, George, wrote in the introduction to the first edition of Bob's posthumously published *Alaska Wilderness*, "Robert Marshall was fascinated by blank spaces on maps and was drawn to them from an early age."[37] Indeed, Bob Marshall never lost that fascination.

Bob Marshall was born in the family brownstone in New York City on January 2, 1901. He was the son of Louis Marshall, a prominent constitutional lawyer and a well-connected leader in Jewish affairs. The family had a cabin in the Adirondacks, and Louis Marshall's passion for the wilderness as well as his determination to speak for the repressed and underrepresented took root early in his son. At fifteen, Bob, as his family always called him, announced that he had decided to become a forester so that he could spend as much time as possible in the woods. During a tragically short but remarkably productive life, Marshall managed to do that quite well, dividing his time between offices in Washington, D.C., and far-flung fields. He graduated from the New York State College of Forestry in 1924, got a master of forestry degree from Harvard the following year, and his Ph.D. from Johns Hopkins in 1930.

Marshall's boyhood dreams of Lewis and Clark quickly gave way to dreams of Alaska. When he first joined the U.S. Forest Service in the summer of 1924, he pleaded for an Alaskan assignment. Instead, he spent time on the trail of his first heroes in northern Montana and Idaho, near what is now called the Bob Marshall Wilderness Area or, like him, simply "the Bob." But Alaska was calling him one way or the other. In the summer of 1929, in between two years of Ph.D. work at Johns Hopkins, Marshall headed north, drawn to the blank spots on the map of the central Brooks Range west of Wiseman. Veteran pilot Noel Wien flew him from Fairbanks to Wiseman—Marshall's first airplane flight.

Technically, he was on vacation, but typically, Bob had a research plan in mind. He planned to study tree growth at what he figured to be the northern limits of the tree line. With miner, hunter, and all-around woodsman Al Retzlaf, Marshall left Wiseman on July 25, 1929, and walked up the dirt road leading toward the Nolan Creek diggings. Climbing Smith Creek Dome that very afternoon, he got his first good view of the enormity of the terrain ahead. The mountaineering bug, which was never very far beneath the surface of his skin, itched, and he later wrote, "Looking northward again, I suddenly realized that probably not a single one of the hundreds of mountains before me had ever been climbed."[38]

To be sure, there had been a handful of hunters and miners in the area before, but Marshall had found no records of any detailed explorations of the upper North Fork of the Koyukuk—certainly no accurate maps had been made. Marshall and Retzlaf crossed Pasco Pass, descended the Glacier River a ways, and then crossed Delay Pass into the valley of the North Fork. They moved up the North Fork past the mouths of the

Tinayguk and Clear Rivers, and then another mountain beckoned. They scampered up Moving Mountain—so named by Marshall because of the evidence of landslides—and took in the view to the north. "Close at hand," Marshall noted, "only about ten miles air line to the north, was a precipitous pair of mountains, one on each side of the North Fork. I bestowed the name of Gates of the Arctic on them, christening the east portal Boreal Mountain and the west portal Frigid Crags."[39]

Northward through the Gates they went. The North Fork bent sharply eastward soon after, but they continued north up Ernie Creek toward a narrow defile between two rows of picketed peaks. Marshall called it the Valley of the Precipices. At the head of the valley, Ernie Creek split into three streams. Retzlaf busied himself hunting sheep, while Marshall explored the eastern branch. The fact that it is still called Grizzly Creek explains what he found. On a return visit the following year, he would plant some spruce seedlings here to see if they might take root.

Returning down Ernie Creek to the North Fork, Marshall became fixated with a slender spire that dominated the upper North Fork Valley to the east of the Gates of the Arctic. He was still a little green to this country, and he named it the Matterhorn of the Koyukuk. Within two years, he had given it a far more suitable name, Mount Doonerak, an Inupiat word for spirit or devil. Marshall calculated that the mountain was in excess of 10,000 feet. More recent surveys have put its elevation at 7,457 feet but done nothing to lessen its grandeur.

After exploring the upper North Fork Valley and getting flooded out of a camp at the confluence of the North Fork and the Clear, Marshall and Retzlaf returned to Wiseman. Marshall had had a grand time, but this first taste of Alaska only whetted his appetite. He wanted to see more country, yet was also fascinated by his encounters with the Native people he had met. The following August, Marshall returned to Wiseman determined to spend a year there studying the townspeople as well as the surrounding geography.

To begin that year, there was another trip up the North Fork, again with Al Retzlaf and also a packer named Lew Carpenter. At the mouth of the Tinayguk, they met up with a local legend, the Ernie of Ernie Creek—Ernie Johnson, one of the toughest and steadiest men in a tough and tenuous country. In the course of a little stroll up the Tinayguk looking for moose, Ernie and Bob quickly formed a strong bond. The following summer the duo made two long trips together to the headwaters of first the Alatna and then the John River.

When Bob Marshall settled down in Wiseman for the winter of 1930–31, everyone in town thought that he was writing about trees. Imagine their shock when each resident received a book titled *Arctic Village* in the mail in the spring of 1933. Meticulous in detail, revealingly frank, and surprisingly personal, the book was about them—almost each and every one of them! No one in town even knew that he had been writing it. Marshall himself joked that his attorney planned to spend the better part of the next two years defending libel suits. That none were filed is probably because people were more content with themselves in those days, even if the view was not always flattering.

Arctic Village quickly became a classic of life in Alaska. If Wiseman residents were surprised at the book's publication, they were even more taken aback the following spring when each adult in town received a check from Marshall for eighteen dollars, their share of half of his royalties from the book. For some, that was a good two weeks' wages in those days.

Marshall, of course, was eager to return to Wiseman and the Brooks Range, but other commitments kept getting in his way. After the publication of *Arctic Village*, he wrote *The People's Forests* and was appointed director of forestry in the Office of Indian Affairs. In 1935, he helped to found the Wilderness Society. In 1937, it was back to the Forest Service as chief of the newly created Division of Recreation and Lands. And there was his Brooks Range mapping. He worked meticulously with the U.S. Geological Survey to ink in some of those blank spots along the Arctic Divide. He certainly added his share of descriptive place-names, but whenever possible he noted the old and established names used by his sourdough companions. Along the way, he wrote almost 100 articles and papers on forestry and wilderness issues. Although he never married, he flirted with the idea several times. Perhaps he thought that such ties might keep him from his beloved woods; or perhaps he sensed his own mortality at an early age.

Finally in March 1937, Marshall wrote to Ernie Johnson: "I can't think of anything more glorious than to be on the trail with you again and exploring some more of what still remains to me the most beautiful country I have ever seen. . . . There is still much exciting country to explore there and it would be too bad not to take advantage of it."[40]

When Marshall managed to return to Alaska in August 1938, Doonerak was beckoning. He flew into Wiseman and found it changing; radios, airplanes, and even cars were making inroads. But the people were remarkably the same—and long over any ruffled feelings caused by the publication of *Arctic Village*. Seventy-eight-year-old Verne Watts did admit that he had been a little sore at Marshall when he was quoted as describing one woman as "so thin that a couple of macaroni sticks would make a pair of drawers for her." Verne got over it, but seventy-seven-year-old George Eaton pulled Marshall aside and said seriously to him, "Of course, Bob, when I was saying how I'd slept with more women than any man in Alaska, I didn't expect you to put it in a book, but I'm a-telling you, it's true."[41]

With Ernie Johnson, Jesse Allen, and Kenneth Harvey, Marshall headed up the North Fork to Pyramid Creek just north of the Gates of the Arctic. He was intent on exploring the western approaches to Doonerak, but it was raining so incessantly that the foursome soon continued north toward Ernie Pass. Marshall detoured up Grizzly Creek to see if the seedlings he had planted in 1930 had taken root. They had not—just too far north of the tree line. Then the party crossed the pass and explored the headwaters of Graylime Creek and the Anaktuvuk River, circling beneath Limestack Mountain in the process. They were inclined to wait for the weather to clear, but fresh snow only 500 feet above the valley floor counseled a retreat. In the process, they survived a harrowing boat wreck on the swollen waters of the North Fork and arrived back in Wiseman only to hear that these had been the heaviest floods in years.

The next year, Marshall got an earlier start. He arrived in Wiseman on the summer solstice, determined once more to reach the summit of Doonerak. Solid Ernie Johnson was busy with a mining operation, but Jesse Allen and Ken Harvey signed on, along with a Kobuk Inupiat named Nutirwik, who was reputed to be the best hunter in the upper Koyukuk. Marshall left the hunting to others and wrote that he preferred to see animals alive, but he was certainly never one to turn down a good rack of lamb.

Now they tried to reach Doonerak from the south, approaching it up Pinnyanaktuk Creek, a name meaning "superlatively rugged" that Marshall had given on a winter trip out of Wiseman in March 1931. Pinnyanaktuk Creek proved aptly named, but reconnoitering the divide at its head, Marshall and Harvey discovered that the main drainage immediately on the southern flanks of Doonerak was not upper Pyramid Creek as they had surmised, but rather the narrow defile of Bombardment Creek. This cascading torrent sliced between Hanging Glacier Mountain and Doonerak and then hooked wildly around the latter's south side.

Marshall and Harvey retreated down Pinnyanaktuk Creek, rejoined Allen and Nutirwik at the mouth of Holmes Creek, and then made a midnight crossing of Holmes Pass into the Pyramid Creek drainage. From the confluence of Pyramid Creek and the North Fork, they circled around Hanging Glacier Mountain via the North Fork and made their way up rocky Bombardment Creek. It was a most strenuous and circuitous route, and one that tried even Marshall's unbounded energy. Marshall, of course, was a legendarily fast hiker, but his penchant for thirty-mile-a-day hikes in the lower forty-eight met with a dose of Alaskan reality in this landscape.

Near the head of Bombardment Creek, they were surprised to discover a relatively large lake, still frozen solid on July 5 at the height of the Arctic summer. Years later, it would be named for Marshall. But the towering view of Doonerak, majestic as it was rising above the lake, did not inspire mountaineering confidence. Harvey, Nutirwik, and Marshall climbed the mountain's northwest ridge to a point Marshall called North Doonerak. Jesse Allen, inveterate expeditioner that he was about most other things, had only one arm and remained in camp because he was not up to dicey rock work.

From the summit of North Doonerak, Marshall pronounced the northwest face of the main peak—barely half a mile away and 2,000 feet above—impossible. So close, and yet so far. But true mountaineers delight in the experience and not the conquest, and Marshall did just that. He and his companions circled around the north side of Doonerak to Amawk Creek and ended the outing with a rewarding climb of 6,945-foot Apoon Mountain before heading down the Hammond River back to Wiseman.

Four months later, Bob Marshall was dead long before his time. He died in his sleep in a Pullman berth en route from Washington, D.C., to New York City to visit his two brothers. He was thirty-eight years old. In life, he had argued in a report to Congress a year before his death that all of Alaska north of the Yukon River, except for a small area immediately around Nome, should be preserved forever as wilderness with no roads and no leases for industrial development. In death, he quickly became the antithesis of development anywhere and the patron saint of wilderness preservation everywhere.

Even before his winter in Wiseman and the bulk of his Koyukuk ramblings, Bob Marshall penned what was to become the rallying cry for future generations of preservationists long after his idyllic wanderings through the Gates of the Arctic. "There is just one hope," wrote Marshall, "of repulsing the tyrannical ambition of civilization to conquer every niche on the whole earth. That hope is the organization of spirited people who will fight for the freedom of the wilderness."[42] Sure, Bob Marshall filled in a few blanks upon the map of the central Brooks Range, but his real significance to Alaska is that his large shadow would continue to fall across the land in the decades ahead, unequivocally defending wilderness for wilderness's sake.

Farmers in the Matanuska

The Great Depression hit Alaska hard, but times had been rough for so long in so many places that some folks didn't seem to notice. In the summer of 1934 with no upturn in sight, President Franklin D. Roosevelt appointed Ernest Gruening, a medical school graduate who instead had practiced journalism and foreign affairs, to be the head of his newly created Division of Territories and Island Possessions. Gruening met with FDR at Hyde Park to receive his instructions and was told by the president that "Alaska needs more people and we ought to do something to promote agriculture. Next spring I would like you to move a thousand or fifteen hundred people from the drought-stricken areas of the Middle West and give them a chance to start life anew in Alaska."[43]

Never one to mince words, Gruening suggested to the president that his new division was really the equivalent of the British colonial office and that a democracy shouldn't have colonies. According to Gruening, Roosevelt replied, "I think you're right. Let's see what you can develop." In the big picture, those words set Gruening on a twenty-five-year campaign for Alaskan statehood, but his specific charge in the summer of 1934—and the only one to come out of that first meeting with FDR—was the establishment of a government-sponsored agricultural colony in Alaska.

As with so many government programs, there was an immediate dollar gap between concept and implementation. With no money budgeted for the plan, Gruening went to see Harry Hopkins, Roosevelt's right-hand man in battling first the depression and later World War II. Hopkins had been present at the Hyde Park meeting and knew Roosevelt's intentions. As head of the Federal Emergency Relief Administration (later the Works Progress Administration), he had both the political muscle and the dollars to make things happen. Sure, Hopkins told Gruening, there was money for agricultural colonies in Alaska, but if the Federal Emergency Relief Administration was going to supply the dollars, it was also going to run the show. So it was that Harry Hopkins, rather than Ernest Gruening, took the lead in Alaska's biggest colonial experiment.

Alaska was not the only place where the New Deal tried to resettle the destitute from poverty-stricken rural areas, but thanks in part to Alaska's special mystique, the proposed project "up north" got considerable press and took center stage. Boosters of the territory were ecstatic. For once the federal government was actually doing something to promote its growth. If these colonies demonstrated that farming could be successful in Alaska, then it could be done on a large scale and many others would stream north. Even *Time Magazine* wrote, "Many an observer has pointed out that past U.S. depressions were relieved by mass migrations to the frontier, [and] that the present depression is uniquely acute because that safety-valve is gone." But there was, the magazine asserted, "one last U.S. frontier: Alaska."[44]

In Alaska, even those traditionally opposed to the federal government's involvement in territorial affairs were guardedly optimistic. After all, these government-sponsored agricultural settlements were supposed to solve all sorts of problems. They would supply required agricultural products to Alaska. They would boost the territory's stagnant population. They would shore up future defense of the area, given the uneasiness of relations with Japan. And they would provide scores of farming families who had been devastated by low prices, drought, and depression with a fresh start. To those with a sense of history, these goals may have been strangely reminiscent of the colonial policies Baranov had advocated for Russian America more than a century before. Whether Hopkins and other Federal Emergency Relief Administration officials heard history's footsteps or thought themselves visionaries in devising the plan is debatable, but the program moved forward rapidly.

The location selected for the settlement was the Matanuska Valley at the head of the Knik Arm of Cook Inlet, forty-some miles northeast of Anchorage. The name Matanuska seems to have been derived from the Russian term for "Copper River people." The Russians applied this name to the Ahtna Natives, and its use in this location at the head of Cook Inlet may have been meant to indicate a route between Cook Inlet and the Copper River—the same route that the Glenn Highway follows today. The Matanuska Valley is about twenty-five miles long and five to ten miles wide. Its gravel deposits and soils are remnants of the moraines of the Matanuska and Knik Glaciers that ground down from the Chugach Mountains and then retreated. In some places, the soil is rich and fertile, but in others—sometimes only yards away—it is sandy and coursed with considerable gravel. (Glaciers are not very systematic in their soil work.)

Weatherwise, the Matanuska Valley isn't exactly the balmy tropics, but neither is it the deep Arctic. The Chugach Mountains to the south block some of the more incessant rains from the Gulf of Alaska and let in the sunshine, while the Talkeetna Mountains to the north keep some of the Arctic cold from spilling south across the Alaska Range. But the Matanuska is not always the "peaceful valley." Some sections are subject to the "Matanuska wind," an icy blast that blows from the northeast usually in winter and spring, and the "Knik wind," a warmer blow from the south that frequently picks up dust from the vast tidal flats around Cook Inlet. In any given year, the growing season averages 110 days with about fifteen inches of rainfall.

Historically, the Matanuska Valley was home to the Tanaina Natives. They established the small village of Knik at the mouth of the Knik River. (*Knik* is the Tanaina word for fire.) About 1903, Joseph Palmer started a little store in the area and ran it for some years. In 1917, the Matanuska Agricultural Experiment Station, one of several such stations operated in the territory by the U.S. Department of Agriculture, was established in the valley partly in response to the need to feed construction crews working on the Alaska Railroad. Some homesteading occurred in the area, but never as anything more than a trickle. There was a quiet ebb and flow of people in the valley as some wore themselves out on its rocky soil and moved on and others arrived to take their place.

As the depression took its toll on just about everything, M. D. Snodgrass, the settlement agent for the Alaska Railroad, and Colonel O. F. Ohlson, the railroad's general manager, did their best to attract newcomers. It was an attempt first and foremost to drum up business for the railroad by filling up some of the abandoned homesteads scattered across the valley. When the Department of Agriculture dropped its support for agricultural experiment stations in 1931, the railroad even supported the Matanuska and Fairbanks stations for a year. But by 1934, there were still only about 100 families living in the entire Matanuska region. Farming methods were geared to subsistence levels, roads were poor at best, and there were the usual complaints levied against the local trading posts for charging high prices for nearly all commodities. Ohlson thought that what the railroad really needed in order to get business moving in the valley was one of the New Deal's new social programs.

In June 1934, Ohlson got his wish when an initial survey of the valley was made to determine its suitability for a colony. The Federal Emergency Relief Administration received the report the following

January, and the next month Roosevelt issued an executive order prohibiting all new homesteading in the valley, effectively reserving its remaining land and all abandoned homesteads to the colony. Now Harry Hopkins had to find the colonists.

Using typical bureaucratic rationale, the federal government decided that people from the colder northern states would be most readily adaptable to Alaskan conditions. So in March 1935, the Federal Emergency Relief Administration directed its state offices in Minnesota, Michigan, and Wisconsin to select 200 families for the Matanuska settlement.

Considering the hordes of the unprepared that had flocked to Alaska during the gold rush, selection standards were quite rigorous. Families were chosen from active county relief rolls. Those of Scandinavian or at least northern European stock were given preference because again they were deemed better adapted to northern climates. The targeted age group was twenty-five to thirty-five, although this was extended in some special circumstances—usually applicants with younger wives. There could be no record of chronic illness in the family, and adults were required to have at least an elementary school education. High school was desirable. Once past these requirements, selected families were given background checks by the state social services divisions. Imagine if some of these standards had been applied to other Alaskan immigrants over the years! In return, the lucky few were to be provided with forty acres, a house, and transportation from their point of origin to Palmer, Alaska—all on easy thirty-year terms.

Hopkins created the Alaska Rural Rehabilitation Corporation under the Federal Emergency Relief Administration to manage the colony until such time as it could establish its own cooperative organization. Each colonist signed a contract with the corporation agreeing to the terms of the migration. The first sixty-seven families with a total of 298 people traveled by special train from St. Paul, Minnesota, to San Francisco and arrived there on April 29, 1935. They boarded the U.S. Army transport *St. Michiel* and found themselves in Seward a week later. But there was a problem. The "house" parts of the deal weren't quite ready. In fact, also at the dock in Seward was the *North Star*, equally fresh from California with an army of laborers on board who were supposed to erect temporary tents for the colony. In the *North Star*'s hold were 1,500 tons of cargo, farm machinery, tools, groceries, and household equipment, all earmarked for the Matanuska Valley.

On May 10, a special train carried the first contingent of settlers from Seward to Anchorage, where Alaska Railroad officials and Anchorage citizens gave them a rousing welcome. Colonel Ohlson was quick to point out in his remarks that the idea for this grand experiment had been President Roosevelt's, neglecting to make any mention whatsoever of his own promotional role. That way, when these newcomers got to Palmer later in the day and found seventy tents pitched in rows alongside the railroad tracks and little else, they would think of FDR and not Colonel Ohlson.

Meanwhile, another group of 135 families from Michigan and Wisconsin had gathered and departed from Seattle on the busy *St. Michiel.* This group arrived in Seward on May 22, and the men hurried to Palmer ahead of their families to be there for the drawing of lots. It wasn't exactly the Oklahoma land rush, but for those down on their luck, it had all of the excited anticipation of it. Two hundred forty-acre lots were drawn at random, and afterward there was a frantic round of trading as settlers switched lots and vied to be near relatives or old neighbors from the States. How good the soil might be throughout a parcel was frequently judged by a couple of turns of the shovel—a tricky proposition along the Knik Arm.

Not all were happy. Despite the promises of sunny skies, the spring and early summer of 1935 proved to be unseasonably wet. Dreary overcast and rain, rain, and more rain left the collection of tents a soggy mess. Scarcely had the settlers arrived when about forty disgruntled colonists protested the conditions in telegrams to President Roosevelt and Alaska territorial governor John Troy. Homes were not built, wells not dug, there were no schools, and there was the perennial complaint about high prices at the government commissary—higher even, it was alleged, than those at the private stores in Palmer, Matanuska, and Wasilla.

Many simply gave up. Six families left within a month, and twenty more by the end of that first summer. The government dutifully paid their passages back to the States, where most returned to the welfare rolls. But progress was being made. By October, 140 houses standing among the 200 lots were ready for occupancy. A community center, warehouse, dormitory, and power plant were complete, and work was under way in Palmer for a hospital. Temporary school arrangements with a circuit-rider teacher had been made, and by the following autumn a new central school building was opened in Palmer. By 1940, about 150 of the colony's 200 tracts were occupied, although 106 of the original colonists had long since returned to the States or left the colony for other Alaskan work.

Meanwhile, the colony was rapidly repopulating itself. With most of its inhabitants young married couples, 130 babies were born to the 200 or so colony families during the first two years. The trend continued, and it was no wonder that Dr. Conrad E. Albrecht, who served as the Matanuska Valley's physician from 1935 to 1941, found that delivering babies took up the bulk of his time.

In the end, the major thing that the colony proved was that such operations were not cure-alls for the ills of the depression. Farming difficulties in the Matanuska Valley between 1935 and 1941 were the same as or worse than those being experienced in the farm belt of the lower forty-eight. Pigs, cows, and sheep were raised, and dairying and truck farming proved the most reliable ventures. But when good crops came in during the fall of 1939, an early October snowstorm buried 80 percent of them. If Matanuska farmers did harvest a good crop, they were limited in their markets. The railroad gave them some access but also exposed them to competition lurking down the line. Seattle suppliers in particular zealously guarded their Alaskan monopoly of shipping goods to Seward and the southeast. Some went so far as to spread the story that Matanuska vegetables lacked taste and were without nutrients because they were grown too rapidly in the long summer days.

Frequently, such agricultural difficulties were compounded, not eased, by the overriding federal administration. Eventually, this administrative structure evolved from the original Alaska Rural Rehabilitation Corporation into a community council that met weekly with the corporation's general manager and then became the Matanuska Valley Civic Association. All vestiges of colonial administration finally disappeared in 1951, when Palmer became an incorporated town.[45]

In the end, what saved the colony and eased its transition into a regular settlement was the 1940 construction of nearby Fort Richardson and Elmendorf Air Base in Anchorage, which offered a job market as well as a farm market. Some critics called the Matanuska colony a social experiment gone awry—just another socialist New Deal scheme of that damn Democrat, Franklin Delano Roosevelt. Others sang FDR's praises and called it the best thing that ever happened to them. One thing is for certain. The towns of Palmer and Wasilla still stand, and as home to Alaska's State Fair, the Matanuska Valley is about as much farm country as one is apt to find in the far north.

Never the Same Again

Mountaineering historians have a propensity to characterize a certain period in the history of a particular mountain or mountain range as its *golden age*. Hence, the golden age of the Alps in the 1860s, the Canadian Rockies in the 1890s, or Yosemite in the 1960s. Trite as the characterization might be, *golden age* describes a period replete with first looks, first maps, first ascents, and a heavy sense that what is being experienced will never be the same again. The duke of the Abruzzi was too young and Mount St. Elias too wild for him to feel that sense, but Bob Marshall felt it in the central Brooks Range. Doubtless, too, that Harry Karstens felt it as he completed his tenure at Denali, having seen the mountain go from unclimbed giant to tourist destination. If there was a golden age of Alaskan mountaineering, it was the 1930s, and the men who stood at its center were Allen Carpé, Terris Moore, and Bradford Washburn.

In the spring of 1930, only four of Alaska's forty-five peaks above 12,000 feet had been climbed—none more than once. The duke of the Abruzzi and his fellow Italians had of course summited St. Elias in 1897. Mount Wrangell and the geography of its volcanic neighbors vexed Robert Dunn enough that he and a companion reached its summit in 1908. Three years later, a diminutive five-foot graduate of Bryn Mawr, Dora Keen, attempted to climb Mount Blackburn. She returned in 1912 for another try and left Kennecott bound for the mountain in the company of seven men. Keen ended up marrying G. W. Handy, the only one of the seven to make it to the summit with her. Later surveys found Blackburn's 16,390-foot northwest summit to be 104 feet higher than the southeast summit attained by Keen and Handy, but considering their thirty-three-day trip out of Kennecott—twenty-two of which were spent without tents and ten without fuel—it seems less than sporting to deny a first-ascent claim. A year later, Hudson Stuck, Harry Karstens, Robert Tatum, and Walter Harper stood atop the South Peak of Denali. Four peaks out of forty-five—five out of forty-six if one counts the sourdoughs on Denali's North Peak—but that left the field pretty well open.

Allen Carpé, a research engineer by training, was widely recognized by his peers to be both a brilliant scientist and an able mountaineer. Carpé cut his teeth climbing in the Alps and the Canadian Rockies before participating in the first ascent of Canada's Mount Logan in 1925. A year later, Carpé, Andy Taylor, a fellow Logan summiter and one of the most experienced packers in the territory, and William Ladd, a Columbia University medical professor, made a reconnaissance of Mount Fairweather but did not reach its summit. Carpé and Taylor joined forces again in 1930, this time in the company of a Harvard-educated economist named Terris Moore. The trio left McCarthy and followed a circuitous route over Skolai Pass and up the Russell Glacier to make the first ascent of 16,421-foot Mount Bona. Carpé hadn't given up on Fairweather, but Bona proved too inviting a prize.[46]

Meanwhile, however, a team led by a twenty-year-old Harvard undergrad by the name of Bradford Washburn had been having a look at Fairweather. The expedition was a learning experience for Washburn, and he came away from it not only with a healthy respect for Alaska's scale but also incurably hooked on its mountains. So in 1931, Fairweather still remained to be climbed when Carpé, Taylor, and Ladd again returned to its slopes. The Bona climb had hooked Terris Moore as well on Alaska's mountains, and he eagerly joined the group.

Twice the foursome was poised for a summit dash high on Mount Fairweather only to be repulsed by the bitterest of weather. Clearly, such high adventure was not without competition, because as they huddled in the fury of the second storm, Carpé told his teammates that if they did not manage to climb Fairweather that year, Bradford Washburn would most surely be back the following year to finish the job. Ladd and Taylor descended to conserve supplies, but when the storm slackened somewhat, Carpé and Moore struggled through fierce winds and two feet of fresh snow to reach the summit. For Carpé it was his third major first ascent (Logan, Bona, and Fairweather), and for Moore—still only twenty-three—it was his second (Bona and Fairweather).

But Carpé and his young protégé had even bigger goals in mind. Since their first meeting at New York City's venerable Explorers Club, they had been intrigued by reports of a mountain in western China that exceeded 30,000 feet. If the reports were true, it would be the highest mountain in the world. Moore was eager to go, but middle-aged sensibilities soon got the better of Carpé. He was nearly forty, married with a family, and deeply involved in his professional work with Bell Telephone

Laboratories. Carpé opted for what appeared to be the safer choice and chose to head for the slopes of Mount McKinley and take part in a cosmic ray research project. Organized by Dr. Arthur H. Compton of the University of Chicago, it was charged with studying cosmic rays (radiation) at high elevations and high latitudes in different parts of the world. One observation station was established at the Kennecott Mine, and another was planned for the head of the Muldrow Glacier at about 11,000 feet on Mount McKinley.

At first glance, the most dangerous part of Carpé's work seemed to be the manner in which he planned to arrive on the mountain. Having talked with Joe Crosson of Alaska Airways about the feasibility of resupplying the party via airdrops, Carpé soon decided to have Crosson fly him and his team directly from the Alaska Railroad at Nenana to the Muldrow Glacier—something that had never before been done. With Carpé on the first flight to the mountain were Edward P. Beckwith, a consulting engineer for General Electric, and Theodore Koven, a twenty-eight-year-old Sierra Club member who had signed on as a scientific assistant.

Loaded weight for the flight quickly became the overriding issue, particularly when Carpé asked Beckwith why his sleeping bag weighed twenty pounds instead of twelve. Sheepishly, Beckwith admitted that he had brought along an extra quilt. The quilt and some other items stayed behind, and the three passengers and their gear got down to Crosson's 1,200-pound maximum weight limit. Landing his single-engine Fairchild equipped with skis on the ice of the Tanana River, Crosson took his party and their gear on board and then headed for the Muldrow. No one, not even Crosson, seems to have suggested that he try the first attempt with a lighter load. Perhaps all were skeptical that there would be a second trip.

As it turned out, Joe Crosson pulled off the landing at almost 6,000 feet near McGonagall Pass without a hitch. The date was April 25, 1932, and the flight was the first glacier landing in Alaska and a harbinger that was to change the face of Alaskan mountaineering. As Beckwith later reported, "Carpé was delighted, and shook hands with Crosson, who took it much as a matter of course, and lit a cigar before leaving the plane."[47]

But now Crosson had to take off. He managed to get airborne but was unable to climb above the surrounding ridges and was forced to make an emergency landing about a mile above the first site. Early the next morning, with the glacier's surface still frozen solid, he tried again and succeeded. To supervise the loading of more gear and the arrival of two other party members, Beckwith flew out with him—apparently the first passenger to be flown off a glacier in Alaska.

A week later, Crosson and fellow pilot Jerry Jones flew Beckwith, his two newly arrived companions, and another load of supplies to the Muldrow, landing without incident at the McGonagall Pass site. Then Beckwith persuaded Crosson to make a flight up the glacier and air-drop boxes of supplies to the camp that Carpé and Koven had established at about 11,000 feet. They found the tent, and Beckwith pushed out the boxes on Crosson's command. It was another first—the resupply of a mountaineering expedition by airplane.

Meanwhile, there were others on the mountain. In 1929, Harry Karstens's successor as superintendent of Mount McKinley National Park had been Harry Liek, a veteran who had served under National Park Service director Horace Albright at Yellowstone. When Albright encouraged Liek to get out into the park more and "do really conspicuous work," Liek determined that the only way to really do that was to make a second ascent of Mount McKinley.[48]

In the spring of 1932, Liek joined with Minneapolis attorney Alfred Lindley, Norwegian skier Erling Strom, and park ranger Grant Pearson to make an attempt on the mountain. Dog teams from the National Park Service ferried gear to the head of the Muldrow, and in the process they helped Carpé and Koven establish their high camp there. One day while cutting steps along the icy spine of Karstens Ridge, the Liek party heard the then unfamiliar hum of an airplane as Crosson and Beckwith made their airdrop delivery to the high camp. Then the group went on to make the second ascent of Mount McKinley on May 7, 1932. Two days later from their high camp at about 17,000 feet, they also climbed the North Peak, becoming the first to ascend both summits.

But all was not right down below. In an arduous eighteen-hour day on May 10, the Liek party descended the Harper Glacier and Karstens Ridge all the way to Carpé and Koven's tent on the Muldrow. Much to their surprise, the tent was empty and there was no sign of the two men. "That something wrong had happened," wrote Erling Strom, "we at once understood."[49] From the men's diaries in the tent, the Liek party hoped that they might have descended to meet up with Beckwith and the others, but the fact that they had left their sleeping bags behind did not bode well.

Despite the full day that they had already put in, Liek, Lindley, Strom, and Pearson continued down the glacier in search of some sign of Carpé and Koven. When they found it, it was the mangled body of

Theodore Koven, evidently the victim of a fall into a crevasse and resulting exposure, perhaps in an attempt to rescue Carpé. Koven's tracks led from an area of the glacier saturated with crevasses. The group looked in vain for Carpé's body amid the maze but soon concluded that further efforts were fruitless as well as dangerous. Why an accomplished mountaineer such as Carpé met his end in such a way is problematic.

Liek, Lindley, Strom, and Pearson tried to evacuate Koven's body down the mountain, but Pearson soon plunged into a forty-foot crevasse, breaking his snowshoes but luckily nothing else. Wisely, they elected to use their rope for belays rather than to pull Koven's body, which they left wrapped in their tent. Now tentless, the four were forced to continue all of the way past McGonagall Pass to their base camp, arriving there forty-one hours after their departure from the Harper Glacier. Liek and Lindley immediately collapsed and fell fast asleep. Pearson started a fire in the stove, and Strom used his last bit of energy to open some tin cans. "Having kept up with Pearson on this entire trip," Strom noted later, "I could not let him get ahead of me this 42nd hour."[50]

Andy Taylor, Carpé's veteran companion of Logan, Bona, and Fairweather, and Merl La Voy, who had been on the mountain with Parker and Browne twenty years before, climbed to the Muldrow Glacier with Grant Pearson that August to retrieve Koven's body. Carpé's body was never found. These were the first two deaths to be recorded on Mount McKinley. On La Voy's recommendation, the two summits on the ridge extending northeast below Karstens Ridge were named Mount Koven and Mount Carpé.

Meanwhile, Terris Moore had gone off to China in search of the highest mountain in the world. On October 28, 1932, after an epic trip, Moore and Richard Burdsall reached the summit of Minya Konka. Rumors to the contrary, it proved to be a "mere" 24,900 feet. (Today, the Chinese call the mountain Gongga Shan, and its official height is 24,971 feet.) Moore read of his mentor's death in an English newspaper in Shanghai. One can only wonder about the fate of these two friends had either gone with the other.

As Allen Carpé met his tragic end on the Muldrow Glacier, Brad Washburn did indeed return to Lituya Bay and Mount Fairweather, but making its second ascent quickly proved not nearly as enticing as a first ascent of nearby Mount Crillon. With Washburn was his good friend and all-around climbing buddy Robert H. Bates. When they failed to reach Crillon's summit, Washburn and Bates returned in 1933 with a party that included H. Adams Carter and Charles S. Houston. Again they came up short, this time barely a half mile from the top on the summit plateau. Finally in 1934 on Washburn's third try, while Bates was academically occupied and Houston was on another mountain, he and Carter succeeded in making the first ascent of Mount Crillon.

While his friends did battle for a third time with Crillon, Charlie Houston set out to climb Mount Foraker. Houston and his father, Oscar, became interested in the mountain after reading books about the early Denali climbs by Belmore Brown and Hudson Stuck. There had been no prior attempt on Foraker, and the area was only crudely mapped. The two Houstons, T. Graham Brown, Chychele Waterston, Charles Storey, and a horsepacker named Carl Anderson headed west from Wonder Lake toward the Foraker Glacier and the mountain's long northwest ridge. Half a century later, what Charlie Houston would remember most about the trip was the ravenous mosquitoes. The group established a base camp and quickly divided into a support team and a climbing team.

The climbers, young Houston, Brown, and Waterston, set up a camp at 9,800 feet and then worked their way up the narrow knife-edge of the northwest ridge to 13,700 feet below a large plateau. On August 6, 1934, the three reached the summit of Foraker's north peak—the true one it turns out—and several days later also climbed the south peak, just to be certain that they had gotten to the highest summit. Two days later, they were back down the ridge at base camp and celebrating Charlie's twenty-first birthday with a quart of whiskey and fresh caribou steaks. His resulting hangover left young Houston swearing never to drink again. He went on to become the dean of mountaineering medicine and a world-class mountaineer, but surprisingly he never returned to Alaska, instead finding plenty of adventure in the Himalayas, including on Nanda Devi in 1936 and K2 in both 1938 and 1953.[51]

Allen Carpé's good-natured comment about Washburn on Fairweather aside, there was a close-knit and collegial relationship among the handful of mountaineers at the center of this golden age. After his Crillon climb, Washburn was asked by Gilbert Grosvenor of the National Geographic Society if there was anything else to do in the Far North. Washburn bit his tongue and did not mention his keen interest in climbing the Yukon Territory's Mount Lucania, because he knew that his friend Walter Wood was organizing a 1935 expedition for just such a purpose. Instead, Washburn proposed that he, Bob Bates, and Ad Carter map a 5,000-square-mile chunk of the central St. Elias Range generally north of Mount Logan. They took veteran packer Andy Taylor along with them—or perhaps more accurately, he took them—and in addition to mapping made an attempt on Mount Hubbard. When Walter Wood and Hans Fuhrer reached the summit of Mount Steele, Lucania's 16,440-foot neighbor to the north, but failed to climb Lucania itself, the 17,147-foot giant was still the tallest unclimbed peak in North America. Washburn and Bates knew what they had to do.

In 1937, Washburn sent bush pilot Bob Reeve a photograph he had taken on his mapping trip and asked Reeve if he could land a climbing team and gear at 8,500 feet on the Walsh Glacier on the southeast side of Lucania. The elevation was 2,000 feet higher than any of Reeve's previous glacier landings with a load, but Reeve wired back: "Anywhere you'll ride, I'll fly. Bob Reeve."[52]

Norman Bright, a star two-miler, and Russell Dow, a veteran of the Crillon trips, were to join Washburn and Bates on the expedition, but they never made it onto the mountain. Reeve took Washburn and Bates in on the first load and upon landing bogged down in a foot of slush that was covered by an icy crust. For a while, it looked as if Reeve's plane was going to become a permanent fixture on the Walsh Glacier. The two climbers dumped their gear on the mountain, and then Reeve proceeded to strip just about everything out of the plane that didn't contribute to its airworthiness. He changed the pitch of the propeller with a few well-delivered blows of a ball-peen hammer and then barely made it into the air. Reeve quickly made it very clear that not only were Bright and Dow not going to be delivered onto the mountain, but that Washburn and Bates were going to have to walk out on their own.

Left to themselves in the middle of thousands of square miles of glacial wilderness, Washburn and Bates could have panicked, but that would have been decidedly out of character. Instead, they spent two weeks relaying loads up to the wide pass between Mounts Lucania and Steele. They made a first ascent of Lucania and then decided that the easiest way out was by climbing north over Mount Steele and descending Walter Wood's 1935 ascent route. Doing just that, they cut the floor out of their tent to reduce weight and then walked out more than sixty miles over mountains, glaciers, and raging rivers, including a long detour to the head of the swollen Donjek River to effect a crossing. Years later, Reeve told Washburn's wife, Barbara, "I wasn't worried about those guys. I figured they'd get out. It was just a matter of how and when, rather than whether."[53] The summit photo of Washburn and Bates atop Lucania, taken with the former's camera tied to an ice ax with a shoelace, shows both men ramrod straight and grinning broadly. Try telling them that it wasn't the golden age!

Washburn had been taking photographs and motion pictures of his climbs since his first forays in New Hampshire's Presidential Range and the French Alps. He was uncommonly good at it. Washburn quickly embraced the airplane as a way to get sweeping panoramas and dramatic mountain views, even if it meant tying himself and his fifty-pound camera into the open doorway of Bob Reeve's airplane. In 1936, Washburn began his long association with Mount McKinley when he made aerial photographs of the massif for an expedition sponsored by the National Geographic Society and Pan American Airways. It was from these photographs that Washburn located the Kahiltna Glacier–West Buttress route that he would pioneer fifteen years later.

Nineteen thirty-eight found Washburn busy on two mountains. First, he joined with Norman Bright, Norman Dyhrenfurth, and Peter Gabriel to make the first ascent of 13,176-foot Mount Marcus Baker, the highpoint of the Chugach Range. Bad weather delayed the expedition in Valdez for almost a month, and as time ticked away, Washburn got nervous over the delay because his next engagement was to meet Terris Moore and his wife, Katrina, for an attempt at Mount Sanford. On July 21, 1938, these two friends stood atop Mount Sanford, making two major first ascents in one season for Washburn.[54]

Washburn's articles and stunning photographs in *National Geographic Magazine* and *Life* quickly cemented the reputation he had built early on by writing about New England and the Alps. In 1939, while still only twenty-nine, he was offered the directorship of the Boston Museum of Science, a position he was to hold for forty-one years. The following year, he married Barbara Polk. He was careful to include plenty of "expedition time" in his contract with the museum, and if he was similarly concerned about his new marriage relationship, he need not have worried. Barbara Washburn gamely honeymooned with him in Alaska, together making the first ascent of Mount Bertha just east of Mount Crillon. In 1941, they were together in a party of five that made the first ascent of 13,832-foot Mount Hayes via a dicey arête below the final summit pyramid. Later, on June 6, 1947, Barbara Washburn became the first woman to climb both peaks of Mount McKinley, participating in only the fifth ascent of the South Peak and the third ascent of the North Peak. They were quite a team.

Not surprisingly, Washburn was also involved in the third ascent of Mount McKinley, and with him were his buddies Bob Bates and Terry Moore. In 1942, Lieutenant Colonel Frank Marchman led the U.S. Army Alaskan Test Expedition to the mountain for an extended test of cold-weather food, tents, and clothing. The seventeen-man party established a camp in the high basin of the Harper Glacier, and on July 23, Washburn, Bates, Moore, and Einar Nilsson reached the summit of the South Peak. The following day, Sterling Hendricks, Albert Jackman, and Peter Webb also reached the summit.

It had been quite a dozen years. Between them, Washburn and Moore had made first ascents of seven major peaks in Alaska and one more in the Yukon: Washburn on Crillon, Lucania, Marcus Baker, Sanford, Bertha, and Hayes; and Moore on Bona, Fairweather, and Sanford. But far more important than first ascents were Washburn's photographs and maps, which brought the mountains of Alaska into the living rooms of millions and became the winter fodder for future generations of climbers.

And this golden-age generation was far from finished. Indeed, they were just getting started. In 1930 on Bona, Terris Moore was twenty-two. That same year on Fairweather, Washburn was twenty. In 1934 on Foraker, Houston turned twenty-one. They were young kids out on the cusp of adventure, and their exploits and legacies would cast a huge shadow over Alaskan mountaineering for the remainder of the twentieth century. Even as new crops of "upstarts" pushed technical standards on seemingly impossible new routes, they did so only after consulting Brad Washburn's photographs. And oh yes, they weren't all young kids. The indomitable Andy Taylor was fifty years old when he climbed Logan, fifty-six when he stepped atop Fairweather, and sixty when he tutored Washburn, Bates, and Carter in the tricks of a sourdough's survival in the wilds of the St. Elias Range. That thought should quicken more than a few middle-aged hearts.

Some might argue that the decade of the 1950s, which followed the Second World War, was the golden age of Alaskan mountaineering. Perhaps. Another fifteen—sixteen if one counts the true summit of Mount Blackburn—of Alaska's forty-five highest mountains were climbed, as well as many lower but more challenging ones. There were old names like Bob Bates on Hubbard and Alverstone and Brad Washburn on Kahiltna Dome; and new names like Fred Beckey on Hunter and Deborah, but things weren't quite the same—neither the mountains nor the climbers nor the freshness. No, the 1930s were the golden age of Alaskan mountaineering.

Postscript to an Era (1915–1941)

Alaskan mountaineering—*like most other activities—was interrupted by the tumult of the Second World War. As the 1930s waned, what energized Alaska, brought it out of its quiet interlude, and coalesced its developing transportation and economic networks was the looming threat of global war. Just as for the rest of America, the drums of world war beat a commonality of purpose in Alaska. There was much to be done, and little time in which to do it. As events rolled toward an irrevocable crescendo, with them came a sense that far more than mountaineering would never be the same again.*

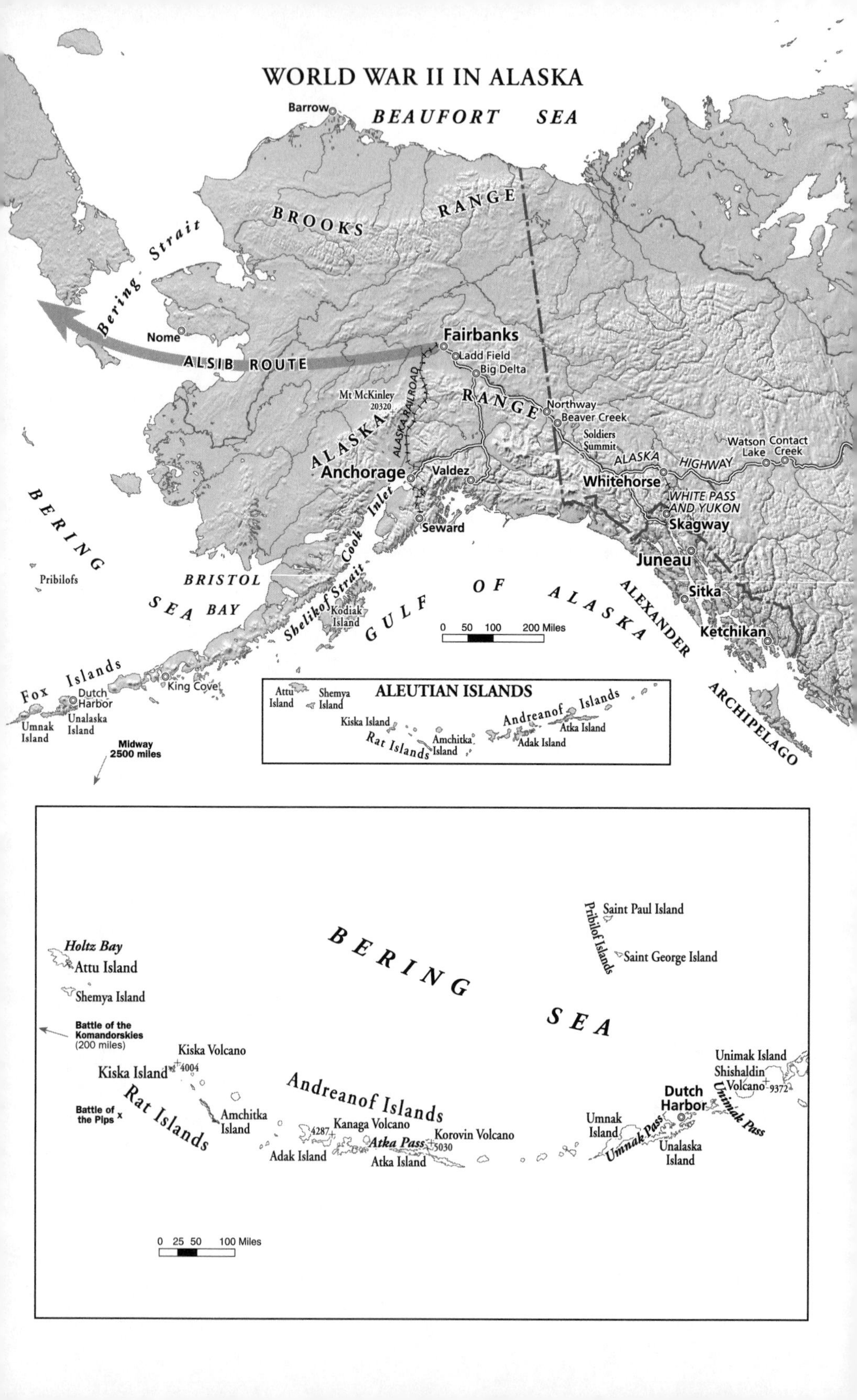
WORLD WAR II IN ALASKA
Barrow
BEAUFORT SEA
BROOKS RANGE
Bering Strait
Nome
Fairbanks
ALSIB ROUTE
Ladd Field
Big Delta
Mt McKinley 20320
ALASKA RAILROAD
ALASKA RANGE
Northway
Beaver Creek
Soldiers Summit
ALASKA HIGHWAY
Watson Lake
Contact Creek
Anchorage
Valdez
Whitehorse
WHITE PASS AND YUKON
Skagway
Cook Inlet
Seward
Juneau
BERING SEA
Pribilofs
BRISTOL BAY
Shelikof Strait
Kodiak Island
GULF OF ALASKA
ALEXANDER ARCHIPELAGO
Sitka
Ketchikan
0 50 100 200 Miles
Fox Islands
Dutch Harbor
King Cove
Umnak Island
Unalaska Island
Midway 2500 miles
ALEUTIAN ISLANDS
Attu Island
Shemya Island
Kiska Island
Rat Islands
Amchitka Island
Andreanof Islands
Atka Island
Adak Island
Saint Paul Island
Pribilof Islands
Saint George Island
Holtz Bay
Attu Island
Shemya Island
BERING SEA
Battle of the Komandorskies (200 miles)
Kiska Volcano
4004
Kiska Island
Rat Islands
Battle of the Pips
Amchitka Island
Andreanof Islands
Kanaga Volcano
4287
Atka Pass
Korovin Volcano
5030
Adak Island
Atka Island
Unimak Island
Shishaldin Volcano
9372
Dutch Harbor
Unimak Pass
Umnak Island
Umnak Pass
Unalaska Island
0 25 50 100 Miles

BOOK SIX

The Forgotten Campaign: World War II in Alaska 1941–1945

Against such a background Americans and Japanese are fighting a strange, unreal, hide-and-seek type of war, battling the elements as much as each other.

—LONNELLE DAVISON, "BIZARRE BATTLEGROUND," 1942

Shortly before midnight *on the evening of February 8, 1904, ten Japanese destroyers quietly slipped into the Russian naval base at Port Arthur (now Lüshun, China) and unleashed a devastating torpedo attack on battleships and cruisers of the Russian fleet. It was a stunning Japanese victory. It was also the result of a surprise attack. Diplomatic relations between the two countries were strained and in the process of being terminated, but no state of war yet existed. Shortly afterward, Alaska's chief game warden, Alfred H. Dunham, speculated on the ensuing Russo-Japanese War in an article about the development of Nome. Dunham wondered "what Alaska would be if it had remained in Russia's hands to the present day. Would Japanese fleets now be bombarding Sitka, Nome, and other Alaskan ports?"*[1]

Thirty-seven years later, early on the morning of December 7, 1941, the Japanese launched another surprise attack without a declaration of war. This time, warplanes of the Imperial Japanese Navy struck U.S. military installations in the Hawaiian Islands and other targets throughout the Pacific. The following day, President Franklin D. Roosevelt termed the events a "date which will live in infamy." For Alaska, it was also to be a date that would result in the most rapid, dramatic, and lasting change in its history.

Nothing altered the face of Alaska more than the Second World War. One may argue that the arrival of the Excelsior *in San Francisco in July 1897 unleashed the Klondike stampede, and that Atlantic Richfield's announcement of oil and gas discoveries at Prudhoe Bay in June 1968 opened the floodgates of another. But the buildup of population, transportation networks, and concomitant infrastructure—all of which remained and was infused with returning GIs at war's end—made December 7, 1941, and the period that followed, the most pivotal in Alaska's history.*

The Rush to Get There

Prior to the Japanese attack on Pearl Harbor, there were only about 2,500 miles of roads in Alaska, and only one major highway, the Richardson, between Valdez and Fairbanks. Ties to the United States amounted to ship service from Pacific Northwest ports and a few Pan American Airways clipper flights. Alaska was, by all accounts, still very isolated from the rest of America. But it had not been caught totally unawares by Japan's aggression.

In July 1940, as the skies over the British Isles were about to explode in the Battle of Britain, Henry H. "Hap" Arnold, by now a major general and chief of the Air Corps of the U.S. Army, made a whirlwind inspection tour of Alaska. His itinerary ranged from Juneau and Sitka, north to Circle, and west to McGrath. Landing in Fairbanks, he reminisced about the flight of bombers he had led to the territory in 1934. That had been a flight to test the limits of new frontiers. Now, Arnold was critically aware that he was about to be called upon to conduct effective military operations upon those very frontiers. "We had spent only a few hours in Alaska," Arnold reported, "before it was evident that it is one thing to decide that national defense requires air bases up near the Arctic Circle, . . . and quite another to accomplish these results."[2]

Much of the army's past experience overseas had been in tropical climates—Hawaii, the Philippines, and the Canal Zone. Arnold recognized that the Arctic was a whole new ball game. Here, cold temperatures, biting winds, and the extremes of belly-deep snow or bottomless mud wore heavily on even the most menial of tasks. Starting aircraft engines in below-zero conditions, forecasting weather along a 2,000-mile front, flying reconnaissance missions in thick cloud cover, building runways, buildings, or just about anything on permafrost were just some of the operational challenges. Then there were the personnel concerns. Soldiers, sailors, and airmen had to be properly clothed with special cold-weather gear, boots, gloves, and goggles, as well as heavy-duty tents, stoves, and vehicles. Anything that was routinely done in temperate climes required more planning, more time, and more durable equipment when done in Alaska.

General Arnold wrote about his Alaskan visit in the October 1940 issue of *National Geographic Magazine*. There was an uneasy dichotomy to the article that probably mirrored the thinking of a great many Americans at the time, including those in Alaska. As was befitting *National Geographic*, there were sections that carried the genial air of a travel piece, replete with easy talk of grand scenery and good salmon catches. But there were other sections that carried an ominous air of inevitability that the war clouds thundering over Europe would soon engulf the rest of the world. Arnold noted that "German and Japanese walking parties have recently been touring Alaska." One photo caption read: "The U.S.S.R. is but a 'Stone's Throw' from the U.S.A." Twenty-five million dollars—then a staggering sum to a territory whose own biennial expenditures for 1939 *and* 1940 were only $6.8 million—were being appropriated, according to the general, for army and navy defense measures in Alaska. "There is no gainsaying," Arnold concluded, "that Alaska will play a vital part in the scheme of national defense."[3]

This national defense mind-set brought to the forefront the issue of linking Alaska with the United States via a road. It was not a new idea. As early as 1929, the United States and Canada had each formed its own International Highway Association. Headquartered respectively in Fairbanks and Dawson, Yukon Territory, their mutual goal was a highway through Canada connecting Alaska with the lower forty-eight and stimulating commerce along the way. By the American association's optimistic estimate, such a 1,350-mile, sixteen-foot-wide gravel thoroughfare from Prince George, British Columbia, to Fairbanks could be built for just over $14 million, provided that ferries and not bridges were employed to cross the Yukon and Pelly Rivers.

Little more than talk came out of these first associations, however, and in 1938, Congress approved the formation of the International Highway Commission. President Roosevelt's appointees to the commission included Congressman Warren Magnuson of Washington State, Ernest Gruening, then still director of the Division of Territories and Island Possessions but soon to be Alaska's territorial governor, and former Alaska territorial governor Thomas Riggs. Canada appointed a similar commission, mostly at the urging of British Columbia's premier, Thomas D. Patullo, who saw the venture as a labor-intensive public works project for his province.

The two commissions met jointly for the first time in Victoria, British Columbia, in July 1939. The Americans favored what came to be called Route A, which ran north from Prince George between the Coast Mountains and the Canadian Rockies, because they envisioned feeder routes linking it to Alaska's southeastern towns. The Canadians championed Route B, which also ran north from Prince George but stayed east of the Rockies to better serve the development of northern British Columbia and the Yukon Territory. Cost estimates for either route had almost doubled since 1929, and when Congressman Magnuson cited national defense considerations as one of the rationales for an appropriation, Secretary of War Henry Stimson replied that the value of the proposed highway as a defense measure was negligible.[4]

Proposed Routes C and D soon joined Routes A and B on the drawing board. Montana, North Dakota, and Alberta representatives lobbied for Route C, because they wanted the highway's southern terminus farther east at Dawson Creek (not to be confused with Dawson in the Yukon Territory) near the British Columbia–Alberta border. Farther east still, Route D was promoted by Canadians seeking to access the growing oil fields at Norman Wells on the Mackenzie River.

In August 1940, with Great Britain hanging on against Nazi Germany mostly on the strength of Winston Churchill's bulldog resolve, President Roosevelt met discreetly with Canadian prime minister Mackenzie King at the tiny border town of Ogdensburg, New York, to discuss the mutual defense of North America. Out of this meeting came the establishment of the Permanent Joint Board on Defense (PJBD), consisting of representatives from each country.

The board's first report was submitted to both governments on October 4, 1940, just as General Arnold's part-travelogue, part-military briefing was hitting the streets in *National Geographic Magazine*. The PJBD report dealt primarily with improvements to the Atlantic coastal defenses of both countries, but it also recommended that a chain of landing fields capable of handling military aircraft, including heavy bombers, be established on a route across northwest Canada between the United States and Alaska. Further elaborated six weeks later, this recommendation called for a series of bases in both Canada and Alaska with cleared approaches, 5,000-foot runways, and facilities for refueling, maintenance, and crew rest.

Early in 1941, the Canadians began work on this series of airstrips that eventually came to be called the Northwest Staging Route. These airstrips included an entirely new field that was to be carved out of the tundra a few miles west of Fort Nelson, some 300 miles north of Dawson

Creek, and an upgraded field at Watson Lake. In Alaska, two new airstrips were to be built at Big Delta, about eighty miles southeast of Fairbanks, and at Northway near the Canadian border.

The Big Delta site was accessible from the Richardson Highway, and the Northway site could be reached via an old mining trail to the headwaters of the Copper River and then by barge down the Nabesna. The sites in northern British Columbia and the Yukon Territory were out in the middle of nowhere. The only way to get the men, materials, and equipment necessary for such heavy construction to those sites was to build a road to them. The closest railroad point to Fort Nelson was the terminus of the Northern Alberta Railway at Dawson Creek, some 400 miles by rail northwest of Edmonton, Alberta. So on February 9, 1941, a Caterpillar D-8 from the Western Construction Company of Edmonton put its blade to the frozen ground and started north from Dawson Creek along a winter trail. No one quite realized it at the time, but they had started to build the Alaska Highway.

The first major challenge was to get across the Peace River near Fort St. John while winter ice still afforded a makeshift bridge. A long convoy made the crossing safely and then followed the bulldozers north. It must have been an amazing sight—modern machinery leading a sinuous line of trucks up a trail that heretofore had seen only sled runners. Slim Byrnes was a veteran trader who encountered the iron machines that winter while mushing his dogs along the trail. When he met the cat train pushing its way north with the endless fleet of trucks bringing up the rear, it was, he said, as if "time went ahead more in a few minutes than it had in a whole lifetime. Like the snap of your fingers, we changed from the old to the new."[5]

The convoy eventually reached Fort Nelson and proceeded to construct the airstrip. It was none too soon. On July 29, 1941, the Permanent Joint Board on Defense reported to its respective governments that the completion of the Northwest Staging Route was "now of extreme importance."[6] Six months before, in one of the most easily understood analogies ever used to articulate a major doctrine of American foreign policy, FDR had quietly explained that if your neighbor's house was on fire, it was only natural that you would lend him your garden hose. You would seek no payment for it, but only to have it safely returned after the fire was out. From such homey thoughts was born the Lend-Lease Program. Not only did that mean "loaning" destroyers, munitions, and other equipment to Great Britain, but it also meant "loaning" aircraft to

America's other "neighbor" in the growing world conflict: the Soviet Union. The shortest route to the Soviet Union, as General Arnold well knew, led across Alaska.

The first flight along the Northwest Staging Route occurred in early January 1942. Thirty-eight planes took off for Fairbanks. The results were a disaster. Only eleven aircraft managed to arrive unscathed. The remaining twenty-seven crashed at various points along the route. Inexperienced pilots and severe winter weather were high on the list of causes, but it was also clear that the great distance between airfields and the limited number of navigation aids along the route were contributing factors.[7]

In the meantime, of course, America's day of infamy had occurred. Many of the reported "sightings" of Japanese planes, ships, and submarines in Alaskan waters during the first few weeks of the war were unconfirmed, but it was nonetheless very clear that Japan's military tentacles were capable of reaching that far. The recent show of Japanese naval power throughout the Pacific, and the belief still held by some senior U.S. Navy officers that America's only hope of defense now lay in shambles along Battleship Row in Pearl Harbor, heightened fears that the entire Pacific would soon be a Japanese lake. If that happened, all sea routes to Alaska would be cut.

It was in this atmosphere that Secretary of the Interior Harold Ickes suggested at a cabinet meeting on January 16, 1942, that the various alternatives for the long-proposed highway to Alaska be reexamined for a safer and more reliable route to getting men and supplies to Alaska. FDR agreed and ordered Ickes to work with War Secretary Stimson and Navy Secretary Frank Knox to develop a recommendation. Stimson, no doubt, was having second thoughts about his earlier comments regarding the military importance of such a route.

Army engineers spread out maps of northwest Canada and looked at the heavy black lines of proposed Routes A, B, C, and D. Local boosterism aside, each alternative had certain benefits as well as detriments, but none of the four routes adequately serviced the airfields of the Northwest Staging Route. As the inaugural staging route flight had just shown, something had to be done to link the remote airfields—for the requirements of speedy defense and resupply as well as navigational assistance to the planes.

When the line of the Northwest Staging Route was overlaid atop the proposed highway routes, the obvious came into focus. Building generally along proposed Route C for 600-plus miles from Dawson Creek to Watson Lake would link the airfields at St. John, Fort Nelson, and

Watson Lake. Building generally along proposed Route A another 500 miles from Whitehorse to Fairbanks via the Northway airfield on the Alaska-Canadian border would link up with Big Delta on the Richardson Highway. A 300-mile line was then drawn across the Cassiar Mountains between Whitehorse and Watson Lake to join the two pieces.

Despite the unknowns of the section from Whitehorse to Watson Lake, the new composite route not only linked the airfields but, more important, had the advantage of construction access at three key points: the railhead at Dawson Creek on the southern end, the White Pass and Yukon Railway from the port of Skagway to Whitehorse two-thirds of the way up the route, and the Richardson Highway from the port of Valdez on the western end. This would enable crews to work simultaneously from both ends and in opposite directions from Whitehorse in the middle. On February 11, 1942, while diplomatic approvals from Canada for such construction were working their way through the Permanent Joint Board on Defense, FDR said "build it."

Not everyone was pleased with the route selection. Alaska territorial governor Ernest Gruening, in particular, threw a fit and wrote a scathing memo to Secretary Ickes calling the route a "hodge-podge" and a "folly."[8] Gruening was still promoting the entire length of Route A in the unrealistic hopes of building roads across the Coast Mountains—something that still has not been done more than half a century later.

Gruening and territorial delegate Anthony Dimond even bulled their way in front of Secretary Stimson—whose main preoccupation by now was saving Australia from the Japanese onslaught. When Stimson explained the overall military advantages of the chosen route, Gruening finally acquiesced and later downplayed his stringent opposition when it came time to write his autobiography. He was mollified in part, Gruening later claimed, because he persuaded the army to build a side road from the new highway to Haines. In fact, the backup of equipment and supplies that soon clogged the Skagway docks and taxed the resources of the White Pass and Yukon had more to do with the army's decision to build the Haines Cutoff than any political lobbying.[9]

The U.S. Army now moved rapidly to spearhead the construction of the Alaska Highway. A scant twenty-six days had passed between Secretary Ickes's first suggestion to review the various road plans and FDR's decision to build it. It just goes to show how quickly things could happen given the exigencies of war. Initial operational plans called for four 1,300-man army engineer regiments, each augmented by a light pontoon company to effect river crossings, to scratch out some semblance of a road along the entire 1,500-mile route—what came to be known as the pioneer trail.

The 18th and 340th Engineer Regiments were ordered to Whitehorse to begin working west and east respectively. The 341st Engineer Regiment was dispatched to Fort St. John to begin improving the winter trail to Fort Nelson. But first the 35th Engineers took advantage of the still frozen winter trail to leapfrog equipment and enough supplies to last the summer ahead to Fort Nelson and to begin trailblazing north from there. At the western end, civilian contractors of the Public Roads Administration (PRA)—another of the New Deal's alphabet soup of agencies—were directed to build east from Big Delta and eventually hook up with the 18th Engineers. Once army engineers had blazed the pioneer trail, additional PRA crews would follow and turn the entire route into a respectable gravel road complete with permanent culverts and bridges.[10]

The military designations of the units hardly give a hint of the men behind them. For one thing, the 340th and 341st Engineer Regiments were brand-new. Not only were the men new to one another, but they were also new to their work. Many were city boys out on their first big adventure. Some had never wielded an ax, let alone felled a tree. Others were assigned to operate heavy equipment because they had driven a tractor on the family farm. All in all, it was a learn-as-you-go operation. But they were the army engineers, and under adverse conditions and in a remarkably short period of time, they managed to do what many looking upon the project from afar thought was impossible. From officers directing the bridging of a creek to cooks figuring out how to keep the pancake batter from freezing in below-zero cold, they learned by doing and—just as important in an isolated land—by improvising.

They were all awed by the endless expanse of the landscape and by the extremes of snow and cold. One lanky young copilot on his first flight to Fairbanks crawled out of his plane upon landing at Watson Lake and announced the obvious with a big grin. "Boy, what a lot of country!"[11]

Of course, once the snow melted, there were mosquitoes. One particular story was told so many times in so many different versions up and down the line that it became a classic. Two mosquitoes, it seems, were discussing a certain soldier. "You take him by the heels," one piped up, "and I'll take him by the head and we'll carry him home to eat." "Oh no," cautioned the other, "if we try that, one of the big fellows will take him away from us! Let's eat him here."[12]

By July 1942, the neophytes had become veterans, and work settled into a routine. The flow of equipment, supplies, and men became more reliable, although the sheer magnitude continued to tax existing supply lines. To ease the bottleneck at Skagway, the U.S. Army leased the White Pass and Yukon for the duration of the war, and experienced railroaders from Railway Engineering Unit 9646-A arrived in Skagway to operate additional locomotives and rolling stock over the heights of White Pass. Photos show flatcar after flatcar loaded with vehicles moving northward up the tracks that still ran smack down the middle of Broadway, Skagway's main street.

Speed and inaccessibility were the two overriding themes of the construction. It had to been done quickly, no matter how remote the terrain. By late September, the gap in the pioneer trail between Fort Nelson and Whitehorse was closed just east of Watson Lake with the linkup of the 340th and 35th Engineers on September 24 at what came to be called Contact Creek. The 340th claimed that it had won the race to Watson Lake, while the 35th boasted that it had scraped out more road—305 miles of it from Fort Nelson. Amazingly, the 340th's regimental band was on hand as the lead bulldozers of each unit touched blades. The army, after all, had its traditions! When the first through truck from Dawson Creek arrived in Whitehorse a few days later, the run had been made mostly for publicity and morale purposes, but it was evidence that the road was taking shape.[13]

Meanwhile, units were pushing to close the gap between Whitehorse and Big Delta. Among the regiments sent to reinforce the initial four units along the route were the 93rd, 95th, and 97th Engineers. In the parlance of the still largely segregated U.S. armed forces, these were "Negro," or "colored," units, whose effectiveness was still questioned by many in the army's officer corps. The 93rd deployed out of Skagway and the 95th out of Dawson Creek, while the 97th blazed the trail eastward up the Tanana for the following PRA civilian crews. When the 97th linked up with the 18th Engineers working west from Whitehorse near Beaver Creek about twenty miles east of the Alaska-Canadian border, it was not only the completion of the highway's pioneer trail but also a symbolic step. By carving the most miles in the final stretch, the 97th had done as much on the ground to silence generations of prejudice as the famed Tuskegee airmen were about to do in the air.

The photo of the linkup near Beaver Creek shows the 97th's Corporal Refines Sims, Jr., from Philadelphia, Pennsylvania, and Private Alfred Jalufka from Kennedy, Texas, of the 18th shaking hands and grinning broadly from atop their dozer blades.[14] Perhaps the most fitting tribute was one penned by reporter Froelich Rainey. "If I were asked to design a monument commemorating the construction of the Alcan [Alaska-Canada] Highway," Rainey wrote, "I would model a 20-ton caterpillar tractor driven by two soldiers, one negro and one white, but so greasy and grimy that the difference in color would be practically imperceptible."[15] Commonality of purpose in the face of great adversity tends to be color-blind.

On October 29, 1942, Secretary Stimson announced that the highway was open and that trucks were rolling along its length. It was a little premature. In truth, there was much that remained to be done. This was, after all, only the pioneer trail. But the road link between Alaska and the lower forty-eight—no matter how tenuous in places—had been made. On November 20, with, as historian Heath Twichell, Jr., wrote, "brass hats and brass bands at fifteen below," Canadian and American representatives braved the cold at Soldier's Summit 162 miles west of Whitehorse above the shores of Kluane Lake and officially opened the road. The ceremony had been delayed several days by unseasonably warm Chinook winds that caused ice jams to take out portions of the Peace River Bridge and the span over the Smith River east of Watson Lake. The weather was nature's bluff, however. The winter of 1942–43 proved to be the coldest in Alaska and the Yukon since 1917.[16]

In the months following the ceremony at Soldier's Summit, the mileage of the Alaska Highway slowly decreased as PRA crews improved the pioneer trail by straightening out switchbacks, blasting cuts and fills, and building permanent bridges. By one estimate, the construction cost of the highway was $138 million.[17] Some thought the figure hideously high and called for an investigation. Others thought it a necessary military expenditure at twice the price. Meanwhile, thousands upon thousands of trucks rolled northward along it with men and materials for Alaska's defense.

Whatever criticism the Alaska Highway was receiving in some quarters for its construction costs, it was getting nothing but praise from pilots ferrying aircraft along the Northwest Staging Route. Inexperienced pilots, frequently unfamiliar with the aircraft they were flying, were all too happy to do a little IFR flying of the I-Follow-Roads type. When mechanical problems forced a pilot either to bail out or to make an emergency landing, the road was a lifeline and afforded ready rescue. In due course, equipment for all-weather navigational aids, maintenance facilities, and auxiliary runways was transported over the road to improve the airfields at Fort Nelson, Watson Lake, and Northway.

After 1942, a large percentage of the air traffic flying north along the Northwest Staging Route was destined for Russia as Lend-Lease shipments. In fact, the route soon became known as ALSIB, the Alaskan/Siberian Ferry Route. This route was much more direct and far safer than the circuitous 13,000-mile loop via the Middle East. ALSIB had three major sections: the flight from the production factory to the ALSIB staging base at Great Falls, Montana, the 2,500-mile leg to Fairbanks, and finally the long series of hops across Siberia to the Russian front. The first stage was often flown by WASPs, Women Airforce Service Pilots, assigned to the ferry groups of the Air Transport Command. At Great Falls, the aircraft were painted with the red star insignia of the Soviet Union, and pilots of the all-male Seventh Ferry Group flew them north to Fairbanks. There, Soviet pilots took over and flew first to Nome for refueling and then on to Siberia.

The first Lend-Lease flight of ten A-20 light attack bombers departed the exchange point at Ladd Field in Fairbanks on September 29, 1942. Only 129 planes were flown to Siberia in 1942, in part because of the logistics of developing the necessary facilities along the route, but also because of the politics of negotiating the transfers with the Soviets. Joseph Stalin was adamant that the transfers be done on American ground with limited contact between the exchanging crews and only after assurances that the aircraft were in tip-top condition. (Neighbors whose houses are on fire are not supposed to be quite so demanding.) The number of aircraft delivered increased significantly in 1943, and by the end of the war in the Pacific, a staggering 7,926 planes had left Fairbanks on "the Red Star Line." These included P-39, P-40, and P-63 fighters; A-20 and B-25 bombers, the latter nicknamed "Mitchell" after the man who had once been military aviation's lone voice in the wilderness; and the venerable C-47, the military version of the DC-3.

ALSIB also provided a relatively safe avenue over which to shuttle Russian and American diplomats and military personnel between Moscow and Washington. American vice president Henry Wallace, Soviet foreign minister Vyacheslav Molotov, and Soviet ambassador Andrei Gromyko were among those who traveled the ALSIB route. Ironically, a few short years later, it would be the fear that Soviet bombers might reverse this course of Allied supply that precipitated a network of early warning stations across Alaska.[18]

In addition to the Northwest Staging Route and the Alaska Highway, there was a third leg to this overall project to better access and defend Alaska. In April 1942, Harold Ickes, who was the petroleum administrator for war as well as the secretary of the interior, told a cabinet meeting that concerted efforts should be undertaken immediately to explore for oil within Alaska in order to ensure a secure source for the territory. Once again, the army was charged with implementing this goal, and, once again, the big map of northwest Canada and Alaska brought the obvious into focus. Suddenly, the Canadian oil field at Norman Wells, which had been discussed in connection with proposed Route D of the highway, swelled in importance.

The solution to decreasing Alaska's dependence on oil arriving by ship, as well as providing fuel for both aircraft on the ALSIB and trucks on the Alaska Highway, was Project CANOL. This undertaking was almost as ambitious, certainly as costly, as the highway itself. It soon had construction teams building a network of oil pipelines across northwest Canada. The main segment was a 600-mile, four-inch pipeline that connected Norman Wells with Whitehorse. Additional wells were sunk at Norman Wells, and by the spring of 1944 the first crude oil reached a refinery that had been constructed at Whitehorse. Another pipeline, CANOL 2, was laid alongside the tracks of the White Pass and Yukon between Skagway and Whitehorse, and still more lines were laid along the highway in both directions from Whitehorse. By the time the Whitehorse refinery was turning out 100-octane aviation fuel in addition to gasoline, however, the war was winding down, and the refinery itself operated for less than a year.

When Missouri senator Harry Truman's Special Committee Investigating the National Defense Program came north to compare costs with results (as it was doing on all war-related spending), CANOL came under extreme criticism. Rather than the seat-of-your-pants, make-do-with-what-you've-got attitude of the military regiments blazing the pioneer trail, the CANOL project was almost awash with resources. But if the CANOL project was not the shining star of World War II construction efforts, it would be remembered when there was talk of other pipelines across Alaska a couple of decades hence.[19]

It had not always been pretty, it certainly did not turn out to be cheap, but what Colonel Heath Twichell, Sr., commanding officer of the 95th Engineers, called the "biggest and hardest job since the Panama Canal" had been built. When the three pieces of the 1,500 miles of the Alaska Highway, the airfields and military bases of the Northwest Staging Route, and the pipelines and refineries of CANOL were put together, 46,000 soldiers and civilians had been involved at a total cost of $500 million. It was the most expensive construction project of World War II.[20] But thanks to these efforts, the race to get there had been won. Alaska could be resupplied and defended no matter whose navy ruled the seas of the North Pacific.

The Darkest Chapter

The fortunes of war in the Pacific from the attack on Pearl Harbor until the Battle of the Coral Sea in early May 1942 lay solidly with the Japanese. Only Jimmy Doolittle's raid on Tokyo from the carrier *Hornet* in April offered the smallest glimmer of Allied hope. But as dark as the prospects were on the battlefront in early 1942, at home an even darker chapter was being written. Later, some would seek to paint over the darkness and excuse it with claims of panic, hysteria, and the exigencies of the moment, but the fact remains that American citizens were rounded up, imprisoned, and denied almost every right in the Constitution largely on the basis of their ethnic origins. It happened to Japanese Americans throughout the United States, but in Alaska this arbitrary singling out also fell upon Alaska Natives.

The basis for these actions was Executive Order No. 9066. Signed by President Roosevelt on February 19, 1942, the order authorized the secretary of war to designate certain military areas from which persons were to be evacuated and then provide "such transportation, food, shelter, and other accommodations as may be necessary." All nonmilitary departments and agencies were required to assist the War Department in carrying out the order by furnishing "medical aid, hospitalization, food, clothing, transportation, use of land, shelter, and other supplies, equipment, utilities, facilities, and services" as might be required.[21]

On its face that seemed reasonable enough, but the order's application left much to be desired. In implementing Executive Order No. 9066, Lieutenant General J. L. DeWitt, the commander of the Western Defense Command headquartered in San Francisco, issued a series of proclamations establishing certain military areas. Then on March 30, 1942, DeWitt issued Public Proclamation No. 5, requiring the evacuation and exclusion of "all alien Japanese, all alien Germans, all alien Italians, and all persons of Japanese ancestry residing or being within the Military Areas." The proclamation allowed half a dozen exceptions for certain alien Germans and Italians—such as those over seventy years of age—but applied to *all* Japanese without exception simply on the basis of ancestry and without regard to American citizenship.[22]

Upwards of 100,000 Japanese Americans were placed under what amounted to house arrest and moved to "relocation" camps around the American West for the duration of the war. In Alaska, ninety-two Issei men—first-generation Japanese immigrants—were rounded up immediately after the attack on Pearl Harbor, separated from their families, and sent to camps in Texas and eventually New Mexico. Their families, numbering about 130, some of whom were American citizens, were sent first to a camp at the Western Washington Fair Grounds at Puyallup, Washington, and then to the more permanent Minidoka, Idaho, internment camp. Among these detainees was a seventeen-year-old boy of mixed Alaska Native and Japanese parentage who had once worked for Bob Marshall in Wiseman. The boy had never met his Japanese father, but he was evacuated nonetheless.[23]

And then there were the Aleuts. They were scattered in tiny villages about the Aleutians from Unalaska to Attu and in the Pribilofs on St. George and St. Paul. Few questioned their American patriotism, but the Aleuts were clearly in the way of the buildup of military bases around Dutch Harbor and any future military operations throughout the Aleutian chain. Many in the military looked upon the Aleuts as a nuisance to be eliminated—from villages impeding the construction of airstrips and other facilities to the complications of social interaction with military personnel. Overlying these concerns was the same decided tone of racism that was being applied to Americans of Japanese ancestry. After much indecision between Alaska's military and civilian authorities, 881 Aleuts were evacuated from the Pribilofs and the villages of Akutan, Biorka, Kashega, Makushin, Nikolski, and Unalaska during June and July of 1942.[24]

The evacuations took place quickly and en masse. On Atka in the middle of the Aleutian chain, eighty-three Aleuts were ordered out of their houses and their village declared off-limits. "They were evacuated while eating breakfast," one officer noted, "and the eggs were still on the table—coffee in cups. A lot of their personal clothing and stuff was still hanging in closets."[25] Within a couple of hours, the village was in flames so that if Japanese troops landed there, they would find nothing of value.

Everywhere, the evacuations were marked by confusion and chaos, poor accommodations and meager rations, and an appalling degree of insensitivity. As 294 Aleuts were herded on board the transport *Delarof* at St. Paul in the Pribilofs, one sailor grabbed a set of heirloom china from an elderly Aleut woman and heaved the carton over the side into the water. The look of horror and misery etched across the woman's face was indicative of the suffering and condescension that all of the evacuees would have to endure for years to come.[26]

Where were they going? There was ample confusion here, too. Most of the Aleuts from the Pribilofs ended up at an abandoned cannery site and adjoining mining camp at Funter Bay on Admiralty Island. Conditions were deplorable. The water system froze in the winter, buildings were ramshackle at best, and the sanitary system consisted of pit toilets poised on pilings set just below the high-water mark. They were flushed with each high tide.

Families from Atka were deposited at a cannery site at Killisnoo on the southern end of Admiralty Island. "We lived in an old herring cannery and those buildings were never meant for winter," then twelve-year-old Nadesta Golley later remembered. "There were no boats to fish. We were just dumped off with the clothes on our backs."[27]

Another isolated, burned-out cannery site at Burnett Inlet on Etolin Island southwest of Wrangell became home to 122 Aleuts from Unalaska. Still others from the smaller villages were sent to an old Civilian Conservation Corps camp at Ward Lake northwest of Ketchikan. All in all, it was a far cry from the words of assistance to be rendered to such evacuees that were contained in Executive Order No. 9066.

Despite such treatment, some young Aleut men were quick to enlist in the services of their country. One, Simeon Pletnikoff of Nikolski, fought in the battle of Attu. Hauled before a provost marshal for impersonating a U.S. soldier, Pletnikoff exclaimed, "What's the matter with you guys? I'm an Aleut."[28] Other Aleut men from among those taken to Funter Bay registered with the Juneau draft board, and seventeen were inducted. Still others from Funter Bay were returned to the Pribilofs in the summer of 1943 for sealing operations. Their families joined them later that fall, but most of the other Aleuts remained in the dismal conditions of Killisnoo, Burnett Inlet, and Ward Lake until the fall of 1945. When they went back to their home islands and villages, they found little if anything intact—homes destroyed, possessions looted, villages leveled. From forced evacuation, to dubious confinement, to desolate homecoming, it is very difficult to imagine that American citizens of the white race would have been treated in a similar manner. It would take a generation before redress was made to the Aleuts as part of the settlement of the claims of the Japanese Americans who were relocated and interned.

While Japanese Americans and Aleuts suffered indignities, the wartime confusion that some used to excuse such lapses in civil liberties was indeed fact. For the first time, the United States and its allies were engaged in a truly global war. A thousand posts on a dozen fronts cried out for attention. In Alaska, the champion of its defense and a leading proponent of its location as a springboard from which to attack Japan was Simon Bolivar Buckner, Jr. "Buck" was a hard-driving West Pointer and proud southerner, who had been born almost a quarter of a century after his famous father surrendered Fort Donelson to Ulysses S. Grant during the Civil War. Arriving in Alaska in July 1940 as a colonel charged with the overall defense of the territory, Buckner was not about to do any surrendering.

His initial command consisted of two infantry companies totaling 276 men and several construction units on temporary duty building new bases at Anchorage and Fairbanks. There were airfields in both locations but no planes. Buck hollered and got a squadron of aging B-18 medium bombers and another of P-36 fighters. Infantryman that he was, Buckner had gotten his wings during World War I and didn't need Billy Mitchell to tell him the importance of airpower, particularly in Alaska. In a year's time his force numbered 6,000 men, and he had his first star. By Pearl Harbor, the Alaskan Defense Command had grown to 21,945 troops stationed in eleven garrisons posted across the territory's thousands of miles, but that still stretched the line pretty thin.[29]

In November 1941, looking for ways to bolster his forces, General Buckner ordered two of his top intelligence officers, Lieutenant Colonel Lawrence V. Castner and Major William J. Verbeck, to form a special commando unit. With four tough sergeants from Chilkoot Barracks at Haines, Castner and Verbeck handpicked a platoon of volunteers that were as gritty and Alaska-wise as any bunch of sourdoughs had ever been. Officially named the Alaska Scouts, they were "long on special skills and short on military discipline, [and] with their rustic parkas, gleaming knives, loose-slung rifles and bearded faces, they were soon known through the territory as 'Castner's Cutthroats.'"[30] The Alaska Scouts were in the vanguard of later landings on Adak and Amchitka and went on to undertake a variety of reconnaissance missions.

Another Buckner protégé, Arctic veteran Major Marvin "Muktuk" Marston, was the army's chief liaison to the Alaska Territorial Guard, a territorial militia variously called the Eskimo Scouts or the Tundra Army. Impressed with the resourcefulness of the Alaska Natives he encountered,

Marston traveled throughout the western half of the territory—much of the time by dogsled—and recruited a network of scouts to serve as an early warning system to watch for Japanese activity along the Arctic coasts. While the Eskimo Scouts saw no combat—and had no connection with Castner's Alaska Scouts—they did their special part to defend Alaska and later became a national guard unit.[31]

What was obvious to General Buckner was that he would have to use every possible advantage if he was going to stem the tide of the Japanese advance. Despite the fact that many thought this land of Arctic cold and snow was no place to fight a war, this was exactly what would have to be done.

No Place to Fight a War

While no one knew it at the time, Japan's visions of Pacific empire would be checked in less than a year after the attack on Pearl Harbor. In May 1942, however, the situation looked terribly grim throughout the thinly stretched Allied front. It ran some 6,000 miles from Alaska and the Aleutians in the north, straight south through the middle of the Pacific and a little atoll appropriately named Midway, around the Japanese-held Gilbert and Solomon Islands, and on west to Australia and New Guinea. Ironically, one of the few who thought that the time of Japan's undisputed triumph would be brief was Fleet Admiral Isoroku Yamamoto, architect of Japan's naval power and mastermind of the attack on Pearl Harbor. Well aware that once engaged, America's industrial and military might would quickly dwarf Japan's, Yamamoto sought to destroy the U.S. Pacific Fleet in one giant battle and use the ensuing victory to broker peace.

But where to fight it? For a time, Yamamoto thought that it would occur in the Coral Sea en route to the conquest of Australia. A four-day running battle there during the first week of May 1942 showed the fury of two navies battling each other with carrier-based planes. The U.S. carrier *Lexington* went to the bottom, and another, the *Yorktown,* was thought by the Japanese to have been sunk. The Japanese claimed victory, but hindsight suggests that what really occurred was a blunting of their drive toward Australia and the first stumble of their expansive advance. So Yamamoto's eyes turned to charts of the North Pacific. Tiny though it was, Midway stuck out like a sore thumb.

Not until after the war did Japan learn for certain that Jimmy Doolittle's B-25s had been launched from an aircraft carrier. At the time, some on the Japanese Imperial Staff thought that they had come from airfields in the Aleutians. Both sides were critically aware that the great circle route put San Francisco 1,000 miles closer to Tokyo via the Aleutians than through Hawaii. In fact, the great circle route between Seattle and Tokyo lay just forty miles north of Dutch Harbor, a protected anchorage on the north shores of Unalaska Island. So when Yamamoto devised a massive assault against Midway, he also planned a coordinated thrust against the Aleutians.

Vice Admiral Boshiro Hosogaya was chosen to lead Japan's northern force, which planned to land troops on Attu, Kiska, and Adak Islands in the western Aleutians and also to attack Dutch Harbor, at the time the only American base of consequence in the Aleutians. These attacks in the Aleutians were calculated to be diversions that would draw American attention and forces away from Japan's primary goal: Midway. Vice Admiral Chuichi Nagumo, who had led the carrier strike force against Pearl Harbor, was assigned to lead a similar carrier force against Midway a day after the Dutch Harbor attack began. Meanwhile, Admiral Yamamoto with the main Japanese battle fleet would lurk nearby in the North Pacific, ready to pounce on the American fleet and annihilate it whether it steamed north to rescue the Aleutians or west to support Midway.

On the morning of June 3, 1942, the Japanese carriers *Ryujo* and *Junyo* steamed out of a thick fog bank less than 170 miles south of Dutch Harbor and launched their torpedo-bombers and accompanying Zeroes. Hosogaya and his task force commander aboard the *Ryujo*, Rear Admiral Kakuji Kakuta, had been hoping for the surprise of another Pearl Harbor. But thanks to a PBY patrol plane spotting the two carriers amid the fog the day before, Dutch Harbor was ready—sort of. Early that morning its installations held an air-raid drill. One of the PBY pilots, Ensign Marshall C. Freerks, recalled that after the drill, "everybody went back to bed and then the Japs bombed us."[32]

For twenty minutes planes from the *Ryujo* bombed and strafed the harbor and shore facilities. The planes from the *Junyo* never made the target. They became lost in the fog—the first of many such occurrences on both sides in the Aleutian campaign—and barely managed to return to their carrier. On the American side, communications between widely separated commands were far from reliable. When the Dutch Harbor radio operator tapped out a frantic and uncoded "About to be bombed by enemy planes," P-40 fighters at Cold Bay, 180 miles to the northeast, were airborne within four minutes and en route to join the fray. Other pilots on alert at Umnak, less than sixty miles to the west, never got the word; they continued to play poker throughout the morning.

When all but one of the *Ryujo*'s planes returned safely with reports of widespread destruction at Dutch Harbor, Admiral Kakuta was elated. The truth of the matter was that things were a mess, but American forces had suffered less than 1 percent casualties and lost nothing of major strategic importance. They were, however, determined to strike back.

Through the short midsummer night, PBYs and other planes searched for the Japanese carriers. During the morning of June 4, as contact was established and then lost and then established again in the swirling fog, a strange assortment of early-model B-17s, twin-engine B-26s, and anything else that would fly—even a PBY hurriedly jerry-rigged to carry a torpedo and bombs—harried the Japanese force. Seas were so rough that for a time Kakuta could not even launch his protective fighters from the wildly pitching carrier decks. His ships suffered no major damage, despite the near miss of a torpedo launched within yards of the *Ryujo*, but the attacks convinced Kakuta that the Americans were far from beaten.

Originally, Kakuta's instructions had been to proceed west to Adak after the Dutch Harbor attack and support Admiral Hosogaya's landings there. In the confusion of the fog and the American attacks, his fleet was crawling westward at barely ten knots when reports of favorable weather back over Dutch Harbor prompted him to reverse course and steam at full speed to launch a second assault against it. Late in the afternoon, seventeen bombers and fifteen fighters—Kakuta dared send only his very best pilots off into the fickle weather—took off from both Japanese carriers and headed for Dutch Harbor. At four in the afternoon they were over their target.

Kate bombers dropped incendiaries and high explosives from above, while Zeroes strafed the shore facilities. The only ship left in the harbor was the aging *Northwestern*, a fifty-year-old steamer that had been beached earlier for use as a barracks. The vessel was something of an Alaskan legend, having run aground and survived sixteen times over the years. The Japanese bombed her and set her afire, but she survived yet again. Most of the rest of Dutch Harbor did likewise, although 750,000 gallons of fuel went up with such a roar that it could be heard on neighboring Umnak. The radio was still not working there, but its P-40 pilots heard the explosion and this time scrambled to the attack.

As this second wave of Japanese aircraft was halfway to its target, Admirals Hosogaya and Kakuta both received urgent messages from Admiral Yamamoto that they should break off the entire attack on the Aleutians and steam south to support the main Japanese fleet off Midway. Kakuta had no choice but to wait and recover his planes, and by that time—despite reports that things had not gone well at Midway—Hosogaya had convinced Yamamoto that some face could be saved by continuing the Aleutian invasion. Yamamoto begrudgingly agreed, but Hosogaya chose to occupy only the western two islands of Attu and Kiska. He feared that Adak lay too close to the newly discovered airfield on Umnak—especially now that its pilots were done playing poker. So after the skies cleared of aircraft over Dutch Harbor on the evening of June 4, Kakuta's carriers picked their way westward into the fog, having lost fifteen men and less than a dozen planes.[33]

In the global picture at the time, the attack on Dutch Harbor seemed a mere footnote. But was it? On the second day of the Dutch Harbor battle, Admiral Nagumo's massive carrier force tangled north of Midway with the American carriers *Enterprise* and *Hornet* and that phantom survivor from the Coral Sea, *Yorktown*. When the sun set in the North Pacific on June 4, 1942, *Yorktown* was crippled for good, but four Japanese carriers, 332 planes, one-third of Japan's combat pilots, and over 3,500 men were on the bottom. It was, as historian Walter Lord characterized it, an "incredible victory" for the United States. Might the outcome have been different had Kakuta's carriers turned south on June 3 to threaten the American flank rather than attacking Dutch Harbor? Captain Hideo Hiraide, chief of the naval press section at Japanese Imperial Headquarters, put a different spin on the entire outcome. "The enormous success in the Aleutians," he reported to the Japanese people, "had been made possible by the diversion at Midway."[34]

So remote were Attu and Kiska that the Americans did not even confirm the Japanese invasions there until a patrol plane flew over Kiska on June 10 and was fired upon by unidentified ships. Before dawn on Sunday morning, June 7, 1,250 Japanese troops had landed on Kiska, which was occupied only by a ten-man weather team. A few hours later, a similar force landed on Attu. Here, in the little village of Chichagof, there were an older white couple and forty-two Aleuts, more than a third of them children, who were eventually sent to a prison camp at Otaru City on Hokkaido for the remainder of the war.

The American response to the landings was unequivocal all along the chain of command, from General Buckner and Rear Admiral Robert A. Theobald on up to General George Marshall and eventually President Roosevelt: American territory had been invaded; drive out the invaders. But given the terrain, the weather, and the great distance from the main bases at Anchorage, on Kodiak Island, and even at Cold Bay and Dutch Harbor, that was far easier said than done.

Colonel William O. Eareckson, "Wild Bill" to his navy buddies and "Eric" to his fellow bomber pilots, personally led flight after flight against Kiska in what historian Brian Garfield called the Kiska blitz. General

Buckner commandeered commercial aircraft temporarily assigned to the Northwest Ferrying Command and organized an airlift of men and materials to Nome in response to an intercepted message that the Japanese intended to strike there next.

In fact, the Japanese had their hands full and were trying desperately to hold on to what they had already captured. They augmented their landing forces on Attu and Kiska and made plans to construct airfields on the captured islands. The occupation wasn't without risk. On the Fourth of July, after Admiral Hosogaya had dispatched a convoy of reinforcements to Kiska, the submarine *Growler* quietly slipped into Kiska Harbor and sent one Japanese destroyer to the bottom and damaged two more.

Back in Japan, the heavily censored Japanese press spoke only of glorious Aleutian victories and the newly won safety of Japan's northern gates. But there was almost unparalleled censorship in the American press as well—live reports on the evening news were still a war or two away. One reason for the news blackout that descended across the Aleutians and much of the war in Alaska was that it was frequently difficult to get a complete and accurate picture of just what the devil was going on across a 1,000-mile front. Witness, were the Japanese attacking Nome or not? More important, of course, there was a general and concerted tendency to downplay reports of any hostile operations that were taking place on American soil, even if that soil was on some rocky islands more than halfway to Siberia.

By the end of the summer of 1942, attention in both Japanese and American high commands was heavily focused elsewhere. The battle for Guadalcanal and the Solomons was raging in the South Pacific, and forces on both sides, particularly naval units, were diverted to that front. In one of the ironies of war, the carrier *Ryujo*, which had launched the first attack on Dutch Harbor, was sunk in the Solomons by a task force commanded by Rear Admiral Thomas C. Kinkaid, who would soon be tapped to head the Aleutian campaign.

As the brief Arctic summer drew to a close, it soon became more apparent than ever that both sides were fighting the weather as much as each other. Against both sides, the weather usually won. Witness the score. During the fall of 1942 in the Aleutians, the United States lost only nine planes in combat, but sixty-three planes went down due to weather and mechanical troubles. "When you could see a hundred feet," recalled Captain Lucian Wernick, a B-17 pilot with the Thirty-sixth Bombardment Squadron, "that was a clear day."[35]

The weather fostered indecision and uncertainty on both sides. The Japanese Imperial Staff couldn't decide whether to fortify Attu and Kiska or abandon them. In September, Admiral Hosogaya ordered the evacuation of all troops on Attu to Kiska in order to concentrate forces and ease supply lines. Though frequently lumped together, the islands were, after all, some 200 miles apart. This evacuation was completed, but then the admiral thought better of the idea and Japanese troops reoccupied Attu a month later. The following spring, American commanders would rue the fact that they hadn't landed on Attu in the interim, but at the time it would have been much too long of a leap.

Meanwhile, the American counterattack had begun in earnest with the occupation of Adak, a rocky enclave in the Andreanof Islands dominated by 3,924-foot Kanaga Volcano. Adak was some 350 miles beyond the current advance base on Umnak. Despite fears of Japanese opposition, the landings on August 30 were unopposed. But where to build an airfield amid such rocky terrain? The flattest spot on the island was the long, narrow tidal flats at the head of Sweeper Cove. The solution was both ingenious and speedy. Army engineers dammed the mouth of the cove to keep out the high tide and in the remarkably short period of ten days had an airstrip ready to receive aircraft, including the Thirty-sixth Bombardment Squadron's B-17s and a squadron of new P-38 Lightning fighters. The facilities on Adak more than halved the distance required for fighter and bomber operations against Kiska. In keeping with the shroud of secrecy over the Aleutian operations, the navy did not report the landings on Adak until almost six weeks later, and then only with reference to "an island somewhere in the Andreanofs."[36]

Next in line was the island of Amchitka, a slender sliver forty miles long and two to three miles wide at the eastern end of the Rat Islands. Amchitka pointed directly at Kiska, just fifty miles away at the western end of the Rats. But Amchitka would have to wait until after the icy gloom of the winter solstice. On January 4, 1943, Admiral Chester Nimitz, commander in chief of U.S. naval forces in the Pacific, put Rear Admiral Thomas C. Kinkaid, Annapolis class of 1908 and a no-nonsense, fight-with-what-you-got sort, in charge of Aleutian operations with instructions to make things happen. Kinkaid was just the opposite of his predecessor, Rear Admiral Robert A. Theobald, who had used his naval units cautiously and sparingly. Of Kinkaid it was said by veterans of his battles in the South Pacific that you could tell where the action was going to be by looking for him.[37]

While air corps bombers continued to harry Japanese supply lines to Attu and Kiska—sinking two freighters loaded with soldiers and supplies off the islands in early January alone—Kinkaid lost no time in taking Amchitka. It was the middle of January, seas were running high, the wind was howling, and the temperature was flirting with zero. It was an Arctic-style hell, but 2,100 army engineers and troops under Brigadier General Lloyd E. Jones did the impossible and secured the unoccupied island. Amchitka was flatter than Adak, but it turned into one great muddy quagmire once the spring thaw came. Despite attacks from Japanese bombers out of Kiska that kept cratering the runway and slowing construction, the field was operational on January 28, in time to receive much-needed P-40 fighters led by Lieutenant Colonel Jack Chennault, the son of Flying Tiger legend Claire Chennault.

Once Amchitka was secure, it was time to tighten the noose around Attu and Kiska. Admiral Kinkaid ordered Rear Admiral Charles H. McMorris to blockade the islands and prevent Japanese resupply convoys from reaching them. McMorris had earned the moniker "Soc," short for "Socrates" because of his scholarly ways, in the Naval Academy's class of 1912 and had gone on to teach English and history there. Now his two-star flag flew from the World War I–vintage light cruiser *Richmond*, which was accompanied by four destroyers and the heavy cruiser *Indianapolis*. Because his force was small and Attu and Kiska were several hundred miles apart, McMorris chose to concentrate his ships southwest of Attu and intercept Japanese convoys as they steamed out of the Kuriles—before they could disperse among the Aleutians.

During late February and early March 1943, McMorris's blockade began to disrupt Japanese supply lines. When ships did manage to elude the blockade and reach Attu, American bombers continued their attacks. They sank the *Chieribou Maru* in the deep waters of Holtz Bay before her cargo of bulldozers intended for airfield construction could be unloaded. On March 10, another Japanese transport slipped into Attu, but during the next week, half a dozen ships turned back in the face of McMorris's force.

At Japanese Imperial Headquarters, the debate over what to do with the island garrisons continued. There were those who had long written off the Aleutians as a no-win situation and were still calling for a complete withdrawal. The other faction looked at the map of the North Pacific and argued that the islands had to be defended because of their proximity to Japan and their value as a wedge between the Soviet Union and Alaska. (Never mind that by now ALSIB was functioning quite safely and effectively 1,000 miles to the north.)

Consequently, Admiral Hosogaya decided to run the American blockade with all of the firepower at his disposal. He loaded three transports until they were bursting at the rivets, stocked more supplies onto the decks of four destroyers, and then gathered four heavy cruisers around them. On March 22, with his flag flying from the cruiser *Nachi*, Hosogaya steamed east from Paramushiro in the northern Kuriles, determined to blast McMorris's blockade apart.

Early on the morning of March 26, 1943, the *Richmond* and her consorts were cruising in a long line about 200 miles west of Attu and 100 miles south of the Komandorskie Islands, where Vitus Bering had met his end more than two centuries before. The squadron now included the aging heavy cruiser *Salt Lake City*, which had recently replaced the *Indianapolis*. PBY patrol planes had been reporting intermittent contact with what appeared to be a major convoy, and soon radar on the destroyer *Coghlan* confirmed the presence of surface ships. But McMorris received a rude surprise. It was not a lightly guarded convoy but the backbone of Japan's northern force. He was clearly outgunned, but the stakes were too high to run because such a major resupply might prolong the war in the Aleutians indefinitely. Besides, McMorris was cut from the same cloth as the man who had sent him out there, Admiral Kinkaid.

McMorris closed his line of ships and swung to the attack, attempting to get at the defenseless transports. Hosogaya responded in turn by shielding the transports and bringing his guns to bear on the *Salt Lake City*. The two biggest cruisers, *Salt Lake City* and *Nachi*, quickly engaged in a running battle of eight-inch and five-inch guns at a range of ten to twelve miles. Each admiral called for air support and expected it to arrive, but there were no available bombers at Paramushiro to aid Hosogaya and Eleventh Air Force bombers at Adak were already loaded with the wrong ordnance for an attack against Kiska. In the time it took to rearm them, the Battle of the Komandorskies would be over.

The battle lasted for over three and a half hours and became the longest continuous gunnery duel between surface ships in modern naval warfare. Finally, the *Salt Lake City* signaled McMorris on the *Richmond*, "My speed zero." Salt water in the main fuel line from numerous near misses had flowed into the *Salt Lake City*'s burners and extinguished them. As the Japanese closed in for the kill, McMorris ordered one destroyer to lay smoke around the crippled ship, and the other three to launch a last-ditch torpedo attack against the oncoming Japanese cruisers.

The tiny destroyers had one-tenth of the displacement of the onrushing cruisers, but they gamely steamed forward to close within range and fire a spread of torpedoes. The opposing ships were still five miles apart when McMorris received a surprise query from the destroyers: "The enemy is retiring to the west. Shall I follow them?"

Ironically, although airpower played no direct role in the battle, the possibility of its appearance had decided the outcome. In the confusion of the battle and smokescreen, Hosogaya did not realize that the *Salt Lake City* was sitting dead in the water. Moments before, the big cruiser had run out of armor-piercing shells and started shooting high explosives, whose white phosphor trails looked somewhat like bombs falling from the sky. Hosogaya ordered his antiaircraft batteries to open up and became convinced that his ships were under attack by bombers from Adak.

As the Japanese fleet, including the three transports, retired to the west, the crew of the *Salt Lake City* got steam back up and fired its remaining salvos. Miraculously, in this last gunnery duel of surface vessels where aircraft played no direct role, casualties were only seven Americans and fourteen Japanese killed. Despite a number of hits on each side, no ships were sunk or permanently damaged. It had been a wild contest, but it was decisive in its outcome. The Battle of the Komandorskies broke the Japanese supply line to Attu and Kiska and put the Japanese navy on the defensive in the North Pacific. And Admiral Hosogaya? He was relieved of his command for failing to finish off the *Salt Lake City*. Admiral McMorris and his men got a hero's welcome upon their return to Dutch Harbor.[38]

No one expected the invasion of Attu to be a cakewalk, but neither did anyone expect it to be the epic that it became. Originally, Admiral Kinkaid's plan had been to keep on marching right down the Aleutian chain. After Amchitka came Kiska. But the U.S. Navy was still spread terribly thin in the Pacific in the spring of 1943. Guadalcanal was finally secure, but the battles for the remainder of the Solomons and New Guinea still raged. When ships were not made available in the numbers that Kinkaid and Vice Admiral Francis W. Rockwell thought sufficient for an invasion of Kiska, Kinkaid promoted a plan to bypass Kiska and use the smaller force to drive the Japanese off less heavily defended Attu once and for all. Operations planners in San Francisco warmed to the idea and soon decreed that Attu was so lightly defended—500 troops at the most—that a single regiment could take it in three days flat.

By this time, General Buckner had close to 150,000 troops under his command spread out across Alaska. But the units had no amphibious landing training, let alone experience, and to strip soldiers from far-flung posts across the territory was not the answer. So the combat-trained Thirty-fifth Infantry Division was recommended for the Attu landings. Both the division's commander, Major General Charles H. Corlett, and its assistant commander, Brigadier General Eugene M. Landrum, had Aleutian experience. Landrum had in fact commanded the occupation of Adak. It seemed a perfect fit, but in the humdrum of global war, the War Department worked in mysterious ways.

Rather than assign the Thirty-fifth Infantry to the Attu landings, planners in Washington gave the job to the Seventh Motorized Division. Until then, the Seventh Division had been training in the California desert for deployment against Erwin Rommel's Afrika Korps. Scrap the tanks, they were told. Instead of a desert-warfare tank division, the Seventh was going to become an Arctic infantry force.

Major General Albert E. Brown, the Seventh's commanding officer, had fought in the trenches of World War I and pulled no punches. Brown took one look at the maps of Attu—the fact that the army, navy, and air corps were all using a different coordinate system was a clue in itself—and snorted, three days, hell. The mountainous terrain alone would be enough to keep his men from crossing the ten-mile-wide island in less than a week—without opposition.

When reconnaissance reports indicated that the estimate of 500 Japanese on Attu was decidedly low and that there might be as many as 1,600 troops dug in on the island, the entire Seventh Division of 10,000 men was mobilized for the invasion. General Buckner was told to assemble the Fourth Infantry Regiment on Adak and hold it in reserve as reinforcements. Meanwhile, the Seventh Division practiced amphibious landings in the surf of California's sunny Monterey Bay and San Clemente Island.

For a while, typical spring weather in the Aleutians grounded even the Thirty-sixth Bombardment Squadron. Frederick Ramputi, a bomber pilot who had first discovered the Japanese ships in Kiska harbor the summer before, recorded in the squadron's log, "No flying this station due to inclement weather." It seems that the 110-knot (126.5 miles per hour) anemometer at Adak had been destroyed in what was the worst storm recorded until then in the Aleutians.[39]

As plans for what was called Operation Landcrab wound into high gear, General Brown picked Captain William H. Willoughby for a special assignment. Willoughby was given free rein to handpick 410 men from among the Seventh's regiments and organize them into a special battalion. Willoughby's Scout Battalion—not to be confused with the Alaska Scouts—was akin to specially trained commandos or later special-forces units. Loaded up with grenades, armor-piercing bullets, light mortars, and plenty of guts, they had the job of landing undetected on the western end of Attu, crossing the island, and seizing the high passes and ridgelines above the Chichagof Valley to block a Japanese escape into the mountains.

Meanwhile, one regiment of 1,500 men would land north of Holtz Bay on the northern side of the island, and another regiment of 2,000 men would land on the south side at Massacre Bay—named for an early encounter between Russian *promyshlenniki* and Aleuts, but soon all too appropriate. These two forces would converge with Willoughby's Scouts on the high ground above the Chichagof Valley and trap the Japanese below them.

When the big storm finally blew on through, Colonel Eareckson's bombers and Admiral McMorris's cruisers pummeled Kiska as a diversion. But somehow Colonel Yasuyo Yamasaki, the garrison commander on Attu, knew that his island, and not Kiska, was to be the target of the assembling invasion fleet. In fact, after months of a news blackout over most operations on the Alaskan front, it seemed as if everyone was talking about the coming campaign. Even radio commentator Walter Winchell told his listeners early in May, "Keep your eye on the Aleutian Islands."[40]

An invasion force of thirty-four ships put to sea from Cold Bay, including the escort carrier *Nassau* and the battleships *Idaho*, *Pennsylvania*, and *Nevada*, the latter the only battleship to get under way at Pearl Harbor. The fleet slipped through the Aleutians and into the Bering Sea, circling wide of Kiska and converging upon Attu from the north. By now, intelligence was estimating Yamasaki's strength on Attu at 2,600 men and confirming that he knew of the coming invasion.

But first there was another battle to be waged with the weather. As the invasion force assembled north of Attu, ships were tossed about like toys by high seas and ferocious winds. The flight deck of the *Nassau* looked like a swimming pool. All land-based planes were grounded. Infantrymen crammed aboard the transports turned six shades of green, and more than one of them thought fondly of those pleasant, practice landings on the beaches of sunny California. D-day had been planned for May 7, but as the storm continued, it was postponed day after day. Finally, in the wee hours of May 11, submarines *Nautilus* and *Narwhal* landed the first half of Willoughby's Scouts in Austin Cove, and the invasion was on.

At first, things were pretty tentative. Numerous rocky shoals hampered the northern force's landing at Holtz Bay. Poor visibility delayed the landings of the southern force until late afternoon. By evening, the two regiments had secured tenuous beachheads, and Willoughby's Scouts were climbing eastward across snow-covered mountainsides. So far, there had been no Japanese resistance. The silence emanating from the low clouds and fog blowing across the hills above the beachheads was eerie. Then all cold hell broke loose.

Colonel Yamasaki had decided that his 2,650-man force could not repel attacks on two beachheads simultaneously, so he grouped his troops along the snow-covered ridges and high ground at the head of the Massacre and Holtz Valleys. As the Americans began to move up the valleys, they were met with a hail of rifle and mortar fire. Three days tops, the planners had said. It took a week of intense fighting just to gain the heights of what came to be called Jarmin Pass at the head of Massacre Valley. Meanwhile, Willoughby's Scouts had clawed their way into position in Yamasaki's rear above Holtz Bay, but most of the men were suffering frostbite and hunger—the thirty-six hours of rations with which they had landed were long gone. "Wild Bill" Eareckson braved the fog to fly sorties and try to direct the fire of the navy ships, but for the most part, it was far too foggy for any kind of air support.

Thousands of miles to the rear, the desk-bound brass couldn't understand what was taking so long. General Brown was relieved of his command and replaced by Major General Eugene Landrum, who had almost drawn the assignment with the Thirty-fifth Infantry in the first place. This Aleutian veteran found no fault with Brown's operational plan and continued to slug it out one foot at a time. There really wasn't any other choice.

By now, all of the Seventh Division and Buckner's Fourth Infantry had been committed to the fray. When the weather cleared momentarily on Wednesday, May 26, Eareckson directed bombing runs against the main Japanese camp in Chichagof Harbor and their positions on the surrounding ridgelines. Atop Fish Hook Ridge at the head of Holtz Valley, Japanese defenders hid in snowdrifts and ice-encrusted crevices and rolled hand grenades down the hillsides on the advancing Americans. Private Joe P. Martinez of Taos, New Mexico, the BAR (Browning Automatic Rifle) man in his squad in Company K of the Thirty-second Infantry Regiment, finally had enough. Martinez led the way up the final ridge spraying bullets right and left. Company K scrambled to the ridge behind him and secured it. Private Martinez won the Medal of Honor—posthumously.[41]

By May 29, the nineteenth day of the battle, Colonel Yamasaki had 800 fighting men left. Surrender was not acceptable in the Japanese warrior's Bushido Code. Yamasaki's only option was to attack. If he could seize the 105-mm howitzers that had been pounding his positions from the head of Massacre Valley, he might be able to turn them around and use them to drive the Americans back to the beachhead. Accordingly, Yamasaki ordered his wounded killed with morphine overdoses and grenades, and moved forward to attack.

Yamasaki advanced against the spot in the American line that was held by Company B of the Thirty-second Infantry Regiment. By coincidence, the company had been ordered back to the battalion kitchen for a warm breakfast prior to resuming their attack. As the company marched off to breakfast, the Japanese struck. For a short time, Yamasaki was heartened. This might work, after all. Japanese troops swept up Engineer Hill toward the howitzers but then encountered heavy hand-to-hand fighting on the ridgeline. Incredibly, as Yamasaki's forces fell back down the slopes and dispersed, about 500 Japanese soldiers committed mass suicide by pulling the pins on their grenades and holding them to their chests. Yamasaki himself took a bullet while leading one last charge. The Americans were stunned.

By Sunday morning, May 30, the Battle of Attu was over except for some mop-up operations. Of the 2,650 Japanese on the island when the fighting started, only 28—none of them officers—remained alive to be taken prisoner. Five hundred and forty-nine Americans were dead. In proportion to the number of Americans engaged in the operation, it was the most costly American battle in the Pacific during the entire Second World War, save for Iwo Jima. Among the thousands of wounded and injured were numerous cases of severe frostbite, exposure, and trench foot.[42]

The lessons of Attu were bitter and many, and they would resonate far beyond this rocky island and have a major impact on American operations throughout both the Pacific and European theaters for the remainder of the war. Never again would American troops go into combat in cold climates—the mountains of Italy, for example—without improvements in footwear, clothes, and other gear. Command structure for joint army-navy operations would be refined for future amphibious landings. Colonel Eareckson's pioneering forward-air-controller role would be adopted.

Operationally, the Attu campaign had caused the Japanese navy to shift forces to the North Pacific in its perpetual quandary of whether to reinforce Attu and Kiska or abandon them. This robbed Japanese naval forces from the Solomons and left the American invasion at Rendova (just west of Guadalcanal) virtually unopposed. Finally, Kinkaid's thought to bypass Kiska evolved into the island-hopping strategy that became the centerpiece of operations in the South Pacific during 1944 and 1945.[43] The Aleutians were no place to fight a war, but American GIs, sailors, and airmen had done it.

Victory

After Attu finally fell, General Buckner and Admiral Kinkaid were eager to make plans for the invasion of Japan from the Aleutians, perhaps as early as the fall of 1943. The Joint Chiefs of Staff balanced the global picture and were more cautious. They set a tentative date of early 1945 for an invasion of Japan's northern islands and told the Alaskan commanders that their first priority remained the capture of Kiska. Given the experience on Attu, everyone knew that the task was not to be taken lightly.

Even in the heat of the battle for Attu, Colonel Benjamin Talley's army engineers had been busy building airfields on the tiny atoll of Shemya, twenty-five miles east of Attu, and in Massacre Valley itself. Shemya had only one thing to recommend it: It was flat. The airfield there was built to accommodate long-range B-29 Superfortresses, then still in the experimental stage but soon capable of carrying destruction directly to Tokyo's doorstep. The Japanese got a taste of what to expect from the new Aleutian bases when B-24 Liberator bombers, replacements for the aging B-17s, refueled at Attu and flew several raids against the major Japanese naval base at Paramushiro. It was the first attack on Japanese soil since the Doolittle raid and the first ever flown from an American land base.[44]

The first steps to invading Kiska itself were to hammer it with air and naval bombardments—when weather permitted—and to reinforce the naval blockade of the island. The largest battle fleet to assemble in Alaskan waters, including the battleships *Idaho*, *Mississippi*, and *New Mexico* and the cruisers *San Francisco*, *Portland*, *Louisville*, *Santa Fe*, and *Wichita*, pounded Kiska with their fourteen-inch and six-inch guns and kept a lookout for Japanese ships trying to enter Kiska Harbor. Meanwhile, the Japanese were planning to do just that.

Not willing to face another Attu, the Japanese Imperial Staff finally decided once and for all to abandon Kiska. It ordered Vice Admiral Shiro Kawase, Admiral Hosogaya's replacement after the Battle of the Komandorskies, to make it happen. As with so many operations in war, particularly war in the Aleutians, it was far easier said than done.

The Japanese had some 6,000 men on Kiska. Kawase was first ordered to effect their evacuation with submarines. Given the limited number of submarines that were still operational in his command—eight—and their limited carrying capacity, the order was problematic. During June, 820 men were evacuated from Kiska by Japanese I-class submarines, but American destroyers sank three of the boats as they departed Kiska, with a loss of some 300 of the evacuees. That left five submarines and 5,183 men still on Kiska. Admiral Kawase knew that he had to try something else.

Kawase left Paramushiro at dawn on July 21, 1943, with four cruisers, an oil tanker, and eleven destroyers. Operationally, his plan was simple. He planned to sail well to the south of Kiska to avoid detection by American patrol planes and then wait for a good fog. Under its cover, he would make a mad dash for Kiska and evacuate the garrison.[45]

The next day, July 22, the American flotilla encircling Kiska received a report from a PBY patrol plane of a radar contact with seven unidentified objects—presumably Japanese ships—heading east to the southwest of Attu. Admiral Kinkaid was determined to stop this force well short of Kiska and ordered the battle fleet to intercept it. Unaccountably, all ships were dispatched away from Kiska Harbor, leaving it guarded only by four PT boats. These boat tossed about the wild seas like corks and were soon withdrawn to Amchitka.

The American battle fleet, led by Rear Admirals Robert C. "Ike" Giffen and Robert M. Griffen (now that must have caused some confusion), steamed southwest and searched in vain for the seven targets. No contact was made, and they were finally ordered by an exasperated Kinkaid to return to station off Kiska and cover its western and southerly approaches. The destroyer *Hull* was belatedly dispatched to patrol north of Kiska to prevent Japanese ships from slipping undetected into Kiska Harbor from that direction.

By the night of July 25, the Giffen-Griffen task force was some ninety miles southwest of Kiska. The night was clear and warm for the Aleutians—a pleasant fifty-two degrees Fahrenheit with a westerly wind of less than ten miles per hour. Four hundred miles to the south, Admiral Kawase was still waiting for a good fog.

Shortly before 1:00 A.M. in the early morning of July 26, radar on the battleship *Mississippi* picked up what appeared to be the seven elusive radar blips about fifteen miles to the northeast. *New Mexico*, *Portland*, *San Francisco*, and *Wichita* quickly recorded similar images. Pulses quickened, and the fleet beat to general quarters, certain the Japanese were making a run for Kiska. A furious flurry of broadsides shattered the calm night as the American battleships and cruisers fired at the radar targets. No visual contact was ever made, but salvo after salvo filled the night sky. The seven targets changed course about twenty degrees and shortly thereafter disappeared from all radar screens. The targets had not returned fire.

At dawn, a catapult plane from the *Wichita* searched the area but found nothing—no debris, no oil slicks, nothing. The American navy had launched a massive attack against . . . what? The theories were diverse. Radar malfunctions were high on the list. Perhaps the radar had picked up the distant peaks on Amchitka and other islands and scrambled the range, making them appear much closer on the screen. Some thought that radar reflections from ionized clouds (never mind that the night was clear) had caused the "sighting." Others were convinced that the radar pips were from some sort of decoys towed by Japanese submarines, or even a fleet of subs themselves running on the surface. Individual radar malfunctions hardly explained the engagement because the multiple contacts from a number of ships all converged about twelve and one-half miles off *Mississippi*'s bow.[46]

The official U.S. Navy version remains that the radar sightings were caused by a bending of radar beams due to atmospheric disturbances. Brian Garfield, writing in 1995 in a new edition of his classic, *The Thousand-Mile War*, suggested, however, that there may be another explanation. Buller's shearwaters are gull-sized seabirds that annually migrate between New Zealand and Alaska. They fly in massed flights and congregate on the surface at night in large groups to feed on plankton. Their mass might have been large enough to reflect a radar image, and doubtless they would have dispersed, or "disappeared," after a few rounds of fourteen-inch shells came crashing in. Who is to say for certain what the target was, but the U.S. Navy won what came to be called the "Battle of the Pips." To paraphrase Robert Service, the Arctic nights have seen strange sights. The Battle of the Pips was one of the strangest, but it was about to have major ramifications.[47]

Because of their dash west of Attu after the first PBY report and their ensuing maneuvers, the American ships were now low on fuel. They retired eastward en masse to rendezvous with the oil tanker *Pecos* to refuel. Meanwhile, Admiral Kawase found his fog bank and followed it northward toward Kiska, right into the area in which the American ships would have been had they not turned east to refuel.

On July 28 under a thick fog, while Kawase on board the cruiser *Tama* stood a rearguard action on lookout for the American fleet, the bulk of the Japanese task force arrived in Kiska Harbor. The troops there had long since blown up their installations, heavy guns, and anything of any conceivable use to the Americans. It took less than an hour to load more than 5,000 men onto the waiting ships. As the force steamed out of the harbor, it passed undetected within 1,500 yards of the American submarine S-33. In a thick fog and without radar, the submarine was blind. Its crew never saw a thing. Four days later, Kawase's fleet was back in Paramushiro without firing a shot, losing a man, or—as it would soon become all too clear—being detected.[48]

Refueled, the American battle fleet returned to station off Kiska, and the noose was tightened further for Operation Cottage—the August 15, 1943, invasion of the island. During the two weeks prior to that date, some officers became a little skeptical about Japanese strength on the island. There was no antiaircraft fire coming from Kiska, although some pilots reported such. Reconnaissance photos showed no movement of trucks or attempts to repair bomb craters. Were the Japanese still there? Tales of elaborate caves on the island suggested to some that they had simply gone underground to await the invasion.

General Buckner was among the increasingly suspicious, and he recommended landing a detachment of Alaska Scouts for a quick look. Kinkaid mulled the recommendation but in the end made the command decision to go ahead with the full-scale invasion. The admiral was under pressure from the War Department to stick to the schedule because of other operations in the Pacific. Besides, if Kiska had been abandoned, it would all make for one immense training exercise.

As the final invasion forces were assembled, a quarter of a million soldiers, sailors, and airmen stood ready throughout the Aleutian chain and the Alaskan mainland bases to support the attack. Adak alone, which had been deserted a year before, was the temporary home of 90,000 servicemen. Among the almost 35,000 combat troops in the early waves of the invasion were veterans of Attu from the Seventh Division and Buckner's Fourth Infantry Regiment. Also involved were the Eighty-seventh Mountain Regiment, which would go on to fight in the mountains of northern Italy as part of the famed Tenth Mountain Division, and 5,300 troops of the Thirteenth Royal Canadian Infantry Brigade. And then there was "the Devil's Brigade." Some 2,500 paratroopers of a crack Canadian-American special services force, which was to be dramatized in a 1968 movie starring William Holden as Colonel Robert T. Frederick, were making their first field deployment together.

When these forces splashed ashore on Kiska on August 15, a half dozen dogs rushed up to greet the advance elements. Ensign William C. Jones recognized one dog as Explosion, a mutt he had given to the ten-man Kiska weather team some fifteen months before. Somehow Explosion had survived the endless air and naval bombardments. The lonely dogs turned out to be the island's only inhabitants. As the troops moved inland, it slowly became obvious that the Japanese were gone. There were booby traps that claimed casualties, and, regrettably, in the clouds and fog there were casualties from friendly fire—a total of 21 dead and 50 wounded. The destroyer *Abner Read* struck a mine offshore and went down with 71 men. In all, there were a total of 306 killed, wounded, and otherwise injured in the occupation of a deserted island.[49]

Some called the invasion of Kiska a gargantuan blunder—a senseless expenditure of men and equipment. Even General Buckner scrutinized the critical press coverage and remarked, "To attract maximum attention, it's hard to find anything more effective than a great big, juicy, expensive mistake."[50] In retrospect, however, that appraisal seems overly harsh. To be sure, the Kiska campaign cost lives and for a good ten weeks after the fall of Attu kept forces in the Aleutians that might otherwise have been used to augment Allied efforts in the Solomons and New Guinea. But in the big picture, the balance had swung 180 degrees in just over fourteen months. Whatever satisfaction the Japanese took from their stealthy escape and the subsequent American efforts against a deserted island was short-lived. A year before, it had been the Japanese that threatened Alaska and the American mainland from the Aleutians. Now the view was reversed, and it was American troops that threatened Japan from there. In the final analysis—despite inevitable snafus along the way—that is how wars are won.

Now the question was, where to from Kiska? In September 1943, the Joint Chiefs of Staff pondered the next move. A strike against Paramushiro and the Kuriles was tempting, particularly given the forces now massed in the Aleutians. Ultimately, however, such an attack never came to pass for a number of reasons.

First, there was the perpetual enigma of what the Soviet Union was planning. The Soviets all too gladly picked up Lend-Lease airplanes in Fairbanks and flew them westward, but Stalin dragged his feet in declaring war on Japan until the final month of the war. Operations against Paramushiro and the Kuriles without Soviet support from neighboring Kamchatka would have been dicey at best. Then there were the lifelines of Japan's war effort. They ran south to the rubber, rice, oil, and strategic minerals of the East Indies. Attacking Paramushiro would do nothing to disrupt them. Finally, of course, there was that other enemy of the Aleutian front: the weather. One could be better prepared for it, but one could not prevail against it. Few relished the idea of more major operations in such conditions of uncertainty. Ironically, of course, the war against the weather continued even in the balmy Central Pacific, where typhoons would wreak major havoc with the 1945 landings at Okinawa.

So the Aleutians were beefed up and became important air and naval bases, but after the recapture of Kiska the main thrust of the war effort in the Pacific shifted southward. With it went many veterans who had cut their teeth on operations in Alaska. The veteran pilots of the Thirty-sixth Bombardment Squadron, who had made a habit of landing heavy bombers in soup so thick that one often couldn't see across the runway, were scattered throughout the air corps to lead newly organized squadrons and training commands. Proving that a victory against an unoccupied island is a victory nonetheless, Admiral Kinkaid was summoned south to command the U.S. Seventh Fleet in support of General Douglas MacArthur's island-hopping campaigns. Colonel Eareckson went to New Guinea to apply the air support tactics he had pioneered in the Aleutians. Behind them all went tens of thousands of soldiers, sailors, and airmen.

General Buckner stayed on in Alaska for a time as theater commander and then assumed command of the new Tenth Army for the final drive against Japan. General Buckner led his troops onto the beaches of Okinawa in the spring of 1945. It was Attu all over again, but on a lengthier, far more gruesome scale. Four days before the two-and-a-half-month battle ended, the man who was as eager as any buck private to get back to Alaska and call it home, Simon Bolivar Buckner, was struck dead by Japanese shrapnel while at a forward observation post.

During the war's final months, Alaska became more and more of a backwater. Attu survived a lone raid by Japanese bombers from the Kuriles—no doubt as payback for raids on Paramushiro—and radar on Adak tracked several hundred of the 10,000 bomb-carrying balloons that the Japanese launched against the American mainland beginning in late 1944. Fighters routinely scrambled and shot them down. But for the most part, boredom joined the weather in the battle for the troops' morale.

It was difficult for the remaining American forces in Alaska to understand that, though far removed from the current headlines, they were nonetheless playing an important strategic role in keeping Japanese forces that might otherwise be deployed farther south tied down in the Kuriles. Although no B-29 raids were ever flown from Shemya, the Superfortresses rained fire on Toyko and its neighboring cities from other island bases. Once the final outcome was fixed after atomic bombs fell on Hiroshima and Nagasaki, Stalin jumped into the war and used its final days to grab the Kurile Islands for the Soviet Union. It was the end of World War II, but the precursor to another era of conflict in which Alaska and the Aleutians would continue to play a major role.

Postscript to an Era (1941–1945)

There is one political footnote *to the Aleutian campaign. In the summer of 1944, President Franklin D. Roosevelt inspected naval bases in the Aleutians and then toured massive Bremerton Navy Yard in Washington State. Along the way, he predicted that many veterans of the Alaskan theater, especially those who had no strong ties elsewhere, would likely return to Alaska and, in his words, "pioneer as their ancestors had."*[51] *Both the comment and its prophecy are fact. A fiction spread, however, that the president's little dog, Fala, had somehow been left behind at one of the Aleutian stops and that FDR had dispatched a destroyer to retrieve him.*

As the 1944 presidential campaign wound into high gear that fall, there were many—even among his most ardent supporters—who wondered whether FDR was up to the task of another campaign, let alone another four years of the presidency. Did he still have his old political magic? How would he respond to the onslaught of Republican attacks? On September 23, 1944—blissfully late for the official opening of a presidential campaign even back then—Roosevelt addressed the Teamsters Union at the Statler Hotel in Washington, D.C. In a speech slated to rank with his "nothing to fear" and "day of infamy" utterances, the old master grandly touched all of the bases and then turned Fala loose against the entire Republican establishment.

"These Republican leaders," Roosevelt declared with mock seriousness, "have not been content with attacks on me, or my wife, or on my sons, [but] . . . they now include my little dog, Fala." He and his family took such attacks as a matter of course, Roosevelt asserted, but Fala, he said, resented them. "You know," continued FDR in his folksy way, "Fala is Scotch, and being a Scottie, as soon as he learned that the Republican fiction writers in Congress and out had concocted a story that I had left him behind on the Aleutian Islands and had sent a destroyer back to find him—at a cost to the taxpayers of two or three, or eight or twenty million dollars—his Scotch soul was furious. He has not been the same dog since."[52] *His audience in the hall and those listening by radio across the country howled in delight. The grand old man still had it. But Alaska, much like Fala, had not been the same land since.*

Even from a distance of more than half a century, in an era of quantum leaps in computer speeds and the Internet, it is difficult to grasp the rapid and profound mobilization that occurred throughout Alaska and all of America in the relatively short time span of World War II. Granted, FDR and others had begun to push America's productivity by 1940. The fact remains, however, that in less than four years—the 1,366 days between December 7, 1941, and the official Japanese surrender on September 2, 1945—America flexed its industrial might and led the world into the modern age. There was no "I'll get back to you next month on that." Things were done immediately with whatever tools were on hand. The fruits of those labors, be they highways, railroads, harbors, or airfields, remained.

As far back as 1940, General "Hap" Arnold had noted on his Alaskan tour that "air bases and airways facilities, emergency landing fields, radio beacons, weather stations, and other air improvements which the Federal Government is introducing in Alaska are equally usable by civil air commerce. They lie along the logical air routes from the Far East to the industrial centers of the United States. They are the airways of the future."[53] *And now that future was about to descend upon Alaska in a big way.*

POSTWAR ALASKA

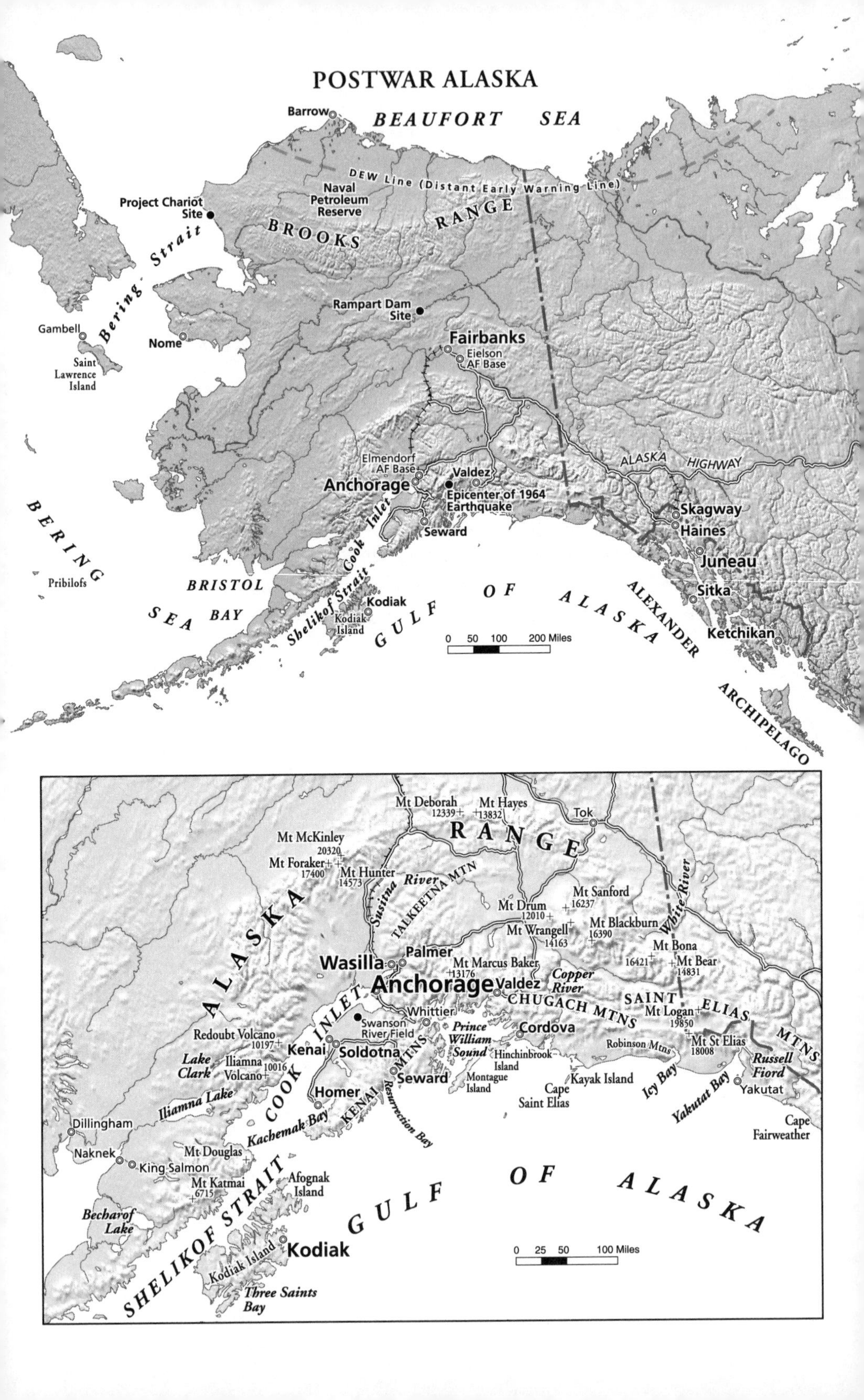

BOOK SEVEN

Postwar Rumblings: Statehood and Earthquake 1945–1964

Eight stars of gold on a field of blue,
Alaska's flag, may it mean to you.

—Marie Drake, "Alaska's Flag"

"I have a girl *back in Wisconsin. We were in high school together," a smooth-cheeked youngster in a private's uniform informed territorial governor Ernest Gruening as the governor met him on one of the roads north of Fort Richardson in 1942. "I like this country," the private continued. "After the war we're going to get married, and I'll bring her up here."*

"There are thousands like him," Gruening later wrote. "They have come to keep free the freest part of all America, . . . [and] many of them, their task complete, will stay. They will homestead. They will hew log cabins from the wilderness. And the love of adventure that is in the American's heart will hold them here."[1]

Ernest Gruening was of the old school that unabashedly waxed poetic in both speech and written word. Poetic visions of America's next frontier aside, however, the governor was absolutely correct in anticipating the new wave of stalwarts who would descend upon Alaska in the two decades after the end of the Second World War. Some of them had been through hell at Attu and Kiska; others had battled wintry weather as fierce as any Japanese charge. But they liked the land. It grew on them. And they were intent upon returning to it. So, too, were others who had only heard its tales.

With them all would come the rumblings of new industries, bigger cities, and increasing refinements in transportation facilities. These rumblings would have lasting consequences, but the postwar years would also be marked by the military rumblings of a new kind of war, the political rumblings of the fight for statehood, and the geologic rumblings of the land itself.

Offspring of Victory

After World War I, one of the hit songs asked the families of returning veterans, "How ya gonna keep 'em down on the farm, after they've seen Paree?" The question was even more poignant after the end of World War II. Upwards of 10 million men and women came marching home, many having been to the four corners of the globe and then some. There were those content to fit back into the America they had left, but to many others the Main Street to which they returned seemed less grand than the one they had spent so many lonely nights dreaming about in far-flung outposts. There was a new restlessness, a new sense of "can-do," a new sense of adventure that saw a generation pull up the roots of its depression childhood and stride boldly toward the future. "In short," as one contemporary writer put it, "the personal horizons of millions of young Americans have been widened."[2]

Horizon was a word that Alaskans—and those who dreamed about Alaska—knew well. Victory was a long way off—and still not certain—when Ernest Gruening first wrote in September 1942 about servicemen returning to Alaska to make their homes. It was a theme to which he returned time and again, including in a January 1944 *Reader's Digest* article that he titled "Go North, Young Man."[3]

Gruening, of course, wasn't the only one to beckon the swarm of servicemen who would soon call Alaska their home. Later in 1944, the Department of the Interior, which still had nonmilitary responsibility for Alaska in its Division of Territories and Island Possessions, published a guide for those interested in Alaska that dealt mostly with land settlement issues. This was followed in 1945 by a general information bulletin on the territory. The publication was supposed to sort fact from fiction, but to those who had spent time in muddy foxholes or stoking coal boilers on angry seas, its warnings about difficulties sounded more like calls to adventure. In its foreword, Secretary of the Interior Harold Ickes came across almost like a newspaper editor in some mining boomtown when he went on and on about the opportunities of "our last big land frontier."

Early in 1946, Ickes even raised the issue of settling refugees from war-torn Europe in Alaska in an article titled "Let's Open Up Alaska!"[4]

Long before Germany and Japan surrendered, hundreds of letters a month flowed into the Division of Territories and Island Possessions seeking information on Alaska. Countless more were posted to towns throughout the territory. Not everyone was ready for the deluge. In Fairbanks, the large number of requests for information pouring into the Fairbanks Chamber of Commerce prompted the following reply postcard:

> In answer to your inquiry about Fairbanks, Alaska:
>
> 1. Pre-war population about 5,000; present population, about 9,000.
> 2. Housing problem presently acute. Building materials not available.
> 3. Basic industry, gold mining, suspended during war.
> 4. Good post-war possibilities in gold mining, agriculture, stock-raising, dairying, increasing housing and hotel facilities.
> 5. Living costs high, but will probably be reduced by local production.
> 6. Probably be opportunity in various services such as auto and airplane.
> 7. Substantial but seasonal tourist business expected after the war.
> 8. Aviation activities expected after war to expand substantially.
> 9. Extent of peacetime development after the war cannot be forecast now.
> 10. Only persons with pioneer instincts will succeed in Alaska. You would have to adjust yourself to pioneer conditions, be willing and energetic in your work and a good citizen of Alaska, to find your niche here.
> 11. Further requests should be accompanied with return postage.
>
> —FAIRBANKS CHAMBER OF COMMERCE[5]

Accompanied with return postage! In other words, come if you must, but pay your own way.

In anticipation of the boom of returning servicemen, Alaska's territorial legislature created the Alaska Development Board in the spring of 1945 at Governor Gruening's behest. Essentially, it was a fledgling statewide chamber of commerce. One of the related pieces of this effort was the book *Opportunity in Alaska*. Published in the fall of 1945 and written by George Sundborg, a newspaperman, technical writer, and planner for various government agencies, it quickly became the indispensable handbook for anyone thinking about moving to Alaska. Not surprisingly, Gruening wrote the foreword.

Gold nuggets were not to be found on the streets of Fairbanks, Gruening asserted, and fortunes would not be made overnight or even in a year or more. (That only happened a generation before in the Klondike or a generation later on the pipeline!) But opportunity, wrote Gruening, "is not lacking," even if "Alaska's opportunities are primarily potential rather than actual." But indeed that potential opportunity was indisputable if for no other reason than Sundborg's statement of the obvious: "Opened by the war, the North is going to stay open."[6]

While not without its moments of boosterism, *Opportunity in Alaska* offered a reasonably objective look at the territory's potential. In the section "A Thousand Careers in Ten Chapters," Sundborg cataloged Alaska's opportunities in forest industries, agriculture, livestock raising, fisheries, mining, out-of-doors (into which was put trapping, hunting, and guiding), construction, transportation and tourism, trade, service and the professions, and a miscellaneous catchall category.

Logging and sawmills were ripe for a boom, Sundborg reported. There were big postwar construction projects planned. Fishing would remain a mainstay. Now that the War Production Board's 1942 order closing down gold mining as a nonessential activity had been rescinded, mining—ever the golden siren—would rebound. And while Sundborg confessed that "Alaska is not essentially an agricultural land," location, location, and location still offered selective opportunities as a photo of dairy cows grazing in the meadows at the toe of the Mendenhall Glacier seemed to prove.

Sundborg had spent some time during the war writing about Project CANOL, so his take on the petroleum industry was particularly interesting. Acknowledging that there were no known Alaska oil fields capable of commercial production, he nonetheless reported survey work being performed near Barrow on the North Slope and a Navy Seabee expedition said to be studying "the feasibility of a 500-mile pipeline from the Barrow field to Fairbanks." Meanwhile, the territory was importing $4 million of petroleum annually, although much of it was destined for military use.

It was obvious to link transportation with tourism. "Whatever it may be after the war," Sundborg wrote, "the Territory will not again be isolated." The future of the Alaska Highway was difficult to predict because it was still under military control, but it didn't take too much foresight to see that route becoming a pipeline of commerce and tourism. There was talk of an automobile ferry linking Ketchikan, Wrangell, Petersburg, Juneau, and Skagway with the highway connections at Haines and Prince Rupert, British Columbia. And there were even some who had not given up on either a highway or railroad route running north along the Rocky Mountain Trench from Prince George—the old Route A of the Alaska Highway. According to one survey by *Time Magazine,* 340,000 people said that they were planning trips to Alaska soon after the war's end. Part of the surge in transportation facilities was a result of the military surplus airplanes that were beginning to fill Alaska's skies.

So much had Alaska's traditional isolation been reduced by its new transportation potentials that a joke made the rounds about a grizzled old prospector who ventured "outside" from his claims deep in the interior for the first time in years. When he returned to his old cronies on the creeks, all he could do was complain about the hustle and bustle of civilization, the heavy traffic, the incessant noise, and the great crush of people. "Goodness, how far did you travel?" exclaimed his friends. "Why all the way to Juneau!"

Among Sundborg's boasts were the claims that there was no hay fever in Alaska and no snakes. Apparently, he said of the latter, the climate does not agree with them. His miscellaneous opportunities ranged from cranberry growing, peat harvesting, and souvenir manufacturing to well drilling and hydroelectric production. He stated unequivocally that the grinding of lenses for eyeglasses had to be done in Seattle, but wasn't so sure about taxidermy. "I do not wish to state categorically that there is no taxidermist in Alaska," he wrote, "but if there is, he is not of sufficient repute to be generally known."[7]

There it was—a land not lacking in opportunity. And when one went through the list, a land not lacking in plenty of issues for future controversy. But as the Second World War ended, it was the opportunities that fired imaginations.

Sometimes the mood of an era can be summed up in one or two seemingly innocuous sentences. Pulitzer Prize–winning novelist Edna Ferber, author of such classics as *So Big, Giant,* and *Cimarron,* wrote her last novel about postwar Alaska. Called *Ice Palace,* it was far from her finest work, but in it she found those two sentences. Of her fictitious Alaskan town, Ferber wrote: "Every third woman you passed on Gold Street in Baranof was young, pretty, and pregnant. The men, too, were young, virile, and pregnant with purpose."[8] Such was Alaska as it gathered the offspring of victory to its bosom and crossed yet another frontier into the postwar boom.

Logging the Forests

Tlingit had been hewing canoes from the trunks of massive red cedar trees for centuries. During the Russian era, Finnish shipwrights built Bishop Innocent's residence from stately Sitka spruce. Timber has always been one of Alaska's most abundant, renewable resources. So why have Alaskans had such a hard time getting their hands on it?

Six days after the Russian flag fluttered down the pole at Sitka and the Stars and Stripes went up, the federal government issued an order banning any private attempts to acquire title to lands in the district—by homestead or otherwise. The order also effectively banned timber cutting on public lands. Since just about everything was public land—Native claims having yet to be addressed—this pretty much locked up the entire store. In the early years of Juneau, there were numerous complaints that these timber restrictions were impeding new construction. Naturally, it cost much more to import lumber than to cut it locally.

Perhaps the most ludicrous example of timber restrictions choking local commerce, or at the very least inflating the cost of local goods, occurred at a salmon cannery at Klawock in 1886. The cannery built its own small sawmill for the limited purpose of producing wooden barrels and boxes in which to ship its fish products. The federal government forbid it to do so and ordered the cannery to import its containers from Seattle and other points in the Pacific Northwest, much to the delight and profit of businessmen there.

Indeed, part of the reason for hamstringing Alaska's timber industry early on seems to have been a powerful northwestern timber lobby that zealously guarded its monopoly on the Alaskan lumber trade, just as did suppliers of other commodities bound for the territory. As the flow of goods from Seattle, Portland, and other coastal towns in the Northwest increased during the 1890s and then exploded with the gold rushes, lumber was a major part of the goods flowing north—a boon to the steamship companies as well as northwestern loggers.[9]

Thus, as Alaska grew, so too did the volume of its lumber imports. They exceeded $1 million in 1916, $1.3 million in 1917, and $1.8 million in 1918, in part because of construction on the Alaska Railroad. By comparison, the output of some fifty small sawmills cutting limited lumber and shingles under permits for local use was valued at only $60,000 in 1917.[10]

The creation of the Chugach and Tongass National Forests in 1907 and 1909 were the first steps in changing this ratio. The National Forest Act of 1897 and the Weeks Act of 1911 both explicitly emphasized wood production as the U.S. Forest Service's main responsibility. Slowly, permits were issued and local sawmills erected. By 1939, twenty-four sawmills mostly in the southeast cut almost 26 million board feet, the largest of those being in Juneau, Ketchikan, Sitka, Wrangell, and Whittier. Fed by wartime demands, production from these operations peaked in 1943 at 62.6 million board feet.[11] Some of this production was clear-grained Sitka spruce that was exported and found its way into the noses and wings of British De Havilland Mosquito bombers.

Lumber for construction is, of course, only one use for timber. Its other major use is as wood pulp, employed in the making of paper, newsprint, plastics, rayon, cellophane, and other assorted products. As early as 1914, a Ketchikan firm announced that the sawmill it was building was but the first step toward a pulp mill operation. That never came to pass, but by 1920, Secretary of Agriculture Edwin T. Meredith was strenuously promoting the use of Alaska's forests for pulp. "The government owes it to Alaska to develop its resources and foster its economic growth," Meredith urged. "The opening up of the forests of Alaska for the development of the paper industry will supply one of the most critical needs of the United States." The head of the U.S. Forest Service, W. B. Greely, turned even more enthusiastic and described Alaska as a "second Norway."[12]

Industry journals and experts echoed the calls and predicted that two or even three pulp mills would soon be in operation in Alaska. That proved overly optimistic. One small pulp mill started operations southeast of Juneau in 1921 and briefly pumped out twenty tons of pulp per day. The pulp was shipped to California, but production and shipping costs proved too expensive and within a couple of years it shut down.

After that, what with depression and war, not much happened in the logging industry—with boards or pulp—until the late 1940s. After World War II, just as after World War I, there was a newsprint shortage, and the paper industry eyed Alaska's forests as a source of pulp. Senator Homer E. Capehart of Indiana, chairman of a subcommittee investigating the newsprint shortage, led a tour of Tongass National Forest during the summer of 1947. The senator left Alaska advocating government action in developing the area's hydroelectric generating capacity and the transportation infrastructure necessary to encourage the future development of pulp mills.

As a result, Congress passed the Tongass Timber Sales Act of 1947. This opened the Tongass to large-scale timber harvests. Essentially, in return for building and operating mills, companies were guaranteed access to billions of board feet of timber over a fifty-year term. Although no bids were initially submitted, Ketchikan Pulp and Paper Company, a firm established in 1948 and jointly financed by Puget Sound Pulp and Timber Company and the American Vicose Corporation, finally secured a fifty-year timber contract on the Tongass and announced plans to build a pulp mill in Ketchikan. Much to Senator Capehart's dismay, however, it was to manufacture the dissolving pulps used in rayon and the like and not pulp for newsprint.

Construction on the mill began in earnest in the spring of 1952. Hundreds of construction workers descended upon Ketchikan. A dam for waterpower was completed in September 1953, and 200 new houses were built for the influx of some 300 families. By June 1954, the mill was in full operation, on its way to cranking out about 100,000 tons of pulp annually. Directly employing upwards of 900 workers, the mill was seen by many as the economic saving grace of southeast Alaska. Just as the oil industry could do no wrong in some quarters a few years later, the timber industry—Forest Service workers, loggers, and mill operators alike—were the toast of Ketchikan and could do no wrong in southeast Alaska in the 1950s.

On July 14, 1954, territorial governor B. Frank Heintzleman, who as a Forest Service employee had written a 1929 report "Pulp-Timber of Southeastern Alaska," gave the dedication speech for the Ketchikan mill. The governor swelled up with boosterism and declared that "for Alaskans, the project marks the first realization of a dream and years of effort to get a major industry started in the vast, untapped virgin forests of the Territory's southeastern reaches."[13] Some Alaskans, however, were quick to object to the smoke and the resulting clear-cuts. Still others, the Tlingit and the Haida, watched with growing frustration as a case demanding compensation for their historic timber rights remained mired in the courts.

With the Ketchikan pulp mill in full operation, the annual timber harvest on the Tongass jumped dramatically. Between 1909 and 1953, the cumulative timber cut on the forest was 1,844 million board feet. In just the five years between 1954 through 1958, almost half that amount—some 865 million board feet—was cut.[14]

Meanwhile, a consortium of nineteen pulp and paper makers, thirteen rayon producers, and two timber companies—nearly all based in Japan—proposed a similar mill near Sitka. The aftermath of World War II had interrupted Japan's traditional timber sources from Manchuria and Sakhalin Island (both now occupied by the Soviet Union), and in the rush to rebuild it turned to other sources. Given the recent war with Japan, there was considerable apprehension in the United States about such a large foreign investment in Alaska. Eventually, the State Department blessed the deal as good not only for Japan's economic recovery but also for "the economic development of a self-sustaining economy in the Territory of Alaska."

The Japanese consortium organized the Sitka Lumber and Pulp Company, obtained a similar fifty-year timber contract, and shipped some lumber as early as June 1955. Long before the Clean Water Act would have required a dredge-and-fill permit, eighty acres of muskeg were buried under nearly 900,000 cubic yards of fill dredged from nearby Silver Bay during the summer of 1957. Much of the mill itself was prefabricated and shipped north on rail car barges. It finally went on-line in November 1959 and was soon shipping 340 tons of bleached pulp per day.[15] By 1960, the Ketchikan and Sitka mills and their logging operations employed a seasonal peak of 2,700 workers.[16] Whatever else was being said about the logging industry, it certainly brought a measure of economic stability to southeast Alaska during the 1950s.

There were, of course, other things being said about the logging industry. First of all, the Tlingit and Haida still claimed most of the Tongass, and as part of the Tongass Timber Sales Act of 1947, forest revenues were being escrowed pending a court decision or settlement. And then there were the conservationists, who quickly took strong exception to the harvesting practice of clear-cutting large tracts of timber.

From the lumber company's and forest manager's points of view in the 1950s, clear-cutting, especially in southeast Alaska, made eminent sense. The two most commercially desirable trees on the Tongass are Sitka spruce and western hemlock. Sitka spruce germinates best in mineral soil that is less prone to erosion, and it prefers sunlight. Not only was clear-cutting the cheapest way of logging these trees, but it provided the best environment for new growth of Sitka spruce. In southeast Alaska, these trees reach prime maturity in about 100 years. After that time, there is a much slower increase in wood volume, and the additional yield is frequently canceled out by rot and wind damage. Thus, ran the timber industry's argument, the greatest yield of wood for pulp uses could be produced with even-aged stands cut at about 100-year intervals.[17]

While conservationists fumed at the practice, clear-cutting also ran squarely into the developing concept of multiple-use land management. Gifford Pinchot's early stewardship of America's forests was not without controversy, but his charge was generally clear: Manage the forests for timber yields. Then along came Bob Marshall and others with the idea that forests should be managed for recreation. Others had priorities of grazing, wildlife protection, or watershed preservation. The mission of the Forest Service evolved to become, as Alaskan planner and educator Robert B. Weeden wrote, "not a custodian of a single economic resource but a steward of diversely endowed public lands."[18]

This evolution led to the Multiple Use–Sustained Yield Act of 1960. It charged the Forest Service with managing its lands by giving equal consideration to five surface-renewable resources: grazing, recreation, timber, water, and wildlife. Each use was to be given equal consideration and have no predetermined order. That said, it was recognized that no piece of land could be all things to all people, so the circumstances of the individual forests, or portions of them, dictated a primary slant toward one or more uses.

In Alaska, the Chugach, having been criticized for its marginal timber resources as far back as its creation, was heavily weighted toward recreation and wildlife. The Tongass was definitely weighted toward timber-related programs and a stable forest products industry. It was difficult, however, to balance even a small measure of recreation, wildlife, and water uses against the timber industry's clear-cutting techniques. In some ways, the multiple-use concept articulated in 1960 was a significant step on the road toward the enactment of the 1964 Wilderness Act. The multiple-use concept did, after all, firmly put to rest the subservience of all other uses to timber management on the nation's forests. Critics of the Wilderness Act would hold just the opposite, of course, and claim that it was the antithesis of multiple use, merely replacing timber as a dominant use with recreation and wildlife.

In retrospect, the logging industry in Alaska might be characterized as the perpetually pregnant boom that never burst wide open. The potential was always there. The promotion was frequently there. And, for a short time following World War II, the industry roared. But then it quickly came face-to-face with the strangely allied concepts of conservation and multiple use. The days when Forest Service rangers and lumberjacks were the toast of Ketchikan because they were the heralds of a new industry were brief.

The death knell to that period came in 1968 when U.S. Plywood–Champion Papers successfully bid on a fifty-year contract and proposed to build a third major pulp mill about thirty miles north of Juneau. In short order, conservationists had the weapons of the National Environmental Policy Act (NEPA) in their arsenal. The Sierra Club led a group charging noncompliance with the NEPA-required environmental impact statement process, which required an analysis of the environmental consequences of any major federal government action. It was to be a delaying tactic that involved considerable litigation and finally saw the company cancel the contract in 1976. By then, the same Tlingit and Haida who had protested so long against the denudation of their ancestral forests were successful in their claims and ultimately successful in the Alaska Native Claims Settlement Act of 1971. One new Native corporation in southeast Alaska promptly selected more than half a million acres of timberland formerly on the Tongass and then proceeded to clear-cut much of it to earn profits for its shareholders.

Cold War Standoff

Despite the wave of returning veterans in the late 1940s, Alaska's economy was once again on the wane. Nothing could match or begin to keep pace with the explosion of population and construction that the territory had witnessed during the war years. There was considerable infrastructure in place, but much of it—airfields, harbors, communications networks, and even roads—amounted to makeshift operations that had been built on the fly during the flurry of the Aleutian campaign. Almost all had been constructed and operated in response to the external pressures of war. The vitality of Alaska's internal, or homegrown, economy of mining, fishing, and logging paled in comparison. Alaska may well have slipped into another era of doldrums except for one thing: The United States was now facing a new enemy and a new kind of war.

Back in 1935, airpower champion Billy Mitchell had testified before Congress that Alaska was the most central place in the world for aircraft. "He who holds Alaska will hold the world," the unconventional general proclaimed. "I think it is the most important strategic place in the world."[19]

Mitchell assumed that he was speaking with regard to the Japanese threat in the Pacific and the coming Second World War, but his prediction would prove even more prescient with regard to the confrontation looming in the postwar era. What William Mungen had predicted on the floor of Congress upon the 1867 purchase of Alaska had come true. Russia, now called the Soviet Union, and the United States were the two great powers on Earth. But they were no longer friends. Just as surely as Churchill's Iron Curtain had dropped across Europe, it had also dropped across the icy waters of the Bering Strait.

Nowhere was this more apparent—or more illogical—than on the islands of Little Diomede and Big Diomede, smack in the middle of the Bering Strait halfway between Cape Prince of Wales in Alaska and Cape Dezhnev on the tip of Siberia. Named for St. Diomede and first charted by Vitus Bering, the two islands straddle the 169th meridian, the dividing line between Russian and American interests delineated in the 1867 treaty. Big Diomede is west—just barely—of the 169th meridian and hence in Russia, while Little Diomede is east—just barely—of the line and thus in the United States. ("Russia" versus "Soviet Union" terminology gets a little complicated as to tense and dates; technically, Russia was a part of the Soviet Union between 1924 and 1991.)

Native families living on Little Diomede had long paddled less than three miles over to Big Diomede to trade, socialize, and even marry. Cape Prince of Wales was, after all, a much more grueling and hazardous journey of almost twenty-five miles across open water in the opposite direction. The international boundary meant very little and the concept of passports even less. Then, during the summer of 1948, Soviet soldiers imprisoned a party of eighteen Little Diomede residents who had visited the big island to trade as per their long custom. The group was held for fifty-two days and grilled extensively about Alaskan defenses before being released. Such summer outings came to an end. The cold war was heating up.[20]

While the United States and its European allies squared off against the Soviet Union and the Eastern Bloc over Berlin—the famous airlift began in June 1948 and lasted fifteen months—the U.S. military quickly determined to strengthen its bases in Alaska. Once again, hundreds of millions of dollars in military spending began to pour into the territory. Construction was under way on Eielson Air Force Base southeast of Fairbanks. Named after aviation pioneer Carl Ben Eielson and built to handle long-range bombers, it was then the largest airfield in the world. Reconstruction efforts were also under way at key World War II facilities, including Fort Richardson and Elmendorf Air Force Base near Anchorage; Fort Greely near Big Delta (named for Arctic explorer Adolphus W. Greely and not newspaperman Horace); and major bases on Kodiak, Adak, and Shemya Islands.[21]

By 1949, the U.S. Air Force was worried that the Soviets' own long-range bombers based across the Bering Strait on the Chukotsk Peninsula could theoretically attack nuclear and other strategic sites in the Northwest, including the Hanford, Washington, atomic bomb plant and Grand Coulee Dam. In August in response to the threat, the air force announced that it was moving the Boeing aircraft plant from Seattle to Wichita, Kansas, to keep it out of harm's way. The announcement raised an immediate and thunderous public outcry in the Northwest. Boeing was Seattle's largest employer. Not only would 25,000 jobs be lost, but at least twice that number of spouses and children would likely also be affected.

The Seattle Chamber of Commerce organized a public hearing to confront the air force about its decision and invited all of the allies that it could muster to attend. Among those was Alaska governor Ernest Gruening. The Seattle Chamber of Commerce was not exactly on the best of terms with the governor because the group strongly opposed Alaska statehood. It did so because Seattle jealously guarded the economic monopolies and commercial clout that the Emerald City had always held over Alaskan trade. This was a crisis, however, and the group was desperate. To his credit, Gruening accepted the invitation.

At the hearing, Secretary of the Air Force Stuart Symington and Lieutenant General Kenneth B. Wolfe, deputy chief of staff for matériel, rationalized the military necessity for the Boeing move by explaining that Russian long-range bombers based in Siberia could fly 2,400 miles and reach targets around Puget Sound. Wichita, on the other hand, was 3,600 miles away and hence beyond their range. Consequently, there was no other option but to move the Boeing plant.

By Gruening's own account, he was outraged at such logic. And what would happen, Gruening demanded, when the Soviets developed a bomber, as they surely would, that *was* capable of flying 3,600 miles? The air force representatives did not have a ready answer, but Gruening—never one to take a backseat—did. He proposed that rather than move the Boeing plant, the far better option was to strengthen Alaskan defenses with a comprehensive radar network capable of early detection of aggressor bombers. Squadrons of jet interceptors based in the territory could then be scrambled to shoot them down long before they crossed Alaska, let alone reached Seattle. If that was done, Gruening concluded with no lack of certitude, the only way for Russian bombers to attack Seattle successfully was by making a wide detour around the Aleutians.

Secretary Symington and General Wolfe left the hearing red-faced, but a few days later the air force reversed its decision on the move. The Boeing plant would stay in Seattle. The Seattle Chamber of Commerce sent Mrs. Gruening a dozen roses and profusely thanked the governor for his shrewd examination of the issue. If there was ever anything that the chamber could do for Gruening in return, he was told, just ask. Sure, replied Gruening unabashedly. Support Alaska statehood. That set the Seattle group squirming, and its polite reply was that the group couldn't adopt such a resolution because it involved a political issue. Alaska eventually became a state, but to the very end the Seattle Chamber of Commerce never endorsed its statehood efforts—the only major chamber of commerce on the West Coast that did not do so.[22]

Shortly after the Seattle hearing, the air force announced that it was spending an initial $50 million to begin construction of a line of radar stations across the northernmost rim of North America that would create just the sort of early warning system that Gruening had advocated. It came to be called the Distant Early Warning Line—DEW Line for short—and when completed it stretched in a 3,000-mile arc from western Alaska to Baffin Island. In a massive mobilization project reminiscent of the Alaska Highway and predictive of Arctic construction projects yet to come, 23,000 U.S. and Canadian construction men spent four years battling weather and hostile terrain to build fifty stations along the line.

Each station boasted a garden of spindly towers laced with a web of wires and one massive gold dome poised atop a building much like a giant golf ball atop a tee. All were linked to a central, joint U.S.-Canadian command headquartered in Colorado Springs, Colorado, that would eventually be called the North American Air Defense Command—NORAD for short. The original defensive concept was to alert jet fighters from Eielson, Elmendorf, and other air bases and intercept the incoming Soviet bombers in the hour it took them to cross the radar line and fly to the closest probable targets in Alaska. By the end of the 1950s, the DEW Line was backed up by two similar warning systems, the Mid-Canada and Pinetree Lines, the latter along the U.S.-Canadian border.

Not everyone was pleased by the DEW Line's incursions into the Alaskan landscape. "They've changed everything by building that," grumbled Alexander Malcolm "Sandy" Smith, a spry ninety-six years old and a veteran of the tortuous Edmonton trail to the Klondike, "The Arctic will never be the same again."[23] Old "Sandy" Smith was right, but the same could have been said of the gold rush, and the same certainly would be said of later incursions.

In fact, the construction of the DEW Line did what the Alaska Pipeline would do except more so a decade hence. It wasn't just the initial construction. It was the inevitable numbers who followed the new roads and trails. "So many people drop in on us up here—mostly prospectors—that it's become a nuisance," complained Vice Admiral Richard H. Cruzen, project chief for the line. "In the middle of nowhere, we're forced to post 'Keep Out' signs."[24] Now it was possible to fly across the North American Arctic on a clear night without losing sight of the lights of a DEW Line station.

Fully operational by 1957, the DEW Line was extended throughout the Aleutians two years later. By then, the folly of the proposed move of

Boeing to Kansas had been proved. Introduced in 1955, Soviet Tu-16 Badger long-range medium bombers had a reported range of 4,250 miles, and Tu-20 Bears with almost twice that range were not far behind. Squadrons of then state-of-the-art F-102 Delta Dagger and F-106 Delta Dart all-weather interceptors sat on twenty-four-hour alert and scrambled at the first report from a DEW Line station of an unidentified target.

Fortunately, boredom and isolation, not Soviet bombers, remained the chief enemies. Nowhere was the boredom and isolation felt more keenly, but the Cold War tension more real, than on St. Lawrence Island. One hundred miles long and twenty miles wide, St. Lawrence guards the southern approaches to the Bering Strait. At Gambell on the island's northwestern tip—well beyond the DEW Line—the U.S. Army and U.S. Air Force set up a forward listening post to monitor Soviet air traffic.

Enlisted men who were Russian-language experts were given a crash course in Soviet military protocols and then shipped to St. Lawrence Island to eavesdrop. All were college graduates with foreign-language training from the likes of Stanford, UCLA, Yale, and Williams. Day after day for a tour of duty of one year, they monitored Soviet radio frequencies and taped the pertinent exchanges for further study by intelligence experts. The international boundary lies just thirty-seven miles to the west of Gambell, and Siberia itself is only fifty miles away. It was a classic Cold War game of cat and mouse.

Periodically, U.S. Navy Neptune reconnaissance bombers flew along the U.S. side of the boundary, and Soviet Mig fighters from an air base at Provideniya rose to shadow them. On one occasion, the Migs got too close for comfort. The Neptune pilot turned sharply to break off the encounter, but mistakenly turned *into* Soviet air space. The Migs opened fire, but the Neptune was able to circle around and make an emergency landing just south of the Gambell post. Officially, the U.S. government protested this "attack" on an unarmed reconnaissance plane. Unofficially, air force personnel showed up at the crash site and quietly removed the four 20-mm cannon mounted in the Neptune's nose as standard equipment.

Most of the battles, however, were with the weather. Ten months out of the year, temperatures on St. Lawrence Island were well below freezing, and the landing strip for the 100-man base was an ice-covered, freshwater lake. Only in July and August did the weekly supply plane out of Nome have to resort to a makeshift runway of steel mats similar to those used for the hurry-up, wartime airfields built throughout the Aleutians. In fact, the very brief week or two of summer was heralded only after Inupiat from the village of Gambell had held their annual Fourth of July sled dog races. In the dead of winter, with only a few hours of daylight, the wind howled incessantly, and temperatures frequently fell to minus-fifty or even minus-sixty degrees Fahrenheit.

When the sun did shine, a few brave souls, including army specialist Al Ossinger, ventured out for hikes up the windblown ridges nearby. Everything, Ossinger later remembered, was white—the sea, the mountains, the base, the landing strip on the lake, everything. The only spots of color to interrupt this white landscape were the sunsets over the mountains of Siberia off to the west. After a few months of such excursions, those with cameras who had been taking slides of their outings hung up a sheet and prepared a grand slide show to lighten up the winter's boredom. Everybody looked forward to seeing each other's handiwork, but it turned out that everybody had taken pictures of the exact same thing: the only spot of color around. The entertainment turned out to be a slideshow filled with sunsets over Siberia.

The magic number in this outpost of fifteen Quonset huts was 365—one year's tour of duty without leave of any kind. Once you arrived, you stayed. Everyone kept a calendar, marked off the days, and could tell you instantly how many days he had to go. Ossinger, a Stanford graduate proficient in German and French as well as fluent in Russian, counted his 365 days on St. Lawrence from September 1955. Stopping briefly at Fort Richardson en route to Nome and then Gambell, Ossinger endured a round of army physicals and then had two perfectly good wisdom teeth removed on the army's orders—just to be sure that he was staying.

Puffins, Arctic foxes, and walruses were plentiful around the Gambell post, but women were nonexistent. That didn't bother one soldier, who had sworn to his girl back home that he wouldn't look at another woman for 365 days. He happily kept his pledge for about eleven months, but then the weekly supply plane brought a surprise. Alaska Airlines flew a DC-3 for the military on charter flights from Nome. The flight's stewardess was supposed to wait in Nome while the pilots made the potentially dangerous hop to Gambell. On this occasion, however, the woman decided to come along just to see the place. She ended up being the toast of the noontime mess, much to the chagrin of the soldier who had to explain his broken pledge.[25]

As military spending in Alaska increased during the 1950s, so too did consumer prices and population. Workers were paid high wages for those days, but they earned them because of the high cost of living, frequently substandard housing accommodations, and harsh working conditions. In 1950, Alaska's population was almost 130,000, with about one out of six residents in the military. Anchorage and Fairbanks—because of both the construction of new bases and the staging of new air defense units—were hit the hardest by the mushrooming population. From 3,495 souls in 1940, Anchorage jumped to an estimated 20,000 by 1950, counting the outlying suburbs. (Then "outlying suburbs" meant the likes of Spenard and Turnagain, not Eagle River!) Incredibly, this same "Anchorage Bowl" grew 52.1 percent between April 1, 1950, and December 31, 1951—shades of earlier booms. Fairbanks increased from 5,600 in 1940 to 11,700 in 1950. Even Seward more than doubled in size to just over 2,000 during the same decade because of improvements to the Alaska Railroad.[26]

This new wave of military spending made a different impact than that during the recent war. True, the dollars were more substantial—as much as $250 million a year—but what they bought was also more long-lasting. And because many servicemen had their families along, the dollars trickled down to the local economy. Office buildings, banks, hotels, radio stations, and schools were built on the boom. Huge apartment complexes, such as the 696-unit Government Hill Apartments in Anchorage, rose from what had been vacant fields, and FHA loans spurred the construction of tract homes in the outlying suburbs.

But it wasn't just the big towns that felt the impact. Airline connections to the lower forty-eight improved, as did intrastate services, principally by Alaska Airlines and Wien Air Alaska. Dock facilities at Seward and Valdez were upgraded. The mileage of Alaska's road system doubled to 5,200 miles, and both the Alaska Highway and the Haines Cutoff began to be open year-round. An eight-inch pipeline carried oil from the docks at Haines 625 miles to Fairbanks. Aside from the old CANOL lines along the Alaska Highway, it was Alaska's first major pipeline. Interestingly enough, it was built to deliver oil to the interior from the sea and not the other way around.

By 1958, the fear of long-range Soviet bombers was compounded by the threat of intercontinental ballistic missiles, ICBMs, which greatly shortened both warning and response times. Their deployment on both sides was an inevitable ratcheting up of the Cold War. In May 1959, construction of one of three initial Ballistic Missile Early Warning Sites (BMEWS) was announced for Clear, about twenty-five miles south of Nenana. The name had nothing to do with hoped-for missile reports but dated back to Clear Site, established by the Alaska Railroad in 1918 and so named because it was clear of the foothills of the Alaska Range.

During the early 1960s, the construction of the Berlin Wall and the Cuban Missile Crisis only heightened tensions along Alaska's defense perimeter. In the end it was a standoff, but the Cold War's impact on Alaska was long-lasting. The military dollars that poured into the territory because of it during the 1950s and early 1960s were the financial underpinnings of Alaska's economy prior to the major oil discoveries. Even a generation later, when President Ronald Reagan was making one of the last characterizations of the Soviet Union as "the evil empire," he was also still calling Alaska "America's first line of defense."[27]

The Forty-ninth Star

Alaska's long battle for statehood was unique among the other thirty-seven states that were not members of the original thirteen colonies. One obvious distinction was its geographic separation, but that was also the case with Hawaii. What really set Alaska's experience apart was its very low population, enormously high percentage of federal lands, and lack of any provisions made over the years—dictatorial or otherwise—for resolving Alaska Native land ownership claims.

Territorial delegate James Wickersham introduced the first bill for Alaska statehood into the U.S. House of Representatives on March 30, 1916. The timing of the measure was no fluke. As early as 1910, Wickersham had written an article for *Collier's* entitled "The Forty-ninth Star," assuming correctly that New Mexico and Arizona would soon become the forty-seventh and forty-eighth states and proposing that Alaska be next. By no small coincidence, March 30, 1916, was the *forty-ninth* anniversary of the U.S. Senate passing the Alaska purchase treaty. Wickersham enthusiastically promoted the connection, but it would take many years and much more than such symbolism to make Alaska a state.[28]

World War I, the twin declines of population and mineral production in the territory, and the staggering blow of the Great Depression quashed all serious talk of statehood for several decades. The newly appointed director of the Division of Territories and Island Possessions, Ernest Gruening, left his first meeting with FDR at Hyde Park in September 1934 determined to change the "colonial" status of the areas he administered, but part of the problem was a lack of local revenues. As Gruening found out firsthand when appointed territorial governor of Alaska in December 1939, without income taxes or some other type of tax revenue from the resources that flowed out of the territory—be they gold, coal, copper, timber, or fish—the territory's administration was wholly dependent upon federal appropriations.

Gruening's first State of the Territory address was to the 1941 territorial legislature, and it included a proposal for a territorial income tax. Not surprisingly, it was dead on arrival. Then came another war. By the time it wound to a close, Gruening had recommended to the 1945 legislature that a referendum on statehood be held at the 1946 general election. The legislature concurred, and it also passed its own resolution in favor of statehood.

Generally, newspapers in the territory were opposed to statehood, in part because they were controlled by absentee interests that favored the freer wheeling and dealing available under Alaska's territorial status. The major exception to this was the *Anchorage Daily Times*. Its publisher, Robert B. Atwood, and his wife, Evangeline, were always strong voices for statehood. President Harry Truman also turned out to be a strong proponent of statehood for Alaska and for Hawaii, too. Truman plugged Hawaii statehood in his 1946 State of the Union address and urged that "similar action be taken with respect to Alaska as soon as it is certain that this is the desire of the people of that great Territory."[29]

Alaskans declared that desire in November when they voted 9,630 to 6,822 for statehood, roughly 58 percent to 42 percent. It was hardly an overwhelming mandate, but critics of statehood looked beyond the percentages to the total number of votes themselves. How could a territory cast less than 17,000 votes in a general election—less than the population of Reno, Nevada—and still hope to become a state? Alaska's entire population in 1940 was only 72,524. Nevada itself, the least populous state in the 1940 census, had a population of 110,247. That didn't sit too well with some folks, especially U.S. senators from the more populous states who were concerned about diluting their influence in the world's most exclusive club. Meanwhile, a Gallup Poll found 64 percent of the American public as a whole in favor of Alaska's admission as a state and only 12 percent against it. The remaining quarter was undecided.[30]

Governor Gruening and other statehood advocates seized on Alaska's vote and this general public sentiment to lobby Congress and anyone else who would listen about how Alaska deserved the forty-ninth star. Gruening traveled to Washington four or five times a year and over the next decade spoke to hundreds of groups about Alaska statehood. One group that became his ally was the National Governors Conference. (These annual meetings of state governors were begun in 1908 by Theodore Roosevelt to promote his conservation agenda.)

Gruening was always a self-assured arm twister, and in 1947 the governors unanimously approved an Alaska statehood resolution. The conference rules required unanimity in all resolutions, so this was no mean feat. Evidently, the governors were less concerned about Alaska's population than their senators. Doubtless they also knew that their resolution was mostly a showy display of democracy and that it was the senators who would ultimately decide the issue.

Gruening succeeded in getting unanimous resolutions passed again at the 1948 and 1949 conferences, but in 1950 there was a lone dissenter, Governor Herman Talmadge of Georgia. Governor Talmadge declared that he was unwilling to support the admission of a state whose two new U.S. senators were unlikely to support the southern states in filibustering civil rights legislation. Gruening rose and addressed the group at some length. The resolutions were the last item of business, and most participants were eager to depart. As Gruening by his own admission droned on and on in a mini-filibuster of his own, one governor after another quietly whispered to Talmadge to withdraw his objections, lest Gruening hold them there indefinitely. Finally, Talmadge switched his vote to "abstain," and the resolution passed. A similar situation occurred in 1951, and at subsequent conferences, Gruening chose not to ask for another resolution for fear of an outright rejection. Instead, he could rightly tell audiences that Alaska statehood had been supported in five consecutive years by the National Governors Conference.[31]

The 1949 territorial legislature finally got to work on the problem of finances and enacted a comprehensive revenue system, including trap taxes, sales taxes for municipalities, and a territorial income tax based on 10 percent of one's federal tax obligations. But it was statehood that became the overriding issue of the postwar period. Local legislative elections were won and lost over it, friendships severed, and the issue hotly debated from Anchorage streets to remote roadhouse bars.

Robert and Evangeline Atwood formed the Alaska Statehood Association, a voluntary citizens association. William Baker of the *Ketchikan Chronicle* carried the statehood torch in the southeast. Among their opponents was Austin "Cap" Lathrop, one of Alaska's self-made millionaires who had diverse interests in real estate, radio stations, movie theaters, coal mines at Healy, and the *Fairbanks Daily News-Miner*. Old Cap was on the high side of eighty and one of the few major businessmen who had both made plenty of money from Alaska and then reinvested his profits back into the territory. Cap wasn't opposed to Alaska—far from it—but he opposed the financial burden that he thought a state government would create on Alaska's small population.[32]

That same year, the territorial legislature passed Senate Bill 49—again a number assigned with great forethought—and authorized funding for the Alaska Statehood Committee. Governor Gruening appointed its first eleven members. Among them were Mildred R. Hermann, a former president of the State Federation of Women's Clubs from Juneau, state senator Frank Peratrovich, a Tlingit from Klawock and a member of the territorial legislature since 1945, and Percy Ipalook, a Presbyterian minister and the first Eskimo to serve in the legislature. Gruening also organized the out-of-state Committee of 100. Heavily drawn from Gruening's literary and political contacts in the lower forty-eight, names on the list with past Alaska connections included General Henry H. Arnold, novelist Rex Beach, mountaineer Belmore Browne, historian Jeannette Paddock Nichols, and the multitalented Bradford Washburn.[33]

Meanwhile, down in the nation's capital, Alaska's territorial delegates had been promoting various statehood measures for some years. In addition to the concern of low population, chief among the issues of any statehood debate was the question of how much land from the public domain would be granted to the new state. The federal government owned 99.8 percent of Alaska's vast territory, a much higher percentage than in any other would-be state over the years. Even in some of the more expansive western states, the public domain had been reduced by massive railroad land grants prior to statehood. That was not the case in Alaska.

When Judge James Wickersham came out of retirement in 1930 and was elected territorial delegate, he thought that it might be more politically palatable to promote a staged process toward statehood, slowly giving the territory more local control, particularly over its natural resources. Wickersham went down to defeat in the Roosevelt landslide of 1932, and Democrat Anthony J. Dimond, a tall and distinguished-looking lawyer and former teacher, took his place. Dimond not only favored statehood but also went to the extreme on the land grant issue. He introduced a statehood bill that required the federal government to convey almost all public domain lands to the new state—something that Secretary of the Interior Harold Ickes and others were clearly not disposed to do. Dimond chose not to run in 1944, and E. L. "Bob" Bartlett took his place.[34]

In 1947, the House Committee on Public Lands held hearings on a statehood bill and it passed the committee, but no floor action was taken. President Truman stuck by his Alaskan commitments, however, and sent a 1948 message to Congress devoted wholly to urging it to grant Alaska statehood. On March 3, 1950, the House of Representatives passed an

Alaska statehood bill with a vote of 186 to 146. Now Alaska statehood was in the hands of the exclusive club of the U.S. Senate.

The Senate Committee on Interior and Insular Affairs held hearings several months later. Among those whom Delegate Bartlett and Governor Gruening persuaded to testify in favor of statehood was Earl Warren, then governor of California. Father Bernard R. Hubbard, the famed "Glacier Priest" who had explored Aniakchak Volcano in the 1930s, was originally expected to testify against the measure. He was beholden to certain anti-statehood salmon interests that had financed some of his expeditions. Two Alaskan priests gave their colleague a stern lecture on the subject, however, and Hubbard's testimony ended up being neutral except to voice concerns for the tax burden on a small population.[35]

The matter was reported out of committee and debated on the Senate floor before being sent back to committee. The issue of what federal lands would go to the new state was at the heart of this defeat. The House version of the bill called for Alaska to receive four sections per township—four square miles out of every thirty-six, or about 11 percent of the total—as state lands from which to support its various administrative, infrastructure, and education requirements. This was a procedure in keeping with the land grant practices long employed in the admission of most western states.

The Senate version of the bill, however, called for the new state to *select* 21.4 million acres (about 6 percent of the total) from anywhere on the public domain, providing the selections did not interfere with existing reservations, such as national parks, forests, and military installations. Title to the mineral rights under those lands—assuming there had been no earlier conveyance of them—was also to pass to the new state. This measure died because a coalition led by southern senators blocked both the Alaska and Hawaii bills from coming to the floor for a vote. As Governor Talmadge had said, some folks didn't like the idea of new states diluting their votes. The most important point to come out of the 1950 session, however, was that the idea of *selecting* lands—as opposed to section allotments per township—was now firmly embedded in the statehood debate.

Meanwhile, war again pulled attention away from statehood concerns when North Korea invaded South Korea. In time, the 1952 Republican presidential candidate, Dwight D. Eisenhower, promised to "go to Korea," but pro-statehood Alaskans were more concerned about his earlier declared sympathies for both Alaska and Hawaii statehood.

Admission of Alaska and Hawaii to statehood, Ike had declared in Denver, Colorado, in September 1950, "will show the world that America practices what it preaches."[36]

Ike went to Korea, and a wobbly armistice would finally be negotiated there. In his first State of the Union address in 1953, the new president vigorously supported statehood for Hawaii but failed to mention Alaska. What had happened? The assumption was that Hawaii would send Republican senators to the Senate. Alaska's new senators, on the other hand, were likely to be Democrats, despite the fact that the Alaska legislature elected with the 1952 Eisenhower landslide was heavily Republican. The Republican majority in the U.S. Senate was a tenuous 48 to 47 with one independent, and apparently, Ike was not inclined to rock the boat. It was 1955 before he mentioned the possibility of Alaska statehood in a State of the Union address and then only as that—a possibility.

Eisenhower's 1952 election did mean that Democrat Ernest Gruening was out of the territorial governor's chair after almost fourteen years. Ike appointed B. Frank Heintzleman, previously Alaska's regional forester and not a statehood supporter, to replace him in April 1953. Gruening settled down in Juneau to write a history of Alaska. *The State of Alaska: A Definitive History of America's Northernmost Frontier* remains an essential Alaskan source, although it is replete with Gruening's heavy bias for Alaskan self-determination and against what he saw as the federal government's long-standing neglect.

Also in 1953, Bob Bartlett introduced one of a half dozen new statehood bills that attempted to make the rounds in that session of Congress. Significantly, Bartlett's measure doubled the federal land grants to more than 40 million acres. Pennsylvania representative John Saylor introduced a bill identical to Bartlett's and then shepherded it through the House Committee on Interior and Insular Affairs. The bill quickly disappeared into the bowels of the House Rules Committee—probably because House Speaker Sam Rayburn was not inclined toward statehood—but the most significant aspect was that along the way the proposed land grant to the new state grew to 100 million acres. In 1954, both House and Senate committees reviewed measures that contained Alaska's right to select this acreage in large blocks. The traditional sections-per-township approach was dead for good.[37]

But Alaskans were getting antsy and some more than a little cranky about the lack of real progress. After all, such political machinations were likely to drag on for years. When no positive action occurred during the

1955 session, Alaskans took matters into their own hands and by a near unanimous vote in the territorial legislature—there was one dissenter on the House side—passed a bill that called for the election of delegates to a constitutional convention. Held at the University of Alaska in Fairbanks beginning in November 1955, and blessed by Ernest Gruening's exhaustive keynote address titled "Let Us End American Colonialism," the convention of fifty-five delegates labored seventy-five days to draft a state constitution. The convention also resurrected an interesting political ploy from America's past.

The concept of what came to be called the Alaska-Tennessee Plan harkened back to the days of Daniel Boone. After Kentucky was admitted to the Union as the fifteenth state in 1792, its neighbors to the south became envious and then irate when what would become Tennessee was not granted similar status. A constitutional convention was called to draft a state constitution, and in 1796 the territorial legislature elected two "senators" and sent them east to Philadelphia. Four months later, they were back in Nashville as bona fide U.S. senators, and Tennessee was the sixteenth state. Other territories promoting statehood adopted similar strategies prior to the Civil War.

On April 24, 1956, Alaskan voters approved both the new constitution and the Alaska-Tennessee Plan to elect two "senators" and one "representative" at the fall general election. The winners were William Egan, a Valdez businessman and early supporter of statehood legislation, and Ernest Gruening as "senators," and Ralph J. Rivers, a Fairbanks attorney and former mayor, as "representative." All three were Democrats. Bob Bartlett supported the plan and kept his official seat as Alaska's territorial delegate, no doubt figuring that statehood was worth the risk of losing his job.

Bartlett need not have worried. The creatively elected Alaska delegation was treated cordially but not given privileges on the floor of either house. A cartoon in the *St. Louis Post-Dispatch* pictured Gruening, Egan, and Rivers camping out in the cold within sight of the Washington Monument. But there was good news in the wind. For one thing, Fred Seaton of Nebraska became secretary of the interior. Alaskans took that as a good sign because Seaton's maiden speech during a short tenure as U.S. senator in 1951 had been in support of Alaska statehood. (Never mind that Ernest Gruening supposedly wrote it for him!) For another, B. Frank Heintzleman, the anti-statehood governor, resigned, and Mike Stepovich, a Fairbanks attorney who favored statehood, was appointed in his place. Close ranks, the new governor said. Let's get this thing done.

The statehood bills that were introduced into the House and Senate in 1957 recognized Alaska's new constitution and thus were "admission" bills, rather than "enabling" bills meant to start a territory on the legal road to statehood. The distinction may seem arcane to all except bookish legal scholars, but the practical result was that the legislation was streamlined—less to argue about. Alaska was growing so rapidly—its inhabitants would almost double from 128,643 in 1950 to 226,167 in 1960—that population was no longer much of an issue. That left federal land grants and Native claims as the big issues.

The House committee version called for 182 million acres to be selected by the new state from the public domain within a twenty-five-year period. The Senate version was similar except that its number was 103 million-plus acres. Then Bartlett was approached with four amendments to the House version. Decrease the land grant acreage to the Senate's number; hold a statehood referendum; permit federal regulation of Alaska's maritime commerce with the other states to continue; and, oh, by the way, let the federal government retain jurisdiction over Alaska's game and fish interests. Wasn't the last one where so much of this had started in the first place?

Bartlett agreed to the first three amendments, and with Speaker Sam Rayburn's belated blessing, the Alaska Statehood Bill passed the House on May 26, 1958, by a vote of 210 to 166. By now, President Eisenhower was also giving the matter his unqualified endorsement. Over in the Senate, majority leader Lyndon Johnson grudgingly decided that he, too, would go along with Mr. Sam. After some last-minute maneuverings that saw the bill's Senate floor manager, Henry M. "Scoop" Jackson of Washington, adopt the House version wholesale, the Senate approved Alaska statehood on June 30 by a vote of 64 to 20, with mostly southerners and a few New Englanders, including Connecticut's Prescott Bush, opposed. A week later, the president signed the bill into law.[38]

At an August primary election in Alaska, the statehood referendum handily passed with 83.5 percent in favor. In November, William Egan was elected governor and Ralph Rivers the state's lone congressman. Bob Bartlett was elected to one of the Senate seats in a cakewalk, but Ernest Gruening faced a tough challenge from territorial governor Mike Stepovich for the other. Campaigning in downtown Anchorage one day, Gruening stuck out his hand to a prospective voter and introduced himself. The man recoiled and exclaimed, "You're Gruening? I'd sooner vote for the devil." The devil would be a tough guy to beat, acknowledged

Gruening, "but if he decides not to run, do you suppose you could switch to me?" The man gave Gruening a steady scowl and spat back, "Why, you son of a bitch, I might vote for you yet." Whether he did or not is unknown, but enough people did to make Gruening, by a narrow margin, Alaska's other senator. He and Bob Bartlett flipped a coin to determine seniority. Bartlett won and became Alaska's senior senator.[39]

On January 3, 1959, President Eisenhower signed the proclamation officially making Alaska the forty-ninth state. The road to statehood had been long and rocky, but considering the land grant situation alone, the wait had been well worth it. Almost 28 percent of the new state's land—103,350,000 acres—was now slated for state ownership. But what about the issue of Native land claims? On that thorny debate, all sides had taken a bye. The concept of Native land claims had been recognized in the statehood legislation, but the specifics were left to be resolved in the future.

Back in 1927, when Alaska statehood seemed—and indeed was—a long way off, a contest was held to design a new flag for the territory. Benny Benson, then a thirteen-year-old Seward schoolboy and an orphaned resident of the Jesse Lee Mission Home, submitted a design and wrote: "The blue field is for the Alaska sky and the forget-me-not, an Alaskan flower. The North Star is for the future of the state of Alaska, the most northerly in the union. The Dipper is for the Great Bear, symbolizing strength."[40] Later, Marie Drake adapted these words and Elinor Dusenbury composed the music for what became Alaska's state song. Benny Benson's flag with eight stars of gold on a field of blue was now represented on the flag of the United States of America as its forty-ninth star.

Oil Boom on the Kenai

Oil is not new to Alaska. In fact, by some estimates it has been underground there roughly 300 million years, even before the North American and Pacific plates began their steady though belabored advance northward from near the equator. Alaska Natives utilized oil seeps in various ways since prehistoric times. Russians discovered seeps near Chinitna Bay on the western shores of Cook Inlet in the early 1800s. Similar seeps in the Baku region of the Caucasus had long nurtured a fledgling bitumen industry, producing a substance roughly akin to asphalt that had been used to strengthen mud bricks as far back as the ancient Sumerians. The Russians were too far from home and commercial markets to develop the resource, however, and much like the yellow gold, this black gold was neglected for a time.

In 1857, as the Russians were trying to rid themselves of Alaska, Canadian James Miller Williams dug an oil well near Oil Springs, Ontario, and set up a crude refinery to distill kerosene for use in lamps. Williams's efforts aside, historians usually trace the modern oil industry to Edwin Drake's first well at Titusville, Pennsylvania, two years later. Using a wooden rig and a steam-operated drill, Drake struck oil at the then unheard-of depth of sixty-nine and one-half feet. As much as thirty-five barrels a day bubbled to within a few feet of the surface. Drake sold the crude for twenty dollars a barrel—not cheap in those pre–Civil War days.

After the war, Samuel van Syckel built the world's first oil pipeline, linking the Titusville field with the railroad fives miles away. A few years later in 1874, a three-inch, sixty-mile pipeline was laid from the oil fields all the way south to Pittsburgh. Armed guards patrolled the new pipeline, not to monitor spills but to discourage sabotage by disgruntled teamsters and railroaders who considered the pipeline a threat to their transportation businesses. Thirty-five hundred barrels a day—then a veritable river of oil—flowed down the pipeline.

The oil industry flourished quickly, fueling the Gilded Age of American industry and bankrolling numerous individual fortunes. Eastern oil fields throughout the Ohio River valley were soon eclipsed by discoveries at Spindletop near Beaumont, Texas, and in the Indian

Territory of Oklahoma. With seemingly limitless quantities of oil close at hand, there was little incentive to search for more in the wilds of Alaska. But as the great gold rushes drew people north, some early wildcatting (drilling in an unproven area) was done in Alaska in attempts to satisfy local needs, principally for kerosene for lamps and heating oil. As with so many commodities coming north in those days, finding oil locally was a heck of a lot cheaper than importing it from John D. Rockefeller's Standard Oil Company.

In 1901, a hole was begun—"spudded" in the terminology of the oil industry—at Katalla east of Cordova. When completed the following year, it became Alaska's first producing oil well. The Chilkat Oil Company was formed, and eighteen wells were eventually developed in the Katalla field, the deepest being some 1,800 feet. In 1911, in the shadow of the Bering coal fields controversy, a refinery was built at Katalla, and over the next twenty-two years it processed a grand total of 154,000 barrels—an amount about equal to one-tenth of the capacity of the future tanker *Exxon Valdez*. When the Katalla refinery burned in 1933, times were tough and demand low. The facility was not rebuilt, and production from the Katalla field ceased.

Elsewhere, some claims had been staked near Chinitna Bay on Cook Inlet as early as 1892. Between 1902 and 1906, six dry holes were drilled in the area by the Alaska Petroleum Company and the Alaska Oil Company. These failures discouraged further activity for a time, but after World War I, which graphically demonstrated that future wars would require great quantities of gasoline and fuel oil, interest returned to Cook Inlet. At various times between 1923 and 1939, a number of wildcat wells were drilled in the Chinitna Bay district, around Kanatak farther down the coast, and up above the Matanuska Valley at Chickaloon. There were some shows of oil, but nothing to cause much excitement or stimulate more drilling.[41]

Meanwhile, the U.S. Navy was continuing its long-established role of poking around the Brooks Range and Alaska's North Slope. Inupiat Eskimos had long been utilizing oil from seeps on Cape Simpson sixty miles east of Barrow to coat scarce driftwood for fires. These seeps were first reported to the navy in 1890 and then "discovered" by Alexander Malcolm Smith in 1917. Incidentally, this was the very same "Sandy" Smith who forty years later would bemoan the changes to the Arctic wrought by the construction of the DEW Line. Standard Oil Company of California sent its own geologists to the Cape Simpson area in 1921.

They managed to stake the first claim on the North Slope, but the climate, terrain, and the great distances involved were not conducive to exploratory drilling.

In the wake of World War I, the world's two major industrial powers, the United States and Great Britain, chose different approaches to ensuring future military oil supplies. While the government of Great Britain invested directly in what later became British Petroleum, the United States set aside large chunks of land that were thought to contain oil as naval petroleum reserves. Private leasing or drilling was not allowed on these reserves, and indeed the government had no immediate intent of drilling upon them itself. The lands were simply set aside as a "reserve"—albeit unproven—in case of emergency or if existing supplies failed.

In 1923, President Warren G. Harding designated 23 million acres between Point Barrow and the Brooks Range as Naval Petroleum Reserve No. 4. It was an area roughly the size of the state of Indiana. Similar reserves were created throughout the western United States, most famously at Teapot Dome in Wyoming. When Secretary of the Interior Albert Fall leased drilling rights at Teapot Dome to private industry cronies despite the reserve, he ended up in prison and the affair gave the already dubious Harding presidency a permanent black eye.

Some geologic surveys were undertaken in Naval Petroleum Reserve No. 4 from 1923 to 1926, but it took World War II to really pique the navy's interest in the area. Between 1944 and 1953, thirty-six test wells and forty-four boreholes for core samples were drilled there at a total cost of $60 million. Supervised by the U.S. Navy and the U.S. Geological Survey, the explorations located three oil fields and six gas fields, but all were of questionable value in terms of commercial development. The largest oil discovery was at Umiat on the Colville River, well toward the eastern boundary of the reserve. The most useful gas find proved to be a small field discovered near Barrow in 1949 that became a source of lighting and heating for the town.[42]

As the postwar building and automobile boom swung into high gear, the end result of oil and gas exploration on Naval Petroleum Reserve No. 4 and throughout Alaska was very disappointing. Despite more than fifty years of drilling, commercial oil production in Alaska was limited to the 154,000 barrels produced from the Katalla field. But though discouraging, those facts didn't stop wildcatters from looking because—just as with the mining frontier—the oil business was one in which "broke one day, rich the next" held true.

One of the areas about which geologists continued to speak favorably was the western side of the Kenai Peninsula along Cook Inlet. But that's about all that they did—talk. This lack of action frustrated key members of Anchorage's inner circle of businessmen. They were particularly concerned about diversifying Alaska's economy in view of the inevitable fact that the tap of federal dollars for military spending was bound to run dry someday. Chief among these city fathers was Robert B. Atwood, publisher of the *Anchorage Daily Times*. Coming north from Springfield, Illinois, in the 1930s, Atwood had married the banker's daughter and transformed a sleepy little paper with a circulation of 650 into a dominant force in Anchorage, as well as a leading proponent of statehood.

In the early 1950s, Atwood, his father-in-law, Elmer E. Rasmuson, who was president of the Bank of Alaska, and about thirty others pooled some funds and began buying leases on the Kenai Peninsula. With much of Alaska still unreserved federal land, it was incredibly easy to do so. All that was required under the Mineral Leasing Act of 1920 to secure a three-year lease on federal land was the payment of a onetime fee—then twenty-five cents per acre. Standard leases covered 2,560 acres, or four square miles. Atwood and his partners were convinced that if they acquired a large enough block of leases, they could persuade an oil company to drill an exploratory hole.

In the process of buying leases, Atwood met up with Locke Jacobs. Coming north from the timber country of Oregon in 1946, Jacobs arrived in Alaska at the age of twenty-two with twenty-eight dollars in his pocket. Like so many others before him, he was determined to make his fortune in mining. Five years later, Jacobs was working part-time as a stock clerk in an army surplus store in Anchorage to make ends meet, but in the meantime he had been bitten by the oil lease bug. After spinning a couple of leases to Shell Oil and doubling the store owner's money, Jacobs was recommended to Atwood. Fueled by the Atwood group's dollars, Jacobs soon accumulated over 300,000 leased acres around the Swanson River on the northwestern Kenai Peninsula. So far, this was a lot easier than hard-rock mining.[43]

With significant acreage assembled, Jacobs went south to sell the prospect. He made the rounds of various oil companies and at best took a lot of ribbing about ice and snow and at worst got quickly shown the door. A few independent geologists were intrigued by Alaska's promise, but big oil companies just didn't think of Alaska as oil country. It was

Seward's Ice Box talk all over again. Jacobs persisted, however, and finally, Richfield Oil Corporation expressed an interest. At the time, Richfield was a small company mired on the third tier of the petroleum industry. It was well below the "seven sisters" of the first rank—Jersey Standard, Texaco, Gulf, Standard of California, Mobil, Royal Dutch Shell, and British Petroleum—and such up-and-comers of the second level as Continental, Phillips, Getty Oil, Atlantic Refining, and Cities Service. Richfield already held some leases around the Swanson River, but it had never done any fieldwork on the Kenai.

To close the deal with Richfield, Jacobs and his Anchorage group offered to option some of their leases for nothing. In the end, Richfield bought them and then announced plans to drill an initial test well on one of its own adjoining leases. But—as frequently is the case in Alaska—there was a complication. The drill site was located on the 1.7-million-acre Kenai National Moose Range, established in 1941 by FDR to protect the Kenai moose herd. The moose range reservation permitted mineral leasing, including oil and gas, but surface access and use fell under the jurisdiction of the U.S. Fish and Wildlife Service. While this was prior to the enactment of the National Environmental Policy Act, which among other things established the protocol of environmental impact statements prior to certain federal action, the Fish and Wildlife Service could and did impose environmental restrictions.

In addition to moose habitat, one of the agency's prime concerns was over trumpeter swan breeding grounds. At the time, trumpeter swans were on the verge of extinction, about to go the way of the passenger pigeon. There were perhaps no more than 100 nesting pairs left in the entire United States. This was a particularly grim statistic because about 80 percent of the world's trumpeter swans nest in Alaska.

Obtaining a permit to drill on the moose range was further complicated by similar environmental concerns over mineral leasing on wildlife refuges in Louisiana. All leasing and drilling permits on refuges nationwide were stalled while Congress held hearings to review the matter. (Much later, in early positioning in the 2001 campaign to open the Arctic National Wildlife Refuge to oil and gas exploration, Louisiana senator John Breaux would hold these Louisiana refuges up as examples of oil and gas development going hand in hand with conservation.) Richfield Oil Corporation was reluctant to get involved in these political battles, but Atwood and his influential friends were not above jumping into the fray. Calling on such friends of Alaska as General Nathaniel Twining, the air force chief of staff who had served in Alaska and had grown to love it as much as Simon Buckner had, and Secretary of the Interior Fred Seaton, Atwood's group launched a concerted lobbying campaign. The end result was that Richfield's permit to drill on the moose range was approved providing that there was no disruption of lakes where trumpeter swans were nesting.

To move Richfield's giant drilling rig into the drill site, a twenty-five-mile road was bulldozed north from the Sterling Highway. Like sharks circling for the kill, other oil companies, including Standard of California, Atlantic Refining, Union, Marathon, and Phillips, watched Richfield's operation and decided that they had better protect their own positions by taking out additional leases in the area. By the beginning of 1957, there were more than 3,000 oil and gas leases throughout Alaska, and Locke Jacobs was doing his best to keep track of them. Five years before, there had been only 139.

Then on July 23, 1957, Richfield's wildcat well along the Swanson River struck pay dirt. It tapped into a large deposit of oil at a depth of 11,131 feet and set off a flurry of activity not unlike the booms of the mining frontier. The Swanson River field was Alaska's first major oil discovery. By the end of the year, the number of leases in Alaska had swelled to over 9,000 and covered nearly 20 million acres. The Atwood group sold its leases to Standard of California, and Standard and Richfield teamed up in a joint operating agreement to develop the field. When the Swanson River field came on-line in 1960, it produced an initial 20,000 barrels of oil a day. Remember the 154,000 barrels over more than twenty years from the Katalla field? A twenty-two-mile pipeline soon linked the field with a shipping terminal at Nikisiki on Cook Inlet. Most of the crude was shipped south in tankers to West Coast refineries, although a small refinery was built near Anchorage to serve the Alaska market.[44]

Before the arrival of the pipeline, Nikisiki had been a sleepy little Tanaina Athabascan village. A few miles to the south at the mouth of the Kenai River, Kenai was bigger, but not by much. The Russians originally settled it as Fort St. Nicholas in 1791. Soldotna was barely more than a crossroads where the road west to Kenai split off from the Sterling Highway running south to Homer. A post office was established in 1949, but that was about it.

The influx of oil field workers changed all of that. Soldotna even got a suburb, Sterling, where the Swanson River Road headed north into the fields. Because so much of Soldotna was built in the wake of the 1950s oil boom, the town and its environs took on a decidedly middle-American—rather than Alaskan—look. Almost fifty years later, more than one traveler winding his or her way through the Kenai Mountains and savoring Alaska is jerked back to reality upon entering Soldotna. This is an oil boomtown that could be anywhere.

Two years after the initial Richfield oil discovery, Union and Marathon made a major gas discovery along the Swanson River Road just north of Sterling. A pipeline was laid to Anchorage, and the town changed over from imported fuel oil to local natural gas to run its economy. By now, Robert Atwood, Locke Jacobs, and their friends were particularly pleased by their early gamble.

With such success, it was only a matter of time before geologists were eyeing Cook Inlet. Offshore oil and gas exploration was still in its formative years, and wide-ranging tides, glacial silt, unpredictable currents, ice floes, fogs, and blizzards did not make Cook Inlet a very enticing classroom. Nonetheless, the first offshore drilling platform was erected in its waters in 1962, and it soon proved that there was gas under the inlet. A year later, a Shell-operated consortium that included Richfield and Standard of California struck oil in Cook Inlet at what was to be called the Middle Ground Shoal Field.

Because of the inherently high costs of operating in Alaska, it became the norm for several companies to share the financial risks of exploration, while designating one of the group to act as operator. In this manner, three new oil fields were brought in during 1965. Union (as operator) and Marathon made the biggest find, the McArthur River Field, in Cook Inlet. Mobil operated for itself and Union in developing the Granite Point Field, and Union also operated for Marathon, Texaco, and Superior Oil in locating the Trading Bay Field. By the beginning of 1968, offshore wells were popping up in Cook Inlet almost as frequently as whales. There were eleven fixed drilling platforms—most with twin rigs—at work looking for more, while fifty producing wells in the four Cook Inlet fields were each averaging more than 2,000 barrels of oil a day. Another forty wells in the Swanson River field were averaging 900 barrels a day each.

When Cook Inlet production peaked at over 220,000 barrels a day later that year, in a mere decade Alaska had gone from zero oil production to become the eighth-ranking oil-producing state in the United States. The Cook Inlet basin alone was estimated to hold reserves of 1.5 billion barrels of oil and 5 trillion cubic feet of gas. The revenue from this bonanza gave the fledgling state of Alaska a solid financial boost. Fifty thousand dollars a day in royalties and taxes poured into state coffers, while millions more flowed into the state's economy, just as Bob Atwood and his fellow lease investors had intended. In the span of a decade, oil and gas had become Alaska's third largest industry behind only fishing and timber.

But just as countless hard-rock miners could testify, reserves in the ground were one thing, dollars in the bank after delivery to smelter or refinery were quite another. By one count, by 1968 the oil industry's total investment in Alaska was approaching half a billion dollars, out of which it had seen a return of only fifty cents on the dollar. "Eighth largest oil-producing state" made for good chamber of commerce talk, but as bore-hole pressures on million-dollar wells declined, payout was frequently some years down the road.[45]

Then, too, the 220,000 barrels per day backing up that boast paled in comparison to the 3 billion barrels per day being generated by Texas, the top oil-producing state. The truth of the matter was that the Cook Inlet and Swanson River fields were small on U.S. scales and even smaller when compared to the international fields where the "seven sisters" were used to operating. One of the seven, British Petroleum, participated in several Cook Inlet wildcats but was put off by the cutthroat leasing climate that permeated such relatively small fields. BP was already looking around for "the big one." While most other majors savored the oil boom on the Kenai, British Petroleum turned its attention north of the Brooks Range.

And the oil did not calm all waters. Indeed, ten years of frenzied construction in a largely pristine wilderness and in the teeming waters of Cook Inlet had their moments of environmental confrontation, but the experiences also raised environmental consciousness levels among both industrialists and conservationists. By and large, Standard of California established rigid standards for its operations on the moose range and at the time got high marks for its efforts, including revegetating drill sites in the lush, willow-rich habitat.

Environmental diligence was much more difficult, but even more critical, in the waters of Cook Inlet. Accidental oil spills, fish kills from seismic explosions, rigs interfering with fishing and migrations all became points of contention. Inevitably, there were moments of extremism on both sides. One of the strangest was when the Fish and Wildlife Service tried to impose a penalty of $25,000 for any moose struck and killed by an oil company vehicle. The oil companies grumbled, yet they were willing to chalk it up to the cost of doing business. But not some local businessmen, who noted that if the law was applied statewide, the government-owned Alaska Railroad would have to pay about $6 million to account for the approximately 250 moose killed on its tracks each year. The penalty went by the wayside.[46]

The Swanson River Road opened up a large area to recreational activities, particularly canoeing, hiking, and fishing. In the debates ahead, the Kenai experience would be held up by the oil industry as an example of how oil development and conservation could work closely hand in hand. At the time, many conservationists seemed to agree. No one seemed to mind the increased recreational use on the Kenai, but whether such improved access was a good thing elsewhere would be strenuously debated in the future.

And what about the trumpeter swans? Not to be confused with their smaller and much more ubiquitous cousins, the tundra swans, the trumpeters have made a miraculous recovery. By the 1990s, there were thousands of nesting pairs throughout Alaska. That is the kind of victory that all should cheer.

A New Meaning of Wilderness

Conflicts between development and conservation were certainly not new to Alaska's history, but as Alaska wrestled with its postwar growth and the many demands of statehood, debates between the two intensified. In part, these debates were fueled by the 103 million acres that the new state of Alaska was authorized to select from the public domain. Another central issue was the one Congress had chosen not to resolve in its grant of statehood—the Native land claims. Overlying both was a continuing discussion of the meaning of wilderness. Sometimes the discussion was philosophical, sometimes it was pragmatic, but always it was emotional.

It is difficult, if not impossible, to ponder the meaning of wilderness in Alaska's Arctic without coming across the footsteps of Olaus and Margaret Murie. Margaret—simply "Mardy" to at least three generations of wilderness lovers who came to revere her—arrived in Fairbanks on a river steamer in the fall of 1911 as a rather precocious nine-year-old. Her stepfather had recently been appointed an assistant U.S. attorney there. In the summer of 1921, after two years away from home at Reed College in Oregon, Mardy returned to Fairbanks and was introduced to Olaus Murie, a budding young scientist with the U.S. Biological Survey, the forerunner of the Fish and Wildlife Service. If it wasn't love at first sight, it didn't take many "sightings" to seal a lifelong partnership.

Three summers later, after becoming the first woman graduate of the University of Alaska, Mardy took another river steamer down the Yukon to marry Olaus at the little town of Anvik, thirty-some miles upstream from Holy Cross. Having just completed a summer studying birds in the Yukon Delta, Olaus was en route to the Brooks Range to continue the early stages of his landmark caribou studies. Given the communications at the time in interior Alaska, it was a leap of faith on both of their parts to pick a rendezvous point well in advance and then arrive there separately, Mardy with a wedding trousseau that included a fur parka and long johns.

Their honeymoon journey up the Koyukuk to Bettles, and then by dogsled to Wiseman, bound them to each other, but it also bound them to the land. On Olaus's 1926 assignment to band geese along the Old Crow River far up the headwaters of the Porcupine, they took along their infant son, Martin, and to the worries of a first-time mother were added hordes of voracious mosquitoes, fast-moving rivers, and a leaky boat. The glow of Olaus and Mardy Murie's Alaskan adventures together never waned, and in fact it fired a passion for wilderness preservation that they brought to their association with numerous wilderness causes and organizations, particularly the Wilderness Society.[47]

The godfather of the Wilderness Society was none other than Bob Marshall. Bob Marshall and the Muries were kindred spirits. The Muries first crossed paths with him in 1933, when Mardy wrote Marshall a letter congratulating him on *Arctic Village*. Marshall avowed as how the Wiseman folks had still been talking about Mardy and her Arctic honeymoon when he arrived in town a decade later. Marshall spearheaded the formation of the Wilderness Society in 1935 and became its chief benefactor. Upon Marshall's death, Olaus became one of the trustees of the Robert Marshall Wilderness Fund. Grants from this endowment provided the Wilderness Society with the funding to promote no less milestone legislation than the 1964 Wilderness Act and, eventually, the 1980 Alaska National Interest Lands Conservation Act (ANILCA).[48]

Olaus Murie was elected president of the Wilderness Society in 1950. About the same time, George L. Collins, a senior planner for the National Park Service, was surveying recreation opportunities in Alaska and chose to spend time on the North Slope. As Collins later told the story, U.S. Geological Survey geologists steered him away from the Naval Petroleum Reserve No. 4 and sent him east of the Canning River. "If you stay east of Pet 4, there's nothing over there until you get to the Mackenzie Delta in the Yukon," geologist John C. Reed told him. "You'll be out of our hair, and that's where the finest relief, the highest mountains in the Alaskan arctic, and the greatest landscapes are."[49]

By the time Collins had spent several field seasons tramping around the Franklin, Romanzof, and Davidson Mountains and floating the likes of the Sheenjek and Kongakut Rivers, he was a firm believer. "This is the finest place of its kind I have ever seen," he wrote. "It is a complete ecosystem, [and] needs nothing man can take to it except complete protection from his own transgression."[50] Somehow this land had to be preserved.

Collins soon proposed that the whole extent of far northeastern Alaska become an Arctic wildlife sanctuary. Olaus and Mardy Murie and the Wilderness Society led a growing constituency of scientists and conservation organizations who came to champion the idea. In 1956, the New York Zoological Society, the Conservation Foundation, and the University of Alaska cosponsored a summer of field studies that Olaus led on the upper Sheenjak River. The slide shows and lectures that he gave around the country in the years after this trip were the impetus to creating an Arctic wildlife preserve.

But what legal form would the preserve take? Murie worried that traditional national park status would simply attract recreation and visitor developments. George Collins was concerned that subsistence hunting by the region's Alaska Natives and some prospecting were at odds with traditional park management. Whatever was done, there was the realization that the natural boundaries of the ecosystem they sought to protect extended well beyond the international border and all the way to the Mackenzie Delta. Wilderness does not recognize political boundaries.

The idea that made the most sense—and was probably the most politically expedient—was the creation of a national wildlife range similar to the preserves established by FDR in 1941 to protect Kodiak Island's bears and the Kenai's moose. Out of a 1957 conference, which George Collins chaired, came a proposal for the Arctic National Wildlife Range. Bureau of Land Management director Edward Woozley, under whose jurisdiction the land fell, and Clarence Rhode, the Alaska regional director of the Fish and Wildlife Service, both supported the move.

But political expedience did not mean political unanimity. When legislation was introduced into Congress to limit mining in the designated area to subsurface rights in an attempt to preserve wildlife habitat, Alaska's new senators blocked the measure, calling it one more federal attempt to dictate Alaskan land use. With no congressional legislation forthcoming, Secretary of the Interior Fred Seaton, who had been won over years before by an Olaus Murie slide show, took executive action and signed Public Land Order No. 2214 establishing the 8.9-million-acre Arctic National Wildlife Range on December 6, 1960, a few weeks before the Eisenhower administration left office. The order closed the range to mining activities but permitted oil and gas exploration and development under certain circumstances. A few months later, an appeal was made to the new secretary of the interior to rescind the order, but Stewart Udall declined to do so.[51]

Tragically, Clarence Rhode died in a plane crash in 1958 in the mountains he was trying to save. Olaus Murie succumbed to cancer in 1963. But the first major step to preserve the land for which they and so many others had labored had been taken. Years later, remembering her time there with Olaus, Mardy Murie would reminisce, "We had a personal interest in the proposed arctic range, and we were determined to save it." At the time, Mardy described their emotions in a letter to Fairfield Osborn, the director of the Conservation Foundation: "We both wept—and I think then we began to realize what a long and complicated battle it had been."[52] But the battle was just beginning.

Meanwhile, nowhere was the debate over wilderness more passionate than at what was still called Mount McKinley National Park. One of the key players here was also named Murie. In the tradition of Charles Sheldon, Olaus's brother, Adolph, had done landmark natural history studies on the park's animals, particularly its wolves. Some people wanted to throw the park gates wide-open, build a huge hotel at Wonder Lake, pave the park road, and even bridge the McKinley River to extend the road toward McGonagall Pass. While the park boasted grand scenery for tourists, Adolph Murie and others remembered that Sheldon's original plea for its creation had been to preserve a complete ecosystem for Dall sheep, caribou, wolves, and moose. The national park existed first and foremost, they argued, to protect the animals. Tourists and improvements to accommodate them should take second place or not be built at all if they impinged on the chief goal.

The first trickle of tourists arrived at McKinley Station when the Alaska Railroad was completed past the park's eastern boundary in 1923. A year later, the park road—built along much the same route that it still follows—was passable for touring cars as far as Savage River, twelve miles from McKinley Station. A grand total of sixty-two visitors were recorded that year, a number that more than tripled to 206 in 1925. Year after year, a few miles at a time, the road was extended westward over Polychrome Pass, across the Toklat, and past Wonder Lake until in 1938 it finally reached Mile 90 at the old mining camp of Kantishna just north of the park boundary. Even in those days, there were arguments between park superintendents and construction crews over disposition of construction debris and proper alignments.[53]

Pre–World War II visitation to Mount McKinley peaked at 2,200 in 1939, with all tourists arriving on the Alaska Railroad. The McKinley Park Hotel, able to accommodate 200 guests, opened that year at the park entrance. Built under the auspices of the Works Progress

Administration, the hotel was decidedly lacking in the grand architecture so evident at other WPA national park projects. Even Secretary of the Interior Harold Ickes remarked with disappointment that the building looked like a factory and that he expected to hear a shift-change whistle blow. During World War II, the army took over the entire operation and ran the hotel as an R-and-R camp. During 1943, as many as 8,000 troops per month disembarked at McKinley Station for a few days of fun and savored their escape from the Aleutians.[54]

Now, with the war over and increased numbers of visitors about to descend, the primary purpose of the park as wildlife refuge or tourist attraction was heatedly debated. Underlying the debate was the prospect of greatly improved access. In 1957, the 170-mile Denali Highway was completed between Paxson on the Richardson Highway and the park entrance. To be sure, it was rough gravel at best, winding and narrow, but it provided private motorists with the first direct access to Mount McKinley. And it was just the beginning. Plans were under way for an all-weather highway linking Anchorage and Fairbanks along the route of the Alaska Railroad—right past the park's front door. (Countless tourists think this road is the "Park" Highway because it leads north from Palmer to the park entrance, but in truth it is officially the George A. Parks Highway, named after Alaska's territorial governor from 1925 to 1933, and it continues to Fairbanks.)

In the years between the two road openings—the Denali Highway in 1957 and the George A. Parks Highway in 1972—there were lengthy and sometimes bitter debates about the future of Mount McKinley National Park, centering on how to preserve Sheldon's vision while accommodating the tidal surge of visitation. What to do? Develop facilities to serve the increased number of visitors or restrict access to preserve the resource? The early 1960s would not be the only time those questions were posed.

Olaus Murie joined his brother in being displeased with plans to widen and pave the park road all the way to Wonder Lake. Olaus well remembered his own early years studying caribou in the park. Even the early road was too much of an intrusion for him. To turn it into a highway capable of handling thousands of cars was bound to create the same "rush-past-the-scenery" clutter that had come to plague such jewels as Yellowstone, Yosemite, and Glacier. The road became the focus of the wilderness-versus-development debate because it was, in the words of park historian William E. Brown, "the umbilical that would feed all other growth."[55]

During 1963, *National Parks Magazine*, the voice of the watchdog

National Parks and Conservation Association, carried a series of articles on the mounting pressures for development. Olaus Murie reminisced about earlier visits in light of the proposed new road improvements. Only a few years before, Olaus mused, "I had a comfortable feeling on this road. The moss, blueberry bushes and dwarf birch came close to the edge. We were not frightening the landscape away from us, as we seem to do on carefully manicured highways."[56]

The balance, of course, was that Alaska's congressional delegation and much of the state were eager to promote tourism. It would be a long educational process on both sides of the issue before the realization that tourism did not necessarily require megahotels at Wonder Lake. As one letter to the editor of *National Parks Magazine* put it, "We wish to be reassured that the Park Service is not selling its soul to the public demand for easy comfort and amusement."[57]

Conservationist Sigurd F. Olson wrote to Adolph Murie what became the prevailing view: "The reason McKinley is such a wonderful game sanctuary is because there are no interior developments beyond those at Denali and Headquarters. Start developing elsewhere and the charm and wildness will be gone." The National Park Service had a great opportunity at Mount McKinley, Olson argued, to do the very thing that should have been done in other parks, namely keep all development outside park boundaries. In the end, even the National Park Service was forced to admit that it had "finally learned" from its experience in parks in the lower forty-eight that "the solution is not to provide more and more roads for more and more automobiles."[58]

By 1966, managing the park road had become an integral component of managing the greater Denali ecosystem. The pavement still ends at Savage River, and the road has been significantly widened no farther than Mile 30 at Teklanika. Once the George A. Parks Highway was completed in 1972, the access solution was to prohibit private autos beyond Savage River and implement a shuttle bus system. The Mount McKinley discussions underscored the importance of access to any wilderness debate. The ramifications of public access would continue to be a wilderness issue that would be even more heatedly debated when a road was built north to Prudhoe Bay a few years hence.

By the way, there is one decidedly Alaskan twist to the Denali road management plan: the road lottery. In the fall after the tourists have gone, the road is opened to a few people in their private autos who have been selected in a lottery from among thousands of entries. In 2001, 400 cars were allowed to drive the length of the park road on each of four days in mid-September. It doesn't exactly take one back to the days of Charles Sheldon, but without the clutter of numerous buses, it does suggest how things were in the "early" days of the road in the 1960s.

As debates over the meaning of wilderness continued, two proposed development projects became lightning rods of controversy. They involved two issues almost guaranteed to coalesce developers and conservationists into opposite camps: nuclear power and dams. Although located in Alaska, the projects generated national debate and set the tone for even more raucous battles in the future. Suddenly, the battle over developing or conserving Alaska was everyone's business.

In 1957, the Atomic Energy Commission launched the Plowshare Program to promote peaceful uses of atomic energy. Some scientists at the University of California quickly recommended to the AEC that large-scale, one-blast-moves-all, earth excavations were a perfect application for the atom's might. Thus was born Project Chariot, a plan to detonate a gargantuan explosion equal to 2.4 million tons of TNT and in the process move, well, just about anything. But where to park the chariot?

AEC scientists were eager to test bombs for such a purpose (low-dose radiation worries were only beginning to be discussed) but were shrewd enough to recognize that they had to wrap the project in an enticing package. What better to cloak it in than economic development? Rather than a mere hole in the ground, Project Chariot proposed to blast out a harbor—or at least a big hole that would sort of look like a harbor—on the Arctic coast somewhere north of Nome. This new harbor, the company line ran, would promote mineral development in the nearby DeLong Mountains, principally oil and coal.

USGS studies admitted that not much was known about the region's geology. Nonetheless, a private consulting firm prepared a glowing report on its mineral potential—without conducting field investigations—and speculated that a new harbor in a smooth strip of coastline at Cape Thompson, just south of Point Hope, would encourage development. In 1958, no less a spokesman for atomic energy than the father of the hydrogen bomb, Edward Teller, headed a delegation of AEC officials and scientists charged with selling Alaskan audiences on the project.

In general, the Alaskan press was enthusiastic. Here was an opportunity for more federal dollars to flow into the new state, as well as a groundbreaking (no pun intended) project that would propel Alaska into the atomic age—then still thought to be much more promising than the future would prove. Business leaders liked the sound of the federal dollars, but even they were a little skeptical about the mineral deposits and the need for a new harbor on the edge of the unruly Chukchi Sea. Point Hope residents, on the other hand, were very concerned about the environmental impact of what was being proposed for their backyard.

The AEC contracted for environmental studies of the Cape Thompson area, but many charged that the process was mere window dressing for plans the AEC had already set in motion. When data collected on Inupiat Eskimo hunting patterns and the biological effects of the blast differed with AEC views, it was ignored. University of Alaska biologist Leslie Viereck resigned from the AEC's environmental review team rather than take part in what he considered a sham. Viereck and others took their concerns to the Alaska Conservation Society. Before long, the Sierra Club also took on the cause. There were simply too many unanswered questions about how radioactivity from the blast would affect everything from drinking water to the nesting grounds of thousands of seabirds to lichen essential to caribou that in turn were essential to the Inupiat—in other words, the entire natural balance of the region.

By 1961, despite the assertion of the *Anchorage Daily Times* that "nuclear power opens amazing new avenues for greater achievements," the 300 residents of Point Hope had attracted powerful allies and set an important precedent. Governor Egan assured Point Hope residents that no detonations would occur until he was convinced that they would cause no harm. Senator Bartlett figured that if the AEC was as confused and insensitive to Alaskan concerns in the project's development stage as it had been in its planning to date, it could take its blasting elsewhere.

But most important, the Association of American Indian Affairs sponsored the first general meeting ever held by Inupiat and Yup'ik Eskimos to discuss Project Chariot and other land use concerns. Village representatives from the Arctic coast to the lower Kuskokwim gathered in Barrow and found a commonality of purpose that strengthened their cultural identity. Talk of splitting an atom fused a heretofore loosely organized culture.

In August 1962, all the while insisting that its studies had found no danger from the proposed blast to the region's inhabitants or its fish or wildlife, the AEC quietly put Project Chariot on the back burner. Later in the 1960s, the commission conducted underground tests on Amchitka

Island out in the Aleutians, where the only inhabitants were seabirds. Loud public outcries were heard with each detonation, but to no avail.

Today, there is just a little circle on the map marking Chariot, Alaska, and the surrounding coastline is part of the Alaska Maritime National Wildlife Refuge. The defeat of Project Chariot showed the power of the environmental coalition, but more important, it demonstrated the power of Alaska Natives to challenge any action when they acted in concert. It was a lesson that would be remembered a few years later.[59]

And then there was the dam. Suffice to say that the Yukon River has always been Alaska's main vein. Salmon swim its length, commerce plies its many arteries, and wildlife and waterfowl find shelter in its verdant wetlands. In the early 1960s, however, a plan was proposed that some said would make the Yukon do even more. Others feared an irrevocable disaster.

About eighty miles northwest of Fairbanks, the Yukon cuts generally southwest through the mountains dividing the Yukon and Tanana watersheds. Downstream is the confluence with the Tanana. Upstream is the watery maze of the Yukon Flats. The flats, by one count an estimated 36,000 lakes, ponds, sloughs, and assorted backwaters, is one of the world's premier wetlands and waterfowl breeding grounds. Just downstream from where the Dalton Highway and the Alaska Pipeline now cross the Yukon is a site in Rampart Canyon that made engineers' mouths water. Here was the perfect dam site.

While this project, too, would attract national combatants, the two chief protagonists were Alaskans. Proving that nothing is cut-and-dried in Alaska—except perhaps smoked salmon—Rampart Dam's principal proponent was Senator Ernest Gruening, a liberal Democrat. State representative Jay Hammond, a moderate Republican, led the antagonists who opposed both the financial cost and the environmental impacts of the proposed project. While the two men had decidedly different backgrounds and philosophies, neither was ever known for biting his tongue on a controversial issue.

Gruening got the ball rolling by obtaining an initial congressional appropriation for a U.S. Corps of Engineers feasibility study. If a 530-foot-tall, 4,700-foot-long dam was built at the Rampart site, this colossus would back up the Yukon and flood an upstream area of 10,000 square miles, inundating roughly 2 percent of the entire state of Alaska and creating a reservoir larger than Lake Erie. Rampart Dam would dwarf Grand Coulee Dam both in size and power-generating capacity and cost an estimated $1.3 billion, still a staggering sum in the pre-Vietnam era. Depending on nature's vagaries and who was crunching the numbers, it would take fifteen to twenty years to fill the reservoir. This, Gruening postulated, was what developing Alaska was all about.[60]

Developers and businessmen rallied to the Rampart cause. Project Chariot and even the DEW Line were small potatoes compared to the huge construction payrolls that would flow into the state if Rampart Dam interrupted the Yukon's flow. The dam would be by far and away the largest single project ever undertaken in Alaska. At Gruening's request, the Alaska state senate unanimously passed a Gruening-drafted resolution supporting the project and urging additional appropriations. The resolution then went to the state house of representatives for its consideration. Since the matter dealt with natural resources, it was assigned to the House Resources Committee, the chairman of which was Jay Hammond. Much to Gruening's frustration, Hammond sat on the matter and refused to bring the resolution to a committee vote. Despite the dam's generally widespread popularity at the time, Hammond was convinced that it had "monumental boondoggle" written all over it—both financially and environmentally.

Gruening's Democrat cohorts and some of Alaska's pro-development newspapers were all too happy to take Hammond to task on the issue. "Lone Naknek Legislator Opposed to Rampart Dam" blared one headline just before the next election. In some places that may have been a political death knell. In independent Alaska, especially the Bristol Bay country, however, it was tantamount to praise. Hammond got more votes because of the headline than he lost and was reelected.

Fellow Republicans, particularly those from Fairbanks where there was strong support for the dam, finally persuaded Hammond to bring Gruening's resolution to a vote and thus stifle cries of heavy-handedness. Hammond brought the resolution to the committee all right, but only after making brief but substantive changes. Essentially, the resolution now called for the construction of Rampart Dam only after all environmental and economic studies were fully funded, completed, and sufficiently analyzed. The amended resolution was *read* before the entire house of representatives, but it was evidently not *heard* because what Hammond termed the "Rampart Regardless" faction passed the new version overwhelmingly—with Hammond among the few dissenters just to preserve the illusion.

The state senate concurred with the amended resolution, and it was signed by the governor and delivered to an unsuspecting Senator Gruening in Washington. When the resolution reached Congress, Secretary of the Interior Stewart Udall reportedly exclaimed, "You mean to tell me the Alaska State Legislature passed this resolution? I can't believe it. It's the first intelligent thing they've had to say about Rampart!"[61]

Gruening, of course, was crimson with rage when he read the resolution's substance. Whether by error or design, his original version was introduced into the *Congressional Record.* But it didn't matter. Clearly, there was going to be much more study of the dam's environmental impacts.

While the Alaska Conservation Society, the Sierra Club, and other groups opposed Rampart Dam, perhaps the most decisive opposition came from the U.S. Fish and Wildlife Service. The agency announced publicly and loudly that the proposed dam would endanger the Yukon's salmon spawning grounds, effectively cutting off the upper third of the river. "Nowhere in the history of water development in North America," its report to the U.S. Corps of Engineers proclaimed, "have the fish and wildlife losses anticipated to result from a single project been so overwhelming."[62] Senator Gruening tried to stifle such government criticism of Rampart with an angry letter to the secretary of the interior, but he didn't get very far with Stewart Udall.

There were, of course, equally vocal pleas in support of the dam. Senator Gruening's administrative assistant, George Sundborg, who by no small coincidence was the author of a book about Grand Coulee Dam, found little of redeeming value in the area to be drowned. "Search the whole world," Sundborg asserted, "and it would be difficult to find an equivalent area with so little to be lost through flooding." Sundborg claimed that the Yukon Flats contained "not more than ten flush toilets." Tongue in cheek, conservation writer Paul Brooks challenged the toilet estimate and claimed it "grossly exaggerated."[63] Such exchanges struck to the core of the debate over how land is valued and for what purposes.

Other supporters of the dam, clearly not understanding the distinction between water and wetlands, said in all seriousness that if ducks loved water, what could be better for ducks than more water? Such speakers had obviously not paused to consider how many ducks were nesting out in the middle of Lake Erie. Such insensitivity to the environmental ramifications of the project only served to weaken legitimate arguments for power-generating capacity.

The Council of Natural Resources opposed the dam on both ecological and economic grounds, and in the end, the Rampart Dam project came to be summarized in one line by Ira Gabrielson, director of the Wildlife Management Institute: "Rampart Dam is synonymous with resource destruction."[64] With such consensus in hand, Secretary Udall issued a recommendation that the Rampart Dam project not be built. Calling Rampart a major turning point in the country's belated concern for environmental abuse, Jay Hammond later wrote, "Perhaps nothing did more to exacerbate environmental awareness in and outside Alaska than the Rampart Dam debate."[65]

It might be added that few debates did more to distill the many shades of conservation and prove that such issues are never black and white. To Gruening, the Yukon Flats was "a vast swamp." To Hammond, it was the shelter for one of the largest wetlands and waterfowl breeding grounds in North America. Each man thought of himself as a conservationist, but there were differing degrees of how much each "conservationist" wanted to change versus how much he wanted to leave alone. What was the value of wilderness? Was the ground worthless because it lacked flush toilets and electricity, or did that very fact make it priceless?

Ten years later, writing in his autobiography, Ernest Gruening was still convinced that he had been right about Rampart Dam. By then, few agreed with him, but Gruening spoke to the many shades of conservation. Recognizing that the issue was fraught with emotionalism on both sides, Gruening declared himself not only a conservationist but a fervent one. "But where I disagree with some of my fellow conservationists, whom I class as extremists," he wrote, "is that their concern omits the essential part of the problem—the human element. They are interested in preserving the habitat of a variety of feathered, furred, and scaled creatures. That's good. But man requires a habitat too, and without a viable economy does he have one?" Conservation and development could and must be reconciled, Gruening went on, "but sometimes these friends of nature are, in their laudable zeal, misguided, too."[66]

Gruening might have stopped there, but in seeking to drive home his point, he only managed to come full circle to prove the differences of the many shades. Why, look at new Lake Powell, he wrote. Its sparkling waters were a vast improvement over the arid depths of Glen Canyon. A similar lake over the "anything but beautiful" Yukon Flats would have been "an enhancement," he concluded.[67] So, unlike so many other rivers, the debate rolls on.

There is an interesting footnote to the Rampart Dam story. Gruening and others were so committed to the Rampart proposal that they rejected conservationists' entreaties to back a much smaller dam and power project in Devils Canyon on the Susitna River in exchange for shelving Rampart. It was a trade-off similar to the one David Brower and the Sierra Club negotiated to save Echo Park in Dinosaur National Monument at the loss of Glen Canyon. But in Alaska's case, proponents clung to the Rampart project until the end.

The Susitna–Devils Canyon project, which in fact made much more economic sense than Rampart and would have provided hydroelectric power to both Anchorage and Fairbanks, did not gather momentum until the 1980s. By then, with Rampart long dead and no other reason to trade, conservationists blocked the construction of what would have been viewed as a necessary part of the Rampart defeat twenty years earlier. The Susitna River through Devils Canyon remains one of Alaska's roughest stretches of white water. The Yukon Flats is now an 8.6-million-acre national wildlife refuge.

Earthquake!

It was Good Friday. Many had been to church or were going later in the evening. Easter was early this year, and throughout much of southcentral Alaska on March 27, 1964, winter still held the land in its cold and icy grip. Around the waters of Unakwik Inlet in the northwestern reaches of Prince William Sound, silence cloaked the mountainsides. Animals, some say, can sense impending natural disasters before they strike. If so, perhaps bald eagles took to the skies to shriek a warning, snowshoe hares looked around nervously, and sea otters paused in their incessant preening. Hibernating brown bears slept on—for the moment. The time was 5:35 P.M.

Alaska's rim of fire is no stranger to earthquakes. Here, the grating forces of the Pacific and North American plates have long caused the earth to rumble, shake, and otherwise adjust itself. The town of Valdez, some forty miles east of Unakwik Inlet, has recorded dozens of major earthquakes and numerous minor ones since records were begun in 1898. By some counts, one out of every ten of the world's earthquakes occurs in Alaska. But no one was prepared for the fury that was unleashed this March evening.

At 5:36 P.M., something snapped along the fault lines beneath the waters of Unakwik Inlet. The resultant shock waves spread in all directions—heaving islands upward, sucking lakes dry, and violently shaking ridgelines and valleys. For three to four minutes the rumbling continued. Registering between 8.4 and 8.7 on the Richter scale, Alaska's Good Friday earthquake was the most powerful seismic event ever recorded in North America. In comparison, the 1906 San Francisco earthquake was a mere tremor.

From the quake's epicenter in Prince William Sound, major destruction spread across more than 100,000 square miles. The earthquake was felt throughout an area greater than 500,000 square miles—roughly twice the size of Texas. The change in the landscape was mind-boggling. Some areas in southeastern Prince William Sound rose as much as thirty-eight feet. In the northwest, land dropped an average of two to three feet, flooding beaches and inundating freshwater marshes with salt water. Some land areas shifted seaward as much as sixty-four feet.[68]

The natural scene was the setting, but it was only a part of the story. Human inhabitants were dumbstruck. Some thought the end of the world had come. Others were certain that despite the DEW Line, the Soviets had managed to drop nuclear bombs on Elmendorf Air Force Base. Strictly in terms of power, that would have been mild by comparison. By one calculation, the earthquake unleashed 200,000 megatons of energy—more than 2,000 times the power of the mightiest nuclear bomb then in U.S. arsenals. Most folks were simply too shocked to reason beyond immediate fears for their own safety and that of their loved ones.

At his house overlooking Turnagain Arm south of Anchorage, J. D. Peters walked into the driveway to greet his wife, who was just coming home. Before he could hug her, the earth opened and the two were separated by a gaping fissure. Peters tried desperately to throw his wife a line to keep her from slipping down the slope. Just as suddenly, the fissure closed back up, and she was propelled into his arms. Throughout the surrounding neighborhood there were frantic screams and outcries as houses broke apart and dumped their occupants—some barefooted, others holding babies—out onto buckling streets and sliding hillsides. Peters dashed into one crumbling house and gathered coats and boots for those fleeing into the cold.[69]

In another section overlooking Turnagain Arm, the house of Lowell Thomas, Jr. dropped like a rock when a large portion of bluff gave way. Mrs. Thomas and their two young children managed to escape and gamely climb up to level ground through a maze of shattered buildings and slippery, snow-covered slopes. Later, waiting at their minister's home to be reunited with Mr. Thomas, they heard the calm voice of the KFQD radio announcer matter-of-factly report that as far as he knew, little actual damage had occurred. "Wait till he sees our street!" eight-year-old Anne Thomas exclaimed. Two neighborhood children were not so fortunate. Twelve-year-old Perry Mead shepherded two younger siblings to safety and then returned to his house for his baby brother. Neither was ever seen again.[70]

Joe Kramer, a veteran Anchorage taxi driver, had seen a lot of crazy driving in his time—alcohol-inspired and otherwise—but when vehicles started fishtailing toward him like bumper cars at some Midwest amusement park, he thought that their drivers had gone berserk. "It was when they started bouncing two feet off the ground," recalled Joe gravely, "that I knew it was more than just the drivers."[71]

Downtown Anchorage looked like a war zone. At the first rumblings, attorney Russell Arnett burst out of the Anchorage Athletic Club's steam bath and started running down Fourth Avenue—stark naked. Other souls tumbled or crawled out of collapsing structures and quickly formed a human chain to keep from falling into dozens of yawning fissures. Arnett joined the chain. No one, not even Arnett, seemed to notice that he was missing his clothes.[72]

Along the north side of Fourth Avenue above Ship Creek and the Cook Inlet tidal flats, the ground sank some twenty feet. Restaurants, bars, liquor stores, and pawnshops disappeared from sight. Most of the Denali Theater sank below street level. Only its neon sign, remarkably intact, remained completely above ground. But just across the street on the south side of Fourth Avenue, buildings suffered only minor damage and a few broken windows. Ironically, a banner advertising the Alaska Methodist University's production of Thornton Wilder's *Our Town*—a play without scenes or sets—still spanned the street.[73]

The quake's selectivity along Fourth Avenue was repeated elsewhere. Half of a floral shop was gone, but in the remainder, arrangements of Easter lilies stood tall and serene in their vases. The walls of the year-old J.C. Penney store collapsed outward, killing a pedestrian who had crouched beneath them and a woman in a passing car. A few blocks away, the fifteen-story Anchorage-Westward, the city's largest hotel, stood unscathed but surrounded by rubble. At the airport, the control tower collapsed, killing a controller, but the terminal building was largely untouched.

All too quickly, darkness fell. Without electricity, heat, or running water, the town hunkered down as temperatures dropped into the teens. As the town slowly dug out in the days that followed, William DeAngelis summed up what a lot of people were thinking. Cradling his little dog—the only thing that he had left—on his lap in one of the relief centers, the old man shook his head and softly sighed, "I'm just settin' here with Beauty, waiting for my social security to come. I don't have anyplace to go."[74]

Miraculously, out of a population of 55,000, there were only nine deaths in Anchorage, but it was a different story out on Kodiak Island and in Seward, Valdez, and the little village of Chenega. The surface rumblings were relatively minor compared to the quiet horror of the killer tsunamis that struck soon afterward. Underwater landslides and water displacements caused by seaward movements of land built huge waves and sent them rolling across Prince William Sound and the Gulf of Alaska at speeds approaching 500 miles per hour.

Anchorage braced for a tsunami, but it amounted to only three feet. The town was spared any more major damage in part because much of it was about 100 feet above sea level. The towns around Prince William Sound, however, weren't so lucky. A monster seventy-foot-tall tsunami roared through Knight Island Passage and inundated the little village of Chenega. Many of its twenty homes were built on pilings. When the first wave swept through, it picked the structures up and carried them away like pieces of driftwood. Twenty-three of the town's eighty inhabitants drowned, thirteen of them children. Only the schoolhouse high on the bluff above the village remained undamaged. Here, school should have been in session.

And Valdez? Well, Valdez was just about gone. Earlier on Friday afternoon, the Alaska Steamship Company's *Chena*, fresh from Seattle, had cruised into Port Valdez and tied up at the town dock. The arrival of the regular supply ship was a popular event, and some of the townspeople were on the long dock watching longshoremen unload her cargo. Among them was twelve-year-old Freddie Christofferson. When the rumbling started, Freddie's companion yelled, "Earthquake!" and took off running. Freddie followed close behind, but when he looked back over his shoulder, he saw the *Chena* picked out of the water as if she were a bathtub toy. All of her mooring lines snapped. Water swept over the dock, and the ship slammed into it with a bone-jarring blow that disgorged a fusillade of exploding timber. Some of the ship's crew were trapped belowdecks by shifting cargo.[75]

The *Chena*'s captain gave three blasts on the whistle and attempted to navigate to the presumed safety of the middle of the harbor. By some counts, the ship bottomed three times as the waves rolled back and forth. As the water flowed seaward, some said that it was as if a giant bathtub were draining. When the water reversed its course, the torrent piled up in the shallows into a giant wave that broke across the dock and swept away dozens of people. In all, thirty people drowned in the tsunamis that steamrollered across the town's tidal flat.

The longshoremen who had been on board the *Chena* unloading the ship demanded that the captain return to the splintered dock and put them ashore. When he refused, they piled into a lifeboat and rowed to the beach through a jumble of unpredictable waves, swirling debris, and jagged ice. Somehow, the *Chena* managed to stay afloat.

On shore, fuel tanks ruptured and soon caught fire. Teenager Helen Irish, who just minutes before had been among those on the dock, cried hysterically as her father left their house to search for her mother. Slowly, most survivors struggled toward higher ground. Many spent the night in the freezing cold high on the slopes of Thompson Pass. They looked down on the ruins that had once been their town, but that now were illuminated only by the fiery glow from blazing oil tanks.

The following morning, many survivors were taken north on the Richardson Highway to Glennallen. There, Helen Irish, still frantic with worry, found her mother safely asleep on an army cot. Others weren't so lucky. One old woman labored painstakingly to fill out the hauntingly yellow Western Union forms that would notify relatives that her husband had been one of those swept out to sea. Because the ground was either still frozen or flooded, other casualties were committed to the deep waters of Port Valdez for burial.

It was much the same in Seward. The afternoon had been dead calm—a rarity without the winds that alternated between the Harding Icefield and Resurrection Bay. Able-bodied seaman Ted Pederson was checking valves on the fuel lines between Standard Oil's bulk storage tanks and the tanker *Alaska Standard*—in those days still delivering oil *to* Alaska. Pederson felt the dock tremble and knew what was happening. "Earthquake!" he yelled, and ran toward his ship just as the dock rose ten feet into the air. The hoses parted and gasoline sprayed in every direction. As the dock fell downward, a giant wave laced with timber and debris swept over Pederson. But he, too, was one of the lucky ones. When he regained consciousness, he was aboard his ship with nothing more serious than a broken leg.

Lenny Gilliland watched as the bulk storage tanks and the *Alaska Standard* seemed to disappear from sight. In their place appeared a huge fireball fed by the wildly spewing gasoline. It looked to Gilliland as if all of the water had suddenly gone out of Resurrection Bay. Gilliland couldn't be sure, but if that was even partially true, he knew that it would return in a big way. As soon as the wild shaking stopped, Gilliland gathered his family into the car and headed for higher ground.

The gasoline from the broken lines on the Standard Oil dock stoked a water-borne inferno in the bay. The *Alaska Standard* plowed through the fiery waters in an attempt to get away from it. Then came the first tsunami. Later, no one could agree on how many there had been, how long after the quake they struck, or how tall they were. But their impact was undeniable. The first wave lifted a wall of fire from the bay and pushed it inland over the Texaco tank farm eight blocks from the waterfront. Water and burning fuel surged across the railroad yards, setting fire to dozens of homes, trailers, waterfront shops, and even the radio station.

Dorothy Horne was alone in her house that doubled as the Railway Express office. She was reading the J.C. Penney catalog when the house started to shake. Remembering that one was supposed to stand in a doorway during an earthquake, Horne did so until the shaking got so bad that she finally stumbled outside. "I remember seeing a train coming down the tracks," she later recalled. "The harbor master had a house right over there. It started falling into the water. Then I saw the boats in the small-craft harbor—it looked like they were trying to climb on top of each other. I looked back where the train was and it was gone."[76]

All night long, flames from the oil tanks lit the sky while the occasional explosion of another storage tank punctuated the stunned silence. Then there was another eerie sight. The pilings from the exploded docks floated in the waters of Resurrection Bay with their waterlogged portions submerged and their relatively dry, tar- and oil-encrusted ends above water and now aflame. These burned like candles and made for a flickering procession of torches as they bobbed with the tide out to sea.

Seward had been scheduled to receive an All-America Cities Award for its industrial and civic improvements. When a reporter offered his sympathies in the face of such a calamity, one Seward resident shrugged and said, "That's all right. That bay out there and the pass behind the town aren't named Resurrection for nothing."[77]

Kodiak was farther away from the quake's epicenter than either Seward or Valdez. It shook less but was subject to the full brunt of the tsunamis that raced across the Gulf of Alaska. When Karl Armstrong, editor of the weekly *Kodiak Mirror*, felt the first tremor, he thought that he had a story. Walking during the quake was like marching across a field of Jell-O. But when Armstrong couldn't reach Anchorage by telephone, he quickly realized that, far more than a story, he had a disaster in the making.

The first tsunami on Kodiak sneaked in quietly—a fast surge of rising water that had no wave crest. Off-duty navy lieutenant Raymond Bernosky was climbing a small hill to tend some traps when he happened to look back to where he had parked his Scout several hundred yards from the beach. The vehicle was suddenly afloat and coming toward him. Studded with floating ice, the waters swept swiftly and silently up the hill and chased Bernosky to higher ground.

Down at Kodiak Harbor, Captain Bill Cuthbert was eating dinner in the galley of his fishing boat, *Selief.* When he felt the eighty-six-foot-long vessel shudder sharply, his first thought was that some greenhorn had rammed him. By the time Cuthbert got on deck, the surrounding 160 crab and salmon boats in the harbor were bucking at their moorings like wild broncos. Then the first wave retreated, and the sea level in the harbor dropped. Cuthbert figured that he could ride out whatever followed.

The second wave hit with a much wilder vengeance, pushing a crest as tall as thirty feet. Townspeople who had not already done so scurried up Pillar Mountain. Captain Cuthbert stayed with his boat. The wave picked up the small boats and tossed them inland at least two or three blocks. Some, including the *Selief,* rode the wave back out as it took an assorted collection of buildings with it. Later when they could laugh again, townspeople would joke, "Come to Kodiak to see the tide come in and the town go out."

The enormous waves continued. Again, the details of how many, how high, and how often were lost in the blur of the night. Cuthbert rode several waves in and out and then ended up aground several blocks inland. The port radio operator was frantically contacting boats, and when she got to the *Selief* and queried her position, Cuthbert drawled, "By dead reckoning, in the schoolyard."[78]

When the waves at last subsided, Kodiak was a shambles of splintered buildings and broken water lines squirting geysers into the air. Most of its fishing fleet lay in a heap, with holes in their hulls and masts and rigging hopelessly entangled. On average, Kodiak slumped five feet. Everyone pitched in to help with the cleanup. One resident couldn't resist telling city manager Ralph Jottes, "I wished you every success when you campaigned to clean up Kodiak, but this is ridiculous."[79] Elsewhere on the island, tracks showed that many of the island's famed brown bears had awakened in a hurry, and rather than wandering around as would be their norm after hibernation, they had made a beeline for the ridgelines.

As bad as the earthquake was, it could have been worse. A total of 115 Alaskans died, and 4,500 were left homeless throughout the state. Another 16 people drowned in tsunamis along the Oregon and California coasts. But the fact that the quake struck on a holiday, and at an hour when many were at home, no doubt eased the death toll. Crowded schools and downtown areas, particularly in Anchorage, might have resulted in many more deaths. At one Anchorage school, an entire wing dropped a dozen feet into a fissure-laced pit, and the roof collapsed on much of the school's other wing. In many respects, it had been a good Friday after all.

The fishing fleets were in port, so the crews were safe even if many of the vessels were tossed about like Matchbox toys and left heavily damaged or destroyed. The Alaska Railroad north of Seward was a tangled mess, and the railroad's dock facilities and yards were in ruins. Preliminary reports by Governor William Egan's office—he himself was from Valdez—estimated total damages of $205 million. In the light of late-twentieth-century hurricane disasters, this figure may seem small, but in 1964 it was roughly twice the state's annual budget and equivalent to $1.14 billion in 2000 dollars. Final figures would more than double the early estimate.

Clearly, a new state still struggling to get its economic house in order could not begin to absorb such a staggering financial loss. Thus, amid such death and destruction, there was a silver lining. The U.S. military in Alaska was among the first to respond, helping others as well as its own. Alaska's U.S. senators, E. L. Bartlett and Ernest Gruening, were in Washington, D.C., when the earthquake struck, but President Lyndon Johnson quickly dispatched a presidential jet to whisk them back to Anchorage along with Edward McDermott, the director of emergency planning. In the weeks ahead, Bartlett and Gruening led the efforts to secure major federal relief for Alaska. Bartlett drafted an emergency relief bill, and Gruening told the senate, "[The] State of Alaska has suffered a catastrophe which, in my reasoned judgment, surpasses in magnitude that suffered by any state of the Union in our Nation's entire history."[80]

Eventually, the federal government provided Alaska and its citizens aid totaling $393 million, of which $66 million was in low-interest loans that were later repaid.[81] Even Ernest Gruening and others used to railing against the federal government's neglect of Alaska were forced to admit that Washington had come through in a big way to help the state. In retrospect, the emergency aid measures were probably the only issue affecting Alaska ever to come before the U.S. Congress that was not the subject of heated debate. Future governor Jay Hammond, then a member of the state legislature, summed up the end result by noting in retrospect that "thanks to a benevolent federal government and the ensuing rebuilding boom, the earthquake ultimately proved to be an economic boon to many Alaskans."[82]

And what of the land itself? Once again, it had demonstrated its dynamic nature. In the years ahead it would once again prove its resilience—up to a point. Salmon spawning beds that were covered with silt recovered, forests that were leveled sprouted new trees, and wetlands briefly inundated with salt water once again became critical bird nesting grounds. But reminders remain about the power of the land.

At the far eastern end of Turnagain Arm, easily visible from the highway between Anchorage and Seward, there is a forest of dead trees. They stand in stark contrast to the fiery hues of the fireweed that carpets the ground. Once there was a town here called Portage. Sure, it was never more than a little railroad town where the Alaska Railroad's lines split to run south to Seward and east through the mountains to Whittier, but it was a town, a living, breathing, kid-hollering, dust-kicking town. The Good Friday Earthquake caused this entire area to sink, and in the process the town of Portage was flooded and destroyed. Unlike Seward, Valdez, and Kodiak, it was never rebuilt. The trees died when the freshwater table dropped and their roots could reach only salt water. Their skeletons remain as proof that the land does not stand still.

Postscript to an Era
(1945–1964)

So Alaska was finally a state, *and if there were those who held any illusions of grandeur about that status, the land itself had spoken to remind one and all that it remained at the basis of all things. In the years ahead, despite the initial promise of oil and gas in Cook Inlet and on the Kenai, Alaska would struggle with the economic realities of its new statehood and the postponed resolution of the claims of its Native peoples. Then Alaska would be rocked by a major new discovery that was destined to have far more aftershocks than any natural event. It would be the gold rush all over again, and then some. Only this time, the gold was black.*

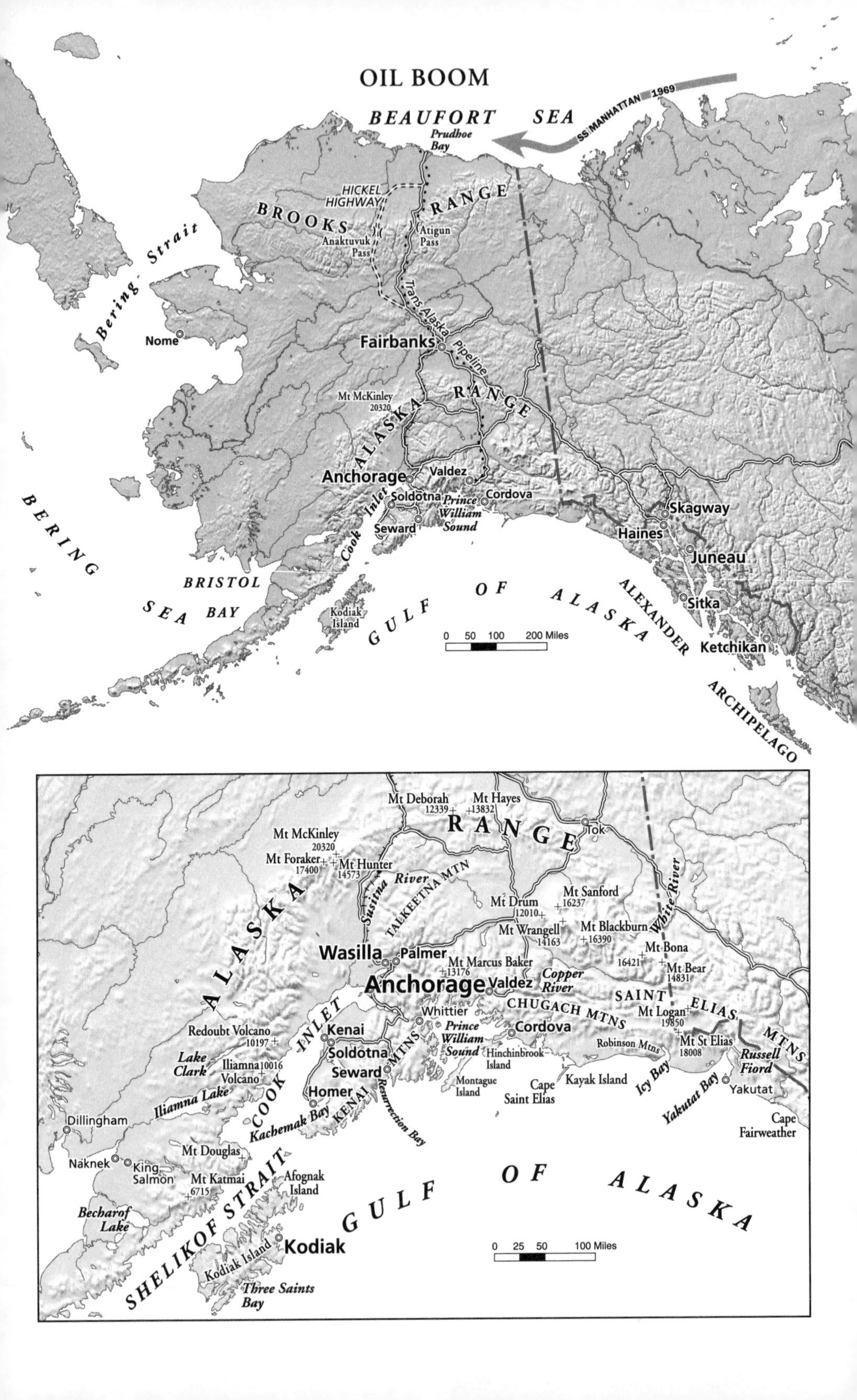
OIL BOOM
BEAUFORT SEA
SS MANHATTAN 1969
Prudhoe Bay
HICKEL HIGHWAY
BROOKS RANGE
Anaktuvuk Pass
Atigun Pass
Bering Strait
Trans Alaska Pipeline
Nome
Fairbanks
Mt McKinley 20320
ALASKA RANGE
Anchorage
Valdez
Soldotna
Prince William Sound
Cordova
Seward
Cook Inlet
Skagway
Haines
Juneau
Sitka
Ketchikan
BERING SEA
BRISTOL BAY
Kodiak Island
GULF OF ALASKA
ALEXANDER ARCHIPELAGO
0 50 100 200 Miles
Mt Deborah 12339
Mt Hayes 13832
RANGE
Tok
Mt McKinley 20320
Mt Foraker 17400
Mt Hunter 14573
Susitna River
TALKEETNA MTN
ALASKA
Mt Drum 12010
Mt Sanford 16237
Mt Wrangell 14163
Mt Blackburn 16390
White River
Mt Bona 16421
Mt Bear 14831
Wasilla
Palmer
Mt Marcus Baker 13176
Anchorage
Valdez
Copper River
CHUGACH MTNS
SAINT ELIAS MTNS
Mt Logan 19850
Mt St Elias 18008
Whittier
Redoubt Volcano 10197
Kenai
Soldotna
Seward
KENAI MTNS
Prince William Sound
Cordova
Hinchinbrook Island
Robinson Mtns
Russell Fiord
Lake Clark
Iliamna Volcano 10016
COOK INLET
Montague Island
Cape Saint Elias
Kayak Island
Icy Bay
Yakutat Bay
Yakutat
Iliamna Lake
Homer
Kachemak Bay
Resurrection Bay
Dillingham
Cape Fairweather
Mt Douglas
Naknek
King Salmon
Mt Katmai 6715
Afognak Island
SHELIKOF STRAIT
GULF OF ALASKA
Becharof Lake
Kodiak
Kodiak Island
Three Saints Bay
0 25 50 100 Miles

BOOK EIGHT

North Again: This Time the Gold Is Black

1964–1980

Hell, this country's so goddam big that even if industry ran wild, we could never wreck it. We can have our cake and eat it too.

—ADVISER TO GOVERNOR KEITH MILLER, 1970

The defining event *of Alaska's first quarter of a century of statehood was triggered by Section 6 of the statehood-enabling legislation. This section permitted the new state to "select" approximately 103 million acres from the public domain in Alaska that had not been otherwise reserved. These parcels would become state lands, and revenues from them—from their sale, lease, or use—were supposed to provide the new state with a secure financial foundation. Among the first selections the new state made were some 2 million acres on the Arctic coast in a roughly 140-mile-wide by 20-mile-deep strip wedged between the federal reservations of the Naval Petroleum Reserve to the west and the Arctic National Wildlife Range to the east. Inupiat Eskimos had lived on this land for centuries, but in the early 1960s they weren't reading the legal notices that announced the state's selection and extinguished all other rights.*

Later, some would say that the state of Alaska had been uncannily prescient in selecting this area. The truth was that the reason behind the selection was far more bureaucratic than it was geologic. Alaska, like other states, was given ownership of submerged lands on the continental shelf within its boundaries. Given the ubiquitous lakes, marshes, and assorted wetlands strung along the Arctic coast and the very low tidal fluctuations of the Arctic Ocean, it was sometimes very difficult to determine where these submerged state lands ended and onshore federal lands began. Because several oil companies were mildly interested in leasing in the area and no one wanted to go to the trouble of surveying a wildly meandering boundary, the state simply selected these 2 million acres as part of its 103-million-acre allotment. At the time, more than a few people were certain that such a selection on the Arctic coast was an utter waste of the state's entitlement.

Before the first quarter century of its statehood was out, these 2 million acres would directly provide the state of Alaska with hundreds of millions of dollars in lease bonuses, landowner production royalties, and severance taxes, not to mention hundreds of millions of dollars more in indirect economic stimuli. The very selection process of which they were a part would also provide the state with the era's two most pervasive and divisive issues, Native land claims and resource development versus wilderness preservation claims. Added to the mix were differing views for using the enormous flood of surplus revenues. One thing was certain: Alaska was no longer a backwater. The rush to the Far North was on again.

Black Gold

When British Petroleum (BP) folded its cards and walked away from the oil-leasing frenzy on the Kenai Peninsula and around Cook Inlet, it went looking for the type of globally significant oil field that it had previously developed. BP had played a major role in opening the Iranian oil fields, as well as others throughout the Middle East. The company was accustomed to investing huge amounts of capital in the pursuit of mega-sized fields. Having weathered Iran's nationalization of the Abadan refinery in 1951 and the disruptions caused by the 1956 Suez Crisis, BP was critically aware that Middle Eastern politics and oil were likely to be entangled for years to come. Clearly, there was no guarantee of uninterrupted oil flows from that region. Others could fight over the Kenai and Cook Inlet. BP went looking for another Middle Eastern–like mother lode in an area uncluttered with major competition and political instabilities.

Alaska's North Slope was just such a place. In the summer of 1963, British Petroleum, in partnership with Sinclair Oil, barged a Canadian drilling rig 2,000 miles down the Mackenzie River to the Arctic Ocean and then westward along the coast to the mouth of the Colville River. The rig was assembled, and once the ground froze later that fall, it was hauled 60 miles south to an area of federal leases that was just east of the old Umiat field on the Naval Petroleum Reserve. Between December 1963 and early 1965, BP crews drilled six wells in this area, but they yielded only marginal shows of oil and gas. As prior discoveries on the petroleum reserve had demonstrated, this far north in Alaska it wasn't good enough just to find oil. It had to flow in substantial commercial quantities that were capable of recovering not only the expenses of exploration and development but also the high transportation costs of moving the crude out of the frozen region and south to refineries.

While these operations went on east of Umiat, BP also conducted seismic surveys closer to the coast around the Colville River delta and eastward to Prudhoe Bay. In September 1964, BP petitioned the state of Alaska to put its newly selected state lands in these areas up for lease. By the end of the year, the state held its first sale of North Slope leases. It decided, however, to offer only the area around the Colville River and, for the moment, exclude those lands around Prudhoe Bay. BP and Sinclair jointly leased 318,000 acres at the sale for an average of less than six dollars an acre. There wasn't much competition.

Six months later, the state added the Prudhoe Bay acreage to another lease sale. Sinclair, discouraged by the results with BP near Umiat, decided not to acquire any additional acreage. This left BP alone to protect its earlier lease positions against new bidding interest from Richfield Oil, Atlantic Refining, and Humble Oil. These companies weren't sure what BP was planning, but BP's worldwide reputation made them suspicious. If BP thought there was value to North Slope leases, then maybe . . . Richfield and Humble jointly acquired 25,600 acres covering the heart of the coastal area immediately surrounding Prudhoe Bay for ninety-three dollars an acre, outbidding BP two to one for these tracts. Prices were going up. Only Atlantic Refining's bid of six dollars an acre suggested a degree of lingering naïveté. BP was still able to secure leases of another 82,000 acres largely ringing the Richfield-Humble leases to the south and west for about sixteen dollars an acre.

Shortly after this second lease sale, BP moved its Canadian rig north from its prospects near Umiat and spudded the Sinclair-BP Colville No. 1 east of the Colville River. The well showed oil-bearing sands, but they failed to deliver a gusher. Sinclair had stayed in the well as a working interest partner but now called it quits for good on the North Slope. Undeterred, BP partnered with Union Oil and drilled the Kookpuk No. 1 to the south. It, too, was dry. Thus by 1967, after eight dry holes, BP was also ready to give up on the North Slope, having been beaten down by high costs and left with nothing to show for them.[1]

Fifty miles east of the mouth of the Colville, not much had changed at Prudhoe Bay since John Franklin struggled west to Beechey Point in his attempted rendezvous with the *Blossom* in 1826. One thing had, however. On this treeless, barren plain, a tall, spindly structure now cast a dim shadow in the face of the fleeting sun of the Arctic winter. How cold was it? Well, it was so cold that . . . If one asked the question, any number of crude descriptions were apt to follow. But cold it was. And windy, too. The wind cut across the blank, white landscape like a knife slammed through a pile of potato chips—snapping, crackling, breaking, and grating against the latticework of girders rising high into the darkness. It was a far cry from Drake's crude wooden derrick at Titusville a century before. The wind muffled the whine of the big diesel motor turning the long drill string. Some 444 twenty-foot sections of pipe disappeared 8,883 feet into the frozen ground. The date was February 18, 1968. A wind-blasted sign identified the well as Prudhoe Bay State No. 1 and listed the operator as the Atlantic Richfield Company.

Atlantic Refining and Richfield Oil merged in late 1965, in part because both companies had operations on the Kenai and interest in the North Slope—even if Atlantic itself had just lost out on the new Prudhoe Bay leases. In March 1966, the new Atlantic Richfield Company (ARCO) and Humble decided to find out what they had bought. They began drilling their first joint well, the Susie Unit No. 1, along the Saganavirktok River just east of Prudhoe Bay. The logistics alone of moving 4,000 tons of equipment to the drill site were mind-boggling. Unlike BP's barge operation, this mobilization involved the largest civilian airlift in Alaskan history until that time. Beefy, four-engine Hercules Skyfreighters, the civilian version of the workhorse C-130s, flew load after load of 20 tons at a time north from Fairbanks to a short airstrip bladed across the tundra. Ten months and $4.5 million later, the Susie was declared a dry hole.

Now what? BP was calling it quits. ARCO and Humble were already out millions. Sinclair was counting itself lucky and smiling smugly in the background. Maybe it was time to walk away from this one. But walking away was not the sort of tactic that had gotten Robert O. Anderson to the chairman's seat at ARCO. As a young CEO at Atlantic Refining, Anderson had pushed the company toward the industry's second tier and secured its place there by engineering the merger with Richfield Oil. Now exploration geologists Harrison Jamison and Rollin Eckis urged him to try "just one more." Knowing that a goodly part of the battle in this Arctic war was just getting into the field, Anderson agreed and ordered the drilling rig from the Susie hole skidded to a new location two miles inland from frozen Prudhoe Bay. Later, Jamison would reminisce that if this "last" one had come in dry, it probably would have been a very long time before major oil exploration was again undertaken on Alaska's North Slope.

Prudhoe Bay State No. 1 was spudded on April 22, 1967. Its drilling engineer was Jim Keasler, a thirty-three-year-old from Missouri who had worked on rigs in Cook Inlet. Eleven days later, the rig was down to 590 feet when drilling was suspended for the summer. This was done both to avoid damaging fragile tundra during the soggy, boggy months and because operations were much more efficient on frozen ground. When winter draped its icy skirt across the plain once again, drilling resumed on November 18.

Steady progress was made, and during December two drill stem tests showed signs of natural gas. The flow of the second test at about 8,600 feet was in excess of 1 million cubic feet of natural gas a day—a good well in a developed field in the lower forty-eight but still questionable for a rank wildcat in Alaska. Then, when the driller tried to rotate the pipe and continue drilling, he found that the lower 900 feet of the string were frozen solid. There was nothing to do except cement off this section and, by inserting a wedge to direct the bit around the plug, redrill the lower section. All of this took considerable time, and it was February 18, 1968, before a third test could be run between 8,750 and 8,883 feet.

As a watery mixture surged up the borehole, one of the workers lit the discharge that was flowing into the containment pit. Whoosh! A huge fireball of white and yellow flames shot fifty feet into the air. Jim Keasler checked the pressure readings and shook his head in disbelief. They worked out to an incredible 22 million cubic feet a day of natural gas. Meanwhile, as geologist Bill Penteller was taking condensate samples from the pit, the fireball quickly changed to a deep orange color. Now, thick black smoke curled across the white landscape. Prudhoe Bay State No. 1 had drilled through a light layer of natural gas and struck oil. Black gold. A lot of it.

Not trusting radio communications with such delicate news, Jim Keasler flew into Fairbanks to phone his supervisor. The fireball from the flaring gas and burning oil was visible for 100 miles. If anyone was looking, the secret was already out. "You mean two million a day, don't you?" district drilling superintendent Lee Wilson asked skeptically. "No, twenty-two," Keasler replied. "And oil, too."

The news shot up ARCO's chain of command. With straight faces, ARCO executives casually called upon British Petroleum. "I say, would you chaps be interested in selling those worthless leases over by Prudhoe Bay?" BP's regional manager in London, Alwyne Thomas, had been around far too long to fall for that one. Thomas had bid on the Prudhoe leases during his days as BP's Alaska manager and had even tried to persuade Richfield to drill a joint well there. A few weeks later, ARCO mumbled an announcement about a well returning "a substantial flow of gas" and then revealed plans to dig another hole seven miles to the southeast, once again along the Saganavirktok River.

As the press beat a path to its corporate door, Atlantic Richfield spoon-fed them the numbers. Yes, a new test between 9,500 feet and

9,800 feet had flowed 1.3 million cubic feet of gas a day and almost 1,200 barrels of oil. No mention was made of Keasler's 22-million-cubic-feet figure or the 40 million cubic feet recorded after that. Finally, on June 25, 1968, ARCO announced that Prudhoe Bay State No. 1 had indeed produced natural gas at 40 million cubic feet a day and crude oil at up to 2,415 barrels a day. It also confirmed that Sag River State No. 1, drilled with the Canadian rig that BP had relinquished on the Colville, had also found oil in the same formation and at the same depth seven miles away. Engineers and accountants whipped out their slide rules and did the math. If that indeed was all one field, it meant a heck of a lot of oil. Fairbanks and the rest of Alaska went wild. No one had seen anything like it since the day the steamer *Portland* docked in Seattle with "a ton of gold." Only this time, the gold was black.[2]

During the summer of 1968, Atlantic Richfield and British Petroleum continued drilling wells around Prudhoe Bay in an attempt to determine the size and boundaries of the newly discovered field. Depending on who was crunching the numbers, initial reserve estimates ranged close to 10 billion barrels. At the then-prevailing market price of between three and four dollars a barrel, you were suddenly, to paraphrase the inimitable Everett Dirksen, "talking about real money."

The rest of the majors in the oil industry practically fell over one another in their attempts to get into the field. Mobil, operating for itself and Phillips, barged a rig and supplies down the Mackenzie River route. Most others, including Texaco, Standard of California, Pan American, and Colorado Oil and Gas, flew rigs in piecemeal on board an endless stream of Hercules transports and just about anything else that could fly. Seven major landing strips operated within a seventy-mile radius. Eventually, the state of Alaska took over Standard of California's strip at Deadhorse and made it the main airfield for the area.[3]

ARCO's respect for the environment, as well as certain operational practicalities in not drilling during the summer months, was quickly forgotten. Operations continued around the clock in the long days of the Arctic summer. To give substance to the soggy tundra, ton after ton after ton of gravel was dug from the Saganavirktok and other rivers until their lower reaches resembled endless gravel pits. What about the salmon runs? some asked. Won't they be affected? Can't talk about that now, was the reply. The boom was on, and seemingly nothing could stop it.

The most hideous example of a boom run amuck that first season was the infamous Hickel Highway. Walter J. Hickel was a real estate developer who had been elected governor of Alaska in 1966. At his urging the following year, the Alaska legislature created the NORTH Commission to promote the development of the North Slope. But even Wally Hickel couldn't have imagined the difference that one year's time would make. It was the overnight boom of the mining frontier all over again.

With all available aircraft filled to the gills and the water route down the Mackenzie or through the Bering Strait annually short-lived at best, Hickel ordered that a winter trail be blazed to the North Slope from the end of the existing road system at Livengood. Located some seventy-five miles northwest of Fairbanks, Livengood wasn't much more than the jumping-off place for the Brooks Range. From there, the route crossed the ice of the frozen Yukon, ran to Bettles on the Koyukuk, and then continued up the John River through the heart of the Brooks Range to Anaktuvuk Pass. Three hundred–plus miles from Livengood, it reached Sagwon on the Saganavirktok River south of Prudhoe Bay. Conservationists were outraged. For once, most everyone agreed with them.

Originally, the plan was to compact the snow into a durable surface, a true winter trail. Instead, construction crews plowed off the snow and bladed away the overburden down to the permafrost, effectively creating a trench. As any Klondiker who had spent time on White Pass could tell you, this quickly made for an endless sea of mud. Even worse, when the surrounding snow melted, the Hickel Highway filled with a torrent of water and quickly became the "Hickel Canal." Attempts to reconstruct the road the following winter resulted only in major reroutings. Now there were two ditches instead of one.

The costs of hauling equipment via this route versus flying it in vary with the telling, in no small measure depending on one's view of the "highway." What is certain is that the route was open for only the months of March and April in the spring of 1969. Long convoys of massive tractor-drawn sleds made the twelve-day journey crawling along at speeds of one or two miles an hour, their drivers taking turns sleeping and pushing onward around the clock.

Conservationists were particularly upset because this gash cut right through an area being promoted as Gates of the Arctic National Monument. Even though Hickel gave the order to blade the route across the range, it was all federal land. But evidently, the federal government wasn't too upset. While big D-9 bulldozers were cutting the swath, newly elected president Richard Nixon appointed Hickel his secretary of the interior. Secretary of State (Alaska had no lieutenant governor then) Keith Miller succeeded Hickel as governor and immediately christened the route the Hickel Highway. It was supposed to be a name given seriously in honor of Hickel's "opening up the North Slope." Subsequent events, however, quickly equated the name with a boondoggle and an example of the worst sort of environmental calamity.[4]

The embarrassment of the Hickel Highway put all on notice that when it came to getting Prudhoe Bay oil to market, there was going to be no easy or universally acceptable solution. Despite growing environmental and land ownership concerns, a pipeline running south was the early, hands-on favorite. It was not, however, the only idea seriously considered that first year. Remember the seemingly endless search for the Northwest Passage? For a time in 1969, the long-held dream of its commercial success was revived. After all, men had built "impossible" railroads across Alaska when the economic prize at the end of the line gleamed brightly enough. Now it looked to some as if the economic prize of oil at Prudhoe Bay was enough of a lure to force open the Northwest Passage at last.

The compelling challenge was somehow to transport at least 2 million barrels of oil a day from Prudhoe Bay and to start doing so as quickly as possible. Reserves in the ground were a fine thing, but they had to be pumped and delivered before they became money in the bank. Even before Prudhoe Bay State No. 1 hit, BP and ARCO-Humble engineers had pondered how they would get North Slope crude to market. Tankers loading right offshore were certainly an immediate answer, and given the ever-increasing tonnage, two ships per day might be able to handle the load.

That kind of talk was music to some ears. Because the archaic Jones Act still required shipments between U.S. ports to be in U.S.-built ships, American shipyards sensed a building boom for tankers. But there were also problems with the tanker plan. The continental shelf extends for miles and miles off the North Slope coast into the Beaufort Sea and renders the waters off Prudhoe Bay relatively shallow. Huge tankers drawing fifty or more feet of water when loaded would have to take on their cargo perhaps as far as twenty-five miles off shore. This would require an underwater pipeline and a floating loading dock that would sit at the unyielding mercy of the Arctic winds and ice. Some proposed to anchor the tankers in the deeper mouth of the Mackenzie and simply pipe the crude 300 miles eastward along the coast.

Those ideas were premised, however, on the fact that huge oil tankers could indeed navigate the Northwest Passage—a feat that it had taken Norwegian Roald Amundsen three years to accomplish in a tiny fishing yawl. Clearly, there had to be a test run before anyone got much more excited. If it proved feasible, the industry would worry about the hues and cries of environmentalists up in arms over the recent *Torrey Canyon* oil spill off the coasts of France and Great Britain.

Thus, Humble Oil, with several million dollars each from ARCO and BP, took the lead in what became the Arctic Marine Expedition. For the test run, Humble chose the largest tanker in the U.S. fleet, the *Manhattan*. Built in 1962 by Bethlehem Steel Company at a cost of $28 million, the ship was 940 feet long with a beam of 132 feet and a draft of over 50 feet. Forty-three-thousand-horsepower steam turbines turned two large propellers to push the ship at a top speed of 17.5 knots. Initially, the *Manhattan* was built as part of a complicated deal that involved Greek shipping magnate Stavros Niarchos. She quickly ended up with Seatrain Lines, however, and was carrying grain to India when Humble chartered her for the Arctic cruise. The oil company promptly ordered major modifications.

Because time was of the essence in making a test voyage during 1969, and because no one shipyard was able to accommodate the entire refitting, the *Manhattan* was literally cut into four sections, worked on in separate East Coast shipyards, and then reassembled. The vessel was strengthened overall, a second hull was built around the engine room, and a sixteen-foot-high, eight-foot-thick steel belt was wrapped around the entire hull—sort of the ultimate in bumper car design. But most impressive of all, the entire bow was replaced with a new section mimicking traditional icebreaker design that allows a ship's prow to ride up onto the ice and then crush it with the weight of the ship. On the *Manhattan*, the prow angled thirty degrees from the keel and featured eight-foot-wide shoulders, or wings, jutting out on either side. These shoulders were designed to open a wide channel at the bow and thus make passage smoother for the rest of the ship.

When her makeover was complete, the *Manhattan* had grown. She was a much bigger, much weightier ship, totaling 1,005 feet in length, 155 feet abeam, and carrying a draft of 52 feet. While her total displacement tonnage had increased from 137,000 to nearly 150,000 tons, the amount of her cargo capacity, or deadweight tonnage in seaman's terms, had actually shrunk from 115,000 long tons (2,240 pounds each) to 105,000 long tons. In terms of oil capacity, that translated into about 700,000 barrels of oil, slightly less than half the capacity of the future *Exxon Valdez*.

The *Manhattan* finally cast off from the shipyards at Chester, Pennsylvania, on August 24, 1969—mighty late in the season for a cruise through the Northwest Passage. In addition to her crew of fifty-four, just about anyone and everyone was on board: oil company representatives, Canadian and American government officials, journalists, scientists, and television cameramen, a total of seventy-two passengers. Whatever happened, it was going to be a well-publicized event. Joined by the Canadian icebreaker *John A. MacDonald,* the *Manhattan* sailed north into Baffin Bay between Greenland and Baffin Island and then turned west through Lancaster and Melville Sounds. The ghosts of Sir John Franklin's *Erebus* and *Terror* were watching.

Out in open waters, the *Manhattan* easily pulverized the relatively thin ice she encountered, but in the more sheltered waters of the sounds, the thicker polar pack ice gave it pause. At the very slow speeds required to negotiate the ice, moving the ship's massive weight became a detriment, particularly because she was able to generate only about 40 percent of her forward horsepower while going astern. The big ship was forced to back up and ram the pack ice time and again, and frequently she plowed into the ice, got stuck, and was extricated by the *MacDonald.*

Off Banks Island, where the Royal Navy's Robert McClure had convinced himself in 1850 that a watery Northwest Passage did in fact exist, the *Manhattan* struck due west into McClure Strait rather than slip southward through Prince of Wales Strait. No surface vessel had ever negotiated that passage. Three days later, the *Manhattan* found out why. She came to a grinding halt amid an endless array of polar ice and barely managed to escape. The two ships backtracked to Prince of Wales Strait and finally broke into open water. On September 17, the tanker anchored off Prudhoe Bay—all in all, a fairly fast, 4,500-mile trip. A helicopter flew a symbolic barrel of oil out to the ship, certainly the easiest part of the experiment.

Originally, there was talk of continuing around Point Barrow and passing south through the Bering Strait. The initial delay in sailing nixed this option, however, as that route was now solidly blocked by pack ice. So *Manhattan* and her escort turned east and retraced their route. On the return journey, ice punched a hole in one of the ship's storage tanks. All that escaped was seawater that the *Manhattan* was carrying for ballast, but the accident must have furrowed at least a few brows.

When the ship made port in New York on November 12, two conclusions were obvious: The *Manhattan* couldn't have made the trip without icebreaker escorts, and no voyages would be possible during the dead of winter. In 1970, the *Manhattan* returned to the Arctic for further trials, but Humble soon ended its charter and the vessel went back to carrying grain. She remains the first and last commercial ship to sail the Northwest Passage.

All told, Humble spent about $50 million on this early tanker experiment, but supertankers loading directly on the Arctic coast weren't the only alternatives to a pipeline being discussed. For a short time, Boeing promoted a giant air tanker version of its jumbo 747. Let's just fly the oil out. Others suggested a fleet of 900-foot-long nuclear submarines, literally filled with oil. Still others proposed extending the Alaska Railroad north from Fairbanks and shipping the oil south in tank cars. All of these forms had limited capacity and incurred substantial transportation costs. Finally, there was cartoonist Mark Wheeler from Ketchikan who showed a bucket brigade of Alaskans stretching from Prudhoe Bay to Valdez. Thus, a pipeline remained the favorite and seemed to make the most economic sense. But whether it would run south to Valdez or east into Canada and directly to Midwest markets was still a matter of heated debate.[5]

Meanwhile, bidding for additional state leases reached a fever pitch at a lease sale held in Anchorage on September 10, 1969. Some would mark the date as the day that Alaska got truly rich. By then, twenty-three wells had been drilled around Prudhoe Bay: seven by ARCO-Humble, nine by BP, four by Mobil-Phillips, two by Standard of California, and one by Hamilton Brothers. Only two were dry holes. In the oil business, those odds were about as good as one was apt to get. Consequently, the tracts up for lease had been crisscrossed and studied umpteen times by rival companies. Bidding competition promised to be both keen and expensive. There was even talk that the state of Alaska's revenue from the lease bonuses alone might reach $1 *billion*—then an absolutely astronomical sum.

Private jets squeezed wingtip to wingtip at Anchorage's Merrill Field, and all available hotel space was solidly booked. Good-old-boy Texans in

Stetsons and cowboy boots, dapper Englishmen in Brooks Brothers suits, New York bankers talking fast and furious, and even a local prankster masquerading as an Arab sheikh moved among the diverse crowd that packed 350-seat Sydney Lawrence Auditorium in downtown Anchorage. All rose for the Pledge of Allegiance and the singing of Alaska's state song, "Alaska's Flag." Many a bidder smiled determinedly at one line in particular: "The gold of the early sourdough's dreams, the precious gold of the hills and streams." Afterward, Alaska's commissioner of natural resources, Tom Kelly, managed to gavel the roar down to a buzz. There were many, he said, who hoped that Alaska could both preserve and responsibly develop its wilderness.

Then each representative—whether of behemoth or hopeful start-up—marched dutifully down the aisle and handed over sealed bids marked only with a tract number. Almost every oil company in the world was represented, either individually or in partnership or joint venture with another or a group. One name was noticeably absent: Sinclair Oil. Recently, it had been acquired by Atlantic Richfield. Had Sinclair stayed in the North Slope game for another round or two, it might have been the other way around.

There were 179 tracts totaling 451,000 acres up for lease, most consisting of 2,560 acres each, or four square miles. Incredibly, 131 of those tracts had been offered at one time or another in the three previous sales of North Slope leases. They were back on the block this time because no one had bid on them before. When all of the bids had been received, the process began. One by one, one tract at a time, the bids were opened and announced.

What was the day going to be like? The BP-Gulf consortium set the tone by taking the first tract with a bid of $15.5 million, or a whopping $6,055 per acre, 1,000 times what BP had first paid for North Slope leases in the fall of 1964. BP-Gulf captured the next five tracts, as well, ending up with six tracts near the mouth of the Colville River for a staggering $97.6 million. In its twenty-two previous state lease sales, including the previous three on the North Slope, the state of Alaska had netted $97.7 million. It was going to be a *good* day.

Tract 57 had everyone on the edge of their seats. It was the closest to the discovery wells. ARCO-Humble bid $26 million in hopes of getting control of this adjacent property. These hopes were quickly dashed when BP-Gulf's bid of $47.2 million was read. There was a collective, audible gasp throughout the auditorium. The Continental-Sun–Cities Service group bid $36.6 million, and Hamilton Brothers $36.8 million. It looked as if BP-Gulf was in the driver's seat. Then a consortium of Mobil, Phillips, and Standard of California bid $72.1 million. The cheers almost blew the roof off the place. They quickly turned to a moment of stunned silence, however, and then pandemonium. The final bid on the tract was announced from Amerada-Hess–Getty: $72.3 million, upwards of $28,000 per acre. By $200,000—less than three-tenths of a percent—Amerada-Hess–Getty controlled a tract on which one estimate put recoverable reserves at 200 million barrels.

By 5:15 P.M. it was over. Bids on 15 outlying tracts were rejected as too low, but the remaining 164 tracts yielded $900,040,000, an average of more than $2,100 per acre. A private jet chartered by the state rushed the cashier's checks directly to the Bank of America in San Francisco so as not to lose a day's interest. Amerada-Hess–Getty was the day's high bidder, having spent $272 million. As part of the day's opening ceremonies, Governor Keith Miller had noted that Alaska would never be the same again. That was true, but it was suddenly $900 million richer. The ghost of William H. Seward must have gotten a good chuckle out of that one.[6]

The Blue Canoes

Alaska's major transportation corridors have frequently been watery ones. Nowhere was this more true than in southeast Alaska. It wasn't just that there weren't any roads or trails. Even if there had been, where would they go? With its towns and villages scattered on a myriad of islands and the Coast Mountains posing an impassable barrier to the east, if you wanted to reach it, leave it, or move about the region, you took to the sea. The only well-established avenues of land access to the southeast were via the historic Tlingit trail over Chilkoot Pass or, later, the White Pass and Yukon Railway across White Pass. A third route, the Haines Cutoff link to the Alaska Highway, was not forged until military necessity required it in 1942. Thus, the southeast has a legacy of Tlingit canoes, military vessels like the venerable tug *Pinta*, and an array of coastal steamers plying its waters.

Once the Alaska Highway was opened to civilian travel shortly after World War II, Haines became a key access point to the southeast. Haines resident Steve Homer soon envisioned a regularly scheduled ferry service between Haines and Juneau. In 1949, Homer went into partnership with Robert Sommers and Associates and started Chilkoot Motorship Lines. The company's one and only vessel was a war-surplus LCT (Landing Craft Tank), similar to those that had unloaded tanks, trucks, and heavy artillery pieces at Normandy and countless amphibious landings throughout the Pacific. Christened the M/V *Chilkoot* (for Motor Vessel), the stocky, 114-foot-long, 32-foot-wide craft initially made one trip weekly between Haines and Tee Harbor, about fifteen miles north of Juneau. Operating only between May and November, the vessel could carry up to thirteen autos, twenty passengers, and a crew of seven at a top speed of a blistering nine knots.

Most folks thought that it was a great idea, but operating costs on the limited run ate up all of Steve Homer's profits and then some. He and his partners turned the service over to the Territorial Board of Road Commissioners in 1951. The converted *Chilkoot* continued to operate until 1957, when the road board replaced it with the M/V *Chilkat*. Doubtless the similarity in names caused some confusion, but there was no mistaking the new vessel. Although roughly the same dimensions as its predecessor, the *Chilkat* had an upper deck behind the wheelhouse and could accommodate thirty additional passengers. The one major similarity with the original LCT was a huge bow ramp for either beach or dock loading. Top speed? Not much improvement there—ten knots.

Throughout the 1950s, the territory of Alaska conducted several transportation studies. In their first election after statehood, Alaska voters in 1960 approved two transportation bond issues, a $3-million expenditure for dock facilities and a seagoing ferry for a Kodiak Island–Kenai Peninsula run, and $15 million for three new vessels and seven dock facilities spread throughout the southeast. These locations were the region's principal towns: Skagway, Haines, Auke Bay serving Juneau, Wrangell, Petersburg, Sitka, and Ketchikan. A terminal operation in Prince Rupert, British Columbia, was also authorized. A million dollars, it seems, went a lot further in those days!

The three new vessels in the southeast went into service in the spring of 1963, and their arrival inaugurated the new Alaska Marine Highway. This was no small achievement. From the highway connections at Haines or Prince Rupert, for the first time there was regularly scheduled, year-round vehicle access to the principal towns along the Inside Passage. Alaska's Department of Transportation designated the system a marine *highway*, because people could take their cars, trucks, or RVs between the southeast's towns almost as easily as they could have driven with a road system.

The marine highway's new vessels were far cries from the squat and sputtering *Chilkoot*. The multidecked *Malaspina* was 352 feet long and 73 feet wide with a draft of 15 feet. The ship could carry 109 vehicles and 500 passengers, a fifth of which could be accommodated in berths. The *Malaspina*'s cruising speed was 16.5 knots. Her sister ships, *Taku* and *Matanuska*, went into service a few months later. All three ships were painted in Alaska's state colors, Arctic blue with yellow gold trim, and locals quickly came to call them the "blue canoes."

A local newspaper summed up the transformation taking place: "Alaska ferry fever broke out and began when the *Malaspina* broke trail the first week of February 1963. The ferry operation created a severe hotel room shortage in Alaska cities and towns. The population as never before was on the move; people and their automobiles free at last from isolation."[7]

But this newfound mobility was not restricted to the southeast. Once the new ships arrived in Lynn Canal, the workhorse *Chilkat* began service between Valdez and Cordova through Prince William Sound, giving Cordova its first highway access via the Richardson Highway out of Valdez. And in 1964, the 240-foot-long *Tustemena* went into service between Kodiak and the Kenai Peninsula. *Tustemena* was shorter than her sisters and had to operate in the rougher seas of the Gulf of Alaska. Her shorter length gave the vessel a decided roll that earned her the nickname "Dramamine Express."

In 1964, the first full year of the marine highway's operation, its five ships carried 83,000 passengers and 16,000 vehicles. Not everyone, however, thought it was a good investment. Despite the big increase in traffic, much of it summer tourists, Norwegian-born Ed Locken, a banker in Petersburg, was not in favor of the public indebtedness required for the bond issues. "The ferry doesn't bring us any business," Locken told a reporter. "The tourist brings his family and his camping equipment with him. He comes in here with a $10 bill and the Ten Commandments and doesn't break either."[8] Later, the same reporter caught up with Senator Ernest Gruening, who, despite the new ferry system, was touring the upper reaches of the Taku River, busily scouting a route for a road east from Juneau to the Canadian highway system—just as he had first proposed in 1939. Some ideas never die, although it has still not come to pass.

Of course, navigating the Inside Passage was not without risk. On January 26, 1964, while en route to Sitka, the *Malaspina* struck submerged rocks in Peril Strait. The ship was taken to Seattle for extensive repairs to her keel plates. The *Matanuska* rammed a troller in Whitestone Narrows near Stika in May of that year and later in the fall tore up her propellers on submerged rocks in Wrangell Narrows. *Taku* got her baptism within days of her April 1963 delivery when she caught a rock while trying to dock in Petersburg on a minus-3.4-foot tide. An eight-by-seventy-foot section of the ship's keel was gouged. Later that summer, while the ship was idling at the Petersburg dock, two young boys sneaked into the *Taku*'s wheelhouse, shoved the engines full ahead, and sent the bow crashing into the support towers that held the counterbalance weights for the loading ramp. The 180-foot-long loading ramp ended up in the water, and it took three months to repair the dock.

By far *Taku*'s worst accident, however, occurred in 1970 when the ship missed the turn into the harbor at Prince Rupert and ran aground while making seventeen knots. The British Columbia ferry *Queen of Prince Rupert* came to the rescue and evacuated passengers and vehicles before the ship began to list steeply at low tide. Hours later, the incoming tide righted the stranded *Taku*, and the ship was pulled off the beach and sent south to Seattle under her own power for repairs.

In 1967, the Alaska Marine Highway extended its system to Seattle. With that additional route, and the ensuing oil rush to Prudhoe Bay, the ferries soon became quite crowded. In turn, the *Tustemena* and then the *Malaspina* were pulled out of service, cut in two, and lengthened with the addition of fifty-six-foot sections added to the middle of each ship. New staterooms and a solarium were added to the *Malaspina*, and the *Tustemena* was outfitted with new stabilizer fins that when coupled with its increased length reduced its roll considerably. The "Dramamine Express" was no more, and by 1979 the ship's service area was to include the Aleutians, a modern-day revival of the historic Aleutian mail route by steamers. By then, the *Matanuska* had also undergone an overhaul to increase its length by fifty-six feet.

In 1970, flush with newfound oil revenues, Alaska voters passed another marine highway bond issue. This one for $21 million was to extend service and build the *Columbia*, destined to reign as the flagship of the blue canoes. At 419 feet in length, the *Columbia* was capable of carrying 158 cars and 1,000 passengers, 318 of them in berths, and making seventeen knots. But even this queen was not immune to an Inside Passage baptism. Put into service during the summer of 1974, the *Columbia* struck a rock in Peril Strait—indeed an aptly named passage—and tore a 100-foot gash in her starboard hull. The next year, the ship made it into John McPhee's *Coming into the Country* by pulling away from the dock in Ketchikan—while still tied up. The boarding ramp was torn loose, but there was no damage to the ship, save for a few red faces in the pilothouse.

As service increased during the 1970s, three smaller vessels were put into service to expand the routes. The *E. L. Bartlett*, named for one of Alaska's first U.S. senators, replaced the veteran *Chilkat* in Prince William Sound in 1969. The 235-foot *Le Conte* was actually built in Sturgeon Bay, Wisconsin, and arrived in Alaska only after a trip down the Great Lakes and St. Lawrence River, through the Panama Canal, and up the West Coast. Finally, the *Aurora*, also built in Sturgeon Bay, was put into service in 1977 to serve the area around Ketchikan, including trips to Hyder, Hollis, and Metlakatla.

In 1989, the Alaska Marine Highway moved its southern terminal north from Seattle to Bellingham, just south of the Canadian border. The move eliminated cruising through the congested waters of Puget Sound and docking in downtown Seattle and cut as much as ten hours off the run to Ketchikan. From Bellingham, it was a straight shot out into the Strait of Georgia past Vancouver and north. The *Matanuska* was the first vessel to sail from the new Bellingham terminal, and appropriately the ship carried the body of ferry system pioneer Steve Homer, who had recently died in a Bellingham nursing home, northward to Haines for burial. Some things do come full circle. The next year, the Alaska Marine Highway carried 400,000 passengers and 110,000 vehicles.

And what of the fate of the stout and reliable *Chilkat*? After being replaced in Prince William Sound by the *Bartlett*, like an old steam engine it was put on the milk runs, first between Juneau and Hoonah and then around Ketchikan. It was finally retired in 1988.

Without a doubt, the Alaska Highway and Alaska's expanding internal road system played key roles in the state's rapid growth in tourism during the 1960s and 1970s. But in no small measure, it was the Alaska Marine Highway that annually provided hundreds of thousands of visitors—whether with RVs or solitary backpacks—access not only to southeast Alaska but also to the state's road system via the Haines Cutoff and later the highway over White Pass. Camping out on the solarium decks of the larger vessels, basking in the twilight glow of the perpetual summer sun, and exploring the various ports of call became a rite of passage for first-time visitors and a strong message of "welcome home" for returning ones. Certainly, the popularity of Alaska's ferries as an adventure as well as a mode of travel contributed to the explosion in cruise ship visits to Alaskan ports a few years down the road.

Many visitors got their first glimpse of Alaska from the decks of the *Columbia*, *Matanuska*, *Malaspina*, or others. U.S. Forest Service interpreters gave history and cultural talks, local bards swapped poems and stories, and Alaska Natives shared the secrets of their handcrafts. The ships became small parts of Alaska before one even set foot in Ketchikan. And while Anchorage, Fairbanks, and other interior and southcentral towns boomed with the pulse of oil, the tourists the Alaska Marine Highway carried gave a much-needed shot in the arm to the economy of southeast Alaska. They poured millions upon millions of dollars into the state's economy and, despite that Petersburg banker's assertion to the contrary, almost always managed to break a ten.

The Long, Long Road to ANCSA

In 1909, Major General Adolphus W. Greely, a veteran Arctic explorer who spent time in Alaska, wrote a book titled *Handbook of Alaska*. Greely boasted of the region's grand scenery and its wealth of natural resources. His enthusiasm for the country was diminished only by the condition of the Alaska Natives, a situation that Greely viewed as "disgraceful to a nation claiming to be civilized, humanitarian, or Christian." Greely went on to ask several pointed questions that would reverberate across Alaska for decades:

> What, if anything does the General Government owe the natives of Alaska, and in what form shall the payment be made? It is a problem great in its moral as well as in its practical aspects. Having largely destroyed their food supplies, altered their environment, and changed their standards and methods of life, what does a nation that has drawn products valued at $300,000,000 owe to the natives of Alaska? Will this nation pay its debt on this account?[9]

Fifty years and more hundreds of millions of dollars later, Greely's questions had still not been answered.

Historically, the three charters of the Russian American Company had decreed in various ways that the company's chief and overriding goal in Alaska was the exploitation of natural resources. It was not concerned with the extension of the czar's dominion over Alaska Natives, except as might be necessary to protect Russian activities along the coasts. That exception had been bad news for the Aleuts and to a lesser degree the Koniag and Tlingit, but many Alaska Natives had no or only minimum contact with the Russians.

The 1867 Treaty of Cession between Russia and the United States devoted barely a paragraph to the future legal status of Alaska's inhabitants. It gave them the option of returning to Russia within three years or, if they remained in the ceded territory, being "admitted to the enjoyment of all rights, advantages and immunities of citizens of the United States," including "the free enjoyment of their liberty, property and religion." There was, of course, one major exception. The provision did not apply to "uncivilized native tribes." While exact census figures do not exist from 1867, by the definition of the time perhaps as many as 98 percent of Alaska's inhabitants fell into this category. These individuals, the treaty proclaimed, "will be subject to such laws and regulations as the United States may, from time to time, adopt in regard to aboriginal tribes of that country." Significantly, the treaty did not seek to extinguish Native titles outright.[10] A century later, there were still those who wished that it had.

When Congress finally took the first step toward furnishing some measure of civilian government for Alaska, the Organic Act of 1884 provided that Alaska Natives were not to be disturbed in their use or occupation of lands claimed by them. The terms under which they might acquire title to such lands, however, were left to future legislation by Congress.[11] In essence, some measure of Native titles had been recognized but no mechanism put in place to perfect them.

One well-meaning, though somewhat misguided, attempt to do so was the Native Allotment Act of 1906. Each Native head of household was allowed to select a 160-acre homestead on nonmineral-bearing ground. The key problem with the act was that 160-acre homesteads might be fine for farmers, but they were hardly conducive to a traditional hunter-gatherer culture. Why should anyone be confined to 160 acres when for generations his ancestors had been using all of the land? Besides, even if Alaska Natives had been inclined toward farming, the homestead process was not very successful in Alaska even among its white population.

The Native Allotment Program was poorly advertised, and applications were submitted for only eighty allotments in the first fifty years after the act. While the Native Allotment Act made no reference to general Native land rights and certainly did not extinguish them, the homestead process seemed to many to be a clear legislative step away from the earlier recognition of historic titles—attempting to mute larger tribal territorial claims by doling out individual homesteads.

The first significant meeting to discuss Native land issues was held in Fairbanks in July 1915 at the invitation of Alaska's territorial delegate, James Wickersham. Fourteen Athabascan representatives, including six Tanana chiefs, asked Wickersham how they could protect their lands throughout the interior from increasing encroachment by whites. One of two ways, Wickersham told the gathering: File for homesteads under the 1906 act or ask the government to create reservations. Neither course was palatable to the assembled chiefs.[12]

That same year, the newly created Alaska territorial legislature took the then extraordinary step of passing a law that established a process whereby Natives could become U.S. citizens. Whether or not the territory actually had the legal right to grant U.S. citizenship is highly questionable, but the passage of the law showed the newfound political power of the Alaska Native Brotherhood. Formed in Sitka in 1912 by nine Tlingit and one Tsimshian, this group was the first organization of Alaska Natives to mobilize and attempt to affect change within the American legal system.

The Alaska Native Brotherhood quickly recognized that the first step in any land claims process was citizenship. Without citizenship, Alaska Natives had no legal standing, and thus any other guarantees that they might receive were unenforceable. Only a few Natives became citizens under the territorial act, and it was never challenged constitutionally, in large measure because Congress soon erased all citizenship questions by passing the Citizenship Act of 1924. This granted citizenship to all "Indians" born within the territorial limits of the United States and applied to Alaska Natives.

But citizenship was only the first step. Also in 1924, William L. Paul, a Tlingit member of the Alaska Native Brotherhood, became the first Alaska Native elected to the territorial legislature. One of the founders of the Alaska Native Brotherhood soon said to young Paul, "William, the land is yours. Why don't you fight for it?"[13]

William Paul answered that question by suggesting that the Alaska Native Brotherhood seek compensation for lands taken from the Tlingit and Haida since the 1867 Treaty of Cession. The federal government's reservation of Tongass National Forest, Glacier Bay National Monument, and the Tsimshian reservation on Annette Island had in fact taken almost all of southeast Alaska. Under pressure from the ANB and perhaps hearing echoes of its promise in the Organic Act of 1884 to someday resolve such matters, Congress passed a 1935 law allowing the Tlingit and Haida to sue the United States and seek payment for these lands. With the right to bring suit established, William Paul—by now an attorney in Seattle—began a legal battle of more than three decades representing Tlingit-Haida interests.

Although the case was not won until 1959, and the value of the compensation not determined until 1968, these Tlingit-Haida claims set a major precedent in the efforts to establish historic Native rights.

While the Tlingit-Haida lawsuit was slowly working its way through the courts, Congress proposed another solution by extending the Indian Reorganization Act of 1934 to Alaska. This act provided for the creation of reservations under the same dubious reservation system that existed in the rest of the country. Like so many proposed solutions to the Native land claims issue, the reservation concept was controversial. Non-Natives feared that large areas of land would be locked up and made inaccessible. Natives, wary of the legacy of the reservation system in the lower forty-eight, feared confinement in small areas without adequate resources, particularly for hunting and fishing.

Nonetheless, between 1941 and 1946 seven reservations were created throughout Alaska by mutual consent—Native petition and government approval. The largest included the villages of Venetie and Arctic Village in the upper Yukon drainage and totaled some 1.4 million acres. The smallest was the village of Unalakleet on Norton Sound with 870 acres. Three villages voted to reject reservation designations. By 1950, however, eighty other villages had submitted petitions to the secretary of the interior requesting reservations exceeding 100 million acres, more than a quarter of the territory. Clearly, the reservation process wasn't going to work.[14]

As the drive for statehood gathered steam during the 1950s, some Alaska Natives feared that Congress might simply use the occasion to extinguish their claims once and for all. After all, despite apparent willingness to listen to the matter over the years, Congress had promised only a *resolution* in the Organic Act of 1884, not a Native victory. Statehood critics held up the land claims uncertainty as just another reason why Alaska wasn't ready for statehood.

In the end, the Alaska Statehood Act renewed the promise of the Organic Act. The new state disclaimed all right and title to any lands held by Indians, Eskimos, or Aleuts, or held in trust for them by the United States. But once again, the act did not define what "right or title" the Natives might have.[15] Other provisions of the statehood act authorized the new state to select 103 million acres of land from the remaining public domain. When the state of Alaska began to select these lands, the fat was finally in the fire.

The first major conflict between Alaska Natives and the regulatory powers of the new state of Alaska and the federal government began innocently enough. One day in 1960, state representative John Nusungingya, an Inupiat Eskimo of Barrow, went duck hunting for food just as Eskimos had done for centuries. He was promptly arrested by federal game wardens for shooting ducks out of a hunting season established by an international migratory bird treaty. Two days later, 138 other Inupiat went out, shot ducks in protest, and then presented themselves to the game wardens for arrest. While charges against all parties were eventually dropped, the incident characterized the difference—and frequent conflict—between the traditional "subsistence" way of life of most Alaska Natives and the regulatory nature of the "white" system.[16] This difference had always existed, but it would be central to the resolution of Native land claims during the 1960s and continue to influence governmental land management decisions to the present day.

Subsistence is a white word that hardly begins to describe a complex lifestyle. In fact to some, *subsistence* even suggests poverty or bare survival. That couldn't be further from the truth. To Alaska Natives, subsistence is much, much more than putting food on the table. It is an interrelated web of people, land, water, wildlife, and the spirit, where that web is the only source of sustenance and materials. Living in harmony with the land is perhaps a much more meaningful description of this lifestyle than subsistence. Until that harmony was disturbed by outside forces, the experience of many Alaska Natives was a very rich and fulfilling way of life. No wonder that it was difficult to put geographic boundaries around such a lifestyle, or that it was confrontational to disrupt a part of the web by suddenly saying, "Sorry, no duck hunting."[17]

The Barrow duck incident, along with growing opposition to infamous Project Chariot, galvanized Inupiat Eskimos to take the first steps toward organizing a regional association. A conference held in Barrow in November 1961 to discuss these common concerns resulted in the formation of Inupiat Paitot (the People's Heritage), the first Alaska Native association formed since the Tlingit established the Alaska Native Brotherhood fifty years before.

Another recommendation to come out of the Barrow conference called for the publication of a newsletter to disseminate information about common concerns and legal processes. It was largely Howard Rock's idea, and delegates thought it such a good one that they persuaded him to become the paper's editor, despite his pleas that he lacked journalism experience. Rock in turn twisted the arm of Fairbanks reporter Tom Snapp to help him and cranked out the first issue of the *Tundra Times* on October 1, 1962. It is difficult to overstate the paper's importance.

For the next turbulent decade, the *Tundra Times* became the definitive source of information on Native issues. It reported the policies, goals, and individual activities of the other regional associations that followed Inupiat Paitot's lead and soon organized, and it gave them a common voice. The paper called attention not only to land issues but also to the whole array of concerns facing Alaska Natives: inadequate education, poor health care, substandard housing, high unemployment, and widespread discrimination.

In short, the *Tundra Times* became the rallying point that discussed shared concerns while urging a higher visibility and cultural pride. Journalism experience or not, Howard Rock put his pen squarely on the central issue in his first editorial. "Since civilization has swept into their lives in tide-like earnestness," Rock wrote, "it has left the Eskimos, Indians, and Aleuts in a bewildering state of indecision and insecurity between the seeming need for assimilation and, especially in the Eskimo areas, the desire to retain some of their cultural and traditional way of life." [18]

Meanwhile, there was increasing friction over the state's land selection process, most notably in the Minto Lakes region northwest of Fairbanks. The area was a favorite with sportsmen because of its abundant moose and waterfowl. It was also the traditional hunting and fishing grounds of about 150 Athabascans of the village of Minto. The state wanted to select the area as part of its statehood allotment, build a road into the lakes, and develop a recreation area. No one bothered to ask the Minto people what they thought of the idea, even though they had made several earlier attempts to file land claims in the area with the Bureau of Land Management.

In February 1963, Richard Frank, the thirty-six-year-old chief of Minto, appeared before a meeting in Fairbanks to question how the state could be so presumptuous as to proceed with such plans without even informing those who lived there. In what became one of the oft-quoted speeches of the Native claims movement, Frank argued eloquently that the state development would ruin his villagers' subsistence way of life and effectively destroy it. "A village is at stake," he said quietly. "Ask yourself this question. Is a recreation area worth the future of a village?" [19]

Howard Rock seized on the Minto story, and the *Tundra Times* spread Richard Frank's words throughout the state. Rock urged villages to protest state land selections that infringed on their historic hunting grounds. Some villages, including Stevens Village and others impacted by the proposed Rampart Dam, filed similar protests to state land selections. Early in 1963, about 1,000 Alaska Natives from twenty-four villages in the Yukon Delta, Bristol Bay, and Aleutians area petitioned Secretary of the Interior Stewart Udall to halt all transfers of state selected lands in areas surrounding their villages until all Native land claims could be resolved. This first request for a "land freeze" was not granted, but Udall did appoint the three-person Alaska Task Force on Native Affairs to study the issue.

Alaska's congressional delegation was largely unsympathetic to these land claims by Alaska Natives, in part because its members had yet to feel the political heat that Howard Rock was beginning to generate through the *Tundra Times*. In the spring of 1966, Senator Ernest Gruening characterized the claims as based on "dubious grounds of aboriginal rights" and urged the federal government to settle them quickly with cash payments. Gruening saw the entire dispute as an impediment to the fruits of statehood that he had labored so long to plant, and he strenuously recommended that the Department of the Interior not only refuse to accept any additional protests against state land selections but also dismiss all pending ones.

Gruening's position drew the ire of an Inupiat Eskimo student at the University of Alaska who was to have much to say about both the substance and the procedure of the Native claims process over the next few years. Willie Hensley wrote to Gruening and sent a copy of his letter to the *Tundra Times*. Reacting first and foremost to the exclusion of Natives from much of the process itself, Hensley told Gruening, "You should have at least made an effort to discover what the Natives desire behind these land claims before creating a prejudiced attitude toward them by your recent statements." It might be a simple and expedient solution for Congress to pay compensation in cash to buy off Native claims, Hensley continued, "but it seems that we should be given the opportunity to voice our opinions on the matter."[20]

Gruening slowly came to moderate his position, first suggesting legislation giving the Court of Claims jurisdiction similar to what it had exercised in the Tlingit-Haida claims. If that didn't work, he promised to introduce legislation settling each dispute on a case-by-case basis. Alaska's lone congressman, Ralph Rivers, thought that either of those solutions would likely take far too long. He favored extinguishing all rights and awarding cash compensation but no land. "What would they do with it?" was his soon infamous question. "They wouldn't use it. It would just lie there." Alaska's senior senator, E. L. Bartlett, figured that granting 1 million acres around the Native villages, plus a cash payment for other lands, would be sufficient to settle the matter.[21]

Meanwhile, Secretary Udall's Alaska Task Force had reported in and noted the obvious failure of the Organic Act of 1884 and its progeny to provide a means by which Natives might obtain title to their historic lands. While the commission noted that it was long past the time when this issue should have been resolved, its recommendations nonetheless failed to recognize the needs of subsistence lifestyles. Its recommendations of prompt grants of 160 acres to individuals, withdrawals of small areas for village expansion, and designation of use areas for traditional hunter-gatherer activities without ownership were not nearly enough in the view of most Natives.

The Alaska Task Force's initial recommendations were opposed by a growing number of new regional associations. The Alaska Native Brotherhood and the Inupiat Paitot were soon joined by the Association of Village Council Presidents organized in southwest Alaska in 1962, the Gwitchya Gwitchin Ginkhye (Yukon Flats People Speak) formed in 1964, and the Cook Inlet Native Association, made up mostly of Natives living in Anchorage. The Tanana Chiefs Conference, which had first come together to meet with Judge Wickersham in 1915, also began to meet regularly. Then a new association was formed in Barrow in 1966. Called the Arctic Slope Native Association, it quickly made claim to some 58 million acres north of the Brooks Range, essentially the entire North Slope, on the basis of aboriginal use and occupancy.

Alternately nurtured and prodded by the *Tundra Times*, these regional associations gained increasing influence, but there was still no one statewide voice for Native issues. As early as 1963, the Alaska Native Brotherhood had proposed some sort of statewide organization, but regional associations just beginning to flex their own political muscle were reluctant to relinquish any of it to a larger group in which their individual voices might be lost.

Indeed, the many different divisions of Alaska Natives were traditionally suspicious of one another, and there was frequently as much friction among them as there was between Alaska Natives collectively and whites. This attitude was overcome in large part due to Howard Rock's continuing admonitions in the *Tundra Times*. All Natives could work together on their common issues without losing their regional identities, Rock urged, and in fact should do so because all Alaska Natives together—roughly one-sixth of the state's population—would form a powerful political force.

Emil Notti, a thirty-four-year-old Athabascan from Ruby and president of the Cook Inlet Native Association, took the initiative to call a meeting of regional associations in Anchorage on October 18, 1966, by coincidence the ninety-ninth anniversary of the transfer ceremony marking Russia's sale of Alaska to the United States. Incredibly, about 250 people representing seventeen Native organizations attended the conference. Out of the three-day event came a board of directors, standing committees, and a growing sense of political power for a statewide organization that would soon call itself the Alaska Federation of Natives. Amazed at the sudden attention the group was receiving from veteran political leaders, one attendee wryly observed that if any delegate paid for his own meal, it was because he chose to dine alone.

Early the next spring, Emil Notti was elected president of the Alaska Federation of Natives, and the organization quickly took a lesson from the political activities of the Arctic Slope Native Association (ASNA). In the state's 1966 primary, ASNA became the first Native association actively to support political candidates for statewide offices. Mike Gravel, a real estate developer from Anchorage, sought ASNA's support and almost defeated Congressman Ralph Rivers in the Democratic primary. The Northwest Alaska Native Association, founded in part by Willie Hensley, soon joined ASNA in endorsing Congressman Rivers's Republican opponent, Howard Pollock, in the general election. Pollock beat Rivers, and the Natives smiled smugly. Perhaps now the ex-congressman knew what they would do with their land. By no small coincidence, Willie Hensley was elected to the state house of representatives.[22]

Before the end of 1966, Secretary Stewart Udall acted on one of the AFN's recommendations and imposed the "land freeze," effectively stopping all transfers of federal lands. Alaska's newly elected governor, Anchorage developer Walter J. Hickel, complained that such action denied the state its entitlements under the Statehood Act, but Udall stood firm and the freeze remained in effect. By May 1967, thirty-nine protests were on Udall's desk, ranging in size from a one-square-mile claim by the village of

Chilkoot to the Arctic Slope Native Association's claim of 58 million acres. Many of the claims were overlapping, and when their acreage was totaled—depending on who was doing the counting and drawing the lines—the area under protest was greater than the entire state.

Governor Hickel was far from an ardent proponent of Native land claims, but he was shrewd enough to recognize that both the state of Alaska and Alaska Natives wanted the same thing: land from the federal government. As long as the arguments lasted, neither was going to get any. The matter had to be resolved before the state could get on with oil and gas leasing; otherwise, Alaska's fledgling oil boom was headed down the tubes. With Hickel's blessing, state attorney general Edgar Boyko met in 1967 with representatives of the Alaska Federation of Natives to suggest a joint approach. This meeting resulted in the creation of the governor's Land Claims Task Force, whose thirty-seven appointees included Alaska Natives, state officials, and Department of the Interior representatives. Hickel appointed state representative Willie Hensley its chairman.[23]

When Hensley's task force reported its recommendations in January 1968, they fell into three key areas: first, that 40 million acres be awarded to Alaska Natives in fee simple (absolute ownership) and not in trust as under the reservation system; second, that cash payments from 10 percent of the income produced from the sale or lease of certain oil and gas rights up to $65 million be awarded; and finally, that the land and cash settlements be allocated to and paid through business corporations to be organized by region. The dicey question of whether the Native land selections would be made before, after, or coincidentally with state selections was not addressed. While there would be numerous versions and much cantankerous debate before a final bill was passed, the basic components of land and dollars distributed to collective corporations rather than to individual Natives would survive the process.

The next month, the Senate Committee on Interior and Insular Affairs held hearings in Anchorage on legislation to enact the recommendations of the Land Claims Task Force. A long list of Alaska Natives from around the state testified about their historic use of the lands surrounding their villages. Their testimony was proud and determined, but there was equally determined opposition on the other side of the table, particularly from miners.

"Gentlemen," George Moerlein, representing the Alaska Miners Association, told the senators, "I submit to you that neither the U.S., the state of Alaska, nor any of us here gathered as individuals owes the Natives one acre of ground or one cent of the taxpayers' money."[24]

Moderates in the process sometimes found that there was hell to pay. Phil Holdsworth, a former state commissioner of natural resources who had recently given up his state job to become a lobbyist for the Alaska Miners Association, questioned whether Moerlein was speaking for miners as a whole. "I don't believe the group will agree with the position that there is no moral or legal right in this matter," noted Holdsworth. "Legal responsibility is always subject to court action but the moral responsibility is certainly there."[25] Maybe. The Alaska Miners Association fired Holdsworth a few days later.

As 1968 dragged on, Congress was preoccupied with the war in Vietnam and civil unrest at home, but the news of oil discoveries at Prudhoe Bay kept pushing Native land claims to the front burner. Alaska Natives showed that they were determined to keep them there by flexing their political muscle in another round of elections. Although Ernst Gruening's position on land claims had moderated, it was not enough to keep Natives from supporting Mike Gravel in the Democratic U.S. Senate primary. Gravel beat Gruening and again with Native support went on to beat Republican Elmer Rasmuson in the general election. Ironically, when Alaska's other U.S. senator, E. L. Bartlett, died shortly after the election, Governor Hickel appointed Ted Stevens, who had lost the Republican primary to Rasmuson, to take Bartlett's place. No one ever said that politics in Alaska was dull.

The November 1968 election also signaled a changing of the guard in the Department of the Interior. Much to the chagrin of Alaska Natives, Stewart Udall was leaving, to be replaced by Richard Nixon's appointee. "Frankly," said Udall just before he left office, "I do not believe we would have made any significant progress on the Native claims issue if we had not held everyone's feet to the fire (or perhaps I should say the ice) with the [land] freeze."[26] Despite pending judicial challenges to his actions, Udall then imposed a "super freeze" that halted all transactions on Alaska's remaining federal lands for a period of two years, time enough, it was hoped, to settle the claims once and for all.

Udall's would-be successor as secretary of the interior was none other than Alaska governor Walter J. Hickel. Even before his nomination was confirmed, Hickel boasted that he could undo anything that Udall had done during his final weeks. That proved a mistake. As his nomination came under fire from environmental groups, Hickel desperately looked around for support from any and all corners, including that of the Alaska Federation of Natives, which not surprisingly had withheld its endorsement. The end result of several weeks of intense negotiations was that Hickel promised to keep the land freeze in effect for two years or until the Native claims were settled, whichever occurred first, in exchange for the AFN's endorsement of his nomination.

The chairman of the Senate Committee on Interior and Insular Affairs, Henry Jackson of Washington, extracted a similar pledge from Hickel as a condition of his confirmation. Hickel promised to consult both Jackson's committee and the comparable committee in the House of Representatives before issuing any modifications of the freeze even to accommodate public rights-of-way for roads, airports, or—significantly—pipelines. In other words, the land freeze was on, and the heat was also on to reach a settlement prior to any other land actions, including pipeline construction. Ironically, even as Hickel testified at his confirmation hearings, work was under way on the Hickel Highway.[27]

In the spring of 1969, a bill wound its way out of Jackson's committee that would have given Alaska Natives $1 billion and about 10 million acres of land. It also proposed one regional corporation on the Arctic Slope, and two statewide corporations to oversee social programs and investments. When it passed the Senate in July by a vote of 76 to 8, Senator Ted Stevens urged Alaska Natives to take it and run—all the way over to the House. But they didn't. The strongest objections were over the amount of land. Ten million acres, they said, wasn't enough.

Soon, it seemed as if everyone had his or her own version of a settlement bill. Alaska state senator Nick Begich got elected to Congress in 1970 in part by saying that he thought 10 million acres was not enough. Colorado's crotchety congressman Wayne Aspinall, chairman of the House Interior Committee, introduced a bill with only 100,000 acres of fee lands and a permit system for subsistence hunting. Former governor William Egan came out of retirement and won the governor's chair back from Keith Miller with a vow to work more closely with the Alaska Federation of Natives on land claims issues. Meanwhile, the U.S. Supreme Court refused to review a lower-court decision upholding the land freeze. So the freeze stood, the oil companies fretted, and the clock ticked on.

Certainly, the AFN and other Native groups had gained significant political clout in Alaska during the 1960s. And, thanks in part to Walter Hickel's high-profile confirmation hearings, the cause of Native land claims had considerable support at the national level. But it was the even greater political clout of the oil companies champing at the bit to transport oil out of Prudhoe Bay that finally cut through the legislative morass and pushed two similar bills out of the House and Senate. At their core was a settlement of 40 million acres of land in fee simple and almost $1 billion.[28]

Many in Alaska—even those who had supported some measure of a land claims settlement—were aghast. "Over the years I have watched the Native land claims grow and grow," wrote one resident of the Kenai to Alaska's congressional delegation. "At first I was very much in favor . . . , but now, like many others, I am beginning to wonder. They say they need the money to catch up with the 20th Century and, at the same time, the land so they can go back to living like they did 200 years ago." In response to a similar letter, Senator Ted Stevens acknowledged the growing division in Alaska but urged the bill's passage, adding, "It's going to destroy us if we don't do it."[29]

So now it would finally be done. In the fall of 1971, freshman congressman Nick Begich played a subtle yet highly effective role in pushing a final compromise package through a conference committee. The House of Representatives adopted the measure 307 to 6 and the Senate by unanimous consent on December 14, 1971. Four days later, President Richard Nixon signed into law the Alaska Native Claims Settlement Act—ANCSA, as it would be forever known. One testament to Begich's talents in the process came the following fall when Democratic majority leader Hale Boggs of Louisiana came to Alaska to campaign for him. Tragically, they both died when the small plane carrying them from Anchorage to Juneau disappeared and was never found.

One writer called the Alaska Native Claims Settlement Act "a complex settlement of a complex situation." That was putting it mildly. In essence, the final bill provided for 40 million acres and $962.4 million to be distributed among twelve regional corporations largely on the basis of population. In exchange, it extinguished all historic aboriginal land titles, all aboriginal hunting and fishing rights, and all previously created reservations, except the Tsimshian reserve on Annette Island. With extinguishment of these aboriginal claims, Alaska Natives obtained fee title to more land than was in the entire Native American reservation system in the lower forty-eight.

The cash payments were to be made over time to the twelve regional business corporations. All eligible Natives had to enroll in one of these corporations to become in effect shareholders and thus participate in the land ownership and cash dividends. To be eligible for enrollment as a Native corporation shareholder, one had to be a U.S. citizen with one-quarter Alaska Indian, Eskimo, or Aleut blood. To some, it was the long-awaited recognition of their ancestral claims and a proud acknowledgment of their heritage. To others, it was the gold rush come again. Many Alaskans who had a Native grandparent, but who had never given the matter much thought, suddenly became quite interested in their heritage.

The corporations would hold title to the lands selected in the process and supervise the incorporation of village corporations within their regional boundaries. The villages in turn would have rights to a certain number of surrounding acres based on the population of the village.[30]

By 1974, all twelve regional corporations with about 76,500 enrolled stockholders were established and operating. Two hundred and three villages statewide were certified as eligible to participate in the regional corporations. Additionally, a federal court ordered the creation of a statewide thirteenth corporation for Natives living outside Alaska to provide for their participation in the money settlement.[31]

Suffice to say, not everyone was happy. In fact, sometimes it seemed as if no one was happy. Created by committee, the final Alaska Native Claims Settlement Act was reminiscent of the story of the committee that when asked to design the perfect animal, came up with the elephant. Three months after the bill's passage, everyone was still arguing about what it meant. At a Senate hearing on amendments, Senator Len B. Jordan of Idaho whispered off the record, "That wasn't much of a bill we wrote." His colleague Lee Metcalf of Montana whispered back, "That was a lawsuit we wrote."[32]

Cries of "Not enough!" or "Native welfare!" continued to be heard. A decade later, William "Spud" Williams, then president of the Tanana Chiefs Conference, still spoke bitterly: "I get tired of these newspaper articles [that say] they gave us 44 million acres and a billion dollars. They didn't give us shit. They stole it [from us], and the only time they were interested in settling it [the claims] was when they found a few barrels of oil."[33]

Perhaps journalist Mary Clay Berry offered the simplest, yet clearest, perspective on the long, long road to the Alaska Native Claims Settlement Act. "Despite its flaws," Berry wrote, "this settlement was an extraordinary agreement in view of the United States' traditional way of settling its Indian problems."[34] Indeed, it was. But there was also a little matter of a last-minute addition to the act, a requirement stuck in subparagraph (d)(2) of Section 17 that Congress receive recommendations for setting aside certain parks, wildlife refuges, wild and scenic rivers, and forests that were deemed in the national interest. In the end, subparagraph "d-2" would throw almost as much controversy over Alaska's public lands in the next decade as the Native land claims had in the previous one. But first there was a pipeline to build.

Working on the Pipeline

It was just one of those coincidental ironies of history that the formal application for the Trans-Alaska Pipeline System was filed with the Department of the Interior on June 6, 1969, the twenty-fifth anniversary of the D-day invasion of Normandy. Over the next few years, however, D-day would prove to be an apt analogy to describe the opening round of what became another heated and lengthy Alaskan battle. Initially, any pipeline across federal lands was required to comply with the right-of-way provisions of the Mineral Leasing Act of 1920. By the time the Native land claims were settled and the land freeze lifted, however, there was also a whole new list of requirements to be met under the recently enacted National Environmental Policy Act of 1969. NEPA, as the act came to be called, was then largely uncharted water. Many of its early precedents and procedures were debated and litigated while oil companies anxiously sought a way to deliver Prudhoe Bay crude.

The Trans-Alaska Pipeline System (TAPS) began late in 1968 as an unincorporated joint venture of the three major companies controlling production from the initial Prudhoe Bay discoveries: British Petroleum (BP), Atlantic Richfield (ARCO), and Humble, soon to be swallowed by Exxon. BP and ARCO each held a 37.5 percent stake, while Humble was in for 25 percent. In February 1969, TAPS announced plans to build a forty-eight-inch pipeline—larger in diameter than had ever been laid—from Prudhoe Bay south roughly 800 miles to the ice-free port of Valdez. The pipeline, eleven pumping stations, and storage and loading facilities at Valdez were estimated to cost $900 million and be capable of initially delivering 500,000 barrels of oil a day. The completion date? Sometime in 1972 seemed reasonable. Forty-eight-inch pipe was ordered from steel mills in Japan, because no American mills could guarantee that size pipe in both the quantity and time frame required.

In September 1969, the big lease sale was held. It was just another coincidence that Alaska's $900-million revenue windfall from the sale equaled initial construction estimates for the delivery system. By now, there were five other companies in the TAPS joint venture. BP and ARCO each gave up 10 percent of the total, which was allocated as follows: 8.5 percent to Mobil,

3.25 percent each to Phillips and Union of California, 3 percent for Amerada-Hess, and 2 percent for Home Oil, although Home later withdrew from the venture.[35]

With the pipe on order and the winter construction season approaching, the consortium was eager to begin work. All that was needed was the right-of-way permit from the Department of the Interior, and Secretary Hickel was under considerable pressure to expedite it. Such action, it will be remembered, was a modification of the land freeze order, and because of the condition imposed on Hickel's Senate confirmation, it required the consent of the interior committees of both the House and the Senate.

Early on, both committees were generally inclined to modify the land freeze to accommodate pipeline construction, and in fact, the Senate interior committee approved modifying the freeze to permit construction of the first phase of the haul road north from Livengood. But before the committees could act on the pipeline right-of-way, an increasing number of environmental issues were raised—some by such groups as the Sierra Club and the Wilderness Society, but others by Hickel's own Department of the Interior.

Undersecretary Russell E. Train had been charged with coordinating a task force to study the impacts of the pipeline on the Arctic environment prior to the issuance of a permit. When the preliminary report was delivered on September 15, 1969, its members professed to have been "guided by the view that oil development and environmental protection are not inconsistent." But to many, the findings seemed to be more of a status report on the application process rather than constructive suggestions on how this could be accomplished. The task force listed a number of problems for which no ready solutions had been found despite a flurry of exchanges with the oil companies.

The biggest technical issue was the effect that hot oil transported in traditionally buried pipe would have on Alaska's permafrost. Prudhoe Bay crude came out of the wellhead at 160 degrees Fahrenheit—hotter than most domestic hot-water heaters. Cooling the oil before sending it down the pipe wasn't much of an option because tests showed that cooling and then reheating Prudhoe Bay crude turned it into tar. Reheating would occur simply by virtue of the friction generated at the eleven pumping stations along the line. Thus, the line was going to be hot no matter what. The hot pipe's impact on the permafrost, some said, was going to be like pushing a hot nail or knife into a block of ice. Heat the pipe, melt the ice, and spill the oil. The Department of the Interior group concluded that the oil industry's plan to bury the pipe just wouldn't work.

Water pollution from construction activities and possible spills, tanker discharges of oily ballast, impacts on wildlife, and potential earthquake zones were also identified as unsolved problems. Then too, of course, there was the pending legal issue of Native land claims and the additional legal issue that TAPS had requested a right-of-way almost twice the width of that authorized under the Mineral Leasing Act of 1920. All in all, it was quite a laundry list of concerns. Meanwhile, the TAPS project manager boasted to the Senate interior committee that the first shipment of Japanese-manufactured pipe was scheduled to be unloaded in Valdez any day.[36]

Secretary Hickel assured Congress that his department had no intention of issuing a right-of-way permit until these issues were addressed, but he also argued that lifting the land freeze in the meantime was merely a procedural step in clearing the way for such future action. Based on this assurance, and Hickel's dubious assertion that lifting the freeze would not unduly impact the Native claims settlement process, both House and Senate interior committees approved modifying the freeze to accommodate the issuance of the pipeline permit. Senate chairman Henry Jackson made it clear, however, that his committee remained very concerned about the environmental impacts of the project. Jackson noted that the new National Environmental Policy Act of 1969—then just out of conference committee and about to be voted upon—required a public statement analyzing the environmental impacts of any federally funded project or any project on federal lands, including an evaluation of alternatives.

Hickel figured that the easiest first step was to issue a right-of-way permit for construction of the haul road that was required to transport the enormous tonnage of equipment and material along the pipeline corridor. But without waiting for such action, the state of Alaska set about reopening the Hickel Highway—essentially grading its second swath—in order to rush equipment north before the spring thaw. This equipment had to be in place, the argument ran, so that when the official permit was issued, construction companies would already be mobilized along the route to construct the permanent haul road—in most instances the third road cut across the tundra.

With Alaska governor Keith Miller champing at the bit to get the road completed, Hickel was playing a game of catch-up in issuing the road permit. Then the legal process brought things to a standstill with a bang. Five Native villages filed suit in federal district court to enjoin Secretary Hickel from issuing the permit on the grounds that they and not the federal government owned some of the lands over which it would pass. On the one hand, Hickel was blindsided because the suit ran counter to certain construction waivers that some villages had signed the previous fall. Evidently, however, these were obtained in part by promises of Native employment on the project. With construction crews jumping the gun and no Native contractors involved, the Natives went to court. It was a busy place.

At the same time, the Wilderness Society, Friends of the Earth, and the Environmental Defense Fund filed suit in the same federal district court to halt the entire TAPS project. They claimed that it violated the right-of-way limits of the Mineral Leasing Act of 1920 and hadn't provided the environmental impact statement required by the newly enacted National Environmental Policy Act. Governor Miller threw up his hands and, apparently having learned nothing to date from the Hickel Highway fiasco, declared with questionable jurisdiction that he would authorize the haul road himself as a state highway. While Alaskans cheered such independence, Secretary Hickel was forced to express public shock and disdain at such overreaching, although he may have winked in Miller's direction as he said it. TAPS for its part wisely decided to wait for a federal permit. No one could have predicted then that the wait would be for years and years.[37]

The Alaska Pipeline permitting process became NEPA's baptism under fire. With no rules or regulations yet in place to implement the law, the Department of the Interior first made an effort to comply with its intent by issuing an environmental impact statement for the haul road. The document was eight pages long. In April 1970, federal judge George L. Hart, Jr., ruled that not only were the various rights-of-way for the pipeline, its extra width, and the haul road one project and as such in violation of the size restrictions of the Mineral Leasing Act, but also that whatever NEPA's intent, an eight-page environmental analysis just didn't cut it. But most significant, Judge Hart also ruled that environmental groups such as the Wilderness Society, Friends of the Earth, and the Environmental Defense Fund did indeed have the legal right—"standing," in the lexicon—to sue in such matters and claim irreparable injury in the face of such federal action.

By the following year, the Department of the Interior's Draft

Environmental Impact Statement had grown to 12,000 pages. By then, despite the D-day coincidence of the permit's filing date, the military meaning of the acronym TAPS was echoing with an uncomfortable finality. The oil companies reorganized TAPS as the Alyeska Pipeline Service Corporation and appointed Humble Oil executive Edward L. Patton as its head. Patton had extensive experience building refineries in both Norway and northern California in the face of a web of governmental regulations and procedures. He would stay in charge until oil finally flowed to Valdez, but when he first took the job, the timetable and even the eventuality of the pipeline itself were still very much in doubt.[38]

Needless to say, the oil companies and many Alaskans were not taking these delays in the best of humor. Ever the promoter, Anchorage publisher Bob Atwood scoffed at the idea that Alyeska would risk losing oil through major spills. "The companies have a common interest with proper conservationists," Atwood told an outside journalist, "but these other kooks—they talk wildly and people seem to listen to them."[39]

Causing particular ire were the conservationists from "outside" who seemed bent on telling Alaskans what was in their best interests. Conceding that there wasn't much anyone could say or do to convince these "environmental crusaders" that the pipeline was going to be anything less than an unmitigated disaster, the *Ketchikan Daily News* editorialized: "These people are not interested in a pipeline under any conditions and they add their voices to professional conservationists who make a practice of opposing any development any place—Sierra Club, Friends of the Earth, etc."[40] No wonder John McPhee wrote of a fellow's bruin encounter and concluded only partially tongue in cheek that "when you are the Sierra Club's man in Alaska, the least of your problems is bears."[41]

By December 1971, the Alaska Native Claims Settlement Act was law, and at least one piece of uncertainty had been removed from the process. Meanwhile, the first feeble attempt at an environmental impact statement (EIS) suggested that the haul road alternatives were between building the road immediately or building it in three years, and between a federal- or a state-controlled right-of-way. But then a U.S. court of appeals ruled that when it came to considering action alternatives, NEPA required a very broad discussion. Not only did this mean that the long-held route of the pipeline generally along the Hickel Highway corridor—in part because 2,542-foot Anaktuvuk Pass was the lowest crossing of the

Brooks Range—must be scrutinized, but also that alternative routes as well as transportation methods must be studied.

Among the alternatives, loading crude directly onto tankers at Prudhoe Bay was pretty much out of the question. The *Manhattan* voyages had proven the folly of that. Pipeline guys tended to draw straight lines on maps. Yet despite this mentality of a straight-shot pipeline south to the Gulf of Alaska—via Anaktuvuk Pass or some other route—there had also been early talk about a route through Canada. This route would take the pipeline eastward from Prudhoe Bay, along the coastal plain of the Arctic National Wildlife Range, and generally southeast up the Mackenzie River valley to the vicinity of Edmonton and the existing pipelines and refineries of both the Pacific Northwest and the greater Midwest. While this route would be at least twice as long, pass through a friendly though foreign country, and also raise considerable environmental concerns, it had numerous advantages.

First and foremost, a trans-Canadian route would eliminate all tanker traffic. The oil would be delivered by pipe directly to refineries. Thus, while the permafrost issues were likely to grow because of the longer route, there would be no oil danger to North Pacific fisheries, particularly in Prince William Sound. A trans-Canadian route would also avoid the earthquake zones that Valdez knew only too well. Finally, but perhaps most economically significant in the long run, a system of land pipelines would also be capable of transporting Prudhoe Bay's estimated 26 trillion cubic feet of natural gas to markets.[42]

But not too many people were thinking about the long run in 1972. The oil companies were tired of sitting on reserves that couldn't be delivered to market. The state of Alaska was watching its $900 million windfall from the 1969 lease sale disappear in the face of growing demands and badly wanted its share of production royalties. The federal government had already paid what some people thought was a ransom to settle the Native land claims and remove that roadblock to the pipeline's construction. It was time for the oil to flow.

The draft EIS initially devoted only two of its 12,000 pages to the alternative of a Canadian pipeline route. Even when forced by court action to expand on this, the Department of the Interior could never overcome the groundswell of support from the oil companies and most Alaskans for an all-Alaska route. These Alaskans demanded that the pipeline avoid Canadian soil with a vigor that was reminiscent of the debates over the routes to the Klondike three-quarters of a century earlier.

Moderates like Alaska's soon-to-be governor Jay Hammond, who tried to walk a middle ground between those who would lock up the land and those who would develop it and ask questions about impacts later, ended up the target of ridicule from both sides. In truth, Hammond supported development that caused little environmental impact and that paid its own way, and he wasn't convinced that the pipeline met either criteria. Hammond tried to promote greater study of the trans-Canada route, but he was shouted down. (How Jay Hammond managed to say all of this, antagonize so many people, and still be elected governor of Alaska in 1974 is a whole other story—as well as a commentary on Alaska politics!) "Winning their specious argument," reflected Jay Hammond years later of the all-Alaska route promoters, "may have been one of the most costly mistakes of the 20th Century."[43]

In retrospect, Hammond and the few others who questioned the trans-Alaska route were proved correct. When West Coast refineries were stretched to capacity by the pipeline's 2-million-barrel-a-day discharge, Alaskan crude ended up being transferred to smaller tankers on the West Coast for transit through the narrow Panama Canal and shipped north to Texas and East Coast refineries. It was definitely the long way around. Many years later—after the 1989 *Exxon Valdez* catastrophe in Prince William Sound and amid cries for development of production from what became the Arctic National Wildlife *Refuge*—there is still no way of transporting Prudhoe Bay's trove of natural gas southward.

But again, that was the long run. The short run was summarized by Secretary of the Interior Rogers C. B. Morton on May 11, 1972, when he announced his intent to issue the pipeline permit. "I am convinced," said the secretary, "that it is in our best national interest to avoid all further delays and uncertainties in planning the development of the Alaska North Slope reserves by having a secure pipeline located under the total jurisdiction and for the exclusive use of the United States."[44] That statement aside, there was another round of litigation, a portion of which found that the project's environmental requirements under NEPA had in fact been met. Then in February 1973, a federal court of appeals ruled that if Alyeska did in fact require a right-of-way for the pipeline wider than that authorized by the Mineral Leasing Act of 1920, its redress must be sought from Congress and not the courts.

Spurred on by the Arab oil embargo and resulting fuel shortages, Congress at last decided to take legislative action. The crucial moment came on July 17, 1973, when the U.S. Senate voted on an amendment offered by Alaska's Mike Gravel to declare that the ponderous requirements of NEPA had been met and simply to authorize the construction of the pipeline. The outcome of the vote was a 49-to-49 tie. For the first time in four years, Vice President Spiro Agnew, then only weeks away from resigning his office, cast the deciding vote in favor of the pipeline. Similar legislation passed the House, and finally on November 16, 1973, Richard Nixon signed the Alaska Pipeline Authorization into law.[45] Almost six years had passed since that heady test of the Prudhoe Bay State No. 1 well.

And so the rush was on again. In 1974, the floodgates opened, and construction workers flowed north on a tidal wave the likes of which had not been seen in Alaska since the gold rush or World War II. Heavy equipment operators, pipe fitters, welders, truck drivers, surveyors, laborers, cooks, bakers, housekeepers, and assorted camp followers all headed north to work on the line. Anchorage and Fairbanks boomed, and Valdez? Well, Valdez was about to undergo a transformation almost as jarring as the 1964 earthquake.

Two big things had changed on the pipeline since the early days of 1969. The first was its route. While Anaktuvuk Pass was indeed the lowest crossing of the Brooks Range, that route required the line to detour westward. Its biggest drawback, however, was a lack of sizable gravel deposits. Good gravel is important to most construction operations, but it is essential to Arctic construction as an insulator and stabilizer on top of permafrost. The tundra quagmires along the abandoned Hickel Highway both north and south of Anaktuvuk Pass still bore mute testimony to that fact.

So a route was surveyed from Prudhoe Bay south up the gravel-laden Sagananvirktok and Atigun River valleys to 4,790-foot Atigun Pass, and then down the Dietrich River and the Middle Fork of the Koyukuk past the old placer mining camps of Wiseman, Coldfoot, and Tramway Bar. Not only did the river valleys on either side of Atigun Pass offer ready gravel sources, but at the time this route seemed to mollify moderate environmentalists, who continued to dream about the Gates of the Arctic National Monument centered in the heart of the Brooks Range around Anaktuvuk Pass.

Farther south the pipeline corridor met the original trace of the Hickel Highway, crossed the Yukon, and continued to Livengood and Fairbanks. South of Fairbanks, the route crossed the Alaska Range at

Isabel Pass and the Chugach Mountains at Thompson Pass before dropping almost 3,000 feet into Valdez, through Keystone Canyon. From Prudhoe Bay to Valdez, the route came to 789 miles.

The other big change centered on the VSM—pipeline shorthand for "vertical support member." VSMs—some 78,000 of them—would become the skeleton across which the four-foot-wide pipe would be stretched like a string of spaghetti. This was the solution to the permafrost problem. About half of the line was to be buried conventionally, but the other half was to be elevated above ground on VSMs. A VSM was an upright piece of eighteen-inch steel pipe installed in pairs and spaced every fifty to seventy feet. Each pair of VSMs was connected by a steel crossbeam. The pipe rode in a saddlelike assembly mounted atop the crossbeam but capable of sliding back and forth as the pipeline expanded and contracted with changing temperatures. The whole system was installed in a trapezoid pattern that zigzagged back and forth between straight legs. This construction also provided flexibility in the event of an earthquake.

VSMs got the hot pipe away from the permafrost, but the VSMs themselves were still in contact with it. Marginal permafrost with temperatures only slightly less than freezing reacts against the intrusion of anything slightly warmer. The VSMs might be heaved out of the ground or swallowed into the permafrost—in either event twisting the pipeline out of alignment. While a variety of footers were tested to alleviate this problem, the final answer was to use thermal VSMs except in places where they rested on bedrock or were sunk in thaw-stable permafrost in the very far north. Thermal VSMs essentially had their own nonmechanical, natural convection system that worked like a radiator to release heat buildup from the VSMs into the atmosphere, thus keeping their underground contact with the permafrost cool.[46]

Twelve pump stations were strategically placed along the 789-mile route to boost the crude over the three main mountain ranges. Four stations were located on the North Slope for the climb over Atigun Pass, another four along the ups and downs through the Yukon Basin to Fairbanks, two for the climb to Isabel Pass, and the final two for the grade north of Thompson Pass.

Work officially began on the new haul road on April 29, 1974, after the state of Alaska approved the final permits for those portions of the road crossing state lands. By the end of September, all of the gravel—some 35 million cubic yards—was down, and grading was completed by mid-November. The 361-mile road from Livengood to Prudhoe Bay along with twenty major bridges was built in 154 days at the cost of around $500,000 per mile. In a moment reminiscent of the linkup of the Alaska Highway at Contact Creek, a ribbon-cutting ceremony was held on a rise above the South Fork of the Koyukuk between Coldfoot and Prospect Creek.

The only missing link was the crossing of the Yukon River. An ice bridge was used during the winter of 1974–75 as personnel and materials continued to roll north. By the following fall, the state of Alaska had completed a permanent crossing, a half-mile-long highway bridge with five massive piers that was partially paid for by Alyeska. In return, the pipeline would be strapped to the bridge and save Alyeska much treacherous excavation work in the river's ever-changing bottom.[47]

Construction of the actual pipeline began in March 1975, using the same forty-eight-inch pipe that had been stockpiled in Seward, Valdez, and other points since its 1969 delivery from Japan. The first section to be laid was the crossing of the Tonsina River about seventy miles north of Valdez. A 1,400-foot-long string of 40-foot sections was welded together, weighted with temporary concrete collars, and then gingerly lowered into a ditch dug across the riverbed. The ditch was backfilled with gravel, and the concrete weights were removed.

The pipe was supposed to be snug in its trench, but with the weights removed, the whole section popped out of its trench like some Halloween ghoul out of a coffin. It was an inauspicious beginning, and it took another three and a half weeks to get the pipe firmly in place. That done, it would be a year and a half before it was discovered in the final testing that this first section of pipe had been crushed during this initial work. By then, the expedient remedy was to abandon that section and lay new pipe just upstream. Thus, the Tonsina crossing—one of some 800-odd stream crossings along the route—had the distinction of being both the first and the last section of the Alaska Pipeline to be completed.[48]

Before the pipeline was completed, upwards of 70,000 men and women could say that they had had some hand in its construction. At the peak, 20,000 workers were directly employed by Alyeska and its subcontractors at one time. The biggest concentration was at the terminal in Valdez, where the construction of dock facilities, storage tanks, a ballast treatment system, control center, and the pipeline itself down from

Thompson Pass kept as many as 4,000 workers busy. Thirty-four two-story barracks, denoted A through HH, were scrunched together like so many mobile homes on one end of the terminal complex. Each building had fifty-two double rooms. It helped to like your roommate.

But the pay was hard to beat. Office help pulled down $500 a week. Guys in the trades were making as much as $1,800 a week—at least double what they would have earned in the lower forty-eight. Overtime was the norm, especially during the long days of summer when six ten-hour days was the rule. Those who stuck it out for six months, twelve months, or more usually went back home with a sizable nest egg, provided that they hadn't spent too much on the few diversions that Valdez had to offer.

"Oil and alcohol," boasted one ample-bellied Texan on one of the subcontract crews, "that's what Alaska runs on." And when either of those commodities failed, there was always companionship to be had for a price. It was uncanny. It really was the gold rush all over again right down to the isolation that gripped Valdez when winter storms closed the road to Anchorage and grounded the planes. One October when the movie of the evening turned out to be *One Day in the Life of Ivan Denisovich,* the irony of slave-prisoners serving indeterminate sentences was not lost on the audience. But at least in Valdez, or anywhere along the line, the workers knew when their time was up. In the meantime, they got fed steaks and paid top dollar. There were, however, only just so many nights in a row that one could get plastered. Psychologically, there were some definite parallels to the gulag, especially when the phones went down or the mail was late.[49]

The centerpiece of the Valdez terminal area was the initial eighteen storage tanks for oil, lining terraces blasted out of bedrock on the mountainside. Each had a capacity of 510,000 barrels, was 250 feet in diameter and 62 feet high, and had a cone-shaped roof designed to withstand heavy snow loads. The tanks were paired off and surrounded with containment dikes capable of holding 110 percent of their volume. Three other, slightly smaller tanks, also cone-roofed, were each 250 feet in diameter, 48 feet high, and with a capacity of 430,000 barrels. This was the BTA—the Ballast Treatment Area.

Tankers without a load ride high in the water and bob about like a matchbox. A good wave might swamp them. The answer was to take on a partial load of seawater as ballast for stability on the empty run north. When the tankers arrived in Valdez to take on a load of crude, the seawater ballast was jettisoned. The problem, of course, was the oil that had been in the tanks from the last run prior to taking on the seawater ballast. No matter how hard the pumps sucked unloading the crude, some of it remained in the tanks to mix with the seawater ballast. Take a 1-million-barrel capacity tanker, leave just 0.5 percent of its load clinging to its tanks, mix with seawater and discharge, and suddenly you have a 5,000-barrel oil spill.

The final right-of-way agreement prohibited the discharge of tanker ballast anywhere, but what to do with it? The solution was an enormous ballast treatment area at Valdez. The seawater ballast of incoming tankers was pumped into a tank and allowed to stand for several hours. The majority of the oil floated to the surface and was skimmed off and mixed with the supply coming down the pipe. The remaining seawater was pumped through a filtering and chemical treatment process until it contained less than ten parts of oil per million. Finally, the water was gradually discharged into Valdez Arm.[50]

Vic Filimon, Alyeska manager of terminal construction at Valdez, summed up the special challenges of Alaska construction this way: "Nothing that we're building here is all that new," Filimon confessed, "but it's the bigness of the place and the strictness of the specifications. Everything here is being built to withstand 90-pound snow loads, 90-mile-per-hour winds, and an earthquake of 8.5 on the Richter scale." There were, of course, the inevitable surprises—what Filimon called "the consistency of the inconsistency of the rock." Bedrock for the foundations of the oil storage tanks was first thought to be about eight feet down. By the time they were through, some foundations were down forty-five feet.[51]

There were man-made hassles as well. The quality of the welds on the pipeline became such a source of confusion and cost overruns that even Congress held hearings on the subject before it was resolved. The problem was one of both actual workmanship and subsequent documentation. There were about 65,000 welds in the 800 miles of pipeline. In normal pipeline construction, it was standard practice to take random x-rays of about 10 percent of the welds, except those at river and highway crossings, which were x-rayed at a 100 percent rate. On the Alaska pipeline, the protocol was 100 percent testing—in other words, x-ray and document each and every weld—all 65,000 of them.

When whistle-blowers began to question both the integrity of some welds and sloppy documentation procedures, finger pointing was rampant. In one instance, 358 x-rays of suspect welds were stolen and never found. In another, a project manager for the testing firm was found dead of questionable causes. Part of the challenge became to determine which welds were flawed and which were only poorly documented. By the time that Alyeska had sorted through the mess, thousands of welds were rechecked and in some cases redone—frequently by uncovering and then reburying pipe. The company spent an estimated $55 million on the weld repair program alone. No wonder that just like almost every other construction project since Noah's ark, the pipeline was over budget—way over budget.[52]

But the pipeline was getting done. By the end of 1976, almost all of the main line was in place. The pump stations were nearing completion, and Alyeska estimated that the terminal facilities at Valdez were 83 percent complete. Along the line, the workforce began to dwindle as sections were finished and another winter set in, but at Valdez crews continued to work through the dark months to ready the installation for the spring tests. Pressure tests were conducted, and in April 1977 Alyeska announced its intention to begin filling the line with oil later that summer. It was not as easy as turning on a spigot.

For starters, the difference in temperature between the cold pipe and the hot oil had to be balanced. In warmer climes, water would have been put into the line ahead of the oil—separated by a plug that oil workers call a "pig"—to purge the line of air. Water in the Alaska pipeline would likely have frozen solid, so nitrogen was used instead. Unlike oxygen, nitrogen won't support combustion and thus is safer to use with hydrocarbons. The nitrogen was warmed and inserted into the line under pressure ahead of the pig. As it moved down the line, it warmed the pipe to a temperature compatible with the oil that followed. At the rate of about one mile an hour, Prudhoe Bay crude began its journey south to Valdez on June 20, 1977.

The trip was not without mishap. At pump station 8 about 500 miles down the line from Prudhoe Bay, nitrogen was injected into the pipe without being heated. The cold cracked a pipe bend. Later, while workers were replacing equipment on a pump, oil sprayed all over the place and ignited. One worker was killed, a number were injured, and pump station 8 was heavily damaged.

In Anchorage, St. Patrick's Catholic Church raised money for its building fund by holding a lottery to guess the exact time that the oil would reach the Valdez terminal. An Anchorage widow won $30,000 with a time of 11:02 P.M. on Thursday, July 28, 1977—thirty-eight days, twelve hours, and fifty-six minutes en route. Incredibly, with the 800-mile line full, there were 9.2 million barrels of oil—roughly eight tanker loads—in the line.[53]

Meanwhile, earlier that spring the tanker *ARCO Fairbanks* had operated in and out of Valdez harbor with loads of seawater ballast, serving as a training ship for dozens of tanker captains, harbor pilots, and Coast Guard personnel. It was a process akin to an airplane pilot taking a check ride and making landings at a new airport. Soon it was for real, and the *ARCO Juneau* took on the first load of Prudhoe Bay crude and sailed south through Valdez Narrows on August 1, 1977.[54]

At the time, the construction of the Alaska Pipeline was the most expensive private construction project in U.S. history. Having been fed by inflation, the technical demands of permafrost construction, an unprecedented array of environmental regulation, and seemingly endless cost overruns, its original $900-million estimated cost ballooned to over $8 billion before it was completed.

Despite this, as the *ARCO Juneau* sailed out of Valdez on its first run, there was a lot of cheering and counting of future profits. The oil companies were *finally* getting their product to market—almost ten years after the original Prudhoe Bay discovery. Royalty and severance tax revenues were about to swell Alaska's state coffers. Alaska businessmen counted their take from the past few years and looked around for the next boom. Only those who had opposed the pipeline on environmental grounds looked at it with resignation and waited for what would happen next.

A Capital Move

Sitka was always the crown jewel of Russian America. Although boasts of "Baranov's Castle" and the "Paris of the Pacific" were more than slightly exaggerated, the town was nonetheless the undisputed capital of Alaska for more than half a century. Little changed in the first decades after the American purchase—there wasn't much competition for the title. But slowly, Joe Juneau's little mining camp—once it finally got its name figured out—began to throw its weight around. The millions upon millions in gold coming out of the Treadwell Mine added persuasive bulk.

Sitka, it was said by those in Juneau, was becoming less and less important to the territory. It was just too far away from the mainstream, too far away from the center of commerce. In 1900, Congress agreed and ordered that the capital of Alaska be moved from Sitka to Juneau once suitable facilities could be arranged there. Understandably, Sitka dragged its feet, but Juneau hastily arranged something "suitable" and the move took place.

Capital moves—be they those of county seats replete with tales of snatching public records in the dead of night, or carefully planned new cities built in virgin wilderness—are nothing particularly unusual in American history. Thus, it seems somewhat silly in retrospect that Alaskans spent so much of the 1970s debating just that. Moving the capital from Juneau was an issue that polarized communities, divided families, and consumed a great deal of time and money. Maybe the whole thing was just an excuse to talk about something besides the pipeline.

Actually, it was quite a bit more than that, and it was the brainchild of publisher and promoter Robert B. Atwood. After celebrating statehood in 1959, Atwood used the *Anchorage Daily Times* to launch a concerted campaign to move the capital from Juneau. Years later, when asked about the genesis of the whole capital move controversy, Atwood boasted gleefully, "I created it!"[55]

Indeed he did, but no one could deny that Alaska's population center had in fact shifted sharply westward after World War II. From 1939 to 1950, Juneau's population increased by only 2 percent and Ketchikan's by 16 percent. Fairbanks, on the other hand, grew 241 percent and Anchorage a whopping 658 percent. Military spending was at the root of the matter, but many Juneau residents worried that Congress or their fellow Alaskans might try to relocate the capital. They breathed sighs of relief when Alaska's new state constitution fixed the capital at Juneau, but then Bob Atwood got wound up.[56]

Juneau, Atwood editorialized, was becoming less and less important to the state. It was just too far away from the mainstream, too far away from the center of commerce. Sound familiar? It did to folks in Sitka. Surprisingly, however, in 1960 and again in 1962, statewide voters decisively rejected two Atwood-backed ballot initiatives to move the capital from Juneau. In part, this may have been because they thought the new state had much better things on which to spend their money than capital moves. But there had also been considerable infighting between Fairbanks and Anchorage over which town would get the plum if Atwood managed to pluck it from Juneau.

In 1974, with Atwood and the *Anchorage Daily Times* still playing the same tune, yet another initiative was petitioned onto the ballot. This time, however, there were three big differences from the earlier proposals. First, the state was awash in oil royalties. Second, the oil boom had skewed population figures even more heavily away from southeast Alaska. Finally, if the move from Juneau was approved, the new capital had to be an undeveloped site, in an as yet undetermined location, not within a thirty-mile radius of either Anchorage or Fairbanks. The latter was the result of an "OK, if we can't have it here, you're sure not going to get it there either" trade-off between the two towns. Although no cost estimates were given, the 1974 measure passed with 57 percent of the vote.

In that same election, Alaska voters reelected Democrat Mike Gravel to the U.S. Senate and gave the governor's chair to Republican Jay Hammond by 221 votes over Alaska institution Bill Egan. Although the *Juneau Empire* later accused Hammond of being "a closet capital mover," the new governor had in fact spoken out against the initiative. Reluctantly, Hammond appointed the Capital Site Selection Committee to make specific site recommendations. For its chairman, Hammond chose Willie Hensley, who had just lost the race for Alaska's sole congressional seat to Republican Don Young. Hensley was generally against the capital move as too costly, but he did admit to being caught up in the excitement of the prospect of building a new town from scratch.

For much of 1975, Hensley's nine-member committee and a bevy of architects, engineers, and planners crisscrossed Alaska, looking at both close-in and remote parcels of state land and talking with hundreds of people. Central location and accessibility were the key reasons given by most Alaskans who favored the capital move, but when push came to shove, not that many could think how they were really inconvenienced with the capital in Juneau. The only real consensus seemed to be that nobody had much good to say about Anchorage—except Atwood and folks in Anchorage. But then that wasn't really anything new. Earl Cook, the Fairbanks representative on the committee, kept hoping for a location north of the Alaska Range, but that didn't happen. Meanwhile, William Corbus, the Juneau representative, smiled to himself every time the drawbacks of any site were discussed.

In the end, the Capital Site Selection Committee focused on the logic of a location somewhere between Anchorage and Fairbanks and recommended three possible choices. The closest to Anchorage was near Willow, about sixty-five road miles north and easily accessible via the Alaska Railroad and the new George A. Parks Highway. The Larson Lake site was half a dozen miles east of Talkeetna, the turnoff to which is about thirty miles farther along the road to Fairbanks than Willow. Talkeetna folks had voted 69 to 42 in favor of moving the capital, but now that it might end up in their backyard, some of the proponents were having seconds thoughts. The third site was near Mount Yenlo, a 3,961-foot mountain overlooking the confluence of the Yentna and Kahiltna Rivers. It was about thirty-five miles northwest of Willow across miles upon miles of braided rivers and wetlands.[57]

The architects, engineers, and planners estimated the development costs of the various sites as follows: Willow, $2.46 billion; Larson Lake, $2.56 billion; and Mount Yenlo, $2.7 billion.[58] Not to worry, said the consultants. They figured that the state would have to pick up only about 20 percent of the tab and that private development would fund the remainder of the costs. Willie Hensley was among those not convinced. He thought that the costs were being too widely ignored. "In Alaska," said Hensley at the time, "too many people seem to think they are floating to Heaven on a sea of oil."[59]

Alaska voters were asked to choose between the Willow, Larson Lake, and Mount Yenlo sites at the 1976 general election. Pleas to include "none of the above" on the ballot were ignored, and voters chose Willow better than two to one over the others.[60] Even though the location was required to be on state land, some private land in the Matanuska and Susitna Valleys doubled in value—at least on paper—between 1975 and 1977 in anticipation of the construction.

The voters had spoken, albeit less than enthusiastically, and now Governor Hammond and the legislature were forced to figure out how to pay for it. One option was simply to open the oil tap and fund the new capital with royalty revenues. The other was to ask voters to approve a bond issue. Hammond and others opposed to the capital move hoped that such a vote might bring a measure of financial awakening to its proponents.

With the "where?" seemingly settled, the issues now became "how much?" and "by whom?" A group calling itself FRANK, an abbreviation for Frustrated Responsible Alaskans Needing Knowledge, lobbied hard to include accurate construction costs on the bond issue ballot. Interestingly enough, however, as the 1978 election neared and emotions continued to run high on both sides, rarely was the capital move put in terms of per capita cost. When all was said and done, the $900 million bond issue was going to cost every man, woman, and child in the state in the neighborhood of $2,500 each.

Meanwhile, another committee was formed to tour the state and seek input on what the new capital should look like. Many Alaskans wanted it to look like Juneau. Heck, many wanted it to *be* Juneau. In a survey conducted as part of the governor's Alaska Public Forum, 314 rural and urban residents statewide were asked, "Where should the state capital be?" Forty-one percent chose Juneau, 15 percent Anchorage, 13 percent Willow, and 7 percent Wasilla. The other votes were undecided or scattered around the state.[61] And Mount Yenlo? Well, not a whole lot had happened up that way since Frederick Cook last passed by in 1906.

Wherever the location and whatever the cost, Sitka state senator Pete Meland termed the whole thing a "Permafrost Brasilia," a pointed reference to Brazil's less than stellar experience carving a new capital out of the Amazon rain forest. In the 1978 general election, most voters agreed. They decisively rejected the $900 million bond issue. Maybe they had done the personal math after all.

What Governor Jay Hammond thought of the whole mess was probably summed up best in his autobiography. Silver-tongued and loquacious about most issues, he barely mentioned the capital move, except to acknowledge that it had festered throughout his eight years in the governor's chair and that in the end his administration was able "to lance this boil."[62]

The Permanent Fund

With the completion of the pipeline, Prudhoe Bay's oil tap was cranked wide-open. Alaska was suddenly rich. Not just sort of rich but filthy rich. The big lease sale of September 1969 had made the state swagger, but that $900 million in revenues was a drop in the barrel compared to the hundreds of millions of dollars that now flowed annually into state coffers from a combination of oil and gas leases, royalties, and taxes. The words *budget surplus* took on a whole new meaning. By one estimate, it was projected to be $2,500 annually for the next thirty years for every man, woman, and child in Alaska. "This is an absolutely unique situation," mused University of Alaska economics professor Arlon Tussing, "in that the state has money it doesn't know how to spend."[63]

In the tradition of the mining frontier, some Alaskans wanted to spend this newfound wealth on a spree—whether it be a new billion-dollar capital, a winter trip to Hawaii, or just about anything in between. Others, including Governor Jay Hammond, took a longer view. Back in his days as the part-time manager of the Bristol Bay Borough, Hammond had once proposed forming an investment corporation to be funded from a fishing-use tax, the vast majority of which was paid by nonresidents. Hammond went on to suggest that after paying the costs of local government, local residents—as shareholders in the corporation—would receive an annual payment akin to a stock dividend for each year they had resided in the borough. Most folks in Bristol Bay looked at him as if he were crazy, and promptly and overwhelmingly voted down the measure.

A decade later, Jay Hammond was governor, and it was the state of Alaska that was awash in excess revenues. Hammond thought that his original idea of a dividend-paying corporation was still a good one. To promote it among Alaska voters, he created a traveling town meeting called the Alaska Public Forum. This became a raucous roundtable for discussing controversial state issues, but Hammond's dividend idea didn't receive much support. Undeterred, the governor introduced a bill in the Alaska legislature to amend the state constitution and create Alaska, Inc. Under Hammond's proposal, Alaska, Inc. would receive 50 percent of all mineral payments and tax dollars for a trust fund, the principal of which could be invaded only by statewide vote. Half of the fund's annual earnings would be distributed as cash dividends to all Alaskans in the form of one share of dividend-paying "stock" for each year of residency since statehood.[64]

While various committees of the state legislature sparred with Hammond's proposal, there was no shortage of other ideas. Most Alaskans had come to agree that the enormous economic bounty being harvested from oil royalties should be used for something less Brazilian than building a new capital, but there the consensus stopped. Senator Bill Ray, a Democrat of Juneau, termed Hammond's dividend proposal "well intentioned, but misguided" and gave it "not much chance of passage."[65] Loans to small businesses and marginal enterprises, outside investment in blue chip stocks, a host of community development projects, and an Alaskan version of the World Bank to finance megaprojects such as the still-simmering Susitna–Devils Canyon Hydroelectric Project, were among the ideas discussed.

One thing was clear. Alaska's initial $900 million lease windfall had been spent rather quickly to catch up on long-neglected infrastructure and public works, education, and social programs. Meanwhile, in less than a decade the state budget had tripled from just under $300 million, and its growth showed no signs of slowing. It seemed that government was expanding to consume whatever revenues were available.

Rather than have that happen, Governor Hammond opined, "I would far rather take that oil money—money from nonrenewable resources—sail it back out to the people in the form of a dividend, and then, through the normal taxing process, provide for the funds for normal state services."[66] In other words, base the state budget on necessary expenditures and not black gold revenues.

When the legislature first presented Governor Hammond with a fund bill, it was for a statutory investment fund, not an invasion-proof trust created by constitutional amendment. Instead of depositing 50 percent of all mineral revenues into the fund, it called for crediting 25 percent of only the leasing and royalty revenues, not severance taxes as well. Alaska, Inc. wasn't mentioned, and these shortcomings made Hammond skeptical despite the proposed name: the Alaska Permanent Fund. In Hammond's mind, this was not a case where half a loaf was better than no loaf. Legislation could be changed depending on what bunch might congregate in Juneau every two years. Hammond wanted something *permanent* in more than just name or at least something so firmly embedded in the state constitution that it would take a vote of the people to change it. Hammond vetoed the measure.

The legislature got the message and redrafted the legislation. On November 2, 1976, Alaska voters approved an amendment to the state constitution creating a truly permanent fund. It was to be the repository for at least 25 percent, but conceivably 100 percent, of all mineral revenues received by the state: lease bonuses and rentals, royalties, oil and gas sale proceeds, and federal mineral revenue-sharing payments. The Permanent Fund was to invest these revenues in an attempt to provide long-term security from the sale of the state's nonrenewable resources. Just how this was to be done and for what purposes the fund's income could be used were left to the governor and legislature to decide. There was no mention of a dividend to individual Alaskans, but the fund itself was a good start.

In his next State of the State address in January 1977, Hammond renewed his proposal for the dividend-paying Alaska, Inc. and urged the legislature to put at least 50 percent of revenues, rather than the 25 percent required by the amendment, into the Permanent Fund. A reluctant gubernatorial candidate in 1974, Hammond ran for reelection in 1978 in large measure because he wanted to see the benefits of the Permanent Fund accrue to Alaska's residents and not remain a large slush fund subject to legislative whims.

In that 1978 election cycle, Hammond managed an unusual political hat trick. He beat the same candidate, former governor Walter J. Hickel, three times: in the Republican primary, in a court-ordered runoff for the Republican nomination, and in the general election when Hickel refused to give up and ran as a write-in candidate. Hickel likened fellow-Republican Hammond's dividend idea to something akin to socialism and preferred to view the surplus as a huge pot from which to fund development projects. By then, Hickel had become something of an unlikely antiestablishment hero for having been fired as secretary of the interior by Richard Nixon, but even pro-development Alaskans still remembered the Hickel Highway.

Two years later, the Alaska Permanent Fund Dividend became law. Essentially, it provided each Alaskan with one "share" of dividend-paying stock for each year of residency in Alaska since statehood. Aside from resembling Hammond's first dividend idea at Bristol Bay, this approach seemed a fair way to compensate old-timers for the leaner years they had put in before the oil boom. The first dividend was set at fifty dollars per "share," or year of residency. But before the state treasury could crank out the checks, two young attorneys freshly arrived in the state and admitted to the Alaska bar filed suit alleging that dividend payments based on length of residency were unconstitutional. A lower Alaska court agreed, but the Alaska Supreme Court reversed the ruling and upheld the law. The law's critics appealed to the U.S. Supreme Court. Meanwhile, earnings continued to accumulate in the fund.

On June 14, 1982, in an 8-to-1 decision, the U.S. Supreme Court struck down the payments based on residency. The court held that such a system violated the Equal Protection Clause of the Fourteenth Amendment because there was no valid state interest to be rationally served by distinguishing between long-term residents and newcomers. Such a distinction, the court said, could lead to other similar classifications. (Associate Justice William Rehnquist was the sole dissenter.)

Anticipating the high court's decision, Hammond and the legislature had recrafted the dividend program to provide that one-half of the annual income of the Permanent Fund be distributed equally to *all* Alaska residents, including children, who had lived in the state for six months. A portion of the remaining income was to be added to principal to act as a hedge against inflation, and the annual payments were to be based on five-year income averaging that would avoid large yearly fluctuations. A few weeks after the Supreme Court decision, more than 400,000 $1,000 checks were distributed to Alaska residents, roughly the amount of three years of dividends that had accumulated during the litigation.[67]

Meanwhile, with Alaska's state coffers overflowing, the legislature had taken the extraordinary step of abolishing state personal income taxes. There were many, including Governor Hammond, who thought this action was feeling just a little too flush. Suspend these taxes or reduce them for a few years perhaps, but abolish them? Woe to the governor and legislature that would ever have to reimpose them should oil revenues drop.

Some thought that they never would, and by the end of the 1970s, Alaska's state government was beholden to oil and gas revenues in a way that dwarfed the dependency of other states. In 1978, for example, oil and gas revenues accounted for 29.2, 19.6, and 15 percent of the state budgets of Louisiana, Texas, and Oklahoma, respectively. In Alaska that year, the figure was 58 percent, and it topped out at a whopping 90 percent in 1980 and remained in the 80 percent range for much of the following decade. At the peak of oil revenues in 1981, more than $10,000 per resident flowed into the state treasury.

But the truth of the matter was that even with the spigot cranked wide-open, there were a number of variables affecting Alaska's oil revenues. The state was the beneficiary of the fivefold price increase for Middle Eastern crude in 1973–74 and the resulting threefold increase in domestic crude. At thirty dollars a barrel, things looked mighty sweet, but such market swings could work both ways. Alaska historian Claus-M. Naske summed up the downside quite simply: "At $22 a barrel, the state of Alaska shudders, and at $15 per barrel it would face a fiscal revolution."[68]

Meanwhile, of course, pipeline pay had forced up other prices. Alaska was an expensive place both to do business and to live. As always, distance from manufacturing centers and a heavy dependence on imports added to the burden. But with annual fund distributions of $404 paid to more than 500,000 residents in 1985, such worries were for the future.

Back in November 1976, shortly after the passage of the Permanent Fund amendment, the *Anchorage Daily Times* reported that "the impact of the new oil money, with its potential to influence life in Alaska for better or worse, has hardly begun to sink into the public consciousness."[69] Perhaps. But once this got into the collective consciousness, it was going to be awfully hard to shake. As the fortunes of oil have ebbed and flowed in the quarter century since then, one thing is certain. Every Alaskan who cashes an annual Permanent Fund check should thank the determined foresight of Jay Hammond and a few of his cohorts for putting something away for a rainy day.

Postscript to an Era (1964–1980)

The closer a historian gets *to the present, the more difficult it is to paint with a broad brush. The sweep of events is less clear, and the individual brushstrokes are more apt to confuse the broader strokes rather than give them detail. Alaska's history has always had complex and frequently divergent themes, but in Alaska after the Good Friday earthquake they became even more so. Many of the events and movements that were set in motion in the 1960s have yet to play themselves out completely, or even sit still long enough almost half a century later to be put on canvas.*

The only constant is change. Did anyone imagine in 1964 how dramatically the payment of a century-old debt would change the face of Alaska? And how loud the outcries of "Too much!" or "Not enough!" would be on both sides of the issue? Or how a last-minute compromise and the innocuous d-2 subparagraph would spawn the acronym ANILCA and a whole new round of debate?

Meanwhile, the oil flowed—and sometimes spilled—and the debate went on. "Whose land?" But then, the question of "Whose land?" has always reverberated throughout Alaska's history, particularly as each successive frontier was pushed upon it.

NATIONAL PARKS IN ALASKA

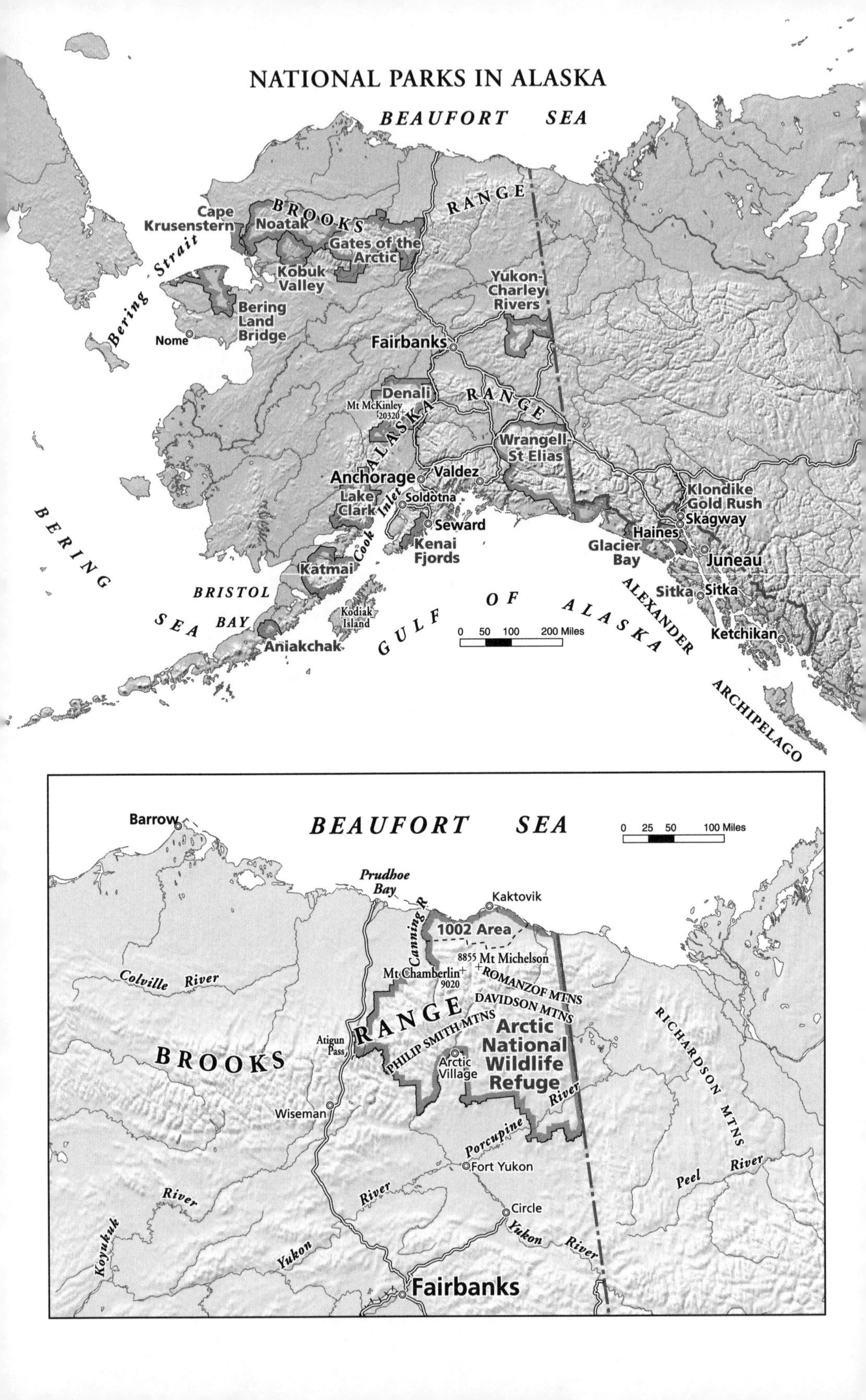

BOOK NINE

Whose Land? Competing Claims 1980–2001

Alaska's grandeur is more valuable than the gold or the fish or the timber, for it will never be exhausted. This value, measured by direct returns in money received from tourists, will be enormous.

—Henry Gannett, *Harriman Alaska Series*, 1901

The question "whose land?" *is not new to Alaska. Indeed, it has been asked with each successive frontier that has been crossed since the earliest peoples slowly plodded eastward across the Bering Land Bridge. But as the decades after the Second World War had shown, as the land filled up with often competing uses, the question "whose land?" was asked with increasing frequency, increasing antagonism, and increasing uncertainty over the future.*

Those who argued for preservation of wilderness were accused of stifling economic growth or even precluding a living wage. Those who promised that they could both develop the land's resources and protect them came up short, especially when a few moments of neglect exposed a system of complacency and let loose a quarter of a million barrels of oil upon the waters of Prince William Sound.

The long-delayed Alaska Native Claims Settlement Act of 1971 resolved some issues but created others. Out of its Section 17(d)(2) came the Alaska National Interest Lands Conservation Act of 1980. Though hardly thought of at the time as an economic measure, ANILCA pointed Alaska on a course toward a new economy where the beauty of the land itself could be harvested again and again. Time would show, however, that even this type of land use was not without its irreversible impacts and uneasy questions.

And when the warnings of one energy crisis were ignored during two decades of largely economic boom and the short end of the cycle returned, it was to Alaska that some in the United States once more looked to ease the burden. "Whose land?" everyone asked. "Whose land?"

d-2 Becomes ANILCA

In a land of superlatives, it is dangerous to bestow one—even after careful consideration. ANILCA, however, seems to deserve a superlative. With perhaps the exception of the clause in the Alaska Statehood Act granting the state the right to select 103 million acres from the public domain for state ownership, no land decision has had, or is likely to have, as significant an impact on the history of Alaska. ANCSA (the Alaska Native Claims Settlement Act) comes very close, but ANILCA—if for no other reason than it affected twice as much acreage—edges it out. Indeed, few acronyms have ever had such a major impact on a landscape.

ANILCA (the Alaska National Interest Lands Conservation Act of 1980) was spawned in the hectic final days of the political maneuverings to resolve the Native claims issues and thus make way for the pipeline. At the instigation of Arizona congressman Morris Udall, brother of former interior secretary Stewart Udall and no stranger to Alaska land issues, Section 17(d)(2) was added to the final version of ANCSA. It authorized the secretary of the interior to withdraw 80 million acres of unreserved public domain for study as possible additions to national parks, national forests, wildlife refuges, and wild and scenic river systems.

Although the "d-2" subparagraph did not contain the operative word *study,* it was assumed that such withdrawals were to be temporary and subject to future recommendations about which lands among these 80 million acres should be accorded some sort of permanent park or forest status. Two time triggers were put in place. The Department of the Interior had two years from the date of ANCSA to recommend the location and size of the lands to be withdrawn, and Congress had an additional five years to act on the recommended withdrawals. Lands initially withdrawn but not included in the secretary's recommendations would revert to the public domain.[1]

Needless to say, there were a number of conservation groups carefully watching this process. These organizations launched a campaign to arouse public interest in the entire Alaskan lands debate. They raised the question "Is Alaska Worth 8 Cents?" as a slogan to promote a letter-writing campaign—eight cents then being the cost of a stamp—to the Secretary of the Interior and members of Congress.

At the two-year deadline for recommendations in December 1973, Secretary of the Interior Rogers C. B. Morton asked Congress to incorporate 63.8 million acres into the national park and wildlife refuge systems. In particular, Rogers advocated the creation of three entirely new and enormously massive national parks. These were Gates of the Arctic in the heart of the Brooks Range, Wrangell–St. Elias between the Copper River and the Canadian border, and Lake Clark at the head of the Alaska Peninsula. Morton also proposed to expand Mount McKinley National Park.[2]

Now it was up to Congress to act, but the Arab oil embargo, the Watergate scandal, and a messy exit from Vietnam all distracted national attention away from Alaska. When attention did focus on Alaska, the major item in the news seemed to be the pipeline. There was, however, plenty of positioning going on off the radar screen. Robert Belous, a National Park Service planner in the Alaska Regional Office in Anchorage, summed it up best in an interview with the *New York Times* in October 1976. "People in the 'lower-48' don't know what's going on in Alaska," Belous claimed. "All they hear about is the pipeline. In fact, there's a big battle in the offing over what to do with the rest of the state. The whole country has something at stake in Alaska."[3]

That last line was undoubtedly true, but there were some, mostly in Alaska, who took strong exception to the concept of "national interest lands" being carved out of their own backyard. Once again, "outsiders" were telling Alaskans what was best for them. At the bottom of this divergence of opinion was the same battle that had been steadily escalating for decades between those who would lock up the land at one extreme and those who would promote almost limitless multiple-use development at the other.

In 1977, legislation was finally introduced from both of these camps to act upon Secretary Morton's recommendations. Congressman Udall carried the conservationist banner and advocated that 116 million acres—roughly one-third of the state and well above subparagraph d-2's 80 million acres—be designated as new park, forest, and refuge units. Resource development (logging, mining, sport hunting, etc.) would be precluded on most of them. At the other end of the spectrum, Alaska Senator Ted Stevens represented the views of many Alaskans, the majority of the state's legislature, and industry in general with a bill that proposed only 25 million acres of new conservation units, all without wilderness designation. It was left to Jimmy Carter's new administration and especially Secretary of the Interior Cecil Andrus to look for middle ground between these two positions.

By this time—well into the process—something very interesting had happened. Section 17 (d)(2) of ANCSA had not decreed 80 million acres of new parklands and wilderness, but only directed the study of *possible* additions up to that amount. However, the decade of the 1970s had been launched with the first Earth Day and then proclaimed the "Decade of the Environment" by no less than Republican Richard Nixon. There was generally growing sentiment—unless one happened to be sitting in some pipeline bar in Valdez—that preserving sizable chunks of Alaska was a good idea. In fact, it may well have been the pipeline construction that did as much as anything to fuel the cause of environmental conservation.

Congressman Udall backed off on some of the acreage in his bill, and on May 19, 1978, the House of Representatives passed an Alaska lands bill. It added about 100 million acres to new conservation units, 66 million of which were to be designated as wilderness. Alaska senators Gravel and Stevens opposed this version, and the Senate reported out a bill with considerably less acreage. When a conference committee was unable to reach a compromise, the measure died for the session.

Opponents of *any* Alaska lands bill noted with smug satisfaction that the five-year time limit for congressional action after Secretary Morton's original recommendations was due to expire in December 1978. If that happened, all of the d-2 lands would be thrown back into the public domain and possibly be subject to development. To prevent this, Secretary Andrus proposed that President Carter use his authority under the 1906 Antiquities Act to designate 56 million acres of the d-2 lands as national monuments.

Thus were created Aniakchak, Bering Land Bridge, Cape Krusenstern, Denali, Gates of the Arctic, Kenai Fjords, Kobuk Valley, Lake Clark, Noatak, Wrangell–St. Elias, and Yukon-Charley Rivers National Monuments, as well as Admiralty Island and Misty Fiords National Monuments on the Tongass National Forest. When Theodore Roosevelt obligingly signed the Antiquities Act in 1906, it is doubtful that even he of the bully pulpit could have imagined this sweeping an exercise of presidential power. Only Congress—and not the passage of time—could reverse such national monument designations.

Now the battle was joined. On one side was a loose coalition of some fifty state and national environmental and conservation organizations. This "Alaska Coalition" mobilized a vocal grassroots constituency to flood Congress with letters and telephone calls. On the other side was Citizens for Management of Alaska's Lands (CMAL). Its pro-development supporters included the U.S. Chamber of Commerce, the National Rifle Association, and the Alaska legislature.

Buoyed by President Carter's monument designations, the House of Representatives once again passed a Udall lands bill. This time, however, the acreage had swelled to 127 million acres, half of which were designated wilderness. The Udall bill sat for well over a year until late in the summer of 1980, when the Senate passed its own version. Through extensive lobbying, Ted Stevens's original, pro-development proposal had grown to encompass 104 million acres of new units with 57 million acres of wilderness—more than four times Stevens's 1977 proposal. It didn't match Udall's final numbers, but it was pretty close.[4]

A few months later, a new brand of conservatism swept Ronald Reagan into the presidency, and there was also to be a new Republican majority in the Senate. Morris Udall and other House Democrats quickly realized that the existing Senate version of the Alaska lands bill was not only "pretty close" but as good as they were likely to get. Accordingly, on November 12, 1980, a lame-duck session of the House of Representatives passed the Senate version, and on December 2, 1980, Jimmy Carter signed the Alaska National Interest Lands Conservation Act of 1980—forever after ANILCA—into law.

In short, ANILCA designated a total of 104.3 million acres as new conservation units—47 million as new and expanded national parks, monuments, and preserves; 53.8 million as wildlife refuges; and 3.5 million spread among recreation areas and wild and scenic river corridors. Among these additions, the Arctic National Wildlife Range was greatly expanded and became the Arctic National Wildlife *Refuge*, yielding yet another Alaskan acronym: ANWR. When these ANILCA additions were added to Alaska's existing national park and forest systems, the total acreage embraced 151 million acres, or about 41 percent of the state.[5]

If some people thought that ANCSA was an invitation to litigation, they should have waited to see ANILCA. The final bill included some 45,000 words. As befitting the landscape it covered, the law was grandiose in vision, sweeping in application, and complex and confusing in detail. ANILCA's purposes focused on two major issues. The first was preservation. New park, monument, and refuge units would preserve "unrivaled scenic and geological values," maintain "sound populations of, and habitat for, wildlife species," safeguard "arctic tundra, boreal forest, and coastal rainforest ecosystems," and preserve wilderness resource values and "related recreational opportunities." It was a tall order made all the more so by the second issue. Consistent with sound principles of fish and wildlife management, the act also sought to continue the opportunity for rural residents to engage in a subsistence way of life.[6]

The maintenance of subsistence lifestyles had, of course, been one of the nagging issues of the Alaska Native Claims Settlement Act. In the decade after ANCSA, however, the dichotomy between preserving traditional cultures and capitalizing on special privileges did nothing but widen. Some Natives religiously stuck to the old ways, hunting and fishing with the same methods that they had employed for generations. Others, flush with dollars from oil revenues, armed themselves with high-powered rifles and roared across the tundra on a fleet of new snow machines. Both groups said that they were practicing subsistence lifestyles. Meanwhile, rural non-Natives out in the bush cried foul over hunting and fishing privileges based on racial distinctions.

As with so many pieces of legislation, ANILCA's initial intent regarding subsistence was noble in purpose. Recognizing the uniqueness of life in rural Alaska, the statute promised the opportunity for traditional subsistence lifestyles for rural residents—Natives and non-Natives alike—and gave the state of Alaska the authority to oversee a unified system of fish and game management on both state and federal lands. In theory, this was supposed to unify wildlife management and ensure that the priorities of species conservation and subsistence harvesting came before sport hunting and commercial uses.

But what is noble in purpose is frequently litigated in the courts. The state of Alaska's administration of subsistence hunting and fishing with its priorities for rural residents was challenged on the basis of a provision in the state constitution prohibiting exclusive or special privileges to take fish or game. The Alaska State Supreme Court ruled that the same privileges of subsistence for rural residents must be extended to urban residents. This ruling was strange but manageable on state lands; however, it placed the mandated state law squarely at odds with the rural subsistence priorities required by ANILCA for federal lands. The end result was that in 1990 federal agencies began to manage subsistence hunting on federal lands.

This was far from the end of the subsistence debate. A Native named Katie John soon sued the federal government, claiming federal subsistence rights to the fish in Alaska's navigable rivers and offshore waters. Her claim ran squarely contrary to the state of Alaska's long-desired and now zealously guarded jurisdiction over its fisheries. In 1995, the federal courts ruled that federal subsistence priorities extended to all waters within the boundaries of the federal reservations. Federal agencies soon began to take this one step further by claiming management for subsistence purposes of fish migrating to federal lands and waters.[7]

"The most troublesome issue of subsistence," noted an *Alaska* magazine article, "is that of allocation—who is to get the fish and game?"[8] Given the court battles over subsistence that continue, that remark proved to be an understatement, but in reality it was just another way of asking "Whose land?"

If the legalese of ANILCA was complex and confusing, its geographic scope was downright mind-boggling. In one fell swoop both the National Park system and the National Wildlife Refuge system doubled in size—not just in Alaska but throughout the United States. No wonder the words *national interest* were part of the statute's title.

Mount McKinley National Park, first championed by Charles Sheldon, shows how deeply ANILCA cut the landscape. Almost 4 million acres were added to the park, tripling its size. While the mountain officially remained "McKinley" in the rolls of the U.S. Board of Geographic Names, Sheldon finally got his wish when the park was redesignated Denali National Park and Preserve.

And Denali was just the beginning. ANILCA created ten new units within the National Park system in Alaska. Wrangell–St. Elias National Park and Preserve became the largest national park in the United States, encompassing some 13.2 million acres—six times the size of Yellowstone. Gates of the Arctic National Park and Preserve was created around Bob Marshall's old stomping grounds—8.4 million acres in the heart of the Brooks Range. Kobuk Valley National Park and Noatak National Preserve promised to safeguard that much land again along two great river basins and the range of the western Arctic caribou herd. Cape Krusenstern National Monument and Bering Land Bridge National Preserve were created to protect the continent's earliest steps.

Two and one-half million acres became the Yukon-Charley Rivers

National Preserve, including the entire watershed of the 106-mile-long Charley River. Lake Clark National Park and Preserve was created amid the chaotic beauty of the meeting of the Alaska and Aleutian ranges with about 4 million acres of park and preserve—over half of which was designated wilderness. Aniakchak Volcano and its roily namesake river became the centerpieces of Aniakchak National Monument and Preserve. Finally, last in size only when measured against its Alaskan rivals, Kenai Fjords National Park was created to preserve the tidewater glaciers and idyllic fjords of the Kenai Peninsula. Additionally, two existing monuments were given national park and preserve status: Glacier Bay, where William S. Cooper first studied plant succession, and Katmai, which Robert Griggs had championed.

Out of ANILCA also came designations of twenty-five wild and scenic rivers, half of which were within the boundaries of national parks, monuments, and preserves. These scenic rivers included the Alatna, John, Kobuk, North Fork of the Koyukuk, and Tinayguk in Gates of the Arctic National Park and Preserve and the Chilikadrotna, Mulchatna, and Tlikakila in Lake Clark National Park and Preserve.

ANILCA also added lands to seven existing refuges in Alaska and created nine new ones, doubling the size of the National Wildlife Refuge system nationwide to 87 million acres, the vast majority of them in Alaska. The smallest Alaskan unit is Izembek Refuge, encompassing most of Unimak Island at the tip of the Alaska Peninsula. Izembek is studded with snow-clad volcanoes, and its broad coastal lagoons are sanctuaries for migrating birds, including the colorful Steller's eider. The largest refuge is the Yukon Delta, roughly the size of South Carolina, and home or stopping grounds for an estimated 100 million birds. The refuges at Kenai and Kodiak, originally proclaimed as ranges to protect moose and brown bears, were among those expanded.

Most of the remaining federal lands in Alaska not directly affected by ANILCA continued to be managed under general multiple-use policies by the Bureau of Land Management. These lands still number about 90 million acres—one quarter of the state. They include the National Petroleum Reserve in Alaska (what was once called National Petroleum Reserve No. 4 or "Pet 4") on the North Slope, the haul road corridor, a sizable chunk of land on either side of the 135-mile-long Denali Highway, and the historically controversial Katalla coal fields.

For all that ANILCA did on the side of conservation, there were some pro-development provisions. Most significant, ANILCA left the door ajar for future production of oil and gas from a portion of the Arctic National Wildlife Refuge. Dubbed the "1002 lands" because they were addressed in Section 1002 of the act, these 1.5 million acres lay along the coastal plain of the Beaufort Sea—east of the Canning River and the state lands at Prudhoe Bay. Some looked at this 100-by-25-mile rectangle and saw another Prudhoe Bay. Others saw an ecosystem that was critical calving grounds for the Porcupine caribou herd, as well as denning habitat for polar bears. The Udall version of ANILCA would have designated this area as wilderness, but when the Senate version was hastily agreed to after the 1980 election, limited oil and gas exploration here was part of the compromise.

ANILCA was very clear on two points about the 1002 lands. First, there had to be extensive wildlife monitoring and other environmental studies—essentially a massive and continuing environmental impact statement—before any exploration plan would be approved. Once approved, "exploratory activity" was limited to surface or seismic exploration. Second, if and when such surface exploration yielded promising results, there could be no leasing or other development—including initial test wells—unless authorized by Congress in separate legislation.[9]

Even if the door to drilling was thus ajar, these requirements seemed to make it almost impossible to force it open without congressional concurrence. But there are always loopholes. The Arctic Slope Regional Corporation, one of the Native corporations authorized by ANCSA, traded some of its selected inholdings within Gates of the Arctic National Park to the federal government in exchange for subsurface mineral rights on the 1002 coast plain. Then, with control of the mineral rights, the Arctic Slope Regional Corporation teamed up with one of its village corporations, the Kaktovik Inupiat Corporation.

The village of Kaktovik lies within the 1002 area boundary and by virtue of earlier selections around the village had what amounted to inholdings within the wildlife refuge. In 1986, the Arctic Slope Regional Corporation partnered with Chevron and British Petroleum to drill an exploratory well on Kaktovik village land within the boundaries of the Arctic National Wildlife Refuge. The results of KIC No. 1 (for Kaktovik Inupiat Corporation) were kept largely confidential, and Congress—sensing an end run—passed legislation to prohibit similar land or mineral interest swaps within ANWR without its approval. Even fifteen years later, the findings in KIC No. 1 remained a closely guarded secret.

Meanwhile, a five-year study completed in 1987 reported twenty-six major geologic structures that had an estimated one in five chance of yielding oil and gas—pretty fair odds in the wildcat terms of the industry. The report conceded that there would be some scarring and temporary disruption of wildlife from seismic explorations, and that these effects would increase considerably if Congress approved a full-fledged leasing and development program. Direct losses for drill sites, mud pits, and ancillary operations were estimated at 5,650 acres, with another 7,000 acres impacted by dust, altered snowmelt and erosion, small oil spills, and freshwater diversions.

The oil industry beamed and boasted that such impacts affected less than 1 percent of the 1002 lands. Conservationists looked at the pockmarks that the bevy of drill sites would cause and shuddered at the report's estimate that up to 37 percent of the area's core caribou calving grounds would be affected. The final word in 1987 was Secretary of the Interior Donald Hodel's concluding recommendation that "an orderly oil and gas leasing program for the entire 1002 area can be conducted in concert with America's environmental goals."[10] That remained to be seen.

ANILCA's legacy is still far from certain. It remains as dynamic and changing as the land itself. What is certain is that in its sweeping land use determinations, it answered the question "Whose land?" with a sort of riddle: "In the final analysis, the land belongs to all of us, and yet to none of us." No doubt that is a legacy that Charles Sheldon, William S. Cooper, Bob Marshall, and certainly Theodore Roosevelt himself would have understood.

The Big One

In late March 1989, as they prepared for another Easter weekend, pre-pipeline Alaskans—the sourdoughs of the twentieth century—were marking the twenty-fifth anniversary of the Good Friday earthquake and remembering the devastation that it had brought. Unbeknownst to them, an event was about to occur that would unleash an irrevocable series of events that would cause almost as widespread destruction. This time, however, the environmental disaster would be man-made. There was little that Mother Nature could do except to watch in horror. This was "the big one."

After participating in the first winter ascent of Denali back in 1967, mountaineer and writer Art Davidson had gone to work for a variety of environmental and Native causes and then served as a natural resources planner for the state of Alaska. Along the way, he was generally opposed to the pipeline, particularly after the presumably safer, all-overland trans-Canadian route had been summarily rejected in favor of the combination pipeline-tanker system. By the time oil was flowing down the line in 1977, Davidson remained skeptical but was somewhat resigned to its existence in light of the umpteen assurances of safety, safety, and more safety that emanated from the oil industry and both state and federal governments. One high-ranking British Petroleum executive later confided to Davidson: "You environmentalists drove us crazy, throwing up all those ecological concerns. We fought you. But if you hadn't slowed us down, we'd have made a real mess of things. Now we have a truly superior oil transport system."[11]

Maybe. To many, the weak link was definitely the tanker leg between Valdez and West Coast refineries. By 1989, there had been major changes to the tanker fleet since the *Manhattan*'s voyages through the Northwest Passage twenty years before. Many of the vessels calling at Valdez were now termed supertankers, capable of carrying twice the capacity of the refitted *Manhattan*. One thing, however, hadn't changed. Most of the tankers, including those only recently built, were still single-hulled.

Early in 1989, British Petroleum's *Thompson Pass* arrived in Valdez to reports that the ship was leaking oil. Under supervision from the U.S. Coast Guard, Alyeska was charged with the primary responsibility for monitoring and responding to spills. As a precaution, Alyeska placed a containment boom around the *Thompson Pass* but then neglected to monitor the situation. When the ship was next checked, oil was overflowing the boom. Additional booms were spread around the ship in a hectic routine that saw two work boats run aground and a third tear up part of the boom while pulling it across some rocks. It took the better part of two weeks to recover 1,700 barrels of oil in the narrow confines of Valdez harbor—relatively undisturbed by the tides. Compared to the 1.6-million-barrel, loaded volume of one of the supertankers, it was like spilling a teaspoonful of oil and then taking two weeks to dab it up with a napkin. For its part, Alyeska called the exercise "a textbook response."

To celebrate the cleanup of the *Thompson Pass* spill, Alyeska oil executives and technicians held a "safety dinner" at the Valdez Civic Center on Thursday evening, March 23, 1989. In another part of town that same evening, Valdez mayor John Devens convened a town meeting to discuss the *Thompson Pass* incident in light of the area's broader concerns about the impact of major spills on tourism and fisheries. Cordova fisherwoman and scientist Dr. Fredricka "Riki" Ott spoke and reminded the group about a 1980 incident when the loaded tanker *Prince William Sound* had drifted out of control for seventeen hours off Naked Island in the heart of the sound. Seventy-knot winds prevented tugs from towing the ship, and it wallowed around aimlessly until its engines were finally restarted, literally just minutes before the tanker would have piled up on a reef. That hadn't been the big one, but Riki Ott warned that such an occurrence wasn't a matter of "if," but only of "when."

Despite such history, Alyeska's spill-response plan—mandated by the pipeline authorization act in much the same way as the ballast treatment system—had suffered from neglect. A highly trained team dedicated to full-time spill response was disbanded in 1982, in part because it was also highly paid and expensive to maintain. And after all, despite the close call on the *Prince William Sound*, not much had happened. Had Alyeska just been lucky, or was the system pretty foolproof?[12]

Down at the pipeline terminal that same evening of March 23, the three-year-old, 987-foot tanker *Exxon Valdez* was being loaded with 1,264,155 barrels of Prudhoe Bay crude in anticipation of her five-day run to a refinery in Long Beach, California. This was a "light" load, well below the ship's 1.6-million-barrel capacity, because she was bound for Long Beach's shallow harbor. Her captain, forty-two-year-old Joseph Hazelwood, was a nineteen-year veteran of Exxon Shipping and had made the run between Valdez and Long Beach many times over the previous decade. This evening, however, Hazelwood was not on board overseeing operations. About 5:00 P.M., Hazelwood and his chief engineer, Jerzy Glowacki, had dropped into the Pipeline Club, one of Valdez's regular watering holes among the oil crowd.

Things were much quieter in Valdez now than during the heady days of the pipeline's construction, but with an average of two tankers a day arriving, loading up, and departing, the pulse of the town and its establishments was still very much measured in oil. Three hours later, after sipping vodka straight up, Hazelwood, along with Glowacki, moved on to the Club Valdez, ordered another drink and a pizza to go, and then caught a cab back to the terminal.

At 9:21 P.M., Hazelwood was on the bridge of the *Exxon Valdez* as her last mooring line was reeled in. Tugs shepherded the long vessel away from the dock, through Valdez Narrows, and into Valdez Arm. For these maneuvers, harbor pilot Ed Murphy had command of the ship, and Hazelwood—against company policy and accepted seamanship—went below after passing the Narrows. Two hours after getting under way, the *Exxon Valdez* was fourteen miles out of port and off Rocky Point, where the harbor pilots routinely relinquished command of a ship back to her captain and returned to port. Murphy ordered Hazelwood called to the bridge, transferred the command to him, and boarded a pilot boat to return to Valdez.[13]

Hazelwood radioed the Coast Guard control center that the *Exxon Valdez* was "hooking up to sea speed," and that her course might veer into the inbound tanker lane to avoid ice floes from nearby Columbia Glacier. The Coast Guard maintained a Traffic Separation Scheme (TSS)—essentially a maritime version of an interstate highway—with one 1,500-yard-wide lane headed into port and another outbound, and a 2,000-yard-wide separation zone between them. The inbound lane was free of traffic, and its use by outbound vessels was not unusual. But given the ice buildup, Hazelwood's subsequent orders to divert into the inbound lane and also engage the ship's autopilot while accelerating to sea speed were all highly unusual. Then Hazelwood did what no ship's captain does when his vessel is navigating a difficult course. He turned command over to Third Mate Gregory Cousins and went below to his cabin.

Cousins was faced with rounding the leading edge of the ice and then making a tight turn back to starboard (right) before the shallows of Bligh Reef appeared on his port (left) side. As long as the *Exxon Valdez* remained in the inbound lane there was no problem, but the tanker sailed straight on through the lane on the last course that Hazelwood had given. By the time that Cousins began to make course changes to starboard, it was too late. Within a couple of minutes of midnight, Maureen Jones, who was standing a lookout watch on a wing of the bridge, reported a blinking red light to starboard. It should have been to port. Cousins ordered another turn to starboard, and then put the helm all the way over. Cousins called down to Hazelwood, who was still in his cabin, with the ominous message, "I think that we are in serious trouble."

At 12:04 A.M., Cousins's thoughts of trouble became fact. The *Exxon Valdez* trembled and then bumped ahead another 600 feet before shuddering to a stop atop Bligh Reef. This finally brought Hazelwood running to the bridge. Rather than shutting down the engines or attempting to reverse them and back off the reef, he kept them running full ahead and ordered a series of right and left turns. For fifteen minutes, Hazelwood kept the throttles full ahead and spun the wheel right and left in an attempt to wiggle the tanker off the reef.

Plowing ahead with a 211,000-ton, deadweight tanker already listing four to five degrees to starboard only made a bad situation worse. Exactly how much these maneuvers added to the resulting disaster will never be known, but the smell of crude now enveloped the ship and control gauges showed that the tanker was rapidly losing oil. Later inspection would show eight of the tanker's thirteen cargo tanks to be ruptured. Hazelwood turned to Chief Mate James Kunkel and uttered the only comment of the evening that would not be challenged: "I guess this is one way to end your career."

A full twenty-three minutes after the *Exxon Valdez* ran aground, Hazelwood finally radioed the Coast Guard control center to report the accident. The duty officer called the captain of the port, Commander Steve McCall, who by a strange coincidence had gone to maritime school many years before with Hazelwood. The duty officer woke McCall with the news. "This is the big one. We have the *Exxon Valdez* aground on Bligh Reef."[14]

While Hazelwood continued to make matters worse by attempting to refloat his vessel with a series of grinding maneuvers, the Coast Guard notified Alyeska, which in turn alerted its spill-response team. With no specially trained crew on twenty-four-hour alert to function much like a team of firemen, the spill-response team now had to be assembled from regular workers. Five long hours after the *Exxon Valdez* ran aground, thirty-nine workers assembled at the Alyeska docks to receive spill assignments. The spill-response barge with booms and oil skimmer equipment that should have been under way within minutes of the incident was sitting idle in dry dock awaiting repairs from a battering it had taken in a winter storm. By then, Coast Guard lieutenant Thomas Falkenstein had reached the stricken tanker and reported that at least 138,000 barrels of oil were already in the water and that 20,000 more were escaping every hour.

Grave as this situation was, as the *Exxon Valdez* spilled her cargo her list increased to starboard. Unless her remaining cargo was transferred quickly, there was the distinct risk that the entire vessel would sink or split in two and disgorge another million barrels of oil. The tanker *Exxon Baton Rouge*, empty and inbound to Valdez for a load of her own, was diverted to the *Exxon Valdez* to await a delicate transfer operation. Meanwhile, nothing was being done to contain the oil. At 2:30 P.M. the next afternoon—more than fourteen hours after the grounding—the Alyeska barge finally chugged onto the scene. Six hours later, the *Exxon Baton Rouge* was alongside the stricken tanker, but the transfer of oil between the two ships did not begin until early Saturday morning.

By Saturday morning, March 25, almost a quarter million barrels of oil were in the water and spreading rapidly throughout eastern Prince William Sound. Chaos reigned. There was confusion between Alyeska and Exxon over who was in charge of the cleanup and heated debate among all parties—including the state of Alaska and area fishermen—over the use of dispersants. Finally, there was the utter magnitude of the spill and the pitiful level of the response. By day three, only 3,000 barrels had been recovered— roughly 1 percent.

It seemed as though things couldn't get worse, but they did. By dawn on Monday morning—day four—a blizzard roared out of the Chugach Mountains and pummeled Prince William Sound with twenty-foot waves and high winds. While containment with miles and miles of boom was by then problematic at best, this storm swept the slick out of the center of the sound, broke it apart like so many multiplying cells, and sent them crashing into the beaches of a myriad of islands. Like shock waves radiating out from the epicenter of an earthquake, the oil spill had spread over 500 square miles by Monday evening. Elsewhere, criminal charges were filed against Captain

Joseph Hazelwood, accusing him of operating a vessel while intoxicated, reckless endangerment, and negligent discharge of oil.[15]

Those inclined to draw a parallel between the spill and the earthquake twenty-five years before were quickly reminded by the oil industry that unlike the quake, there had been no loss of human life and, while a mess to be sure, the physical destruction was not as violent. That wasn't terribly comforting if one was a sea otter slowly sinking below the waves with fur matted with oil or one of an untold number of seabirds vainly trying to beat one last flap of wings to escape what had become black death. And if one was a Cordova fisherman (or fisherwoman like Riki Ott), the spill made you sick in the pit of your stomach, and then it made you damn angry. You went out and did whatever you could to make whatever difference you could, no matter how small it seemed at the time.

Cordova fishermen mobilized a fleet of fishing boats and other vessels the likes of which had not been seen since the evacuation of Dunkirk. This "mosquito fleet" sped to lay booms to protect the salmon hatcheries in the western reaches of Prince William Sound and then started recovering oil with everything from "honey-dipper" shovels to fish suction pumps jerry-rigged to suck up oil. All across the sound, it was the same story. People did what they could with what they had.

While some skimmed oil, others tried to help the victims. Volunteers tried to save flocks of plovers, puffins, cormorants, kittiwakes, and other seabirds, shuddering each time they heard a new flight of unsuspecting arrivals. Eagles swooped down for fish and came up coated in thick black muck—if they came up at all. Dr. Ken Hill, a young veterinarian in Cordova, concentrated his efforts on rescuing sea otters. Knight and Latouche Islands along Montague Strait were eerie death zones, he reported. "You see these gorgeous mountains in sunlight, but there is this silence. Just like a silent spring."[16]

Into this mess steamed help from an unlikely source. The Russians were returning to Alaska. Despite bureaucratic grumbling about a Russian vessel entering American waters, the Russians offered use of their 435-foot *Vaydaghubsky*, the world's largest combination oil-skimming vessel, harbor dredge, and fire-fighting boat. The *Vaydaghubsky* had an onboard storage capacity of almost 50,000 barrels and could suck up that much in less than a day—provided the spill was concentrated in a small area or contained by booms. Unfortunately, by the time the Russian vessel steamed into Resurrection Bay on April 19, much of the oil had spread out to numerous beaches, and the big ship's efficiency was greatly reduced. It was a pity that the *Vaydaghubsky*—or an American version of her—hadn't been in Valdez on the night of the spill. More than one Coast Guard official looked over the state-of-the-art ship and wondered which side had won the Cold War after all.[17]

Throughout much of the first few days, Exxon corporately—and most everyone else—seemed in a daze. Despite the oil industry's collective promise of safety and a rapid cleanup if there was a problem, no one was prepared for a spill of this size. Confusion over who was in charge among Exxon, Alyeska, the Coast Guard, and the Alaska Department of Environmental Conservation fueled indecision and added to delays. Much to his credit, Frank Iarossi, president of Exxon Shipping, was bound for Valdez from Houston within twelve hours of the grounding and stayed for weeks amid death threats and talk of bullet-proof vests for Exxon employees. Anger and recrimination on all sides flowed even farther than the oil.

And the oil was flowing much farther than any had initially thought. When news of the spill first reached the National Park Service's regional office in Anchorage, there was anguish, but also some sense of relief that it hadn't happened along a national park shoreline. The relief, however, was short-lived. Within a week of the spill, oil was flowing out of Prince William Sound—straight for the pristine landscapes and wildlife havens of Kenai Fjords National Park, the Katmai coast, and the ubiquitous islands of the Alaska Maritime Wildlife Refuge.

All along the way, there were wishful reports that perhaps the oil wouldn't reach any farther. On Kodiak Island, Charles Christensen, an Aleut and mayor of tiny Larsen Bay, didn't believe them. "All of us were being told that the oil wouldn't get down this far, but fishermen know the waters. They know the winds and the tides. Once the oil started working its way to Shelikof Strait, everybody realized it was coming."[18]

Valdez, of course, had been ground zero—almost at the epicenter of another jarring experience. The town's population doubled overnight. A dozen flights a day mushroomed to between 300 and 400, with a peak of 687 on March 30. It was the pipeline boom all over again as cleanup crews were hired and local services were taxed to their limits. But, as always, Valdez knew how to survive a boom no matter how egregious its cause. One Valdez cabdriver whose business suddenly tripled was quoted as saying: "This means millions for us. This town is going to grow."[19]

There were others, of course, who were very concerned that along with wreaking environmental havoc, the spill would also wreak economic havoc on the state's burgeoning tourist industry. Images of dead seabirds and otters, oily beaches, and a seemingly endless path of thick, black goo filled the evening networks and the nation's newsmagazines. Alaska's chambers of commerce winced but could do little about national reporting. However, when the state's own *Alaska* magazine—long a voice for trumpeting the scenic wonders and encouraging visitation—devoted its May 1989 issue to detailed coverage of the spill, some in the tourism community went ballistic—even going so far as to cancel advertising space in the magazine.

How, *Alaska* publisher Ron Dalby asked, could the magazine do anything but cover this cataclysmic event? His staff, better than any other, had been in the trenches and knew firsthand the loss. In the end, Dalby and his staff received widespread journalistic praise for both their work and their independence. And as far as tourists were concerned, if anything, *Alaska*'s reporting of the plight of Prince William Sound served to coalesce a growing constituency of those determined to see that this never happened again.

Exxon was only too eager to funnel money to the Alaska Visitors Association to downplay the effects of the spill and encourage tourists not to cancel their plans. The oil industry as a whole was only too aware of the bigger stakes. The spill was only ten days old when an editorial in the *Oil and Gas Journal* noted that "industry, not just Exxon, will pay for the *Exxon Valdez* spill. The first casualty will be some, if not all, of the recent progress toward leasing of the Arctic National Wildlife Refuge Coastal Plain. . . . Superior performance," the editorial went on, "is the only entrée to environmentally sensitive areas."[20]

It didn't matter, nor did it get much press, that almost 7 billion barrels of crude had been transported out of Valdez on some 9,000 tankers over the prior eleven years without a major mishap. Opponents of the pipeline had long said that a major spill would happen. The oil industry had promised that it would not. Now the price would be just as high for being right as for being wrong.

In the midst of the emotional outrage that was heaped upon oil companies in general and Exxon in particular, there came a determined but well-reasoned condemnation from an unlikely source. The Oil, Chemical, and Atomic Workers Union, representing some 40,000 oil workers nationwide, announced opposition to drilling in the Arctic National Wildlife Refuge until Congress enacted a comprehensive national energy policy. "I'm for drilling," acknowledged union vice president Robert Wages, "but there has to be some sanity to the process, and there doesn't seem to be any. Saying you shouldn't drill in the Arctic National Wildlife Refuge ought not be viewed as religious heresy by the industry." Acknowledging that such a position "is certainly not job producing," Wages maintained: "We don't believe it's job destructive either. We'd like Congress to take up the nation's energy policy now, before the ghost of the *Valdez* is gone. Because as long as that ghost is there, there's not going to be drilling up there. When that ghost disappears there will be a subtle, quiet, subterranean move to make sure that they are permitted to drill."[21]

The ghost of the *Exxon Valdez* has faded, but it is definitely not gone. Throughout the summer of 1989, Exxon surveyed more than 3,500 miles of shoreline and cataloged 1,100 individual segments. Then it mobilized 11,000 workers to clean some 1,500 miles of shoreline with everything from high-pressure hot-water hoses to paper towels. Later research showed such high-pressure blasting to be detrimental because it killed many small intertidal organisms. As cleanup costs exceeded $1 billion, the question became "how clean is clean?" More than five years after the spill, some areas were still being treated with an experimental biodegradable hydrocarbon cleaner dubbed PES-51. This mega–spot remover was designed to free oil molecules from both porous and nonporous surfaces and allow them to rise to the surface, where the oil was trapped by booms and collected by skimmers.[22]

As complicated and controversial as the cleanup was, it was straightforward when compared to the endless rounds of legal maneuverings. The state of Alaska and the United States settled all criminal and civil claims against Exxon in the fall of 1991 in a three-part agreement. On the criminal side, Exxon was fined $150 million, the largest fine ever imposed for an environmental crime, but then the governments forgave $125 million of that fine in recognition of Exxon's cleanup efforts. Exxon later put these cleanup costs at $2.1 billion. The balance of the criminal fine ($25 million) was divided among the North American Wetlands Conservation Fund and the national Victims of Crime Fund. Additionally, Exxon paid $100 million as criminal restitution for damages caused to fish, wildlife, and beaches, and these funds were divided between the state of Alaska and the federal government.

Finally, there was a civil settlement wherein Exxon agreed to pay $900 million in ten annual payments, plus an additional $100 million for subsequently discovered damages. Roughly a quarter of the $900 million went to reimburse state and federal governments for spill-response and cleanup efforts, and the balance was used to fund the *Exxon Valdez* Oil Spill Trustee Council. Made up of state and federal representatives, it was charged with supervising and funding long-term habitat restoration and marine science monitoring programs. Because the council created significant endowments from these funds, these efforts continued after Exxon delivered its last settlement check in 2001. Thus, the questions "how clean is clean?" and "how significant are the long-lasting impacts?" are likely to be asked for many years to come.[23]

The settlement that Alaska and the United States reached with Exxon was, however, only the tip of the litigation iceberg. A multitude of civil cases filed against Exxon by boroughs, towns, Native corporations, and private parties are still on appeal in the federal courts. At issue for resolution and division is some $5 billion in punitive damage claims that have been levied against Exxon. Meanwhile, an Alaskan jury found Captain Joseph Hazelwood not guilty of the charge of operating a vessel while under the influence of alcohol, but guilty of negligent discharge of oil. He was fined $50,000 and ordered to begin serving a five-year program of community service in 1999.

In the aftermath of the *Exxon Valdez* spill, Alaska governor Steve Cowper issued an executive order requiring two tugs to escort every loaded tanker from Valdez all the way out through Prince William Sound to Hinchinbrook Entrance. The order also required Alyeska to revise its oil spill contingency plan. As the plan evolved through the 1990s, an oceangoing tug and a 210-foot Escort Response Vessel (ERV) were routinely assigned to escort loaded tankers through the sound. The ERVs were equipped to tow or otherwise assist the tankers and to carry immediate spill-response equipment. Once again there is a full-time, around-the-clock response team on duty in Valdez. One thing hasn't changed since that Good Friday in March 1989. The majority of tankers calling at Valdez are still single-hulled, although Congress has enacted legislation requiring that all tankers in Prince William Sound be double-hulled by 2015.

And what of the infamous *Exxon Valdez*? Refloated and towed off Bligh Reef, the vessel was taken to a Japanese shipyard for repairs. It was renamed the *Sea River Mediterranean* and then put to work hauling oil in the Atlantic. It is prohibited from ever returning to Prince William Sound.

Toward a New Economy

Historically, Alaska's economy has been successively dominated by one major industry: first furs, then mining, military spending, and oil. When the spark plug of the time misfired, economic fortunes declined until reignited by another behemoth. As the 1980s passed, the oil industry in Alaska lost some of its luster. The *Exxon Valdez* fiasco certainly didn't help matters, but the economic root of the problem lay much deeper. In the mid-1980s, the price of oil dropped sharply worldwide. What had been thirty-dollar-a-barrel oil was suddenly selling for half that a few months later.

Meanwhile, oil production from Prudhoe Bay, while still significant, was nonetheless showing signs of the normal decline curve of an established field. The number of corporate players on the North Slope was also decreasing because of a variety of mergers. By 2000 after the dust had settled, two giants, BP Amoco and Exxon Mobil, stood atop the pile. Clearly, oil was going to remain a potent force in the state's economy, but over the preceding two decades other industries had joined it in moving the state toward a new economy.

Since before World War II, logging and fishing had long been looked to with the hope that they would fill the growing economic void left by the decline of hard-rock mining. Despite a brief heyday in the 1950s and 1960s, the postwar logging boom quickly skidded to a halt when confronted with new multiple-use land management practices and a growing aversion to clear-cuts. The pivotal turning point was the 1976 cancellation of U.S. Plywood–Champion's fifty-year timber contract and its plans to build the southeast's third pulp mill north of Juneau. The two other fifty-year contracts were terminated early, and the pulp mills at Sitka and Ketchikan closed in 1993 and 1997, respectively.

To be sure, there is still logging being done on the Tongass—too little to sustain the industry locally say some, too much to sustain the forest perpetually say others. Louisiana-Pacific continues to operate two sawmills, but even these scaled-back operations have come under attack.

Some logging critics are environmentally green outsiders—the same type of perceived interlopers that "real" Alaskans have been trying to ship back to the lower forty-eight for decades. Other critics, however, speak from much closer to home.

Ex-lumberjack Vern Ably and former pulp-mill worker Wayne Weihing, for example, are no outsiders. Ably spent thirty years in the timber industry, and Weihing put in twenty-one years at the Ketchikan Pulp Mill. Between them, they have become the poster boys of the anti-logging movement. "I don't advocate zero cut," said Weihing, "but I do advocate a sustainable cut." Ably remembered cutting a "cathedral-like grove of ancient spruce trees" near Ketchikan near the end of his career. "Before we went in," he recalled, "I took a walk through the forest and my breath was just taken away it was so beautiful. A week later, it was absolutely nuked."

Not surprisingly, the Alaska Forest Association, a timber-industry trade group based in Ketchikan, calls the two former loggers environmentalist pawns. Arguing that the area's high rainfall supports sustainable yields even with clear-cutting, another local, Scott Habberstad, said, "When you cut them down, it's exactly like harvesting corn."[24] As always, the issue is hardly black and white, but whatever new economy Alaska is moving toward, logging is playing less and less of a role in it.

If the economic promise of logging as a pillar of Alaska's economy has gone largely unfulfilled, fishing is a different story. The salmon industry has had its ups and downs over the years, affected by both nature's cycles and the overharvesting or conservation edicts of a particular period. During one of the salmon downturns in 1972, Alaska restricted commercial salmon fishing to 12,500 permit holders. Four years later, Congress passed the Magnuson-Stevens Act, which created a 200-mile offshore fishing zone open to foreign as well as domestic vessels, but subject to U.S. regulations. Under the conservation measures, commercial salmon catches went from 22 million fish in 1973 to 132 million fish in 1984.

Halibut have enjoyed a similar comeback. American and Canadian halibut fleets virtually fished out the North Pacific before the two countries adopted a halibut conservation treaty in 1923. The species slowly revived and has become an important staple to the fishing fleets of many southcentral and southeast communities. As Homer's annual Halibut

Derby proves, some of these fish can exceed 500 pounds and are second only to salmon as a mainstay of Alaska's fishing industry.[25]

Crabs have also been an important part of Alaska's commercial fishing scene, but this species, too, has had to contend with cycles. When Congress put an end to all sealing in the Pribilofs during the early 1980s, the largely Aleut population on St. Paul and neighboring St. George Island turned to crabbing. A decade later, St. Paul was booming as one of the busiest, most modern crab-processing ports in the world. Then in 1999, marine biologists found that the number of male snow crabs in the Bering Sea had fallen to an all-time low. The Alaska Department of Fish and Game attributed the decline to overfishing and imposed an 85 percent cut in the annual crab catch. St. Paul's economy was struck a staggering blow. Only those few fishermen who managed to take part in the halibut season got a brief respite from the economic gloom.[26]

Salmon fishing itself has not been immune to such cycles. In 1997, Bering Sea runs of sockeye and chum salmon fell to a twenty-year low. Because of the economic impact on hundreds of small businesses, Alaska governor Tony Knowles declared an economic disaster in Bristol Bay and the Kuskokwim River drainage and asked for federal assistance under the Magnuson-Stevens Act. The low runs were 78 percent below average and caused a drop in revenues of almost $100 million. Returns to the sea in previous years had been normal, but biologists surmised that unusually high temperatures and calm seas during the spring months had somehow dramatically affected salmon distribution and migration. A similar situation occurred in 1998.[27]

In general, however, fishing is now firmly entrenched with oil as one of the pillars of Alaska's economy. In 1999, a record 5.6 billion pounds of fish and shellfish—about twenty pounds for every man, woman, and child in the United States—was harvested in Alaskan waters. Incredibly, this accounted for 55 percent of U.S. seafood production, almost four times more than the next largest producing state. Fishing provided income for an equivalent of 20,000 full-time jobs, but because so much of the industry employs seasonal and part-time work, as many as 75,000 individuals received all or part of their income from either commercial fishing or seafood processing.[28]

On the sport fishing side of things, the Alaska Department of Fish and Game estimated that in 1997, nearly half a million anglers wet a line in Alaska. About half of them were residents and the other half visitors. If the old adage about days spent fishing not counting against one's life span is true, each of these half million anglers averaged five "free" days a year, accounting for a total of 2.6 million person-days fishing.[29]

Oil. Fishing. So what was left to round out the "Big Three" of Alaska's new economy? Henry Gannett first spoke of Alaska's scenery in economic terms in his report of the 1899 Harriman expedition. "Alaska's grandeur," wrote Gannett, "is more valuable than the gold or the fish or the timber, for it will never be exhausted. This value, measured by direct returns in money received from tourists, will be enormous."[30]

From Harriman's elite excursion to giant cruise ships docking in southeast ports, tourism has grown to become the third leg of Alaska's current economic triangle. Indeed, tourism is the only private sector industry that has grown continuously since Alaska became a state. In 1999, 1.4 million visitors poured $800 million into the state's economy. Three out of four of them said that they had come primarily to sightsee and view wildlife.[31] Apparently, Henry Gannett's prediction had come true.

But what Gannett did not fully appreciate—and what is still not completely understood a century later—is that tourism, too, causes environmental change. Hundreds of thousands of people have an impact. Indeed, in some areas dozens of people can have an impact. Tourism, just like any other industry, must balance its use of the land's resources with the sustainability of those resources. Successfully doing so is not only critical to Alaska's future economic well-being but also to its overall quality of life for visitors and residents alike.

When the Sitka pulp mill closed in 1993, naysayers predicted that the town would lose 900 jobs and 1,900 people. In the three years after the closure, the town lost only about a quarter of that, and then employment and population began to increase. During the same period, real estate prices increased substantially. Ketchikan weathered the closure of its mill in much the same fashion. Evidently, there were forces pushing both of these towns toward a new economy. In both cases increased tourism was at the forefront, and at the root of this new rush was the explosion in the number of cruise ships plying Alaskan waters.[32]

Alaska's long summer days have always meant a fever pitch of activity, but when cruise ships are in port, local businesses are literally open at midnight. From May until September, the local economies of Ketchikan, Juneau, Sitka, Skagway, and Seward revolve around the arrivals and departures of cruise ships. It is not unusual to find shops closed for the afternoon when the docks are empty, but sporting a sign "Open at 4:00 A.M." if a ship is due to dock with the tide in the wee hours. The Ketchikan Visitors Bureau estimated that the number of cruise passengers arriving in Ketchikan more than doubled from 236,000 in 1990 to 565,000 in 2000. No wonder that former logger Wayne Weihing looked out his window in Ketchikan and said, "That cruise ship right there represents our future."[33]

Maybe. At some levels, cruise ships bring the economic benefits of tourism without much of the infrastructure burden. A ship docks, and thousands of people get off, take in the scenic wonders, spend some money, and then get back on and sail away—frequently all within a matter of a few hours. Except to provide a few flush toilets, visitor services are usually limited to local-color entertainment, museums, restaurants, and an endless string of gift shops.

At other levels, many have concerns that there is a finite limit to tourism's bounty. Some old-timers say that the ships have changed the southeast towns for good—and for the worse. Others grapple with the frenzy of having to make a year's income in a few short months. Still others must manage the impacts that even these floating hotels have on the regions they cruise. Governor Tony Knowles called a special session of the Alaska legislature in May 2001 just to deal with the issue of wastewater dumping by cruise ships.

Nowhere is cruise ship management more contentious than in the waters of Glacier Bay National Park and Preserve. Almost 450,000 people visited the park in 1999. The vast majority never set foot on land. They watched glaciers calve from multi-storied cruise ships that at first glance have little impact. But park managers are concerned about the impacts of propeller noise and wake disturbance on both humpback whales and Steller sea lions and of smokestack emissions on rare plants, nesting birds, and other animals.

The cruise ship lines have consistently lobbied to increase the annual number of vessels permitted into Glacier Bay. From only a few dozen visits in the 1970s, this number grew to 107 visits in 1993. Then the National Park Service adopted the Vessel Management Plan (VMP), which permitted 219 scheduled visits by cruise ship in 2001. The National Parks and Conservation Association challenged these higher limits on the grounds that the National Park Service had failed to provide a thorough environmental impact statement prior to implementing the new plan. On appeal, the U.S.

Ninth Circuit Court of Appeals agreed and ordered that the increases not take effect until the completion of an EIS.[34]

Glacier Bay's juxtaposition of tourism and wilderness is a reminder that ANILCA was not an end in itself, but rather a continuation of the ongoing debate over the true meaning of wilderness. Perhaps one of the most telling examples of this is sprawling Wrangell–St. Elias National Park. Discoveries of copper at Kennecott and gold at Chisana showed that its rugged geography was no barrier if one was determined enough to breach it. Then for years the area slumbered under the auspices of the Bureau of Land Management. A few mountaineers climbed its peaks, a few trophy hunters roamed its valleys, and people went about their subsistence livelihood undisturbed. The land, the wilderness, was relatively secure, relatively unchanging.

ANILCA made the region a national park and thrust aside its veil of anonymity. People thought, Hey, if it's a national park, we'd better go check it out. Suddenly in a few short years, roads were improved, new bridges were built, private mining property was acquired, and visitor services were expanded. Thus, even on this scale of wilderness, there are tough questions. How many backpackers a year are too many? How many landings of planes in quiet lakes? Who dares to suggest another road? For now, the vastness of the scale disperses this use and helps insulate the land, but what of another decade or two from now? Are the Wrangells, the Brooks Range, and other regions living arguments that the best thing for true wilderness is just that—wilderness?

Once again, there is no black-and-white answer. These debates will continue and indeed intensify as tourism not only remains a part of the triumvirate of Alaska's economy along with oil and fishing, but also is likely to pass both a few years down the road.

Some might argue that over the years Alaska has had one other major industry: reacting to disasters and/or crises. The Second World War fueled the construction of infrastructure that even the mining boom hadn't. The Cold War brought bigger bases, long-term assignments, and the construction bonanza of the DEW Line and BMEWS stations. The earthquake was hell, but an awful lot of building went on in its aftermath. And the pipeline? Well, only some thought that a disaster, and even they couldn't refute the fact that its construction and the development of Prudhoe Bay had been Alaska's economic main vein for the better part of a decade.

Certainly, there could be no argument that as Alaska moved toward a new economy in the last quarter of the twentieth century, the Native corporations created by ANCSA were key players in every aspect of it. The regional corporations have become multifaceted conglomerates with interests ranging among oil, timber, fishing, construction, lodging, communications, health care, and education. Remember the Arctic Slope Native Association that some viewed as an upstart nuisance in the 1960s? Its ANCSA successor, the Arctic Slope Regional Corporation, is now the largest Alaska-owned corporation in the state, employing more than 1,000 of its Inupiat shareholders and grossing more than $850 million a year from a variety of business interests.

The largest Native corporation geographically is Doyon, Limited, headquartered in Fairbanks and embracing some 200,000 square miles (1.3 million acres) of Athabaskan land at the heart of Alaska's interior. The other Native regional corporations are the Bering Straits Regional Corporation on St. Lawrence Island and east of Nome; the NANA Regional Native Corporation in the northwest around Kotzebue; Calista Corporation in the Yukon and Kuskokwim Deltas; the mostly urban Cook Inlet Region around Anchorage; the Bristol Bay Native Corporation in the southwest around Bristol Bay; the Aleut Corporation in the Aleutians and Pribilofs; Chugach Natives around Prince William Sound; Sealaska Corporation in the southeast; Koniag on Kodiak Island; and Ahtna along the upper Copper River.

In the early years after ANCSA, those Natives who rushed to embrace the new corporate culture—or at least to profit from it—were frequently called "Brooks Brothers" Natives by both admirers and detractors. A generation later, it is impossible to take more than a step or two in Alaska and not have some interaction, directly or indirectly, with the business interests of the Native corporations. The questions of "whose land?" have certainly not all been resolved, but Alaska Natives have finally taken their rightful place as integral cogs in the state's economy.

Oil, fishing, tourism. For the moment, they make up the core triangle of Alaska's economy. Look behind the veil of each. The success of each is much more dependent on the success of the others than any individual segment would like to admit. Commercial fishing is likely to continue as a mainstay. Oil and gas will remain important, but as one of a triad and not the omnipotent force it was in the 1970s. Tourism and recreation will continue to grow, but even tourism will at some point be limited by the carrying capacity of the land.

A 2001 study prepared for the Alaska Conservation Foundation by the University of Alaska at Anchorage outlined the limits of that carrying capacity in terms of conflicts over access to and use of public lands. "Alaskans will need to think through and discuss," the report concluded, "what forms of tourism access are most compatible with resource protection, a healthy economy, and their own quality of life."[35] Thus, as the state moves toward a new economy, Alaskans will have to continue to strive to answer the question "whose land?" more than ever.

ANWR—2001

There is a certain parallel to be drawn between the Canning River of Alaska's North Slope and some of history's more famous watery lines of demarcation. West of the Canning is oil country—the Alaska state lands surrounding Prudhoe Bay. East of the Canning is wilderness—the federal lands of the Arctic National Wildlife Refuge. To be sure, it is far more complex than that generalization, but as the twentieth century came to a close, the armies of development and wilderness marshaled their legions of lawyers, lobbyists, media consultants, and volunteers and figuratively dug in on either side of the Canning. Some vowed that when the battle was joined, it would be a fight to the finish.

The Canning River rises from glaciers on the south side of Mount Chamberlain deep in the heart of the Franklin Mountains. In the course of his 1826 coastal exploration, Sir John Franklin named the river for George Canning, the British secretary of state for foreign affairs who had negotiated with the Russians over Alaska's eastern boundary. Joined by its Marsh Fork, the river flows north out of the mountains and then across the broad coastal plain to the Beaufort Sea, a total of some 125 miles.

Between the Canning and the Canadian border 130 miles to the east, no less than fourteen other rivers follow a similar yet varied course from mountains to plains to sea. Their names challenge non-Inupiat tongues, but their Anglo translations give a clear picture: Katakturak, "river where you can see a long way"; Okpilak, "river with no willows"; and Kongakut River, "farthest away river." The Hulahula River is an exception and was named by whalers homesick for warmer climes. The Philip Smith Mountains at the heads of these rivers divide more than just the waters of the Yukon and the North Slope.[36]

The coastal plain is the summer calving ground of the Porcupine caribou herd, one of four major caribou herds that migrate throughout Alaska's North Slope and Canada's Yukon Territory. As many as 30,000 calves are born on the coastal plain each spring. The caribou share these calving grounds with polar bears, musk oxen, nesting snow geese, and a plethora of migratory birds. The coastal plain is also the home of the Inupiat Eskimos of the village of Kaktovik, population 260 in 2000.

Generalities are dangerous in an ever-changing land, but the Kaktovik Inupiat Village Corporation and most of its citizens have long supported oil and gas drilling—urged on by the pro-development experiences of their parent regional corporation, the Arctic Slope Regional Corporation. South of the Philip Smith Mountains, it is a different story.

Arctic Village, population 130, is one of about fifteen villages of the Gwich'in Athabaskans, who call themselves "the people of the caribou." The Gwich'in Athabaskans have not been weaned from traditional subsistence lifestyles by the lure of oil. They have long opposed oil and gas development on the coastal plain and remain particularly fearful that such operations would disrupt the seasonal migrations of the caribou upon which they and much of the rest of the land's inhabitants depend. In earlier times, the Gwich'in crossed the mountains by sled and on foot to trade with the Inupiats. Now the debate over oil divides them.

The combatants in the war to drill on ANWR have faced each other across the Canning River ever since the passage of ANILCA. In March 1989, with Republican George H. W. Bush freshly elected to the White House, Senate Republicans moved to act on the 1987 recommendations of the ANILCA-mandated EIS. The Senate Energy and Natural Resources Committee passed a bill authorizing a leasing and exploration program for ANWR's Section 1002 lands. A week later, the *Exxon Valdez* disemboweled herself on Bligh Reef. As images of blackened beaches and oil-encrusted wildlife filled television screens and magazines, the measure was never reported to the floor for a vote. Two years later, these images had receded somewhat, and in the aftermath of the Persian Gulf War, President Bush's National Energy Bill authorized drilling in ANWR. A filibuster by Senate Democrats kept the measure from coming to a vote.

There was, however, a battle brewing that would not be postponed indefinitely. There was too much at stake—on both sides. An editorial in the *Oil and Gas Journal* after the 1992 election left no doubt at the stakes. "The next battle over ANWR," the journal opined, "will be the fight for everything. . . . Whatever the odds, industry—*all of it*—must fight this battle to the end."[37]

In 1995, flush from a takeover of the House of Representatives, Republicans prepared to take up the battle again and included a provision for ANWR drilling in the federal budget. President Bill Clinton vetoed the entire budget and expressed his intention to veto any other bill that would open ANWR to drilling. So Alaska's Republican congressional delegation of Senators Ted Stevens and Frank Murkowski and Representative Don Young, who had all been vigorously pursuing such legislation, chose to bide their time along with the rest of those on the west bank of the Canning.

In 1998 during this legislative standoff, the U.S. Geological Survey updated its assessment of in-ground oil reserves that it had originally compiled for the 1987 EIS. As the battle to drill or not to drill escalated over the next few years, both oil industry and environmental groups would pick and chose from among the report's numbers to suit their purposes.

Supporters of drilling claimed that there were 16 billion barrels of oil to be recovered. This number was in fact at the extreme high side of the report and represented only a 5 percent probability of technically recoverable oil across the entire assessment area, including Kaktovik Inupiat Village Corporation lands and offshore state lands. In other words, the absolute best one could hope for.

Foes of drilling claimed that there were only 3 billion barrels of oil to be recovered. This number was at the extreme low end and was in fact rounded downward from 3.4 billion barrels. It represented a 95 percent probability of technically recoverable oil only on federal lands and only in that part of Section 1002 nearest the Canning River. In other words, 3 billion barrels was just about as sure of a thing as one was likely to come by in the oil business.[38]

Other than geologic variables, there were economic variables—"economically" recoverable oil as opposed to "technically" recoverable oil. Counting costs of exploration, development, production, and transportation, no economically recoverable reserves could be marketed from Section 1002 if prices were less than $13.00 per barrel. As the market price went up, so too did the amount of economically recoverable reserves.[39]

The bottom line of the USGS report was that there was more oil than previously thought in ANWR and that it was heavily concentrated in the western part of Section 1002. That left the variable of market price. In 1998, the average West Coast price for Alaska crude was $12.54 a barrel. Two years later, the price had climbed dramatically—suspiciously, some would maintain—to $30.00 a barrel.

As the 2000 presidential campaign swung into high gear, oil prices continued to climb, reaching $37.22 a barrel in September. President Clinton ordered a symbolic release from the nation's Strategic Petroleum Reserves.

Democrat candidate Al Gore drew a firm line at the Canning and spoke out emphatically against ANWR drilling. Former oil men George W. Bush and his running mate, Dick Cheney, were equally adamant in their support for drilling on 1002 lands.

Suddenly, the battle over ANWR was front-page news. With the scrutiny, however, came some embarrassing news for the oil companies. In early December 2000, amid contested presidential election returns, a Coast Guard report charged Alyeska with repeated safety violations at the Valdez terminal. Reportedly, unsupervised workers were attaching cables to anti-sparking equipment designed to prevent errant electrical discharges from igniting fumes as oil was loaded on board tankers. This was sloppy at best; Valdez, its 4,500 residents, and millions of barrels of oil blown sky-high at worst. For its part, Alyeska acknowledged the need for more oversight and promised new procedures.[40]

By the following spring, BP Amoco workers were themselves questioning the safety of some of the operations at Prudhoe Bay. Just days before Senator Murkowski led Bush interior secretary Gale Norton on a much-publicized tour of Prudhoe Bay, state inspectors found that almost one-third of the safety shutoff valves on one drilling platform had failed to close properly. Additional tests showed that 10 percent of the safety valves throughout the Prudhoe Bay complex failed to pass state inspections.[41] Such reports were reminiscent of the complacency that had crept into the spill-response system in the years prior to the *Exxon Valdez*. One couldn't help wondering, Was this the lackadaisical calm before another storm?

There certainly was a political storm brewing, and while Republicans generally favored ANWR drilling and Democrats opposed such action, the political lines were not clear-cut. Alaska's Democratic governor, Tony Knowles, himself a veteran of the oil fields before owning Anchorage restaurants, was solidly behind drilling, in large measure because he knew only too well from where came a significant portion of the state's revenues. A few other Democrats, such as Louisiana's senator John Breaux, leaned in favor of drilling, with Breaux citing his own state's history of drilling on wildlife refuges along the Gulf of Mexico. Republican senators Olympia Snowe and Susan Collins of Maine were among those who opposed ANWR drilling.

To Alaska's longtime Republican incumbents, the latter was just another example of easterners telling Alaskans what was best in their own backyard. Alaska likes to rotate its governors, but the men it has elected to Congress seem to have been given lifetime appointments. Frank Murkowski, chairman of the Senate Energy and Natural Resources Committee, is the junior member of the group, although he is in his fourth term in the Senate. Alaska's senior senator, Ted Stevens, never wanted to be anything but Alaska's senator. His fifth six-year term expires in 2003, although he has served since first appointed in 1968. Congressman Don Young has been elected fifteen times, serving since a special election in 1973, and being seriously challenged only once—by Valdez mayor John Devens in the wake of the *Exxon Valdez* publicity. All three have usually won reelection by wide margins, proving their popularity as well as their staying power.

How Alaska's political winds will blow in the future is less certain. In 2000, a statewide poll conducted for the Alaska Conservation Alliance (ACA) asked whether "the Arctic National Wildlife Refuge should be protected from oil drilling." Forty-five percent supported protection, while 49 percent opposed it. Said ACA director Mary Core, "We in Alaska are not of one mind on drilling for oil in the Arctic Refuge, as our congressional delegation would have others believe."[42] Perhaps. But in a state that still draws a sizable breath from the fortunes of oil, it is difficult to imagine that Alaska would elect anyone to Congress who was opposed to ANWR drilling. But who knows? Given the advancing ages of Alaska's congressional "lifers," Alaska may soon see new faces, if not new views.

If the political lines over ANWR are not clear-cut, neither is the international boundary with Canada. The Porcupine caribou herd also travels through a large portion of the upper Yukon Territory. Ivvavik and Vuntat National Parks were established there in 1995 as cooperative ventures between Parks Canada and the Inuvialuit and Vuntut Gwich'in First Nations to protect the eastern end of this ecosystem that does not know political boundaries. *Ivvavik* is in fact an Inuvialuit word meaning "a place for giving birth."

The rest of the Yukon Territory has a low-key economy largely rooted in the mining heritage of the Klondike, but east of the Richardson and Mackenzie Mountains, the Northwest Territories have flirted with oil and gas production since the first discoveries at Norman Wells provided an impetus for Project CANOL. Oil and gas discoveries have been made in the Mackenzie River delta and offshore in the Beaufort Sea, although distances and transportation difficulties have stymied commercial production. But here, too, the clock is ticking, and Canadians watch the continuing debate over ANWR with interest, particularly as it may impact on their own natural gas reserves.

Drilling in ANWR wouldn't in and of itself solve the decades-old quandary of what to do with the natural gas that rushes to the surface along with the oil. Prudhoe Bay's gas reserves—except minor amounts used to fuel local operations—continue to be reinjected into the ground because there is no economic way to transport them. When natural gas prices soared in 2000 along with oil prices, this once again spurred talk of a natural gas pipeline from the North Slope overland through Canada to Alberta.

Never mind the inevitable environmental battles, Canadian politicians quickly lined up on the choice of routes. At stake are the economic advantages of a major construction project, as well as a boost to local natural gas development. Some, such as Northwest Territories premier Stephen Kakfwi, lobby for the long-discussed route east from Prudhoe Bay and then up the Mackenzie River, a route that could link up with the shut-in wells in the Mackenzie Delta. Others, including Alaska governor Tony Knowles and Yukon premier Pat Duncan, want a gas pipeline to follow the haul road south to Fairbanks and then proceed along the Alaska Highway through Whitehorse. Just to confuse the issue, a third route has been suggested running roughly between the two—economically benefiting both territories but affecting a fair chunk of the Brooks Range and Mackenzie Mountains.

In Alaska, BP Amoco, Exxon Mobil, and Phillips Petroleum have formed a group reminiscent of the early days of TAPS to study the feasibility of the various routes. In Canada, a group of their Canadian affiliates and other oil companies have formed a similar consortium. To fan the boom, reserve estimates as high 100 trillion cubic feet of gas in Alaska and 70 trillion cubic feet in the Northwest Territories have been thrown about. USGS estimates of what is actually economically recoverable production are considerably less.[43]

The Yukon and Northwest Territories largely dozed through the last pipeline debate. In fact, Northwest Territories premier Stephen Kakfwi railed against the 1970s pipeline, but now the First Nations member talks of a Native-owned pipeline providing jobs and economic opportunities for the territory's mostly Native inhabitants. The much greater irony—not lost on most—is that a more thorough study and less provincial attitude on both sides of the international border during the early 1970s might have resulted in natural gas and oil flowing for a quarter of a century along an overland trans-Canadian pipeline route. Doubtless this would not have been without its share of turmoil and dissension, but it would have avoided such disasters as the *Exxon Valdez* spill.

And so the question remains, whose land? Environmentalists call the Arctic National Wildlife Refuge "pristine" but tend to forget that construction of DEW Line stations along the coastal plain during the 1950s not only brought changes but also forced the village of Kaktovik to be relocated. Oil companies boast about minimum-impact ice roads but tend to overlook photos that show the scars of winter seismic trails from 1984 still visible almost twenty years later.

So for now the line has been drawn at the Canning River. Never, says the environmental coalition. Once we cross it, says the oil industry. But it is not as simple as that. Pristine, untouched, a national treasure, says one side. Minimum impact, environmentally friendly, "a puny piece of ground," says the other.[44] Both sides are not without their exaggerations. All of us have a stake in the outcome.

Postscript to an Era, Prologue to the Next

So, whose land? *There are no easy answers. Wherever and whatever the environmental battlegrounds, protagonists on either side of a particular issue have almost always posed them in black and white. Their resolution rarely is.*

In the summer before his death, John F. Kennedy wrote the introduction to Secretary of the Interior Stewart Udall's book, The Quiet Crisis. *Kennedy urged: "We must do in our own day what Theodore Roosevelt did sixty years ago and Franklin Roosevelt thirty years ago: we must expand the concept of conservation to meet the imperious problems of the new age."*[45]

Some of the dominant issues of our own new age—this next frontier—appear obvious and urgent to all: whether or not to drill for oil in ANWR, build a natural gas pipeline across the Brooks Range, or maintain the balance of the salmon harvests. Other issues are subtler but have just as long-term ramifications. What is the ultimate carrying capacity of the land—whether in downtown Ketchikan when umpteen cruise ships dock or amid the yellow snow on the West Buttress of Denali?

A whining turbo-prop *flying far too low flashes through the Gates of the Arctic—well below the rocky arêtes of Bob Marshall's twin pillars of Boreal Mountain and Frigid Crags. Yes, those on board will say, we've been above the Arctic Circle. Gates of the Arctic? Yeah, we saw them.*

Down below, ten days into a Brooks Range crossing, a solitary group of five backpackers can only stare skyward and gawk. Even the grandest of scales can be threatened. Who is to say, however, that the five pairs of footprints are not also an intrusion, a disruption of Nature's balance? Who is to say that they and those who will follow them in increasing numbers are not inexorably changing the land? Are a few dozen oil wells on a few thousand acres out of 19.5 million more of a threat to the land's sustainability than half a million tourists at Denali or new bridges and improved roads stabbed straight into the heart of the Wrangell Mountains?

There are no easy answers. Those who profess to have them either have never seen the land or have failed to try to understand it. What is certain is that whatever answers are applied to this land today, their impacts will be increasingly profound—and increasingly irrevocable. Whether one cheers the change or bemoans it, Alaska stands poised once again to cross the next frontier.

EPILOGUE

Alaska—a Sense of Scale

I'm alone somewhere between Wonder Lake and the moraines of the Muldrow Glacier. The silence is deafening. So this is what Robert Service meant when he wrote of "the silence that bludgeons you dumb." It is so very still. Then something stirs a few decibels. At first, it is only a faint buzz. Then the buzz starts to pulsate until it grows louder and turns into an unmistakable, if somewhat monotonous, rhythm. Honk-honk-honk-honk-honk ad infinitum.

Where is it coming from? Finally, looking far to the north, I see faint, thin lines coming toward me high in the sky. The sound crescendos, and the lines become long Vs of many dots. Suddenly, I understand. Sandhills. Sandhill cranes beating their way south above autumn's reddish tint. Once the giant formation is above me, it is the sound, not the silence, that is deafening. Then, as slowly as they came, the hundreds of wings disappear to the south. Once again, there is nothing but silence.

Scale. That's what this land is all about.

May it always be so.

NOTES

Book One: The Land before Time

1 *National Geographic Atlas of the World*, 7th ed. (Washington, D.C.: National Geographic Society, 1999). The seventeen countries larger than Alaska are: Iran (636,296 square miles), Libya, Indonesia, Mexico, Saudi Arabia, the Democratic Republic of the Congo, Algeria, Sudan, Kazakhstan, Argentina, India, Australia, Brazil, China, the United States, Canada, and Russia (6,592,692 square miles). If one counts Greenland, it would fall between Saudi Arabia and the Democratic Republic of the Congo.

2 Charles Sheldon, *The Wilderness of Denali* (1930; reprint, New York: Charles Scribner's Sons, 1960), p. 8.

3 Donald H. Richter, Danny S. Rosenkrans, and Margaret J. Steigerwald, "Guide to the Volcanoes of the Western Wrangell Mountains, Alaska—Wrangell–St. Elias National Park and Preserve," U.S. Geological Survey Bulletin 2072 (Washington, D.C.: U.S. Government Printing Office, 1995).

4 Israel Russell, "Mount St. Elias Revisited," *Century*, June 1892, p. 200.

5 Marilyn George and Linda Slaght, "The Stikine Icefields: Frozen Rivers Shaping The Land," USDA Forest Service publication, n.d.

6 Robert Service, "The Spell of the Yukon," in *Collected Poems of Robert Service.* (New York: Dodd, Mead, 1940), p. 4.

7 Inga Sjolseth Kolloen, "Crossing the Chilkoot Pass," entry for May 27, 1898, unpublished diary in the Klondike Gold Rush National Historical Park archives, Skagway, Alaska.

8 Jan Halliday, *Native Peoples of Alaska, a Traveler's Guide to Land, Art, and Culture* (Seattle: Sasquatch Books, 1998), pp. x–xi.

9 Steve J. Langdon, *The Native People of Alaska,* 3rd ed. (Anchorage: Greatland Graphics, 1993). pp. 14–24.

10 Marilyn Knapp, *Carved History: The Totem Poles and House Posts of Sitka National Historical Park* (Anchorage: Alaska Natural History Association, 1992); Halliday, *Native Peoples of Alaska*; Langdon, *The Native People of Alaska*; and Claus-M. Naske and Herman E. Slotnick, *Alaska, a History of the Forty-ninth State,* 2nd ed. (Norman: University of Oklahoma Press, 1987).

Book Two: Lifting the Veil

1 Raymond H. Fisher, *Bering's Voyages: Whither and Why* (Seattle: University of Washington Press, 1977), pp. 22–23.

2 Georg Wilhelm Steller, *Journal of a Voyage with Bering* (1743; reprint, Palo Alto: Stanford University Press, 1993), p. 78.

3 Quoted in *Glacier Bay, Official National Park Handbook* (Washington, D.C.: U.S. National Park Service, 1983), p. 13.

4 Steller, *Journal of a Voyage with Bering;* Corey Ford, *Where the Sea Breaks Its Back: The Epic Story of Early Naturalist Georg Steller and the Russian Exploration of Alaska* (Boston: Little, Brown, 1966; reprint, Anchorage: Alaska Northwest Books, 1992).

5 John Dunmore, *Who's Who in Pacific Navigation* (Honolulu: University of Hawaii Press, 1991), pp. 31, 175, 192; David Lavender, *Land of Giants: The Drive to the Pacific Northwest, 1750–1950* (Garden City, N.Y.: Doubleday, 1956), pp. 4, 14–15, 31–39; Donald J. Orth, *Dictionary of Alaska Place Names,* Geological Survey Professional Paper, 567 (Washington: U.S. Government Printing Office, 1967), pp. 32–33.

6 Thomas Vaughan et al. *Voyages of Enlightenment: Malaspina on the Northwest Coast, 1791/1792* (Portland, Oreg.: Oregon Historical Society, 1977), p. 3.

7 Ibid., p.11.

8 John Kendrick, *Alejandro Malaspina: Portrait of a Visionary* (Quebec: McGill Queens University Press, 1999).

9 From Vancouver's log, quoted in Dave Bohn, *Glacier Bay: The Land and the Silence* (San Francisco: Sierra Club, 1967), p. 40.

10 Dunmore, *Who's Who in Pacific Navigation,* pp. 64–67, 80, 177, 196–197, 254–256; Lavender, *Land of Giants,* p. 20; Orth, *Dictionary of Alaska Place Names.*

11 John Dunmore, *Pacific Explorer: The Life of Jean-François de la Pérouse, 1741–1788.* (Annapolis: Naval Institute Press, 1985), p. 221.

12 Ibid., p. 228.

13 Ibid., p. 229.

14 Ibid.

15 From La Pérouse's log, quoted in Bohn, *Glacier Bay,* p. 28.

16 Dunmore, *Pacific Explorer,* p. 232.

17 Steve J. Langdon, *The Native People of Alaska,* 3rd ed. (Anchorage: Greatland Graphics, 1993), p. 24.

18 Orth, *Dictionary of Alaska Place Names,* p. 14.

19 Dunmore, *Who's Who in Pacific Navigation,* pp. 145–46; Orth, *Dictionary of Alaska Place Names,* p. 16.

20 Claus-M.Naske and Herman E. Slotnick, *Alaska, a History of the Forty-ninth State,* 2nd ed. (Norman: University of Oklahoma Press, 1987), pp. 30–32; David J. Nordlander, *For God and Tsar, a Brief History of Russian America, 1741–1867* (Anchorage: Alaska Natural History Association, 1994), pp. 7–9.

21 Hector Chevigny, *Lord of Alaska: Baranov and the Russian Adventure* (New York: Viking Press, 1944), p. 235.

22 Dunmore, *Who's Who in Pacific Navigation*, pp. 145–46; Orth, *Dictionary of Alaska Place Names*, p. 16.

23 Richard A. Pierce, "Georg Anton Schaffer," in *Alaska and Its History*, ed. Morgan B. Sherwood (Seattle: University of Washington Press, 1967), pp. 71–81.

24 Hector Chevigny, *Russian America: The Great Alaskan Venture, 1741–1867* (Portland, Oreg.: Binford and Mort Publishing, 1979), p. 208–9.

25 Dexter Perkins, *A History of the Monroe Doctrine* (Boston: Little, Brown, 1963), p. 29.

26 General Mariano G. Vallejo to Governor Juan B. Alvarado, July 2, 1841, quoted in the Fort Ross Visitor Center exhibits, Fort Ross, Calif.

27 Naske and Slotnick, *Alaska,* pp. 47–50.

28 Dunmore, *Who's Who in Pacific Navigation*, pp. 205–8; Orth, *Dictionary of Alaska Place Names*, p. 25.

29 Pierre Berton, *The Arctic Grail: The Quest for the Northwest Passage and the North Pole, 1818–1909* (New York: Viking Penguin, 1988), pp. 91–95, 151, 221–23.

Book Three: Seward's Folly

1 Murray Morgan, *Confederate Raider in the North Pacific: The Saga of the C.S.S. Shenandoah, 1864–65* (Pullman, Wash.: Washington State University Press, 1995), p. 212–13.

2 Ibid., p. 253 (quoting from Cornelius Hunt's account).

3 Morgan B. Sherwood, *Exploration of Alaska, 1865–1900* (Fairbanks: University of Alaska Press, 1992), p. 16.

4 Ibid., p. 21.

5 Ibid., p. 15.

6 William H. Dall, *Alaska and Its Resources* (Boston: Lee and Shepard, 1870), p. 122.

7 Ibid., p. 100.

8 Ibid., p. 293.

9 U.S. House Executive Document 177, 40th Cong., 2nd sess., 1867, pp. 138–39.

10 Julius Pratt, *A History of United States Foreign Policy* (Englewood Cliffs, N.J.: Prentice Hall, 1955), p. 327.

11 Alfred Hulse Brooks, *Blazing Alaska's Trails* (1953; reprint, Fairbanks: University of Alaska Press), pp. 257–58.

12 Richard E. Welch, Jr., "American Public Opinion and the Purchase of Russian America," in *American Slavic and East European Review* (1958): 481–94.

13 Jan Halliday, *Native Peoples of Alaska, a Traveler's Guide to Land, Art, and Culture* (Seattle: Sasquatch Books, 1998), p. 134.

14 Steven. T. Zimmerman, "Northern Fur Seal," Wildlife Notebook Series, Alaska Department of Fish and Game, 1994.

15 Karl Schneider, "Sea Otter," Wildlife Notebook Series, Alaska Department of Fish and Game, 1994.

16 Robert G. Athearn, "An Army Officer's Trip to Alaska in 1869," *Pacific Northwest Quarterly* 40, no. 1 (January 1949): 57.

17 William R. Hunt, *Alaska: A Bicentennial History* (New York: W. W. Norton, 1976), p. 52.

18 Sherwood, *Exploration of Alaska,* p. 44.

19 Hunt, *Alaska,* p. 56.

20 Pratt, *A History of United States Foreign Policy*, p. 358, quoting from J. B. Moore, *History and Digest of the International Arbitrations to Which the United States Has Been a Party* (Boston: Houghton Mifflin, 1909).

21 Corey Ford, *Where the Sea Breaks Its Back: The Epic Story of Early Naturalist Georg Steller and the Russian Exploration of Alaska* (Boston: Little, Brown, 1966; reprint, Anchorage: Alaska Northwest Books, 1992), pp. 197–99.

22 C.E.S. Wood, "Among the Thlinkets in Alaska," *Century,* July 1882, p. 332.

23 John Muir, *Travels in Alaska* (1915; reprint, San Francisco: Sierra Club Books, 1988), p. 95.

24 Ibid., p. 120

25 Ibid.

26 Ibid., p. 154.

27 Samuel Hall Young, *Alaska Days with John Muir* (1915; reprint, Salt Lake City: Peregrine Smith Books, 1990), p. 121.

28 Ibid., p. 147.

29 Ibid., p. 157.

30 J. Arthur Lazell, *Alaskan Apostle: The Life Story of Sheldon Jackson* (New York: Harper and Brothers, 1960), p. 65.

31 William Dall, "Late News from Alaska," *Science,* July 31, 1885), p. 96.

32 Ted C. Hinckley, "Sheldon Jackson and Benjamin Harrison," *Pacific Northwest Quarterly* 54 (April 1963): 66–74.

33 Melody Webb, *Yukon: The Last Frontier* (Lincoln: University of Nebraska Press, 1993), p. 178. (From *Annual Report of the Commissioner of Indian Affairs to the Secretary of the Interior for the Year 1887* [Washington, D.C.: U.S. Government Printing Office, 1887], p. xix.)

34 Ibid. (From "Report of the Commissioner of Education," in *Report of the Secretary of the Interior for the Fiscal Year Ending June 30, 1885,* vol. 4 [Washington, D.C.: U.S. Government Printing Office, 1886], p. lxii.)

35 Hinckley, "Sheldon Jackson and Benjamin Harrison," pp. 66–74.

36 John G. Brady letter to W. B. Morris, May 6, 1878, quoted in Ernest Gruening, *The State of Alaska: A Definitive History of America's Northernmost Frontier* (New York: Random House, 1954), p. 502.

37 Sherwood, *Exploration of Alaska*, pp. 98–99.

38 Ibid., p. 77.

39 Ibid., p. 99.

40 Frederick Schwatka, *A Summer in Alaska in the 1880s* (1885; reprint, Secaucus, N.J.: Castle, 1988), p. 11.

41 David Neufeld and Frank Norris, *Chilkoot Trail: Heritage Route to the Klondike* (Whitehorse, Yukon Territory: Lost Moose, 1996), p. 21.

42 Schwatka, *A Summer in Alaska in the 1880s*, p. 136.

43 Neufeld and Norris, *Chilkoot Trail*, p. 47.

44 Sherwood, *Exploration of Alaska*, p. 104.

45 Ibid., p. 105 (Quoted from Allen's diary, 1883–84, Allen Papers, Library of Congress.)

46 Heath Twichell, Jr., *Allen: The Biography of an Army Officer, 1859–1930* (New Brunswick, N.J.: Rutgers University Press, 1974); Allen to Dora Johnston, January 10, 1883, Allen Papers, quoted in Sherwood, *Exploration of Alaska*, p. 108.

47 Henry T. Allen, "An Expedition to the Copper, Tanana and Koyukuk Rivers in 1885," reprinted by the Alaska North Publishing Company as v. 15, no. 2 of *The Alaska Journal*, 1985, p. 44.

48 Ibid., p. 45.

49 F. S. Pettyjohn, "Publisher's Comments," *Alaskana* 5, no. 1 (December 1977): p. 2.

50 Robert Dunn "Finding a Volcano and Wiping a 16,000 Feet [sic] Mountain from the Map of Alaska," *Outing*, December 1902, pp. 321–32.

51 Alfred Hulse Brooks, "A Reconnaissance in the White and Tanana River Basins, Alaska, in 1898," in *Twentieth Annual Report of the USGS* (Washington, D.C.: U.S. Government Printing Office, 1898–99), part 7, p. 297.

52 Robert N. De Armond, *The Founding of Juneau* (Juneau: Gastineau Channel Centennial Association, 1967), p. 69.

53 Ibid., p. 83.

54 Robert N. De Armond, *Some Names around Juneau* (Sitka, Alaska: Sitka Printing, 1957), p. 36.

55 De Armond, *The Founding of Juneau*, pp. 112–17. (The production figures for the Treadwell complex are in David B. Wharton, *The Alaska Gold Rush* [Bloomington, Ind.: Indiana University Press, 1972], p. 95.)

56 Sherwood, *Exploration of Alaska*, p. 81.

57 William Williams, "Climbing Mount St. Elias," *Scribner's Magazine*, April 1889, p. 387.

58 Frederick Schwatka, "Two Expeditions to Mount St. Elias." *Century*, April 1891, pp. 871–72.

59 Williams, "Climbing Mount St. Elias," pp. 387–403.

60 Israel Russell, "An Expedition to Mount St. Elias, Alaska," *National Geographic Magazine*, May 29, 1891, p. 152.

61 Jonathan Waterman, *A Most Hostile Mountain: Re-creating the Duke of Abruzzi's Historic Expedition on Alaska's Mount St. Elias* (New York: Henry Holt, 1997), pp. 49–50.

62 Mirella Tenderini and Michael Shandrick, *The Duke of the Abruzzi: An Explorer's Life* (Seattle: The Mountaineers, 1997), p. 35.

63 Ibid., p. 37.

64 Chris Jones, *Climbing in North America* (Berkeley: University of California Press, 1976), p. 54.

Book Four: Go North

1 Melody Webb, *Yukon: The Last Frontier* (Lincoln: University of Nebraska Press, 1993), pp. 59–60.

2 Ibid., pp. 61–62.

3 Ibid., p. 68.

4 Alfred Hulse Brooks, *Blazing Alaska's Trails* (Fairbanks: University of Alaska Press, 1973), p. 328.

5 Pierre Berton, *Klondike: The Last Great Gold Rush, 1896–1899* (Toronto: McClelland and Stewart, 1972), p. 15.

6 Webb, *Yukon,* pp. 86–87.

7 Ibid., p. 88.

8 Brooks, *Blazing Alaska's Trails*, p. 334.

9 Ibid., p. 335.

10 *Seattle Post-Intelligencer*, July 17, 1897.

11 Diary of William Erwin Craig, March 23, 1898—North of Canyon City on the Chilkoot Trail, in archives of Klondike Gold Rush Historical Park, Skagway, Alaska.

12 Diary of Inga Sjolseth Kolloen, March 29, 1898, in archives of Klondike Gold Rush Historical Park, Skagway, Alaska.

13 Jack London, "Which Make Men Remember," in *The God of His Fathers: Tales of the Klondike and the Yukon* (1902; reprint, London: Everett, no date), p. 68.

14 Diary of Stewart L. Campbell, February 21, 1898, in archives of Klondike Gold Rush Historical Park, Skagway, Alaska.

15 Ibid., March 8 and 9, 1898.

16 Ibid., March 22 to June 3, 1898.

17 Ibid., August 5, 1899.

18 Berton, *Klondike*, p. 194.

19 Ibid., p. 200.

20 William R. Abercrombie, "Copper River Exploring Expedition, 1899," 56th Cong., 1st sess., Senate document 306, p. 15.

21 Arthur Arnold Dietz, *Mad Rush for Gold in Frozen North* (Los Angeles: Times-Mirror Printing and Binding House, 1914).

22 Berton, *Klondike*, p. 211.

23 Ibid., p. 212.

24 Diary of E. B. Wishaar, January 3, 1898, in archives of Klondike Gold Rush National Historical Park, Skagway, Alaska.

25 Stan Cohen, *Gold Rush Gateway: Skagway and Dyea, Alaska* (Missoula, Mont.: Pictorial Histories Publishing, 1986), p. 87.

26 David Neufeld and Frank Norris, *Chilkoot Trail: Heritage Route to the Klondike*, (Whitehorse, Yukon Territory: Lost Moose, 1996), p. 55.

27 Diary of William Erwin Crain, March 12–13, 1898, in archives of Klondike Gold Rush National Historical Park, Skagway, Alaska.

28 Neufeld and Norris, *Chilkoot Trail*, pp. 57, 62.

29 *Dyea Trail*, January 12, 1898.

30 Roy Minter, *The White Pass: Gateway to the Klondike* (Fairbanks: University of Alaska Press, 1987), pp. 160–63.

31 Cohen, *Gold Rush Gateway*, p. 64.

32 Neufeld and Norris, *Chilkoot Trail*, p. 92.

33 Diary of Harley Tuck, May 30, 1898, in ibid., p. 89.

34 William H. Goetzmann and Kay Sloan, *Looking Far North: The Harriman Expedition to Alaska, 1899* (New York: Viking Press, 1892), p. 85.

35 Ibid., p. 152.

36 Henry Gannett, "General Geography," *Alaska Harriman Series*, vol. I (New York: Doubleday, 1901).

37 L. H. Carlson, "The Discovery of Gold at Nome, Alaska," *Pacific Historical Review* 15 (September 1946): 259–78.

38 David B. Wharton, *The Alaska Gold Rush* (Bloomington, Ind.: Indiana University Press, 1972), p. 187.

39 Carlson, "The Discovery of Gold at Nome, Alaska," pp. 259–78.

40 George Edward Adams, "Cape Nome's Wonderful Placer Mines," *Harper's Weekly*, June 9, 1900, p. 529.

41 Ibid., p. 531.

42 Ibid., p. 530.

43 Eleanor B. Caldwell, "A Woman's Experience at Cape Nome," *Cosmopolitan*, November 1900, p. 81.

44 Ibid., p. 83.

45 Alfred H. Dunham, "The Development of Nome," *Cosmopolitan*, February 1905, p. 469.

46 Ibid., p. 468.

47 Wharton, *The Alaska Gold Rush*, pp. 206–7.

48 Terrence Cole, *Crooked Past: The History of a Frontier Mining Camp: Fairbanks, Alaska* (Fairbanks: University of Alaska Press, 1991), p. 75.

49 Falcon Joslin, "Railroad Building in Alaska," *Alaska-Yukon Magazine*, January 1909, p. 247; Duane Koenig, "Ghost Railway in Alaska: The Story of the Tanana Valley Railroad," *Pacific Northwest Quarterly* 45 (January 1954): 8–12.

50 Koenig, "Ghost Railway in Alaska," pp. 8–12.

51 Terris Moore, *Mount McKinley: The Pioneer Climbs* (College, Alaska: University of Alaska Press, 1967), p. 9. (Originally published in *New York Sun*, January 24, 1897.)

52 Ibid., p. 14–15.

53 Ibid., p. 26.

54 Bradford Washburn and David Roberts, *Mount McKinley: The Conquest of Denali* (New York: Harry H. Abrams, 1991), p. 25.

55 Ibid., p. 26–27.

56 Ibid., p. 45.

57 Belmore Browne, *The Conquest of Mount McKinley* (New York: G. P. Putnam's Sons, 1913), p. 71.

58 Moore, *Mount McKinley*, p. 75.

59 Browne, *The Conquest of Mount McKinley*, pp. 342, 344.

60 Ibid., p. 359.

61 *The Alaskan* (Cordova, Alaska), June 16, 1906.

62 Lone E. Janson, *The Copper Spike* (Anchorage: HaHa, 1993), pp. 98–99. (Originally published in 1975 by Alaska Northwest Publishing.)

63 William R. Hunt, *Mountain Wilderness: Historic Resource Study for Wrangell–St.Elias National Park and Preserve* (Anchorage: National Park Service, 1991), p. 146.

64 Melody Webb Grauman, "Kennecott: Alaskan Origins of a Copper Empire," *Western Historical Quarterly* 2 (April 1978): 207.

65 M. J. Kirchhoff, *Historic McCarthy: The Town That Copper Built* (Juneau: Alaska Cedar Press, 1993), p. 47. (Originally in *McCarthy Weekly News*, August 10, 1918.)

66 Ibid., p. 82.

67 Ibid., p. 88.

68 H. W. Brands, *TR: The Last Romantic* (New York: Basic Books, 1997), pp. 623–25.

69 Ernest Gruening, *The State of Alaska: A Definitive History of America's Northernmost Frontier* (New York: Random House, 1954), p. 130.

70 Ibid., p. 132

71 Ibid., p. 135.

72 Ibid.

73 George E. Baldwin, "Conservative Faddists Arrest Progress and Seek to Supplant Self-Government with Bureaucracy," *Alaska-Yukon Magazine*, February 1912, p. 45.

74 Charles Sheldon, *The Wilderness of Denali* (New York: Charles Scribner's Sons, 1960), pp. 24–25.

75 Ibid., p. 103.

76 William E. Brown, *Denali: Symbol of the Alaskan Wild* (Virginia Beach, Va.: Donning, 1993), pp. 79, 90–91.

77 *Reports of the Secretary of the Interior for the Fiscal Year Ended June 30, 1918*, vol. I (Washington, D.C.: U.S. Government Printing Office, 1919), p. 126.

78 Brown, *Denali*, p. 129. (Mather letter to Sheldon of January 27, 1921, University of Alaska, Fairbanks Archives, Sheldon Collection, Box 2.)

79 Kim Heacox, *The Denali Road Guide: A Roadside Natural History of Denali National Park* (Denali Park: Alaska Natural History Association, 1990), p. 31.

80 J. E. Spurr, "A Reconnaissance in Southwestern Alaska in 1898," in *Twentieth Annual Report of the USGS* (Washington, D.C.: U.S. Government Printing Office, 1900), part 7, pp. 59, 91–92.

81 Jean Bodeau, *Katmai National Park and Preserve, Alaska* (Anchorage: Alaska Natural History Association and Greatland Graphics, 1992), p. 83.

82 Robert F. Griggs, "The Valley of Ten Thousand Smokes: National Geographic Society Explorations in the Katmai District of Alaska," *National Geographic Magazine*, January 1917, p. 33.

83 Ibid, pp. 64–65.

Book Five: Interlude

1 William H. Wilson, *Railroad in the Clouds: The Alaska Railroad in the Age of Steam, 1914–1945* (Boulder, Colo.: Pruett Publishing, 1977), p. 7.

2 Duane Koenig, "Ghost Railway in Alaska: The Story of the Tanana Valley Railroad," *Pacific Northwest Quarterly* 45 (January 1954): 11.

3 Wilson, *Railroad in the Clouds,* p. 18.

4 *Congressional Record*, 63rd Cong., 2d sess., 1913, vol. 51, part 1:76; Wilson, *Railroad in the Clouds*, p. 25.

5 "Alaska's New Railway," *National Geographic Magazine*, December 1915, p. 567.

6 Stephen R. Capps, "A Game Country without Rival in America," *National Geographic Magazine*, January 1917, pp. 73, 83.

7 William E. Brown, *Denali: Symbol of the Alaskan Wild* (Virginia Beach, Va.: Donning, 1993), pp. 135–36.

8 "Report of the Governor of Alaska," *Reports of the Secretary of the Interior for the Fiscal Year Ended June 30, 1918*, vol. II (Washington, D.C.: U.S. Government Printing Office, 1919), p. 563.

9 William M. Bueler, *Roof of the Rockies: A History of Colorado Mountaineering* (Evergreen, Colo.: Cordillera Press, 1986), pp. 209–10.

10 Dave Bohn, *Glacier Bay: The Land and the Silence* (San Francisco: Sierra Club, 1967).

11 *Juneau Daily Empire*, April 28, 1924.

12 Bohn, *Glacier Bay*, p. 96.

13 Ibid., p. 97.

14 Ibid., p. 89.

15 Ernest Gruening, *The State of Alaska: A Definitive History of America's Northernmost Frontier* (New York: Random House, 1954), p. 245.

16 Ibid., p. 74.

17 "Report on the Salmon and Salmon Rivers of Alaska, with Notes on the Conditions, Methods, and Needs of the Salmon Fisheries," in *Bulletin of the U.S. Fish Commission* for 1889, quoted in Richard A. Cooley, *Politics and Conservation: The Decline of the Alaska Salmon* (New York: Harper and Row, 1963), p. 72.

18 Cooley, *Politics and Conservation*, p. 73.

19 Ibid., pp. 75–82.

20 Ibid., p. 93.

21 Ibid., pp. 104–13.

22 Ibid., pp. 121–24.

23 Jean Potter, *The Flying North* (New York: Macmillan, 1947), pp. 144–45.

24 Steven C. Levi, *Cowboys of the Sky: The Story of Alaska's Bush Pilots* (New York: Walker and Company, 1996), p. 9.

25 Ibid.

26 St. Clair Streett, "The First Alaskan Air Expedition," *National Geographic Magazine*, May 1922, p. 552.

27 Potter, *The Flying North*, pp. 29–30.

28 Melody Webb, *Yukon: The Last Frontier* (Lincoln: University of Nebraska Press, 1993), p. 262.

29 Levi, *Cowboys of the Sky*, p. 35.

30 Webb, *Yukon*, pp. 234–35.

31 Potter, *The Flying North*, p. vii.

32 Claus-M. Naske and Herman E. Slotnick, *Alaska, a History of the Forty-ninth State*, 2nd ed. (Norman: University of Oklahoma Press, 1987), pp. 107–8.

33 Gruening, *The State of Alaska*, p. 432.

34 Bryan B. Sterling and Frances N. Sterling, *Will Rogers and Wiley Post: Death at Barrow* (New York: M. Evans, 1993), p. 133.

35 John M. Kauffmann, *Alaska's Brooks Range: The Ultimate Mountains* (Seattle: The Mountaineers, 1992), p. 49.

36 Ibid., p. 55.

37 Robert Marshall, *Alaska Wilderness: Exploring the Central Brooks Range*, 2nd ed. (Berkeley: University of California Press, 1970), p. xxiii.

38 Ibid., p. 7.

39 Ibid., p. 12.

40 Ibid., p. xxviii.

41 Ibid., pp. 114–15.

42 Robert Marshall, "The Problem of the Wilderness," *Scientific Monthly*, February 1930, pp. 141–48.

43 Ernest Gruening, *Many Battles: The Autobiography of Ernest Gruening* (New York: Liveright, 1973), p. 181.

44 *Time Magazine*, May 6, 1935, p. 17.

45 Clarence C. Hulley, "Historical Survey of the Matanuska Valley Settlement in Alaska," *Pacific Northwest Quarterly* 40 (October 1949): 327–40; and Orlando W. Miller, *The Frontier in Alaska and the Matanuska Colony* (New Haven: Yale University Press, 1975).

46 Allen Carpé, "The Ascent of Mount Bona, Alaska," *Alpine Journal* 43 (1931): 69–74.

47 Terris Moore, *Mount McKinley: The Pioneer Climbs* (College, Alaska: University of Alaska Press, 1967), p. 125.

48 Brown, *Denali*, p. 160.

49 Moore, *Mount McKinley*, p. 135.

50 Ibid., p. 137.

51 Charles Houston, "Denali's Wife," *American Alpine Journal* 2, no. 3 (1935): 285–97.

52 Robert H. Bates, *The Love of Mountains Is Best* (Portsmouth, N.H.: Peter E. Randall, 1994), p. 73.

53 Jonathan Waterman, "Brad Washburn, Renaissance Mountain Man," *Alaska Geographic* 25, no. 3 (1998): 19.

54 Bradford Washburn, "The Ascent of Mt. St. Agnes" [Mount Marcus Baker] *American Alpine Journal* 3, no. 3 (1939): 255–64; Terris Moore, "Mt. Sanford: An Alaskan Ski Climb," *American Alpine Journal* 3, no. 3 (1939): 265–73.

Book Six: The Forgotten Campaign

1 Alfred H. Dunham, "The Development of Nome," *Cosmopolitan*, February 1905, p. 472.

2 H. H. Arnold, "Our Air Frontier in Alaska," *National Geographic Magazine*, October 1940, p. 487.

3 Ibid., pp. 487, 491, 504; "Report of the Treasurer of the Territory of Alaska," January 1, 1939 to December 31, 1940.

4 Heath Twichell, *Northwest Epic: The Building of the Alaska Highway* (New York: St. Martin's Press, 1992), pp. 14–15, 18–19, 20.

5 Ibid., pp. 43–45.

6 Ibid., p. 46.

7 Ibid., p. 53.

8 Ibid., p. 89.

9 Ernest Gruening, *Many Battles: The Autobiography of Ernest Gruening* (New York: Liveright, 1973), pp. 305–6.

10 Twichell, *Northwest Epic*, pp. 53–59.

11 Froelich Rainey, "Alaska Highway an Engineering Epic," *National Geographic Magazine*, February 1943, p. 166.

12 Ibid., p. 143.

13 Twichell, *Northwest Epic*, pp. 208–9.

14 Ibid., pp. 213–14.

15 Rainey, "Alaska Highway an Engineering Epic," p. 167.

16 Twichell, *Northwest Epic*, pp. 214–19, 223.

17 Ibid., p. 253.

18 Brian Garfield, *The Thousand-Mile War: World War II in Alaska and the Aleutians* (Fairbanks: University of Alaska Press, 1995), pp. 196–98; Twichell, *Northwest Epic*, pp. 278–87.

19 Twichell, *Northwest Epic*, pp. 157, 164, 272–73.

20 Ibid., pp. xiii–xiv.

21 Executive Order No. 9066, "Authorizing the Secretary of War to Prescribe Military Areas," February 19, 1942.

22 Public Proclamation No. 5, Headquarters, Western Defense Command and Fourth Army, Presidio of San Francisco, California, March 30, 1942.

23 Ronald K. Inouye, "For Immediate Sale: Tokyo Bathhouse—How World War II Affected Alaska's Japanese Citizens," in *Alaska at War, 1941–45: The Forgotten War Remembered,* ed. Fern Chandonnet (Anchorage: Alaska at War Committee, 1995), pp. 259–63.

24 Dean Kohlhoff, *When the Wind Was a River: Aleut Evacuation in World War II* (Seattle: University of Washington Press, 1995), p. 69.

25 Ibid., p. 70.

26 Ibid., pp. 70–71.

27 Ibid., pp. 89, 98.

28 Ray Hudson, "Aleuts in Defense of Their Homeland," in Chandonnet, *Alaska at War, 1941–45*, p. 163.

29 Twichell, *Northwest Epic*, pp. 46–47.

30 Garfield, *The Thousand-Mile War*, pp. 169–70.

31 Chris Wooley and Mike Martz, "The Tundra Army—Patriots of Arctic Alaska," in Chandonnet, *Alaska at War, 1941–45*, pp. 155–60.

32 Garfield, *The Thousand-Mile War*, p. 29.

33 Ibid., pp. 31–55.

34 Walter Lord, *Incredible Victory* (New York: Harper and Row, 1967), p. 286.

35 Garfield, *The Thousand-Mile War*, pp. 143, 153, 189.

36 Ibid., pp. 176–77, 179.

37 Ibid., p. 210.

38 Garfield, *The Thousand-Mile War*, pp. 219–34; John Bishop, "My Speed Zero," *Saturday Evening Post*, February 5, 1944, pp. 26–28, 70.

39 Ibid., pp. 253, 257–58, 260–62.

40 Ibid., p. 271.

41 Ibid., p. 325.

42 Ibid., p. 333.

43 Ibid., p. 334.

44 Ibid., p. 348.

45 Ibid., pp. 358–61.

46 Ibid., pp. 366–70.

47 Ibid., pp. 417–19.

48 Ibid., pp. 370–72.

49 Ibid., pp. 373–76, 380–85.

50 Ibid., p. 385.

51 Orlando W. Miller, *The Frontier in Alaska and the Matanuska Colony* (New Haven: Yale University Press, 1975), p. 182.

52 Doris Kearns Goodwin, *No Ordinary Time: Franklin and Eleanor Roosevelt: The Home Front in World War II* (New York: Simon and Schuster, 1995). p. 548.

53 Arnold, "Our Air Frontier in Alaska," p. 504.

Book Seven: Postwar Rumblings

1 Ernest H. Gruening, "Strategic Alaska Looks Ahead," *National Geographic Magazine*, September 1942, p. 315.

2 George Sundborg, *Opportunity in Alaska* (New York: Macmillan, 1945), p. 6.

3 Ernest H. Gruening, "Go North, Young Man," *Reader's Digest*, January 1944, pp. 53–57.

4 Orlando W. Miller, *The Frontier in Alaska and the Matanuska Colony* (New Haven: Yale University Press, 1975), p. 183.

5 Sundborg, *Opportunity in Alaska*, pp. 11–12.

6 Ibid., pp. vii, 10.

7 Ibid., pp. 5, 32, 34, 157, 160, 164, 171, 175, 182, 221–22.

8 Edna Ferber, *Ice Palace* (New York: Doubleday, 1958), p. 7.

9 Richard A. Cooley, *Alaska: A Challenge in Conservation* (Madison, Wis.: University of Wisconsin Press, 1966), pp. 19–20; Ernest Gruening, *The State of Alaska: A Definitive History of America's Northernmost Frontier* (New York: Random House, 1954), p. 64; *The Alaskan* (Sitka, Alaska), February 13 and February 27, 1886.

10 Robert B. Weeden, *Promises to Keep* (Boston: Houghton Mifflin, 1978), p. 108; Gruening, *The State of Alaska*, p. 213.

11 Sundborg, *Opportunity in Alaska*, p. 70.

12 David D. Smith, "Pulp, Paper, and Alaska," *Pacific Northwest Quarterly* 66, no. 2 (April 1975): 62.

13 Ibid., p. 68.

14 Claus-M. Naske and Herman E. Slotnick, *Alaska, a History of the Forty-ninth State*, 2nd ed. (Norman: University of Oklahoma Press, 1987), p. 159.

15 Smith, "Pulp, Paper, and Alaska," pp. 69–70.

16 Miller, *The Frontier in Alaska*, p. 200.

17 Weeden, *Promises to Keep*, p. 105.

18 Ibid., p. 110.

19 Naske and Slotnick, *Alaska*, p. 122.

20 Audrey Morgan and Frank Morgan, "Alaska's Russian Frontier: Little Diomede," *National Geographic Magazine*, April 1951, p. 551.

21 Naske and Slotnick, *Alaska*, p. 131.

22 Ernest Gruening, *Many Battles: The Autobiography of Ernest Gruening* (New York: Liveright, 1973), pp. 360–62.

23 Howard La Fay, "DEW Line: Sentry of the Far North," *National Geographic Magazine*, July 1958, p. 129.

24 Ibid., p. 144.

25 Al Ossinger, interview by author, Lakewood, Colo., February 23, 2001. (Ossinger was stationed on St. Lawrence Island, 1955–56, U.S. Army.)

26 Naske and Slotnick, *Alaska*, pp. 134–35.

27 Ibid., pp. 138–39.

28 James Wickersham, "The Forty-ninth Star," *Collier's*, August 6, 1910, p. 17.

29 *House Journal*, 79th Cong., 2nd sess., January 21, 1946, p. 26.

30 Naske and Slotnick, *Alaska*, p. 148.

31 Ernest Gruening, *The Battle for Alaska Statehood* (College, Alaska: University of Alaska Press, 1967), pp. 5–8.

32 Naske and Slotnick, *Alaska*, p. 147.

33 Gruening, *The Battle for Alaska Statehood*, p. 13.

34 Naske and Slotnick, *Alaska*, p. 144.

35 Gruening, *The Battle for Alaska Statehood*, p. 46.

36 "Alaska, Hawaii Statehood Seen as Sign to World," *Denver Post*, September 17, 1950.

37 Naske and Slotnick, *Alaska*, p. 153.

38 Ibid., pp. 155–57.

39 Gruening, *Many Battles*, pp. 404–5.

40 Will Swagel, *Alaska's Flag* (Sitka, Alaska: Sitka Historical Society, 1994), p. 5.

41 Bryan Cooper, *Alaska: The Last Frontier* (New York: William Morrow, 1973), pp. 98–100.

42 Ibid., pp. 101–2.

43 Ibid., pp. 106–8.

44 Ibid., pp. 109–2.

45 Ibid., pp. 114–5.

46 Ibid., p. 112.

47 Margaret E. Murie, *Two in the Far North* (New York: Alfred Knopf, 1962; reprint, Anchorage: Alaska Northwest Books, 1997).

48 James M. Glover, *A Wilderness Original: The Life of Bob Marshall* (Seattle: The Mountaineers, 1986), pp. 162, 272.

49 John M. Kauffmann, *Alaska's Brooks Range: The Ultimate Mountains* (Seattle: The Mountaineers, 1992), p. 97.

50 Ibid., p. 100.

51 Public Land Order 2214, *Federal Register*, December 9, 1960 (F.R. Doc. 60–11519), pp. 12598–99.

52 Debbie S. Miller, "A Pioneer Visit: Mardy Murie and the Arctic Refuge," *Alaska Geographic* 20, no. 3, (1993): 45.

53 William E. Brown, *Denali: Symbol of the Alaskan Wild* (Virginia Beach, Va.: Donning, 1993), pp. 135, 192.

54 Ibid., pp. 157, 186–89.

55 Ibid., p. 200.

56 Olaus Murie, "Mount McKinley, Wilderness Park of the North Country," *National Parks Magazine*, April 1963, p. 7.

57 Henry S. Francis, Jr., "Some Views Concerning the Development of Mount McKinley National Park," *National Parks Magazine*, September 1963, p. 18.

58 Brown, *Denali*, pp. 202–3.

59 Paul Brooks, *The Pursuit of Wilderness* (Boston: Houghton Mifflin, 1971), pp. 61–74; Naske and Slotnick, *Alaska*, pp. 193–96; William R. Hunt, *Alaska: A Bicentennial History*. (New York: W. W. Norton, 1976), pp. 135–40; *Anchorage Daily Times*, June 29, 1961.

60 Gruening, *Many Battles*, p. 496.

61 Jay Hammond, *Tales of Alaska's Bush Rat Governor* (Fairbanks: Epicenter Press, 1994), p. 141.

62 Brooks, *The Pursuit of Wilderness*, p. 91.

63 Ibid., pp. 83, 90.

64 Gruening, *Many Battles*, p. 498.

65 Hammond, *Tales of Alaska's Bush Rat Governor*, p. 138.

66 Gruening, *Many Battles*, p. 500.

67 Ibid.

68 Joseph Kurtak, *Of Rock and Ice: An Explorer's Guide to the Geology of Prince William Sound, Alaska* (Anchorage: USDA, no date), p. 24.

69 Maynard Parker, Don Moser, and Jim Hicks, "Fury of the Quake—200,000 Megatons," *Life*, April 10, 1964, pp. 32C–32E.

70 Mrs. Lowell Thomas, Jr., "An Alaskan Family's Night of Terror," *National Geographic Magazine*, July 1964), pp. 142–56.

71 William P. E. Graves, "Earthquake! Horror Strikes on Good Friday," *National Geographic Magazine*, July 1964, p. 120.

72 Parker, Moser, and Hicks, "Fury of the Quake," p. 32E.

73 Kim Rich, "Shattered Dreams," excerpt from *Johnny's Girl*, in *The Last New Land* (Anchorage: Alaska Northwest Books, 1996), p. 697.

74 Parker, Moser, and Hicks, "Fury of the Quake," p. 32D.

75 Naske and Slotnick, *Alaska*, p. 179.

76 Parker, Moser, and Hicks, "Fury of the Quake," p. 32C.

77 Graves, "Earthquake! Horror Strikes on Good Friday," p. 135.

78 Ibid., pp. 129–30.

79 Ibid., p. 117.

80 *Congressional Record*, 88th Cong., 2nd sess., 1964, vol. 110; part 5:6494.

81 Gruening, *Many Battles*, p. 465.

82 Hammond, *Tales of Alaska's Bush Rat Governor*, p. 142.

Book Eight: North Again

1 Bryan Cooper, *Alaska: The Last Frontier* (New York: William Morrow, 1973), pp. 126–28.

2 Ibid., pp. 76–83.

3 Ibid., p. 140.

4 Tom Brown, *Oil on Ice: Alaskan Wilderness at the Crossroads* (San Francisco: Sierra Club, 1971), pp. 42–45.

5 Ibid., pp. 86–96; Cooper, *Alaska*, pp. 159–65; Claus-M. Naske and Herman E. Slotnick, *Alaska, a History of the Forty-ninth State*, 2nd ed. (Norman: University of Oklahoma Press, 1987), pp. 255–56.

6 Mary Clay Berry, *The Alaska Pipeline: The Politics of Oil and Native Land Claims* (Bloomington, Ind.: Indiana University Press, 1975), pp. 99–100; Cooper, *Alaska*, pp. 146, 177–79; Naske and Slotnick, *Alaska*, p. 250.

7 Stan Cohen, *Highway on the Sea: A Pictorial History of the Alaska Marine Highway System* (Missoula, Mont.: Pictorial Histories Publishing, 1994), p. 12.

8 W. E. Garrett, "Alaska's Marine Highway," *National Geographic Magazine*, June 1965, p. 791.

9 A. W. Greely, *Handbook of Alaska* (New York: Charles Scribner's Sons, 1909), pp. 176, 187.

10 Cession of Alaska, Article III, in *Treaties and Other International Agreements of the United States of America, 1776–1949*, vol. 11, p. 1218; 15. Stat. 539.

11 U.S. Senate, *Alaska Native Claims Settlement Act of 1971*, Report to Accompany S. 34, 92d Cong., 1st sess., 1971, Senate Report No. 92–405, p. 89; quoting *Organic Act of 1884* at 23 Stat. 24.

12 Arnold, *Alaska Native Land Claims*, pp. 81–82.

13 Berry, *The Alaska Pipeline*, p. 45.

14 Arnold, *Alaska Native Land Claims*, pp. 86–88.

15 *U.S. Statutes at Large*, 72 Stat. 339, Sec. 4 (1958).

16 Arnold, *Alaska Native Land Claims*, p. 95.

17 David Avraham Voluck, "First Peoples of the Tongass: Law and the Traditional Subsistence Way of Life," in *The Book of the Tongass* (Minneapolis: Milkweed Editions, 1999), pp. 89–93.

18 "Why Tundra Times?" *Tundra Times*, October 1, 1962.

19 Arnold, *Alaska Native Land Claims*, p. 101.

20 *Tundra Times*, April 22, 1966. (Copy of letter from Hensley to Gruening.)

21 Arnold, *Alaska Native Land Claims*, p. 104.

22 Berry, *The Alaska Pipeline*, pp. 46–47.

23 Arnold, *Alaska Native Land Claims*, p. 119.

24 Senate Committee on Interior and Insular Affairs Hearing, 90th Cong., 2nd sess., February 8, 9, 10, 1968, p. 189.

25 Ibid.

26 Arnold, *Alaska Native Land Claims*, p. 124.

27 Berry, *The Alaska Pipeline*, pp. 60–61.

28 Arnold, *Alaska Native Land Claims*, pp. 137–44.

29 Berry, *The Alaska Pipeline*, pp. 174–76.

30 Public Law 92–203, 92nd Cong., 1st sess., H.R. 10367, December 18, 1971.

31 Arnold, *Alaska Native Land Claims*, p. 164.

32 Berry, *The Alaska Pipeline*, p. 9.

33 Don Hunter, "Williams Challenges Convention," *Council*, January 1982. (Reprinted from the *Anchorage Daily News*.)

34 Berry, *The Alaska Pipeline*, p. 9.

35 Cooper, *Alaska*, pp. 165–66.

36 Berry, *The Alaska Pipeline*, pp. 103, 108–10.

37 Ibid., pp. 115–19.

38 Robert Douglas Mead, *Journeys Down the Line: Building the Trans-Alaska Pipeline* (New York: Doubleday, 1978), pp. 154–55.

39 Cooper, *Alaska*, p. 105.

40 *Ketchikan Daily News*, February 18, 1971.

41 John McPhee, *Coming into the Country* (New York: Farrar, Straus, Giroux, 1976), p. 81.

42 Berry, *The Alaska Pipeline*, pp. 215–16; Mead, *Journeys Down the Line*, p. 81.

43 Jay Hammond, *Tales of Alaska's Bush Rat Governor* (Fairbanks: Epicenter Press, 1994), p. 174; John Hanrahan and Peter Gruenstein, *Lost Frontier: The Marketing of Alsaka* (New York: W. W. Norton, 1977), p. 6; and Berry. *The Alaska Pipeline*, p. 217.

44 Berry, *The Alaska Pipeline*, p. 237.

45 *Trans-Alaska Pipeline Authorization Act*, Public Law 93–153, November 16, 1973; Berry, *The Alaska Pipeline*, pp. 258, 272.

46 James P. Roscow, *Eight Hundred Miles to Valdez: The Building of the Alaska Pipeline* (Englewood Cliffs, N.J.: Prentice-Hall, 1977), pp. 112–13.

47 Mead, *Journeys Down the Line*, pp. 209–10, 223.

48 Ibid., pp. 228, 375–76.

49 Potter Wickware, *Crazy Money: Nine Months on the Trans-Alaska Pipeline* (New York: Random House, 1979), pp. 3, 30, 47, 147, 214.

50 Mead, *Journeys Down the Line*, pp. 201–2; Roscow, *Eight Hundred Miles to Valdez*, p. 120.

51 Roscow, *Eight Hundred Miles to Valdez*, p. 174.

52 Ibid., pp. 165–67; Wickware, *Crazy Money*, p. 61.

53 "$30,000 for a Lucky Lady," *Anchorage Daily News*, July 29, 1977.

54 Roscow, *Eight Hundred Miles to Valdez*, pp. 176, 199–204.

55 Hanrahan and Gruenstein, *Lost Frontier*, p. 44.

56 Naske and Slotnick, *Alaska*, p. 148.

57 McPhee, *Coming into the Country*, pp. 111, 119, 121.

58 Hanrahan and Gruenstein, *Lost Frontier*, p. 247.

59 McPhee, *Coming into the Country*, p. 126.

60 "Willow Site Wins Voters' Preference," *Anchorage Times*, November 3, 1976.

61 Hanrahan and Gruenstein, *Lost Frontier*, p. 250.

62 Hammond, *Tales of Alaska's Bush Rat Governor*, p. 238.

63 Kay Brown, "Alaska Confronts the Dilemma of Being Rich," *Anchorage Times*, November 21, 1976.

64 Hammond, *Tales of Alaska's Bush Rat Governor*, pp. 247–48.

65 Hanrahan and Gruenstein, *Lost Frontier*, p. 198.

66 Ibid., pp. 196–97.

67 Hammond, *Tales of Alaska's Bush Rat Governor*, pp. 225, 248, 251–53; *Zobel v. Williams* 457 U.S. 55 (1982); Alaska State Constitution, article 9, section 15.

68 Naske and Slotnick, *Alaska*, pp. 272–73, 282, 284.

69 Brown, "Alaska Confronts the Dilemma of Being Rich."

Book Nine: Whose Land?

1 Public Law 92–203, section 17.

2 Mary Clay Berry, *The Alaska Pipeline: The Politics of Oil and Native Land Claims* (Bloomington, Ind.: Indiana University Press, 1975), pp. 250–52.

3 Boyce Rensberger, "Protection of Alaska's Wilderness New Priority of Conservationists," *New York Times*, October 31, 1976.

4 Alaska Conservation Foundation, "Twenty Years after ANILCA," *Dispatch*, spring 2000, pp. 1–3; Claus-M. Naske and Herman E. Slotnick, *Alaska, a History of the Forty-ninth State*, 2nd ed. (Norman: University of Oklahoma Press, 1987), pp. 230–33.

5 Naske and Slotnick, *Alaska*, pp. 235–36.

6 16 USC 51, section 3101; Public Law 96–487, title I, section 101, December 2, 1980; 94 Stat. 2374.

7 David Avraham Voluck, "First Peoples of the Tongass: Law and the Traditional Subsistence Way of Life," in *The Book of the Tongass*, ed. Carolyn Servid and Donald Snow (Minneapolis: Milkweed Editions, 1999), pp. 108–10; 16 USC 51, section 3111.

8 Jim Rearden, "Subsistence: A Troublesome Issue," *Alaska*, July 1978, p. 6.

9 16 USC 51, section 3142 and 3143; Public Law 96–487, title 10, section 1002 and 1003, December 2, 1980, 94 Stat. 2452.

10 U.S. Department of the Interior, *Arctic National Wildlife Refuge, Alaska, Coastal Plain Resource Assessment: Report and Recommendation to the Congress of the United States*, April 1987.

11 Art Davidson, *In the Wake of the Exxon Valdez* (San Francisco: Sierra Club Books, 1990), p. xiii.

12 Ibid., pp. 5, 7–9.

13 Ibid., pp. 9–11.

14 Rick Hagar, "Huge Cargo of North Slope Oil Spilled," *Oil and Gas Journal*, April 3, 1989, pp. 26–27; "Alaskan Cleanup Campaign Pressed," *Oil and Gas Journal*, April 17, 1989, pp. 20–22; Davidson, *In the Wake of the Exxon Valdez*, pp. 12–19.

15 "Political, Economic Fallout Spreads from *Exxon Valdez* Crude Oil Spill," *Oil and Gas Journal*, April 10, 1989, pp. 13–16; Davidson, *In the Wake of the Exxon Valdez*, pp. 28, 37, 48, 55.

16 Davidson, *In the Wake of the Exxon Valdez*, p. 149.

17 "Alaskan Cleanup Campaign Pressed," p. 20; Davidson, *In the Wake of the Exxon Valdez*, pp. 181–82.

18 Davidson, *In the Wake of the Exxon Valdez*, pp. 239, 289.

19 Fred Bayles, "Cities View Spill Ravages Differently," *Anchorage Times*, April 11, 1989.

20 "*Exxon Valdez* Disaster Leaves Industry with Much to Repair," *Oil and Gas Journal*, April 3, 1989, p. 17.

21 Majorie Anders, "Oil Union Wants Sanity in Drilling" *Anchorage Times*, May 5, 1989.

22 John L. Eliot, "New Chemical Digs Deep into *Exxon Valdez's* Oil," *National Geographic Magazine*, August 1994, p. 132.

23 *Exxon Valdez* Oil Spill Trustee Council, ten-year report, 1999 (available at www.oilspill.state.ak.us); author's telephone interview with Craig Tillery, trustee and Alaska assistant attorney general, May 16, 2001.

24 Jim Carlton, "Ex-Lumberjacks Lead Fight to Save Tongass Forest," *Wall Street Journal*, December 4, 2000.

25 Harry Ritter, *Alaska's History: The People, Land, and Events of the North Country* (Anchorage: Alaska Northwest Books, 1993).

26 Jim Carlton, "On This Alaska Island, Survival Is More than Just a TV Game," *Wall Street Journal*, January 15, 2001, p. A1, A8.

27 "NMFS Declares Commercial Fisheries Failure in Alaska's Bristol Bay/Kuskokwim Salmon Fishery," *United States Department of Commerce News*, NOAA 97–R167, November 6, 1997; "Declaration of Economic Disaster in Bristol Bay and in the Kuskokwim River Drainages," state of Alaska, Tony Knowles, governor, July 18, 1997.

28 Steve Colt, "The Economic Importance of Healthy Alaska Ecosystems," Institute of Social and Economic Research, University of Alaska, Anchorage, January 2001, p. 11.

29 Ibid., p. 13.

30 Henry Gannett, "General Geography," *Alaska Harriman Series*, vol. I (New York: Doubleday, Page, 1901), p. 277.

31 Colt, "The Economic Importance of Healthy Alaska Ecosystems," p. 19.

32 Ibid., p. 47.

33 Carlton, "Ex-Lumberjacks Lead Fight to Save Tongass Forest."

34 *National Parks and Conservation Association v. Babbitt*, 9th Circuit Court of Appeals, 99–36065, February 23, 2001; e-mail from David Nemeth, chief of concessions, Glacier Bay National Park and Preserve, May 21, 2001.

35 Colt, "The Economic Importance of Healthy Alaska Ecosystems," p. 48.

36 Debbie S. Miller, "An Arctic Dream," *Alaska Geographic* 20, no. 3 (1993): 12–13.

37 "Clinton Presidency Hikes ANWR Stakes," *Oil and Gas Journal*, November 9, 1992, p. 25.

38 U.S. Geological Survey, "Arctic National Wildlife Refuge, 1002 Area, Petroleum Assessment, 1998, Including Economic Analysis." USGS Fact Sheet FS–028–01, April 2001, p. 4.

39 Ibid., p. 6.

40 Jim Carlton, "Alyeska Pipeline May Face Fines amid Charges of Poor Oversight," *Wall Street Journal*, December 4, 2000.

41 Jim Carlton, "In Alaskan Wilderness, 'Friendlier Technology' Gets a Cold Reception," *Wall Street Journal*, April 13, 2001; Jim Carlton, "Alaska Finds 10 percent of BP's Safety Valves in Huge Prudhoe Bay Oil Field Fail Tests," *Wall Street Journal*, April 26, 2001.

42 Alaska Conservation Foundation, "Monumental Choice for the Arctic National Wildlife Refuge," *Dispatch*, fall 2000, pp. 3–4. (Newsletter.)

43 Tamsin Carlisle, "Soaring Natural-Gas Prices Revive Canadian Pipeline Contest," *Wall Street Journal*, January 22, 2001, p. A18.

44 Editorial, "Hard, Cold Facts," *Wall Street Journal*, March 30, 2001, p. A14.

45 Stewart Udall, *The Quiet Crisis* (New York: Holt, Rinehart and Winston, 1963), p. xiii.

BIBLIOGRAPHY

Books

Anthony, Carl Sferrazza. *Florence Harding: The First Lady, the Jazz Age, and the Death of America's Most Scandalous President*. New York: William Morrow, 1998.

Arnold, Robert D. *Alaska Native Land Claims*. Anchorage: Alaska Native Foundation, 1976.

Bates, Robert H. *The Love of Mountains Is Best*. Portsmouth, N.H.: Peter E. Randall, 1994.

Beckey, Fred. *Mount McKinley: Icy Crown of North America*. Seattle: The Mountaineers, 1993.

Berry, Mary Clay. *The Alaska Pipeline: The Politics of Oil and Native Land Claims*. Bloomington, Ind.: Indiana University Press, 1975.

Berton, Pierre. *The Arctic Grail: The Quest for the Northwest Passage and the North Pole, 1818–1909*. New York: Viking Penguin, 1988.

———. *Klondike: The Last Great Gold Rush, 1896–1899*. Toronto: McClelland and Stewart, 1972.

Bodeau, Jean. *Katmai National Park and Preserve, Alaska*. Anchorage: Alaska Natural History Association and Greatland Graphics, 1992.

Bohn, Dave. *Glacier Bay: The Land and the Silence*. San Francisco: Sierra Club, 1967.

Brands, H. W. *TR: The Last Romantic*. New York: Basic Books, 1997.

Branson, John, ed. *Lake Clark—Iliamna, Alaska, 1921: The Travel Diary of Colonel A. J. Macnab*. Anchorage: Alaska Natural History Association, 1997.

Broke, George. *With Sack and Stock in Alaska*. London, 1891.

Brooks, Alfred Hulse. *Blazing Alaska's Trails*, 2nd ed. 1953; reprint, Fairbanks: University of Alaska Press, 1973.

Brooks, Paul. *The Pursuit of Wilderness*. Boston: Houghton Mifflin, 1971.

Brown, Tom. *Oil on Ice: Alaskan Wilderness at the Crossroads*. San Francisco: Sierra Club, 1971.

Brown, William E. *Denali: Symbol of the Alaskan Wild*. Virginia Beach, Va.: Donning, 1993.

Browne, Belmore. *The Conquest of Mount McKinley*. New York: G. P. Putnam's Sons, 1913.

Bueler, William M. *Roof of the Rockies: A History of Colorado Mountaineering*. Evergreen, Colo.: Cordillera Press, Inc., 1986.

Chandonnet, Fern, ed. *Alaska at War, 1941–45: The Forgotten War Remembered*. Anchorage: Alaska at War Committee, 1995.

Chevigny, Hector. *Lord of Alaska: Baranov and the Russian Adventure.* New York: Viking Press, 1944.

———. *Russian America: The Great Alaskan Venture, 1741–1867.* Portland, Oreg.: Binford and Mort Publishing, 1979.

Cohen, Stan. *Gold Rush Gateway: Skagway and Dyea, Alaska.* Missoula, Mont.: Pictorial Histories Publishing Company, 1986.

———. *Highway on the Sea: A Pictorial History of the Alaska Marine Highway System.* Missoula, Mont.: Pictorial Histories Publishing, 1994.

Cole, Terrence. *Crooked Past: The History of a Frontier Mining Camp: Fairbanks, Alaska.* Fairbanks: University of Alaska Press, 1991.

———. *E. T. Barnette: The Strange Story of the Man Who Founded Fairbanks.* Anchorage: Alaska Northwest Publishing, 1981.

Cook, Frederick A. *To the Top of the Continent.* New York: Doubleday, 1908.

Cooley, Richard A. *Alaska: A Challenge in Conservation.* Madison, Wis.: University of Wisconsin Press, 1966.

———. *Politics and Conservation: The Decline of the Alaska Salmon.* New York: Harper and Row, 1963.

Cooper, Bryan. *Alaska: The Last Frontier.* New York: William Morrow, 1973.

Dall, William H. *Alaska and Its Resources.* Boston: Lee and Shepard, 1870.

Davidson, Art. *In the Wake of the Exxon Valdez.* San Francisco: Sierra Club Books, 1990.

De Armond. Robert N. *The Founding of Juneau.* Juneau: Gastineau Channel Centennial Association, 1967.

———. *Some Names around Juneau.* Sitka, Alaska: Sitka Printing, 1957.

De Filippi, Filippo. *The Ascent of Mount St. Elias by H. R. H. Prince Luigi Amedeo de Savoia, Duke of the Abruzzi.* Trans. L. Villari. London, 1900.

Deitz, Arthur Arnold. *Mad Rush for Gold in the Frozen North.* Los Angeles: Times-Mirror Printing and Binding House, 1914.

Dunmore, John. *Pacific Explorer: The Life of Jean-François de la Pérouse, 1741–1788.* Annapolis: Naval Institute Press, 1985.

———. *Who's Who in Pacific Navigation.* Honolulu: University of Hawaii Press, 1991.

Ferber, Edna. *Ice Palace.* New York: Doubleday, 1958.

Fisher, Raymond H. *Bering's Voyages: Whither and Why.* Seattle: University of Washington Press, 1977.

Ford, Corey. *Where the Sea Breaks Its Back: The Epic Story of Early Naturalist Georg Steller and the Russian Exploration of Alaska.* Anchorage: Alaska Northwest Books, 1992. (Originally published in 1966 by Little, Brown.)

Fradkin, Philip L. *Wildest Alaska: Journeys of Great Peril in Lituya Bay.* Berkeley: University of California Press, 2001.

Garfield, Brian. *The Thousand-Mile War: World War II in Alaska and the Aleutians.* Fairbanks: University of Alaska Press, 1995. (Originally published in 1969 by Doubleday.)

Glover, James M. *A Wilderness Original: The Life of Bob Marshall.* Seattle: The Mountaineers, 1986.

Goetzmann, William H., and Kay Sloan. *Looking Far North: The Harriman Expedition to Alaska, 1899*. New York: Viking Press, 1982.

Goodwin, Doris Kearns. *No Ordinary Time: Franklin and Eleanor Roosevelt: The Home Front in World War II*. New York: Simon and Schuster, 1995.

Greely, A. W. *Handbook of Alaska*. New York: Charles Scribner's Sons, 1909.

Grinnell, George Bird. *Alaska 1899: Essays from the Harriman Expedition*. Seattle: University of Washington Press, 1995.

Gruening, Ernest. *The Battle for Alaska Statehood*. College, Alaska: University of Alaska Press, 1967.

———. *Many Battles: The Autobiography of Ernest Gruening*. New York: Liveright, 1973.

———. *The State of Alaska: A Definitive History of America's Northernmost Frontier*. New York: Random House, 1954.

Halliday, Jan. *Native Peoples of Alaska, a Traveler's Guide to Land, Art, and Culture*. Seattle: Sasquatch Books, 1998.

Hammond, Jay. *Tales of a Bush Rat Governor*. Fairbanks and Seattle: Epicenter Press, 1994.

Hanrahan, John, and Peter Gruenstein. *Lost Frontier: The Marketing of Alaska*. New York: W. W. Norton, 1977.

Haycox, Stephen, James Barnett, and Caedmon Liburd, eds. *Enlightenment and Exploration in the North Pacific, 1741–1805*. Seattle: University of Washington Press, 1997.

Heacox, Kim. *The Denali Road Guide: A Roadside Natural History of Denali National Park*. Denali Park: Alaska Natural History Association, 1990.

Hood, Mary H. *A Fan's Guide to the Iditarod*. Loveland, Colo.: Alpine Publications, 1996.

Hunt, William R. *Alaska: A Bicentennial History*. New York: W. W. Norton, 1976.

———. *Mountain Wilderness: Historic Resource Study for Wrangell–St.Elias National Park and Preserve*. Anchorage: National Park Service, 1991.

———. *North of Fifty-three Degrees: The Wild Days of the Alaskan-Yukon Mining Frontier, 1870–1914*. New York: Macmillan Publishing, 1974.

Janson, Lone E. *The Copper Spike*. Anchorage: HaHa, 1993. (Originally published in 1975 by Alaska Northwest Publishing Company.)

Johnson, S. P., ed. *Alaska Commercial Company, 1868–1940*. San Francisco, 1940.

Jones, Chris. *Climbing in North America*. Berkeley: University of California Press, 1976.

Kalani, Lyn, Lynn Rudy, and John Sperry, eds. *Fort Ross*. Jenner, Calif.: Fort Ross Interpretive Association, 1998.

Kauffmann, John M. *Alaska's Brooks Range: The Ultimate Mountains*. Seattle: The Mountaineers, 1992.

Kendrick, John. *Alejandro Malaspina: Portrait of a Visionary*. Quebec: McGill Queens University Press, 1999.

Kirchhoff, M. J. *Historic McCarthy: The Town That Copper Built*. Juneau: Alaska Cedar Press, 1993. (Originally in *McCarthy Weekly News*, August 10, 1918.)

Klein, Maury. *The Life and Legend of E. H. Harriman*. Chapel Hill, N.C.: Chapel Hill Press, 2000.

Knapp, Marilyn. *Carved History: The Totem Poles and House Posts of Sitka National Historical Park*. Anchorage: Alaska Natural History Association, 1992.

Kohlhoff, Dean. *When the Wind Was a River: Aleut Evacuation in World War II*. Seattle: University of Washington Press, 1995.

Kurtak, Joseph. *Of Rock and Ice: An Explorer's Guide to the Geology of Prince William Sound, Alaska*. Anchorage: USDA, no date.

Langdon, Steve J. *The Native People of Alaska*. 3rd ed. Anchorage: Greatland Graphics, 1993.

Lavender, David. *Land of Giants: The Drive to the Pacific Northwest, 1750–1950*. Garden City, N.Y.: Doubleday, 1956.

Lazell, J. Arthur. *Alaskan Apostle: The Life Story of Sheldon Jackson*. New York: Harper and Brothers, 1960.

Levi, Steven C. *Cowboys of the Sky: The Story of Alaska's Bush Pilots*. New York: Walker, 1996.

Lorant, Stefan. *The Glorious Burden: The American Presidency*. New York: Harper and Row, 1968.

Lord, Nancy. *Green Alaska: Dreams from the Far Coast*. Washington, D.C.: Counterpoint, 1999.

Lord, Walter. *Incredible Victory*. New York: Harper and Row, 1967.

Marshall, Robert. *Alaska Wilderness: Exploring the Central Brooks Range*. 2nd ed. Berkeley: University of California Press, 1970.

———. *Arctic Village*. 1933; reprint, Anchorage: University of Alaska Press, 1993.

McGinnis, Joe. *Going to Extremes*. New York: Alfred A. Knopf, 1980.

McPhee, John. *Coming into the Country*. New York: Farrar, Strauss, Giroux, 1976.

Mead, Robert Douglas. *Journeys Down the Line: Building the Trans-Alaska Pipeline*. New York: Doubleday, 1978.

Miller, Debbie S. *Midnight Wilderness: Journeys in Alaska's Arctic National Wildlife Refuge*. San Francisco: Sierra Club Books, 1990.

Miller, Orlando W. *The Frontier in Alaska and the Matanuska Colony*. New Haven: Yale University Press, 1975.

Minter, Roy. *The White Pass: Gateway to the Klondike*. Fairbanks: University of Alaska Press, 1987.

Moore, Terris. *Mount McKinley: The Pioneer Climbs*. College, Alaska: University of Alaska Press, 1967. (Originally published in *New York Sun*, January 24, 1897.)

Morgan, Murray. *Confederate Raider in the North Pacific: The Saga of the C.S.S. Shenandoah, 1864–65*. Pullman, Wash.: Washington State University Press, 1995. (Originally published in 1948 by E. P. Dutton under the title *Dixie Raider: The Saga of the C.S.S. Shenandoah*)

Muir, John. *Travels in Alaska*. San Francisco: Sierra Club Books, 1988. (Originally published in 1915.)

Murie, Margaret E. *Two in the Far North*. Anchorage: Alaska Northwest Books, 1997. (Originally published in 1962 by Alfred Knopf.)

Naske, Claus-M., and Herman E. Slotnick. *Alaska, a History of the Forty-ninth State*. 2nd ed. Norman: University of Oklahoma Press, 1987.

National Geographic Atlas of the World, 7th ed. Washington, D.C.: National Geographic Society, 1999.

Nelson, Richard K. *Make Prayers to the Raven: A Koyukon View of the Northern Forest*. Chicago: University of Chicago Press, 1983.

Neufeld, David, and Frank Norris. *Chilkoot Trail: Heritage Route to the Klondike*. Whitehorse, Yukon Territory: Lost Moose, 1996.

Nordlander, David J. *For God and Tsar, a Brief History of Russian America, 1741–1867*. Anchorage: Alaska Natural History Association, 1994.

Orth, Donald J. *Dictionary of Alaska Place Names*. Geological Survey Professional Paper, 567. Washington, D.C.: U.S. Government Printing Office, 1967.

Paolino, Ernest. *The Foundations of the American Empire: William Henry Seward and U.S. Foreign Policy*. Ithaca, N.Y.: Cornell University Press, 1973.

Perkins, Dexter. *A History of the Monroe Doctrine*. Boston: Little, Brown, 1963.

Potter, Jean. *The Flying North*. New York: Macmillan, 1947.

Pratt, Julius. *A History of United States Foreign Policy*. Englewood Cliffs, N.J.: Prentice-Hall, 1955.

Ritter, Harry. *Alaska's History: The People, Land, and Events of the North Country*. Seattle: Alaska Northwest Books, 1993.

Roscow, James P. *Eight Hundred Miles to Valdez: The Building of the Alaska Pipeline*. Englewood Cliffs, N.J.: Prentice-Hall, 1977.

Satterfield, Archie. *Chilkoot Pass: A Hiker's Historical Guide to the Klondike Gold Rush National Historical Park*. Seattle: Alaska Northwest Books, 1994. (Orignally published in 1973.)

Schwatka, Frederick. *A Summer in Alaska in the 1880s*. Secaucus, N.J.: Castle, 1988. (Originally published in 1885.)

Servid, Carolyn, and Donald Snow, eds. *The Book of the Tongass*. Minneapolis: Milkweed Editions, 1999.

Sheldon, Charles. *The Wilderness of Denali*. New York: Charles Scribner's Sons, 1960. (Originally published in 1930.)

Sherwood, Morgan B. *Exploration of Alaska, 1865–1900*. Fairbanks: University of Alaska Press, 1992. (Originally published in 1965 by Yale University Press.)

———, ed. *Alaska and Its History*. Seattle: University of Washington Press, 1967.

Stejneger, Leonhard, *Georg Wilhelm Steller: The Pioneer of Alaskan Natural History*. Cambridge, Mass.: Harvard University Press, 1936.

Steller, Georg Wilhelm. *Journal of a Voyage with Bering*. Palo Alto: Stanford University Press, 1993. (Originally published in 1743.)

Sterling, Bryan B., and Frances N. Sterling. *Will Rogers and Wiley Post: Death at Barrow*. New York: M. Evans, 1993.

Stuck, Hudson. *The Ascent of Denali*. Seattle: The Mountaineers, 1977. (Originally published in 1914.)

Sundborg, George. *Opportunity in Alaska*. New York: Macmillan, 1945.

Tenderini, Mirella, and Michael Shandrick. *The Duke of the Abruzzi: An Explorer's Life*. Seattle: The Mountaineers, 1997.

Twichell, Heath. *Northwest Epic: The Building of the Alaska Highway*. New York: St. Martin's Press, 1992.

———.*Allen: The Biography of an Army Officer, 1859–1930*. New Brunswick, N.J.: Rutgers University Press, 1974.

Ungermann, Kenneth A. *The Race to Nome*. New York: Harper and Row, 1963.

Vaughan, Thomas et al. *Voyages of Enlightenment: Malaspina on the Northwest Coast, 1791/1792*. Portland, Oreg.: Oregon Historical Society, 1977.

Waddell, James I. *C. S. S. Shenandoah: The Memoirs of Lieutenant Commanding James I. Waddell*. Ed. James D. Horan. Annapolis: Naval Institute Press/Bluejacket Books, 1996. (Originally published in 1960 by Crown Publishers.)

Washburn, Bradford. *Exploring the Unknown: Historic Diaries of Bradford Washburn's Alaska/Yukon Expeditions*. Kenmore, Wash.: Epicenter Press, 2001.

Washburn, Bradford, and David Roberts. *Mount McKinley: The Conquest of Denali*. New York: Harry H. Abrams, 1991.

Waterman, Jonathan. *High Alaska: A Historical Guide to Denali, Mount Foraker, and Mount Hunter*. New York: American Alpine Club, 1988.

———. *A Most Hostile Mountain: Re-creating the Duke of Abruzzi's Historic Expedition on Alaska's Mount St. Elias*. New York: Henry Holt, 1997.

Webb, Melody. *The Last Frontier: A History of the Yukon Basin of Canada and Alaska*. Albuquerque: University of New Mexico Press, 1985.

———. *Yukon: The Last Frontier*. Lincoln: University of Nebraska Press, 1993.

Weeden, Robert B. *Promises to Keep*. Boston: Houghton Mifflin, 1978.

Wendt, Ron, ed. *Goldfields of Nome, Alaska*. Wasilla, Alaska: Goldstream Publications, 1996. (Compilation of USGS reports 1900–1913.)

Wharton, David B. *The Alaska Gold Rush*. Bloomington, Ind.: Indiana University Press, 1972.

———. *They Don't Speak Russian in Sitka: A New Look at the History of Southern Alaska*. Menlo Park, Calif.: Markgraf Publications Group, 1991.

Whymper, Frederick. *Travel and Adventures in the Territory of Alaska*. London, 1868.

Wickersham, James. *A Bibliography of Alaskan Literature, 1724–1924*. Cordova, Alaska: Cordova Daily Times Print, 1927.

Wickware, Potter. *Crazy Money: Nine Months on the Trans-Alaska Pipeline*. New York: Random House, 1979.

Wilson, William H. *Railroad in the Clouds: The Alaska Railroad in the Age of Steam, 1914–1945*. Boulder, Colo.: Pruett Publishing, 1977.

Young, Samuel Hall. *Alaska Days with John Muir*. Salt Lake City: Peregrine Smith Books, 1990. (Originally published in 1915.)

Articles

Adams, George Edward. "Cape Nome's Wonderful Placer Mines." *Harper's Weekly*, June 9, 1900, pp. 529–31.

Alaska Conservation Foundation. "Monumental Choice for the Arctic National Wildlife Refuge." *Dispatch*, fall 2000. (Newsletter.)

"Alaskan Cleanup Campaign Pressed." *Oil and Gas Journal*, April 17, 1989, pp. 20–22.

"Alaska's New Railway." *National Geographic Magazine*, December 1915, pp. 567–89.

Armstrong, Ted. A. "Alaskan Oil." *Oil and Gas Journal*, August 22, 1966, pp. 77–84.

Arnold, H. H. "Our Air Frontier in Alaska." *National Geographic Magazine*, October 1940, pp. 487–504.

Athearn, Robert G. "An Army Officer's Trip to Alaska in 1869." *Pacific Northwest Quarterly* 40, no. 1 (January 1949): 44–64.

Baldwin, George E. "Conservative Faddists Arrest Progress and Seek to Supplant Self-Government with Bureaucracy." *Alaska-Yukon Magazine*, February 1912, pp. 44–46.

Beadle, G. W. "Up Doonerak—An Arctic Adventure." *Living Wilderness*, winter 1952–53, pp. 7–12.

Caldwell, Eleanor B. "A Woman's Experience at Cape Nome." *Cosmopolitan*, November 1900, pp. 81–86.

Campbell, Charles S. "The Anglo-American Crisis in the Bering Sea, 1890–91." *Mississippi Valley Historical Review* 48 (December 1961): 393–414.

Capps, Stephen R. "A Game Country without Rival in America." *National Geographic Magazine*, January 1917, pp. 69–84.

Carlisle, Tamsin. "Soaring Natural-Gas Prices Revive Canadian Pipeline Contest." *Wall Street Journal*, January 22, 2001.

Carlson, L. H. "The Discovery of Gold at Nome, Alaska." *Pacific Historical Review* 15 (September 1946): 259–78.

Carlton, Jim. "Alaska Finds 10 percent of BP's Safety Valves in Huge Prudhoe Bay Oil Field Fail Tests." *Wall Street Journal*, April 26, 2001.

———. "Alyeska Pipeline May Face Fines amid Charges of Poor Oversight." *Wall Street Journal*, December 4, 2000.

———. "Ex-Lumberjacks Lead Fight to Save Tongass Forest." *Wall Street Journal*, December 4, 2000.

———. "In Alaskan Wilderness, 'Friendlier Technology' Gets a Cold Reception." *Wall Street Journal*, April 13, 2001.

———. "On this Alaska Island, Survival Is More than Just a TV Game." *Wall Street Journal*, January 15, 2001.

Carpé, Allen. "The Ascent of Mount Bona, Alaska." *Alpine Journal* 43 (1931): 69–74.

"Clinton Presidency Hikes ANWR Stakes." *Oil and Gas Journal*, November 9, 1992, p. 25.

"Completing the Government Railroad in Alaska." *Railway Age*, April 1922, pp. 813–17.

Dalby, Ron et al. "Paradise Lost." *Alaska Magazine*, June 1989, pp. 20–35.

Dall, William. "Late News from Alaska." *Science*, July 31, 1885, pp. 95–96.

Davison, Lonnelle. "Bizarre Battleground—the Lonely Aleutians." *National Geographic Magazine*, September 1942, pp. 316–17.

Dixon, Bob. "Experts View State's Fiscal Future When Oil Revenues Begin to Decline." *Alaska Business and Industry*, January 1983, pp. 10–11.

Dolitsky, Alexander B. "The Alaska-Siberia Lend-Lease Program." In *Alaska at War, 1941–45: The Forgotten War Remembered,* ed. Fern Chandonnet, pp. 333–39. Anchorage: Alaska at War Committee, 1995.

Dunham, Alfred H. "The Development of Nome." *Cosmopolitan*, February 1905, pp. 465–72.

Dunn, Robert. "Conquering Our Greatest Volcano." *Harper's Monthly Magazine*, March 1909, p. 497.

———. "Finding a Volcano and Wiping a 16,000 Feet [sic] Mountain from the Map of Alaska." *Outing*, December 1902, pp. 321–32.

Eliot, John L. "New Chemical Digs Deep into *Exxon Valdez's* Oil." *National Geographic Magazine*, August 1994, p. 132.

"*Exxon Valdez* Disaster Leaves Industry with Much to Repair." *Oil and Gas Journal*, April 3, 1989, p. 17.

Fialka, John J., and Chip Cummins. "Oil Drilling in Alaska Looms as Legislative Test for Bush." *Wall Street Journal*, January 12, 2001.

"Formal Completion of the Alaska Railroad." *Railway Review*, July 1923, pp. 131–32.

Francis, Henry S., Jr. "Some Views Concerning the Development of Mount McKinley National Park." *National Parks Magazine*, September, 1963, pp, 18–19.

Garrett, W. E. "Alaska's Marine Highway." *National Geographic Magazine*, June 1965, pp. 775–819.

Gilbert, Benjamin Franklin. "The Confederate Raider *Shenandoah*." *Journal of the West* 4 (April 1965): 169–82.

Grauman, Melody Webb. "Kennecott: Alaskan Origins of a Copper Empire." *Western Historical Quarterly* 9, no. 2 (April 1978): 197–211.

Graves, William P. E. "Earthquake! Horror Strikes on Good Friday." *National Geographic Magazine*, July 1964, pp. 112–39.

Griggs, Robert F. "The Valley of Ten Thousand Smokes: An Account of the Discovery and Exploration of the Most Wonderful Volcanic Region in the World." *National Geographic Magazine*, February 1918, pp. 115–70.

———. "The Valley of Ten Thousand Smokes: National Geographic Society Explorations in the Katmai District of Alaska." *National Geographic Magazine*, January 1917, pp. 13–68.

Grosvenor, G. H. "The Harriman-Alaska Expedition in Co-operation with the Washington Academy of Science." *National Geographic Magazine*, June 1899, p. 225.

Gruening, Ernest. "Alaska Proudly Joins the Union." *National Geographic Magazine*, July 1959, pp. 42–83.

———. "Go North, Young Man." *Reader's Digest*, January 1944, pp. 53–57.

———. "Strategic Alaska Looks Ahead." *National Geographic Magazine*, September 1942, pp. 281–315.

Hagar, Rick. "Huge Cargo of North Slope Oil Spilled." *Oil and Gas Journal*, April 3, 1989, pp. 26–27.

Haulman, Daniel L. "The Northwest Ferry Route." In *Alaska at War, 1941–45: The Forgotten War Remembered*, ed. Fern Chardonnet, pp. 319–25. Anchorage: Alaska at War Committee, 1995.

Heacox, Kim. "Contested Waters." *National Parks*, July–August 1984, pp. 25–26.

Hinckley, Ted C. "John G. Brady and the Assimilation of Alaska's Tlingit Indians." *Western Historical Quarterly* 11 (January 1980): 37–55.

———. "Sheldon Jackson and Benjamin Harrison." *Pacific Northwest Quarterly* 54 (April 1963): 66–74.

———. "Sheldon Jackson as Preserver of Alaska's Native Culture." *Pacific Historical Review* 33 (November 1964): 411–24.

———. "Sheldon Jackson, Presbyterian Lobbyist for the Great Land of Alaska." *Journal of Presbyterian History* 40 (March 1962): pp. 3–23.

Hodgson, Bryan. "Alaska's Big Spill: Can the Wilderness Heal?" *National Geographic Magazine*, January 1990, pp. 5–43.

Houston, Charles. "Denali's Wife." *American Alpine Journal* 2, no. 3 (1935): pp. 285–97.

Hudson, Ray. "Aleuts in Defense of Their Homeland." In *Alaska at War, 1941–45: The Forgotten War Remembered*, ed. Fern Chandonnet, pp. 161–64. Anchorage: Alaska at War Committee, 1995.

Hulley, Clarence C. "Historical Survey of the Matanuska Valley Settlement in Alaska." *Pacific Northwest Quarterly* 40 (October, 1949), 327–40.

Ickes, Harold. "Let's Open Up Alaska." *This Week*, February 3, 1946, pp. 4–18.

Inouye, Ronald K. "For Immediate Sale: Tokyo Bathhouse—How World War II Affected Alaska's Japanese Citizens." In *Alaska at War, 1941–45: The Forgotten War Remembered*, ed. Fern Chadonnet, pp. 259–263. Anchorage: Alaska at War Committee, 1995.

Joslin, Falcon. "Railroad Building in Alaska." *Alaska-Yukon Magazine*, January 1909, pp. 247–50.

Kerr, Mark Brickell. "Mount St. Elias and Its Glaciers." *Scribner's Magazine*, March 1891, p. 361.

Koenig, Duane. "Ghost Railway in Alaska: The Story of the Tanana Valley Railroad." *Pacific Northwest Quarterly* 45 (January 1954): 8–12.

La Fay, Howard. "DEW Line: Sentry of the Far North." *National Geographic Magazine*, July 1958, pp. 128–46.

Lambart, H. F. "The Conquest of Mount Logan." *National Geographic Magazine*, June 1926, pp. 597–631.

Martin, George C. "The Recent Eruption of Katmai Volcano in Alaska." *National Geographic Magazine*, February 1913, pp. 131–98.

McCarthy, Terry. "War over Arctic Oil." *Time*, February 19, 2001, pp. 24–29.

Mears, Frederick. "Construction Progress on the Alaskan Railroad." *Railway Age*, June 21, 1918, pp. 1458–60.

Miller, Debbie S. "An Arctic Dream." *Alaska Geographic* 20, no. 3 (1993): 10–24.

———. "A Pioneer Visit: Mardy Murie and the Arctic Refuge." *Alaska Geographic* 20, no. 3 (1993): 40–45.

Mitchell, John G. "Oil on Ice: Economic Boon, Environmental Disruption—Alaska Weighs the Problem." *National Geographic Magazine*, April 1997, pp. 104–31.

Moore, Terris. "Mt. Sanford: An Alaskan Ski Climb." *American Alpine Journal* 3, no. 3 (1939): 265–73.

Morgan, Audrey, and Frank Morgan. "Alaska's Russian Frontier: Little Diomede." *National Geographic Magazine*, April 1951, pp. 551–62.

Muir, John. "Discovery of Glacier Bay." *Century*, June 1895.

Murie, Olaus. "Mount McKinley, Wilderness Park of the North Country." *National Parks Magazine*, April 1963, pp. 4–7.

Parker, Maynard, Don Moser, and Jim Hicks. "Fury of the Quake—200,000 Megatons," *Life*, April 10, 1964, pp. 26–33.

Perry, K. W. "Volcanoes of Alaska." *National Geographic Magazine*, August 1912, p. 824.

Pierce, Richard A. "Georg Anton Schaffer." In *Alaska and Its History*, ed. Morgan B. Sherwood, pp. 71–81. Seattle: University of Washington Press, 1967.

"Political, Economic Fallout Spreads from *Exxon Valdez* Crude Oil Spill." *Oil and Gas Journal*, April 10, 1989, pp. 13–16.

"Progress of Government Railway in Alaska." *Railway Age*, April 20, 1917, pp. 828–32.

Rainey, Froelich. "Alaska Highway an Engineering Epic." *National Geographic Magazine*, February 1943, pp. 143–68.

Rearden, Jim. "Subsistence: A Troublesome Issue." *Alaska*, July 1978, pp. 4–6, 83–88.

Rennick, Penny, ed. "The Copper Trail." *Alaska Geographic* 16, no. 4, (1989): 5–55.

———. "A Future in Question." *Alaska Geographic* 20, no. 3, (1993): 68–77.

Rensberger, Boyce. "Protection of Alaska's Wilderness New Priority of Conservationists." *New York Times*, October 31, 1976.

Russell, Israel. "The Expedition of the National Geographic Society and the U.S. Geological Survey." *Century*, April 1891, pp. 872–84.

———. "An Expedition to Mount St. Elias, Alaska." *National Geographic Magazine*, May 29, 1891, pp. 59–64.

———. "Mount St. Elias Revisited." *Century*, June 1892, pp. 190–203.

Schwatka, Frederick. "Mountaineering in Alaska. *Alpine Journal* 13 (November 1886): 90–93.

———. "Two Expeditions to Mount St. Elias." *Century*, April 1891, pp. 665–84.

Scidmore, Eliza R. "Discovery of Glacier Bay, Alaska." *National Geographic Magazine*, April 1896, p. 141.

Shenitz, Helen A. "Alaska's 'Good Father.'" *Russian Orthodox Journal* 30 (February 1957): 6–10.

Smith, David D. "Pulp, Paper, and Alaska." *Pacific Northwest Quarterly* 66, no. 2 (April 1975): 61–70.

Streett, St. Clair. "The First Alaskan Air Expedition." *National Geographic Magazine*, May 1922, pp. 499–552.

Sutherland, Mason. "A Navy Artist Paints the Aleutians." *National Geographic Magazine*, August 1943, pp. 157–76.

Thomas, Mrs. Lowell, Jr. "An Alaskan Family's Night of Terror." *National Geographic Magazine*, July 1964, pp. 142–56.

Thompkins, Stuart R. "Drawing the Alaskan Boundary." *Canadian Historical Review* 26 (March 1945): 1–24.

Topham, Harold. "An Expedition to Mount St. Elias, Alaska." *Alpine Journal* 14 (August 1889): 366.

Vevier, Charles. "American Continentalism: An Idea of Expansion, 1845–1910." *American Historical Review* 65 (January 1960).

Voluck, David Avraham. "First Peoples of the Tongass: Law and the Traditional Subsistence Way of Life." In *The Book of the Tongass*, ed. Carolyn Servid and Donald Snow, pp. 89–118. Minneapolis: Milkweed Editions, 1999.

Washburn, Bradford. "The Ascent of Mt. St. Agnes." [Mount Marcus Baker] *American Alpine Journal* 3, no. 3, (1939): 255–64.

———. "The Conquest of Mount Crillon." *National Geographic Magazine*, March 1935, pp. 361–400.

———. "Over the Roof of Our Continent." *National Geographic Magazine*, July 1938, pp. 78–98.

Waterman, Jonathan. "Brad Washburn, Renaissance Mountain Man." *Alaska Geographic* 25, no. 3 (1998): 8–21.

Welch, Richard E, Jr. "American Public Opinion and the Purchase of Russian America." *American Slavic and East European Review* 17 (1958): 481–94.

Williams, William. "Climbing Mount St. Elias." *Scribner's Magazine*, April 1889, pp. 387–403.

Wood, C. E. S. "Among the Thlinkets in Alaska." *Century*, July 1882, p. 332.

Wooley, Chris, and Mike Martz. "The Tundra Army—Patriots of Arctic Alaska." In *Alaska at War, 1941–45: The Forgotten War Remembered*, ed. Fern Chandonnet, pp. 155–60. Anchorage: Alaska at War Committee, 1995.

Government Documents

Abercrombie, William R. "Copper River Exploring Expedition, 1899." *Compilation of Narratives of Exploration in Alaska*. Senate Reports, 56th Cong., 1st sess., no. 1023, 382.

Allen, Henry T. *Report of an Expedition to the Copper, Tanana, and Koyukuk Rivers*. Washington, D.C.: U.S. Government Printing Office, 1887. (Allen's report was also published in slightly different form in "A Military Reconnaissance of the Copper River Valley, 1885." *Compilation of Narratives of Exploration in Alaska.* Senate Reports., 56th Cong., 1st sess., 1900. no. 1023, Serial 3896, 411–88.)

Schneider, Karl. "Sea Otter." Wildlife Notebook Series, Alaska Department of Fish and Game, 1994.

Spurr, J. E. "A Reconnaissance in Southwestern Alaska in 1898." In *Twentieth Annual Report of the USGS,* part 7, pp. 31–264. Washington, D.C.: U.S. Government Printing Office, 1900.

U.S. Geological Survey. "Arctic National Wildlife Refuge, 1002 Area, Petroleum Assessment, 1998, Including Economic Analysis." USGS Fact Sheet FS–028–01, April 2001.

U.S. House Executive Document 177. 40th Cong., 2nd sess, 1867, pp. 124–89. (Charles Sumner's lengthy speech and supporting documentation.)

Weidlich, Laurie M. "Game Fishes of Alaska." Alaska Department of Fish and Game, 1996.

Zimmerman, Steven. T. "Northern Fur Seal." Wildlife Notebook Series, Alaska Department of Fish and Game, 1994.

Interviews, Reports, and Unpublished Manuscripts

Borneman, Walter R. "Irwin: Silver Camp of the Ruby Mountains." Unpublished master's thesis, Western State College of Colorado, 1975.

Colt, Steve. "The Economic Importance of Healthy Alaska Ecosystems." Institute of Social and Economic Research, University of Alaska, Anchorage, January 2001.

Exxon Valdez Oil Spill Trustee Council. Ten-year report, 1999. (Available at www.oil spill.state.ak.us)

Ossinger, Al. Interview by author. Lakewood, Colo., February 23, 2001. (Ossinger was stationed on St. Lawrence Island, 1955–56, U.S. Army.)

Tillery, Craig. Interview by author. May 16, 2001. (Tillery is a trustee and Alaska assistant attorney general.)

Young, Frank. Interview by author. Georgetown, Colo., May 8, 2001. (Young was land examiner in Fairbanks, 1971–81, for the Bureau of Land Management.)

ACKNOWLEDGMENTS

Having made no bones about the fact that this is a broad overview of Alaska history, I am particularly grateful to the historians who have broken the trail before me. All sources are acknowledged in the bibliography, but a few are so important as to deserve special mention here. First and foremost is Donald J. Orth's *Dictionary of Alaska Place Names.* I cannot imagine any writer of the Alaskan landscape sitting down at a keyboard without it. Orth is not infallible, but he is indeed indispensable. Morgan Sherwood's seminal *Exploration of Alaska, 1865–1900,* Brian Garfield's groundbreaking *The Thousand-Mile War: World War II in Alaska and the Aleutians,* and Claus-M. Naske and Herman E. Slotnick's steady stalwart, *Alaska, a History of the Forty-ninth State,* have been equally indispensable.

Close behind are Melody Webb's *Yukon: The Last Frontier,* Heath Twichell's *Northwest Epic: The Building of the Alaska Highway,* and Mary Clay Berry's *The Alaska Pipeline: The Politics of Oil and Native Land Claims.* While not lacking in personal opinion, Ernest Gruening's *State of Alaska* has been another building block, as have the autobiographies of two of Alaska's most forceful personalities: Gruening himself and his fellow governor Jay Hammond.

Special accolades must go to those who read portions of the manuscript and offered critical advice: National Park Service historian John Branson, University of Alaska (Fairbanks) historian Ronald J. "Burr" Neely, Jr., author and historian Heath Twichell, Alaska attorney Mary Gilson, natural resources attorney Thomas F. Cope, mountaineer and attorney Mark Hingston, and my occasional coauthor, Lyndon J. Lampert. Eric Janota, whose day job is with National Geographic Maps, lent his cartographic skills to the maps. Portions of the "Crest of the Continent," "Copper, Kennecott, and One Heck of a Railroad," and "The Trails of '98" chapters previously appeared as part of National Geographic's Alaska Destination CD-ROM series.

Many dedicated government agency men and women assisted me over the years in my Alaska research and travels, both with this book and earlier projects. First and foremost is Diane Jung, chief of interpretation of the Alaska Region, National Park Service, who opened many doors for me starting in 1994. And these folks: at the Alaska Regional Office of the National Park Service, regional director Bob Barbee and Wendy Davis; at Denali National Park and Preserve, Clare Curtis, Lisa Eckert, Dan Greenblatt, and Jon Mann; at Kenai National Wildlife Refuge, Candace Ward and Emily Dekker-Fiala; at Kenai Fjords

National Park and Preserve, Anne Castellina, Maria Gillett, Glenn Hart, and Jim Pfeiffenberger; on Chugach National Forest, Susan Rutherford, Anne Jeffery, David Allen, and Karen O'Leary; at Wrangell–St. Elias National Park and Preserve, Margaret J. Steigerwald; in the Alaska Region of the U.S. Forest Service, Neil Hagadorn, Sarah Iverson, and especially Kristi Kantola; on Tongass National Forest, Sandy Skrien, Joni Packard, and Rob Morgenthaler; at Sitka National Historical Park, Carol Burkhardt, Lynda Lancaster, and Jenny Baird; at Glacier Bay National Park and Preserve, David Nemeth, Kris Nemeth, and Rosemarie Salazar; at Klondike Gold Rush National Historical Park, Cathleen Cook, Karl Gurcke, and Paul Lofgren; at Gates of the Arctic National Park and Preserve, Dave Mills and Bob Maurer; at Lake Clark National Park and Preserve, Lee Fink, John Branson, and Janna Walker.

The Alaska Natural History Association is one of the best national park cooperating associations in the country. Many thanks to its current executive director, Charles Money, and longtime past executive director, Frankie Barker. Past ANHA staff who were of considerable assistance to me over the years are Gina Soltis and Jenny Harris (Denali), Martha Massey (Portage Glacier/Chugach National Forest), Melody Jamieson (Glacier Bay), and Michael Troina (Ketchikan/Tongass National Forest).

Veteran Alaskan bush pilots who have flown me in and out of places include Mark Lang (Lake Clark area), Jay Jespersen (the Brooks Range), and Gary Green (the Wrangells).

Libraries and museums are great resources for facts, but one can get the "feel" of the scene only by traveling across it. Clearly, that has always been my favorite part of writing. Many thanks to the people who joined me on the more memorable of these outings (or let me come along as the case might be—thanks, Bill and Mary Kay!) and shared in the fun:

Watching caribou atop Chitistone Pass during a July snowstorm with Bill and Mary Kay Stoehr.

Bouncing along on the permafrost roller coaster on the Richardson Highway with David Barwin.

Hiking the Chilkoot Trail, with Bill, Mary Kay, and Greg Stoehr, Maria Gillett, Frankie Baker, Mary Gilson, Cathy Cook, Diane Jung, and Wendy Davis.

Floating the Chilikadrotna River and hiking the Telaquana country with Bill and Mary Kay Stoehr, Janna Walker, historian and woodsman extraordinaire John Branson, and Bill ("Doc" when he's in the bush) Schneider.

And one of the best trips of my life, our Brooks Range crossing with Bill and Mary Kay Stoehr, Kemp Battle, and "Doc" Schneider.

Others who have offered advice, given tours, or provided special hospitality over the years include Bob and Carol Barbee in Anchorage, Jack and Roma Reakoff and their son, Jesse, in Wiseman, Mark and Sandy Lang at Alaska's Lake Clark Inn; Lee and Cilla Robbins, along with Birch, Amber, and Heather, and

Mary Jane Randolph on Raspberry Island; Monroe Robinson and his lovely daughter, Chelsea, who stole the show from me one Fourth of July at Twin Lakes when she recited "The Shooting of Dan McGrew" from memory as I prepared to read it.

I cannot thank Kemp Battle enough for introducing me to my competent agent, Alexander C. Hoyt, and Alex in turn for introducing me to my insightful editor, Hugh Van Dusen. The three of us were on the same page from the beginning, and I look forward to many more projects with both.

There are three other special people who, while not directly related to this history of Alaska, are nonetheless responsible for much of my development as an author and historian. Professor of history Duane Vandenbusche took me in hand as a freshman at Western State College in Gunnison, Colorado, in the fall of 1970 and over the next five years nurtured my skills and channeled my enthusiasms into histories of railroads across Marshall Pass and the mining camp of Irwin. Olive R. Gifford, longtime Western State College librarian, entrusted me with the keys to the special collections and unlocked the world of primary sources. Finally, Eleanor M. Gehres, for a quarter of a century the head of the Western History Department of the Denver Public Library, took a special and continuing interest in me, a would-be guidebook author, as I researched the history of Colorado's 14,000-foot peaks.

Most important to me is my son, Russ—still "Rusty" when we started taking Alaska trips together when he was eleven. He's my only son—as he is quick to remind me—and together we have ferry-hopped our way around the Inside Passage, taken the car train from Whittier to Portage, and mountain-biked through the Kantishna Hills below Denali. Russ, among my fondest memories are dragging you away from the video games at the Juneau Airport so that we could fly to Glacier Bay—your first flight in a small plane; riding bikes from McCarthy to Kennecott—you had more fun on the way down; gingerly crossing the Million Dollar Bridge and driving the railroad grade outside Cordova—I can still hear the strains from *Raiders of the Lost Ark;* and watching you pull in your first halibut off Raspberry Island. I love you, bud.

INDEX

Entries in *italics* refer to maps and tables.